SIGNAL PROCESSING

Proceedings of the NATO Advanced
Study Institute on Signal Processing
with particular reference to Underwater Acoustics
Held at Loughborough, England
Under the Auspices of
The Loughborough University of Technology

Edited by

J. W. R. GRIFFITHS and P. L. STOCKLIN

*Department of Electronic and Electrical Engineering,
Loughborough University of Technology, Loughborough, England*

and

C. VAN SCHOONEVELD

Physics Laboratory, TNO, The Hague, Netherlands

1973

ACADEMIC PRESS LONDON AND NEW YORK
A Subsidiary of Harcourt Brace Jovanovich, Publishers

ACADEMIC PRESS INC. (LONDON) LTD.
24/28 Oval Road,
London NW1

United States Edition published by
ACADEMIC PRESS INC.
111 Fifth Avenue
New York, New York 10003

Library of Congress Catalog Card Number: 73-1472
ISBN: 0 12 303450 7

Printed in Great Britain by
William Clowes & Sons Limited
London, Colchester and Beccles

Contributors

N. ABRAMSON. University of Hawaii, Honolulu, Hawaii 96822, U.S.A.

M. H. ACKROYD. Department of Electronic and Electrical Engineering, University of Technology, Loughborough, Leicestershire, U.K.

J. M. ALSUP. Naval Undersea Research and Development Center (Code 6021), San Diego, California 92132, U.S.A.

D. J. ANGELAKOS. College of Engineering, Electronics Research Laboratory, University of California, Berkeley, California 94720, U.S.A.

G. H. ASH. Admiralty Underwater Weapons Establishments, Portland, Dorset, U.K.

S. W. AUTREY. Hughes Aircraft Company, Fullerton, California 92634, U.S.A.

W. J. BANGS. Department of Engineering and Applied Science, Yale University, Bectan Center, New Haven, Connecticut 06520, U.S.A.

A. BARBAGELATA. SACLANT ASW–Nata, Via. S. Bartolomeo 400, La Spezia, Italy.

H. O. BERKTAY. Dept. Electronic and Electrical Engineering, University of Birmingham, P.O. Box 363 Birmingham B15 2TT, U.K.

G. BIENVENU. Thomson-CSF, Chemin de Tracaila, 06–Cagnes-sur-Mer, France.

T. G. BIRDSALL. Cooley Electronics Laboratory, The University of Michigan, 262 Cooley Building, Ann Arbor, Michigan 48105, U.S.A.

R. E. BOGNER. Electrical Engineering Department, Imperial College, Exhibition Road, London SW7 2BT, U.K.

A. VAN DEN BOS. Technical University of Delft, c/o Lorentzweg 1, Delft, Netherlands.

B. M. BROWN. Tracor Inc., 6500 Tracor Lane, Austin, Texas 78721, U.S.A.

R. DE BRUINE. Technical University of Delft, c/o Lorentzweg 1, Delft, Netherlands.

F. BRYN. Norwegian Defence Research Establishment, NDRE Div. U, P.O. Box 115, 3191 Horten, Norway.

W. BUHRING. ASTRO-Forschungsinstitut fur Fjnk und Mathematik, 5307 Wachtberg-Werthhoven BRD, Germany.

v

D. BUTLER. Plessey Marine Research Unit, Wilkinthroop House, Templecombe, Somerset, U.K.

G. W. BYRAM. Naval Undersea Research and Development Center (Code 6021), San Diego, California 92132, U.S.A.

H. COX. Marine Physical Laboratory, Scripps Institution of Oceanography, University of California, San Diego, California 92152, U.S.A.

D. J. CREASEY. Department of Electronic and Electrical Engineering, University of Birmingham, P.O. Box 363, Birmingham B15 2TT, U.K.

R. DIEHL. Krupp Atlas Elektronik Bremen, D28 Bremen, Postfach 8545, Germany.

J. J. DOW. Tracor Inc., 6500 Tracor Lane, Austin, Texas 78721, U.S.A.

J. A. EDWARDS. Department of Electronic and Electrical Engineering, University of Birmingham, P.O. Box 363, Birmingham B15 2TT U.K.

R. M. FORTET. University of Paris, Laboratory Probabilites, Tour 56, Squai St. Bernard, Paris 5, France.

R. K. P. GALPIN. Plessey Telecommunications Research Ltd., Taplow Court, Maidenhead SL6 0ER, Berkshire, U.K.

R. B. GILCHRIST. Naval Ships Systems Command (Att. PMS 302), Washington D.C. 20360, U.S.A.

C. GIRAUDON. CIT-Alcatel, 1 Avenue Aristide Briand, 94 Arcueil, France.

H. F. HARMUTH. The Catholic University of America, Washington D.C. 20017, U.S.A.

D. HARVEY. Plessey Marine Research Unit, Wilkinthroop House, Templecombe, Somerset, U.K.

PER. HEIMDAL. Norwegian Defence Research Establishment, NDRE Div. U, P.O. Box 115, 3191 Horten, Norway.

W. HILL. Admiralty Underwater Weapons Establishment, Portland, Dorset, U.K.

E. HUG. SACLANT ASW Research Center, 400 Viale san Bartolomeo, La Spezia 19100, Italy.

J. JOHNSEN. Norwegian Defence Research Establishment, NDRE Div. U, P.O. Box 115, 3191 Horten, Norway.

C. D. JOHNSON. Transducer Materials and Reliability Branch, Naval Underwater Systems Center, Fort Trumbull, New London, Connecticut 06320, U.S.A.

J. R. KENEALLY. Admiralty Underwater Weapons Establishment, Portland, Dorset, U.K.

R. LAVAL. Underwater Acoustic Research Group, SACLANT-ASW Research Centre, Via S. Bartolomeo 400, La Spezia, Italy.

R. LAUVER. Bell Telephone Laboratories, Whippany, New Jersey, U.S.A.

I. G. LIDDELL. Admiralty Underwater Weapons Establishment, Portland, Dorset, U.K.

W. S. LIGGETT, JR. Raytheon, Submarine Signal Division, 5 Highland Road, Tiverton, Rhode Island 02878, U.S.A.

E. B. LUNDE. Norwegian Defence Research Establishment, NDRE Div. U, P.O. Box 115, 3191 Horten, Norway.

G. M. MAYER. Transducer Materials and Reliability Branch, Naval Underwater Systems Center, Fort Trumbull, New London, Connecticut 06320, U.S.A.

H. MERMOZ. Laboratoire de Detection Sousmarine, 83 Le Brusc, France.

D. MIDDLETON. Applied Research Laboratories, University of Texas, P.O. Box 8029, Austin, Texas 78712, U.S.A.

J. C. MOLDEN. Defence Research Establishment Atlantic, P.O. Box 1012, Dartmouth, Nova Scotia, Canada.

P. H. MOOSE. Naval Undersea Research and Development Center, (Code 602), California 92132, U.S.A.

D. NAIRN. Admiralty Underwater Weapons Establishment, Portland, Dorset, U.K.

A. M. NOLL. Executive Office of the President, Office of Science and Technology, Washington D.C. 20506, U.S.A.

L. W. NOLTE. Department of Electrical and Biomedical Engineering, Duke University, Durham, North Carolina 27706, U.S.A.

N. L. OWSLEY. U.S. Naval Underwater Systems Center, Code TD111, New London, Connecticut 06320, U.S.A.

F. PICHLER. Hochschule Linz, A 4020 Linz Auhof, Austria.

T. D. PLEMONS. Applied Research Laboratories, University of Texas, P.O. Box 8029, Austin, Texas 78712, U.S.A.

A. R. PRATT. Department of Electronic and Electrical Engineering, University of Technology, Loughborough, Leicestershire, U.K.

C. N. PRYOR. Signal Processing Division, Physics Ordnance Laboratory, Silver Spring, Maryland 20910, U.S.A.

H. A. REEDER. Tracor Inc., 6500 Tracor Lane, Austin, Texas 78721, U.S.A.

A. A. G. REQUICHA. SACLANT ASW Research Centre, Via S. Bartolomeo 400, 19026 La Spezia, Italy.

E. J. RISNESS. Admiralty Underwater Weapons Establishment, Portland, Dorset, U.K.

J. M. ROSS. Defence Research Establishment, Atlantics, P.O. Box 1012, Dartmouth, Nova Scotia, Canada.

R. DE RUITER. Technical University of Delft, c/o Lorentzweg 1, Delft, Netherlands.

C. VAN SCHOONEVELD. Physics Laboratory NDRO, Oude Waalsdorper-weg 63, The Hague, Netherlands.

P. M. SCHULTHEISS. Department of Engineering and Applied Science, Yale University, Bectan Center, New Haven Connecticut 06520, U.S.A.

R. SEYNAEVE. SACLANT ASW Research Center, 400 Viale San Bartolomeo, La Spezia 19100, Italy.

J. A. SHOOTER. Applied Research Laboratories, University of Texas, P.O. Box 8029, Austin, Texas 78712, U.S.A.

M. L. SOMERS. National Institute of Oceanography, Wormley, Nr. Godalming, Surrey, U.K.

J. M. SPEISER. Naval Undersea Research and Development Center (Code 6021). San Diego, California 92132, U.S.A.

G. A. VAN DER SPEK. Physics Laboratory TNO, Oude Waalsdorperweg 63, The Hague, Netherlands.

P. L. STOCKLIN. Department of Electronic and Electrical Engineering, Loughborough University of Technology, Loughborough, Leics. LE11 3TU, U.K.

F. J. M. SULLIVAN. Technisch Physische Dienst TNO-TH, Postbus 155, Delft, Netherlands.

R. S. THOMAS. Defence Research Establishment, Atlantic, P.O. Box 1012, Dartmouth, Nova Scotia, Canada.

D. E. TREMAIN. College of Engineering, Electronics Research Laboratory, University of California, Berkeley, California 94720, U.S.A.

B. P. TH. VELTMAN. Technical University of Delft, c/o Lorentzweg 1, Delft, Netherlands.

P. VERLOREN. Technical University of Delft, c/o Lorentzweg 1, Delft, Netherlands.

M. J. L. VERNET. Thomson-CSF, Chemin de Travails, 06 Cagnes sur Mer, France.

G. VETTORI. SACLANT ASW Research Centre, Viale S. Bartolomeo 400, 19026 La Spezia, Italy.

G. VEZZOSI. Laboratoire d'Etudes des Phenomenes Aleatories, Batiment 210, 91405 Orsay, France.

F. P. PH. DE VRIES. Physics Laboratory N.D.R.O., Oude Waalsdorperweg 63, The Hague, Netherlands.

H. J. WHITEHOUSE. Naval Undersea Research and Development Center (Code 6021), San Diego, California 92132, U.S.A.

F. WIEKHORST. FHP—Werthhoven Germany, 5307 Wachtberg Werthhoven, Germany.

S. S. WOLFF. Department of Electrical Engineering, The Johns Hopkins University, Baltimore, Maryland 21218, U.S.A.

Preface

This volume contains the papers presented at the NATO Advanced Study Institute on Signal Processing with Particular Reference to Underwater Acoustics, held 21 August–1 September, 1972 at the Loughborough University of Technology, Loughborough, Leicestershire, U.K. The purpose of the Institute organizers was to include, by means of both tutorial and research papers, all major technical fields which constitute signal processing in order to summarize in a single collection the state of the art in each field and to encourage communication between the fields. The international character of the meeting, with participants from twelve countries, contributed much to achieving this purpose. We believe these factors are well reflected in the contents.

Fields represented include spectrum analysis, numerical processing methods, acoustics and propagation, detection and estimation, adaptive processing and normalization, and displays. In many cases a summarized discussion of the paper also appears; this is intended to assist the reader in clarifying questions which he himself might have.

Both the scope and the technical treatments provide a comprehensive and timely text for professional scientists and engineers, as well as graduate students in engineering, communication and signal processing.

The Editors take pleasure in acknowledging the splendid cooperation of the following individuals and organizations, without whose help the Institute would not have been possible: Dr. T. Kester, Scientific Affairs Division NATO; Mr. Ralph A. Martin, Vice-President and General Manager, Submarine Signal Division, Raytheon Company; Loughborough University of Technology.

The consistently outstanding work of the publishers, Academic Press Inc. (London), removed much of the onus for author and editor alike, in the preparation of the book.

Finally, to the authors of the individual papers, who have worked so competently and conscientiously to provide this extremely important collection, we express our profound thanks.

J. W. R. GRIFFITHS
P. L. STOCKLIN
C. VAN SCHOONEVELD

Loughborough, June, 1973

Contents

List of Contributors v
Preface ix

Spectrum Analysis

Time-dependent Spectra: The Unified Approach 1
  MARTIN H. ACKROYD
The Cepstrum and Some Close Relatives 11
  A. MICHAEL NOLL
Walsh Functions—Introduction to the Theory 23
  F. PICHLER
Applications of Walsh Functions in Communications:
    State of the Art 43
  HENNING F. HARMUTH
Spectral Analysis of a Class of Random Signals 63
  ROBERT M. FORTET
Nonlinear Transformations of Random Processes 77
  NORMAN ABRAMSON
Minimum Detectable Signal for Spectrum Analyser Systems . . . 79
  C. NICHOLAS PRYOR
Spectrum Analysis of Reverberations 97
  JARL JOHNSEN
Some Statistical Properties of Nonstationary Reverberation . . . 117
  TERRY D. PLEMONS, JACK A. SHOOTER AND
  DAVID MIDDLETON
Some Remarks on the Use of Auto-correlation Functions with the
    Analysis and Design of Signals 131
  B. VELTMAN, A. VAN DEN BOS, R. DE BRUINE,
  R. DE RUITER AND P. VERLOREN
An adaptive Pre-whitening Filter for Radar Clutter Suppression . . 141
  WALTER BÜHRING

Numerical Processing Methods

Time Domain Filter Methods 143
 R. E. BOGNER
The Fast Fourier Transform and its Implementation 165
 D. BUTLER AND G. HARVEY
Interactive Signal Processing With a Minicomputer System and its
 Application to Sonar Research 183
 ROBERT SEYNAEVE, EDWARD HUG AND
 GIANCARLO VETTORI
Sound Propagation Digital Analysis System 205
 A. BARBAGELATA

Acoustics and Propagation

Sound Propagation Effects on Signal Processing 223
 R. LAVAL
Space-Time Relationships in Echo Wave Fields 243
 F. WIEKHORST
Detection of Underwater Sound Sources by Microwave Radiation
 Reflected from the Water Surface 255
 D. E. TREMAIN AND D. J. ANGELAKOS
Passive Ranging Techniques 261
 PER HEIMDAL AND FINN BRYN
Wave Front Stability in the Ocean 271
 EVEN BORTEN LUNDE
Shallow Water Acoustics Related to Signal Processing 281
 R. S. THOMAS, J. C. MOLDON AND J. M. ROSS
Automatic Fixed Site Propagation Tests 299
 T. G. BIRDSALL
Nonlinear Acoustics 311
 H. ORHAN BERKTAY

Detection and Propagation

Passive Sonar: Fitting Models to Multiple Time Series 327
 W. S. LIGGETT, JR.
Relation of Decision and Physical Wavefield Spaces: Concept and
 Example 347
 PHILIP L. STOCKLIN
Practical Limits of Time Processing 357
 RICHARD B. GILCHRIST
Detection of Extended Targets Against Noise and Reverberation . 363
 G. A. VAN DER SPEK

Some Theoretical Considerations Concerning Time Statistics in
Signal Detection 375
   A. R. PRATT
Robust Many-input Signal Detectors 401
   S. S. WOLFF AND F. J. M. SULLIVAN
Signal Processing in Reverberant Environments 413
   PAUL H. MOOSE
A Comparison of Several Data–Rate–Reduction Techniques for
Sonar 429
   J. J. DOW, B. M. BROWN AND H. A. REEDER
Graphical Comparison of Broadband and Tone-pulse Sonar
Transmissions 449
   D. NAIRN
Signal Processing Device Technology 457
   G. W. BYRAM. J. M. ALSUP, J. M. SPEISER AND
   H. J. WHITEHOUSE
Sampled-analogue Signal Processing 477
   R. K. P. GALPIN
Acoustic Signal Processing in Fast Unmanned Underwater Vehicles 489
   E. J. RISNESS

Array Processing

Results on Active Sonar Optimum Array Processing 495
   C. GIRAUDON
Design of Arrays to Achieve Specified Spatial Characteristics over
Broad Bands 507
   S. W. AUTREY
Broadband Hydrophone Arrays for Use With Explosive Sound
Sources 525
   ARISTIDES A. G. REQUICHA
Applications of Holographic Interferometry in Underwater
Acoustics Research 545
   C. D. JOHNSON AND G. M. MAYER
Holographic Processing of Acoustic Data Obtained by a Linear
Array 561
   R. DIEHL
Space–Time Processing for Optimal Parameter Estimation 577
   WILLIAM J. BANGS AND P. M. SCHULTHEISS
A Recent Trend in Adaptive Spatial Processing for Sensor Arrays:
Constrained Adaptation 591
   NORMAN L. OWSLEY
Enhancement of Antenna Performance by Adaptive Processing . . 605
   G. BIENVENU AND J. L. VERNET

Adaptive Processing and Normalization

Sensitivity Considerations in Adaptive Beamforming 619
  HENRY COX
Adaptive Processing: Time-varying Parameters 647
  LOREN W. NOLTE
What is Optimality for an Adaptive Detection System? 657
  G. VEZZOSI
Digital Logarithmic Normalization of Sonar Signals:
    Serial Processing 671
  C. VAN SCHOONEVELD
Digital Logarithmic Normalization of Sonar Signals:
    Batch Processing 691
  F. P. PH. DE VRIES
Normalization and Optimal Processors. 705
  W. HILL, G. H. ASH AND J. R. KENEALLY

Display Processing

Pattern Recognition. 715
  I. G. LIDDELL
Post Detection Information Processing 729
  ROBERT M. LAUVER
Three-Dimensional Displays 745
  D. J. CREASEY AND J. A. EDWARDS
Some Recent Results With a Long Range Side Scan Sonar 757
  M. L. SOMERS

Author Index 769

Time-dependent Spectra: The Unified Approach

MARTIN H. ACKROYD

Loughborough University of Technology, Leicestershire, England

1. Introduction

For the study of linear time-invariant systems or stationary random processes one can work in the time domain or in the frequency domain. One is not concerned with working in both domains at once. However as soon as non-stationarity becomes important, it is no longer sufficient to work in one domain alone. It then becomes necessary to study the distribution of the energy of a signal in a time-frequency plane.

On the evidence of our own ears it seems clear that the energy of a signal can be considered as being distributed in a time-frequency plane for we can perceive not only the pitch of a musical note but also the instant at which it is sounded. In fact, short-time spectra or spectrograms [1] were first used in speech research. Apart from the study of hearing there are other areas where concepts which involve both time and frequency simultaneously are used. For example, in radar a signal is subject to both a time delay and a frequency (Doppler) shift according to the range and radial velocity of the target. In communication the "instantaneous frequency" of the signal is varied as a function of time. In the analysis of random signals from non-stationary sources, a spectral description can involve both time and frequency.

Very many concepts which involve time and frequency both at once have been introduced into signal theory by many different authors, D. Gabor [2] being one of the first. Although these many diverse ideas are necessarily inter-related in some way, it is only recently that it has become evident just what the relations are. The central concept seems to be the time-frequency energy distribution of a signal, defined by $e(t, f) = s(t)S^*(f)\exp(-j2\pi ft)$ where $s(t)$ is the wave form of the signal and $S(f)$ is its Fourier transform. Each concept which involves both time and frequency turns out to be related in a simple way to the time-frequency energy distribution of a signal. It still remains for a complete account to be given of the unified approach to concepts involving time and frequency. However, A. W. Rihaczek's account [3] unifies several of these concepts in

1

terms of $e(t, f)$. W. D. Mark [4] has given a comprehensive discussion of the spectral analysis of non-stationary random processes and of short-time spectral analysis, which is based, effectively, on the use of the expected value of $e(t, f)$. A. A. Kharkevich [5] gives an interesting and readable early approach toward a unified theory.

2. The Time–Frequency Energy Distribution

The function $e(t, f) = s(t)S(f) \exp(-j2\pi ft)$ seems to have the properties that are required of a function which describes the distribution of the energy of a signal in the time-frequency plane. For example, if this function is integrated over all frequencies, the squared magnitude of the signal's waveform results. This of course describes the distribution of the energy of the signal as a function of time;

$$\int e(t, f)\, df = |s(t)|^2.$$

Similarly, integration over all time gives the energy density spectrum of the signal;

$$\int e(t, f)\, dt = |S(f)|^2.$$

Integration over the whole time-frequency plane yields the total energy of the signal;

$$\iint e(t, f)\, dt\, df = E.$$

Some other functions too have these properties. The justification for choosing the present form of $e(t, f)$ is that it bears simple relations to other concepts of interest and it also has a simple physical interpretation. [3, 6]

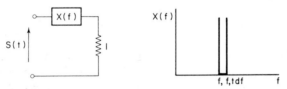

FIG. 1. Physical interpretation of $e(t, f)$.

The time-frequency energy distribution can be interpreted in a quasi-physical way by the artifice of a reactive circuit element which is imagined to have infinite reactance at all frequencies except for an infinitesimal band from f_1 to $f_1 + df$ where it has zero reactance, as depicted in Fig. 1. If the signal waveform is applied as a voltage across a circuit consisting of this circuit element in series with a unit resistor then $Re\ e(t, f_1)\, df$ gives the power entering the circuit at time t. It is thus reasonable to interpret

$Re\ e(t, f)$ as representing the power per unit bandwidth at time t and frequency f. The fact that $Re\ e(t, f)$ may take negative values does not contradict the laws of circuit theory; it merely corresponds to previously stored energy flowing from the reactive circuit element. Only the net flow of energy into the circuit need be positive. The magnitude of the complex quantity $e(t, f)\ df$ indicates the volt-amperes.

The time-frequency energy distribution involves the whole history of the signal, since its definition includes the Fourier transform of the waveform. It is to be noted that it is just as meaningful to deal with the time-frequency energy distribution of a signal as it is to deal with the Fourier transform of its waveform—no more and no less.

3. Functions of Time and Frequency

(a) *The ambiguity function*

The ambiguity function $\chi(\tau, \nu)$ is important in radar and sonar for it describes the response of a correlation receiver to the signal it matches after it has been subjected to a delay and a Doppler shift ν. The ambiguity function, which can be interpreted as being an autocorrelation function for shifts in the time *and* in frequency, can be defined by

$$\chi(\tau, \nu) = \int s(t)s^*(t + \tau) \exp(-j2\pi\nu t)\ dt$$

or, equivalently, by

$$\chi(\tau, \nu) = \int S^*(f)S(f + \nu) \exp(-j2\pi f\tau)\ df.$$

It was shown by Stutt [7] and by Levin [8] that the ambiguity function can be represented as being the inverse double Fourier transform of the time-frequency energy distribution;

$$\chi(\tau, \nu) = \iint e(t, f) \exp{-j2\pi(\nu t + f\tau)}\ dt\ df.$$

This relation is an interesting parallel to the Fourier transform relation between the energy density spectrum of a signal and its autocorrelation function for time shifts only. In the present case there is a double Fourier transform relation between the energy distribution in the time-frequency plane and the autocorrelation function for shifts of both time and frequency.

Possibly the principal use of this formula is in illustrating what a very special function the ambiguity function is. Its double Fourier transform, $e(t, f)$, is a function of which the principal cross-sections, $e(0, f)$ and $e(t, 0)$ themselves form a Fourier pair (apart from a scale factor). It is most unlikely that an arbitrarily chosen function will have this property and thus an arbitrarily chosen function is very unlikely to be realizable as an ambiguity function.

A formula of greater practical use is one, also discovered by Stutt [7], which says that the squared magnitude of an ambiguity function is the double autocorrelation function of the time-frequency energy distribution;

$$|\chi(\tau, \nu)|^2 = \iint e(t, f)e^*(t + \tau, f - \nu)\, dt\, df.$$

This formula often proves useful when the magnitude of an ambiguity function is to be sketched. $e(t, f)$ can usually be sketched directly from a knowledge of the waveform and its Fourier transform. It then remains to sketch the magnitude of the ambiguity function by performing the double autocorrelation either mentally or with the aid of a piece of transparent material which can be displaced by various values of ν and τ.

(b) Instantaneous frequency

The concept of instantaneous frequency is used in the theory of frequency modulation and it has also been applied to speech studies. With a real narrow band signal the instantaneous frequency is taken, loosely, as being one half the number of zero crossings per unit time in an interval which is short compared with the reciprocal of the signal's bandwidth. With a complex signal the instantaneous frequency $f_i(t)$ is defined as the rate of change of the instantaneous phase;

$$f_i(t) = 1/2\pi\, d/dt\, \arg s(t).$$

The instantaneous frequency is related to $e(t, f)$ in a simple way which does not depend on any assumption that the signal is narrowband or that $f_i(t)$ varies slowly [9]. It is easily verified that $f_i(t)$ is in fact that normalized first moment of a cross-section of $e(t, f)$ taken in the f-direction at each point in time;

$$f_i(t) = \int fe(t, f)\, df / \int e(t, f)\, df.$$

This relation shows that the instantaneous frequency is the "centre frequency" of the signal at a given instant even when the signal is not narrowband or where the instantaneous frequency fluctuates rapidly.

A very similar relation holds for the group delay function of a signal, defined by

$$\tau_g(f) = -1/2\pi\, d/df\, \arg S(f).$$

(c) Time-dependent spectra of nonstationary random processes

The time dependent autocorrelation function of a real nonstationary random process can be defined by

$$r(t, \tau) = \overline{s(t)s(t + \tau)}$$

where the bar denotes an ensemble average. There are other equivalent descriptions of the process corresponding to the Fourier transform of

$r(t, \tau)$ taken with respect to either or both variables. If $r(t, \tau)$ is Fourier transformed with respect to τ, a time-dependent power spectrum results.

$$R(t, f) = \int r(t, \tau) \, e^{-j2\pi f\tau} \, d\tau$$

$$= \int \overline{s(t)s(t + \tau)} \, e^{-j2\pi f\tau} \, d\tau.$$

Provided that the functions involved are docile enough, the order of integration and ensemble averaging can be interchanged to yield the time-dependent spectrum of the random process as the ensemble average of the time-frequency energy distributions of the sample functions of the random process;

$$R(t, f) = \overline{s(t)S^*(f) \exp(-j2\pi ft)}$$

$$= \overline{e(t, f)}.$$

If $R(t, f)$ is integrated over all frequency the expected value of the signal power as a function of time is obtained

$$\int R(t, f) \, df = \overline{S(t)^2}$$

while if it is integrated over all time and frequency the ensemble average of the signal energies is obtained (assuming it is finite).

$$\iint R(t, f) \, dt \, df = \bar{E}.$$

With some types of nonstationarity, the time-dependent power spectrum can be averaged with respect to time to yield a time-average power spectrum which can usefully be treated as the power spectrum of a stationary random process [5]. For example, a sine wave which has been amplitude modulated by a stationary random process constitutes a nonstationary random process. However, it is satisfactory for most purposes to average the time-dependent spectrum over one whole cycle of the sine wave and to deal with the resulting time-independent spectrum as if the modulated signal were stationary.

4. Short-Time Spectral Analysis

Short-time spectrum analysis is an important method for studying nonstationary signals such as in passive sonar, for example. Short-time spectra or spectrograms are commonly produced either by *sections in time* or *sections in frequency*.

(a) Sections in time
With this method a bank of bandpass filters with differing centre frequencies is used. The output of each filter is square-law rectified and is then subjected to further lowpass filtering. Figure 2 shows the overall scheme.

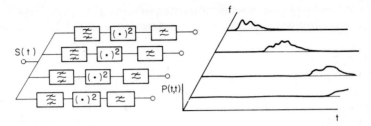

FIG. 2. Short-time spectrum analyser producing sections in time.

The output of each lowpass filter defines a "section in time" of a short-time spectrum, $P(t, f)$. The question arises as to how such a short-time spectrum is related to the time-frequency energy distribution of the signal. The answer is that $P(t, f)$ represents a version of $e(t, f)$ that has been smoothed out to some degree in the t-f plane [10]. If each bandpass filter has a transfer function which consists of the same lowpass prototype $H(f)$ translated to its centre frequency f_1, and if lowpass filtering after the square-law rectifiers is not used, the output of the channel with centre frequency f_1, can be expressed

$$P(t, f_1) = \left| \int S(f) H(f - f_1) \exp(j2\pi f t) \, df \right|^2.$$

This expression has the form of a cross-ambiguity function and can be expressed as a double convolution where $e(t, f)$ and $e_h(t, f)$ are respectively the time-

$$P(t, f) = \iint e(\tau, \nu) \, e_h(\tau - t, f - \nu) \, d\tau \, d\nu$$

frequency energy distributions of the signal and the impulse response of the lowpass prototype filter whose transfer function is $H(f)$. The short-time spectrum measured by a bank of similar bandpass filters is thus a version of $e(t, f)$ smoothed by convolution with the time-frequency energy distribution of the impulse response of the bandpass filter prototype. If lowpass filters following the square-law rectifiers are used, they effect a further smoothing in the time direction only.

This amounts to a further double convolution of the measured short-time spectrum with a function which has zero width in the frequency direction. The finally measured short-time spectrum thus represents a version of $e(t, f)$ which has been smoothed by a double convolution with a function which depends on the bandpass and lowpass filters used.

The width of the smoothing function is determined, in the frequency direction, solely by the bandwidths of the bandpass filters used. The width of the smoothing function in the time direction has a minimum value which is of the order of the reciprocal of the bandwidth of the bandpass filters. Its width in the time direction can be increased from this minimum value to any desired width by increasing the response time (reducing the cutoff frequency) of the filters which follow the square-law rectifiers. The extent of the smoothing function which characterizes a short-time spectrum analyser has no upper limit in either the time or frequency directions. It has a minimum effective area in the time frequency plane of the order of unity, since the impulse response of the lowpass prototype of the bandpass filters has a minimum duration–bandwidth product of the order unity. This is just another way of stating the well-known fact that a filter-bank analyser cannot simultaneously provide good resolution in both time and frequency.

(b) Sections in frequency

An alternate and now commonly used method of short-time spectral analysis involves the following procedure. The waveform of the signal is multiplied by a window function $w(t)$ which has been translated to a point t_1. The Fourier transform of the product of the signal waveform and the window function is computed. The squared magnitude of this Fourier transform is then taken to define a cross-section of the short-time spectrum taken in the frequency direction at time t_1. Each cross-section can, if required, be subjected to smoothing in the frequency direction by convolution with an appropriate function of frequency.

Short-time spectral analysis performed by computing sections in frequency can also be interpreted as producing a version of $e(t, f)$ that has been subject to a double convolution. This time, $e(t, f)$ is convolved with the time frequency energy distribution of the window function $w(t)$, together with a convolution with a function of zero width in the time direction, if the individual sections have been subject to further smoothing. In designing such a spectral analysis, T, the width of the window function $w(t)$ is chosen to give the required resolution in the time direction. The resolution in the frequency direction is then of the order of $1/T$. If further smoothing in the frequency direction is required this is effected by smoothing each cross-section. If a resolution better than T in the time direction and $1/T$ in the frequency direction is required a compromise is necessary.

The interpretation put on the result of a short-time spectrum analysis depends on one's point of view. If the signal is regarded as being a deterministic waveform $P(t, f)$ can be viewed as being a version of $e(t, f)$ which has been subject to distortion by the effect of the double convolution produced by the measuring scheme. This distortion or smearing may well be considered desirable since it removes small rapid fluctions in $e(t, f)$ which are of no interest. If the signal is considered to be a sample function from a nonstationary random process then the measured short-time spectrum $P(t, f)$ can be considered as providing an estimate of the time-dependent spectrum, $R(t, f)$, of the random process. Unfortunately there is little more than intuition available to guide one in choosing the window or filter transfer function widths and shapes to produce a good estimate of $R(t, f)$.

5. Conclusion

The time-frequency energy distribution $e(t, f)$ proves to be a useful concept in relating various ideas in signal theory which involve both time and frequency. It is useful in the study of both deterministic and non-stationary, random signals. Hopefully, the use of this function will lead to a better understanding of the problem of designing procedures for the spectral analysis of nonstationary signals. It has already led to new methods for designing radar and sonar signals.

REFERENCES

1. Koenig, W., Dunn, H. K. and Lacy, L. Y. (1946). The sound spectrograph. *J. Acous. Soc. Am.* **18**, 19.
2. Gabor, D. (1946). Theory of communication. *J. IEE*, 93 pt. 3, 429–457.
3. Rihaczek, A. W. (1968). Signal energy distribution in time and frequency. *Trans. IEEE* IT-14, 369–374.
4. Mark, W. D. (1970). Spectral analysis of the convolution and filtering of non-stationary stochastic processes. *J. Sound Vib.* 11, 19–63.
5. Kharkevich, A. A. (1960). "Spectra and Analysis" (Translated from the Russian), Consultants Bureau, New York.
6. Ackroyd, M. H. (1970). Instantaneous and time-varying spectra—an introduction. *Radio Electr. Eng.* **39**, 145–152.
7. Stutt, C. A. (1964). Some results on real-part/imaginary-part and magnitude/phase relations in ambiguity functions. *Trans. IEEE* IT-10, 321–327.
8. Levin, M. J. Instantaneous spectra and ambiguity functions. *Trans. IEEE* IT-10, 95–97.
9. Ackroyd, M. H. (1970). Instantaneous spectra and instantaneous frequency. *Proc. IEEE* **58**, 141.
10. Ackroyd, M. H. (1971). Short-time spectra and time-frequency energy distributions. *J. Acous. Soc. Am.* **50**, 1229–1231.

DISCUSSION

P. L. Stocklin: You mentioned, in your talk, the uncertainty principles involved in short-time spectrum analysis and in ambiguity functions. Can these be explained in terms of the $e(t, f)$ function?

Reply: Gabor's uncertainty principle says that the product of the r.m.s. duration of a waveform and the r.m.s. width of its Fourier transform cannot be less than unity. The minimum effective area of any $e(t, f)$ function in the *tf* plane is thus about unity. The impulse response of the lowpass prototype filter of a filter bank is, like any signal, subject to this restriction. When we do a short-time spectral analysis, we thus smear the $e(t, f)$ function of the signal by convolution with a function which has minimum area unity.

P. L. Stocklin: I had wondered if you might go along the lines of Gabor's logons in expanding the signal into elementary functions, each occupying minimum area in the time-frequency plane.

Reply: The $e(t, f)$ function is undoubtedly related in some way to the expansion of a signal in elementary functions but this has not been investigated.

P. M. Schultheiss: I wonder if you might comment on the notion of instantaneous frequency, which you have defined as the derivative of the instantaneous phase, when it is applied to a stationary random process and if you could comment on the difficulties which may arise. In particular, moments of the instantaneous frequency may not be finite.

Reply: I am sorry but I have no useful comments to make here. The paper by V. I. Bunimovich (Fluctuating processes as oscillations with random amplitudes and phases, *Zh. Tekh. Fiz.* (*USSR*) 19, No. 11, (1949), pp. 1231–1259. Office of Technical Services translation OTS 62-32700) reports a study of the statistical properties of the instantaneous frequency of Gaussian random waveforms.

J. W. R. Griffiths: One is always concerned about what can be done to produce a required ambiguity function or how we can "push the sand around the box", to use Woodward's analogy. Can the constraints on the ambiguity functions be cast in terms of the $e(t, f)$ function?

Reply: At the present time the significance of the known constraints on ambiguity functions are poorly understood. [Added later] For every constraint on the ambiguity function there must be some corresponding constraint on the $e(t, f)$ function, of course, because they are double Fourier transforms. It may well be profitable to consider the corresponding constraints on $e(t, f)$ with the hope of gaining new insights into ambiguity function properties.

The Cepstrum and Some Close Relatives

A. MICHAEL NOLL

Executive Office of the President, Office of Science and Technology, Washington, D.C., U.S.A.

1. Periodic Signals

(*a*) *Model*

The periodic signals treated in this paper are assumed to have a representation as the output of a time-varying linear filter with a quasiperiodic signal with time-varying period as input. The impulse response, $h(t)$, of the filter is assumed to vary very little with respect to time over a few periods, and the period of the excitation or source signal, $e(t)$, is assumed constant over a few periods. The resonant modes of the linear filter become peaks in the transfer function $H(\omega)$. The fundamental-frequency determination problem is given the signal, $f(t)$, determine the period of the excitation signal.

(*b*) *Rationale for autocorrelation analysis*

As mentioned previously, a periodic signal will be assumed to have a constant period over the analysis interval. If the periodic signal is infinitely repeated, a periodic signal with a spectrum consisting of delta functions spaced at the fundamental frequency results. The effect of a finite analysis interval with weights $w(t)$ for $|t| \leqslant T$ and 0 elsewhere is a convolution of the Fourier transform of $w(t)$ with the impulse tràin. The result is a "periodic" spectrum with spectral ripples.

To prevent confusion between the usual frequency components of a time function and the "frequency" ripples in the logarithm spectrum, Bogert and Tukey have used the paraphrased word "quefrency" in describing the "frequency" of the spectral ripples. Quefrencies have the units

NOTE: The cepstrum and some close relatives were the result of joint research efforts by the author and Dr. Manfred R. Schroeder while both were employed by Bell Telephone Laboratories, Incorporated in Murray Hill, New Jersey. Dr. Schroeder is presently Professor of Physics and Director of the III Physical Institute of the University of Göttingen in Germany.

The bulk of the material in this paper has been published previously by the author in the Journal of the Acoustical Society of America and the Proceedings of the Symposium on Computer Processing in Communications.

of cycles per hertz or, simply, seconds. Likewise, the "period" of the special ripples is called the repiod.

Since the usual method of determining the period of a periodic signal is to take its Fourier transform, the Fourier transform of the power spectrum should have a peak corresponding to the repiod of the power spectrum, and this peak should occur at a quefrency equal to the period of the original signal. All of this might sound strange, but it is nothing more than an intuitive development of the autocorrelation function based upon determining repiodicities in the power spectrum

(c) Rationale for cepstrum analysis

The excitation signal, $s(t)$, is the source of the periodicity, but this signal passes through the filter so that the final signal $f(t)$ is the convolution of the excitation $e(t)$ and the filter impulse response $h(t)$; $f(t) = s(t)*h(t)$. But then the power spectrum $|F(\omega)|^2$ of $f(t)$ equals the product of the power spectra of $s(t)$ and $h(t)$; $|F(\omega)|^2 = |S(\omega)|^2 |H(\omega)|^2$. Thus the autocorrelation function is the convolution of the inverse Fourier transforms of $|S(\omega)|^2$ and $|H(\omega)|^2$, and for many signals the transfer function $H(\omega)$ is such that the resulting convolution gives multiple peaks. The convolution becomes, in effect, a confusion! However, if the logarithm of the power spectrum is examined, the two spectra become additive; $\log|F(\omega)| = \log|S(\omega)| + \log|H(\omega)|$. As depicted below (Fig. 1), the log spectrum of the filter impulse response causes low quefrency ripples in the signal spectrum. The log spectrum of the excitation contributes repiodic fine structure to the signal spectrum. The power spectrum of the log spectrum therefore consists

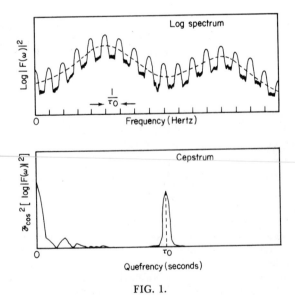

FIG. 1.

of a peak at a quefrency corresponding to the repiodic fine structure and low quefrency information corresponding to the filter response. These two effects are additive, and since they occupy different quefrency ranges they are separately distinguishable. Thus, simply by taking the logarithm of the power spectrum a new function has been derived. This new function is called the cepstrum, again borrowing Bogert and Tukey's terminology, and is formally defined as the power spectrum of the log power spectrum;

$$C(\tau) = |\,\mathscr{F}\,[\log|F(\omega)|^2]\,|^2 = |\,\mathscr{F}_{\cos}[\log|F(\omega)|^2]\,|^2,$$

where $\mathscr{F}$ and $\mathscr{F}_{\cos}$ represent the complex Fourier transformation and Fourier cosine transformation, respectively.

2. The Cepstrum

The cepstrum $C(\tau)$ has been formally defined as the power spectrum of the logarithm power spectrum. Since the log power spectrum is an even function, this definition is equivalent to the square of the cosine transform of the log power spectrum, or

$$C(\tau) \equiv \left\{ \int\limits_{0}^{\infty} \log|F(\omega)|^2 \cos(\omega\tau)\, d\omega \right\}^2.$$

The fundamental frequencies of many signals change with time; therefore a series of cepstra for short time segments of the signal are required. This is accomplished by multiplying the time signal by a function that is zero outside some finite time interval. The function performs something like a window through which the time signal is viewed, and its effects are discussed later in more detail. As shown in Fig. 2, the time-limited signal is spectrum analysed once to obtain the log spectrum and then again to produce the cepstrum. A new portion of the time signal then enters the window and is similarly analysed to produce another cepstrum. This process, when performed repetitively, results in a series of short-time cepstra. The time window, if desired, could also look at overlapping portions of the signal.

3. Some Close Relatives

(a) The clipstrum
The notation $\langle f(t) \rangle$ is introduced to indicate the operation of infinite peak clipping of the function $f(t)$. The subscript LQ is used to indicate the operation of low-quefrency liftering. The new function, called the

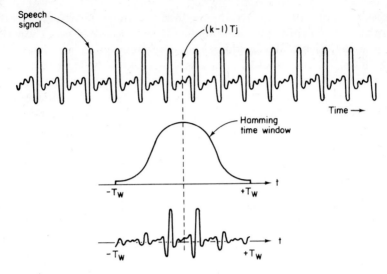

FIG. 2.

"clipstrum" in honour of all the peak clipping, is obtained by performing the following operations:

(1) the time signal $f(t)$ is infinitely peak clipped

$$g(t) \equiv \langle f(t) \rangle;$$

(2) The logarithm of the power spectrum of $g(t)$ is calculated

$$L(\omega) \equiv \log | \mathscr{F}[\langle f(t) \rangle] |^2;$$

(3) $L(\omega)$ is low-quefrency liftered in the frequency domain by a lifter with a hamming impulse response over a 600-Hz frequency interval. This operation is performed by first convolving the hamming frequency function with the log power spectrum thereby obtaining a smoothed log power spectrum, which is then subtracted from the log power spectrum. Expressed algebraically,

$$L_{LQ}(\omega) = L(\omega) - L(\omega)*h(\omega);$$

where

$$h(\omega) = \begin{cases} 0.54 + 0.46 \cos{(\omega/600)}; & |\omega| \leqslant 600\pi, \\ 0; & |\omega| > 600\pi, \end{cases}$$

(4) the liftered log spectrum is infinitely peak clipped

$$I(\omega) \equiv \langle L_{LQ}(\omega) \rangle;$$

and

(5) The Fourier cosine transform of $I(\omega)$ is calculated and squared, thereby producing the final result,

$$C(\tau) \equiv \{ \mathscr{F}_{\cos}[I(\omega)] \}^2.$$

Putting all of the preceding into one equation gives the following expression for the clipstrum of $f(t)$:

$$C(\tau) \equiv \{ \mathscr{F}_{\cos}[\langle \log |\mathscr{F}[\langle f(t) \rangle]|^2 - \log |\mathscr{F}[\langle f(t) \rangle]|^2 * h(\omega) \rangle] \}^2,$$

where $\mathscr{F}$ indicates the complex Fourier transform while $\mathscr{F}_{\cos}$ is the cosine transform only.

At first thought, all of this might seem very weird, but there is actually good reason to expect the clipstrum to perform reasonably well as a fundamental frequency detector. Infinite peak clipping does not eliminate the harmonic fine structure in the spectrum; on the contrary, additional harmonics are usually introduced. Thus, the log power spectrum of an infinitely peak-clipped signal should still possess a frequency spacing between the harmonics equivalent to the fundamental frequency. However, the log spectrum might consist of positive terms only and also possess some low-quefrency components so that liftering is required so that only the harmonic fine structure remains. The liftered spectrum can then also be infinitely peak clipped, and its Fourier cosine transform should have a peak corresponding to the harmonic spacing in the clipped, liftered spectrum.

(b) The harmonic product spectrum

The harmonic peaks in the log spectrum are harmonic multiples of the fundamental frequency. Therefore, compressing the frequency by integers and adding the resulting frequency-compressed version of the log spectrum would give a large peak at the coincidence of all the harmonics as shown in Fig. 3.

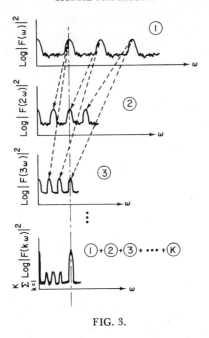

FIG. 3.

Stated mathematically

$$\log \pi(\omega) \equiv \sum_{k=1}^{K} \log |F(k\omega)|^2 = \log \prod_{k=1}^{K} |F(k\omega)|^2.$$

Taking the antilogarithm results in the harmonic product spectrum

$$\pi(\omega) \equiv \prod_{k=1}^{K} |F(k\omega)|^2.$$

The intuitive reasoning for the method is that the harmonic peaks in the log spectrum add coherently while the other portions of the log spectrum are uncorrelated and add noncoherently. The frequency compression results in a sharper final peak.

Since the low-quefrency structure of the log spectrum only adds confusions when compressed to form the harmonic product spectrum, these low quefrencies are removed. This is done by smoothing the spectrum with a hamming function and then subtracting the smoothed spectrum from the original spectrum before calculating the product spectrum. If $w(\omega)$ represents the frequency-domain counterpart of the hamming time window, i.e.

$$w(\omega) = \begin{cases} 0.54 + 0.46 \cos \omega/(2B); & |\omega| \leqslant 2\pi B \\ 0; & |\omega| \geqslant 2\pi B \end{cases}$$

then the low-quefrency liftered log spectrum $L_{LQ}(\omega)$ is

$$L_{LQ}(\omega) = \log|F(\omega)|^2 - w(\omega) * \log|F(\omega)|^2,$$

and the harmonic product spectrum becomes

$$\pi_{LQ}(\omega) = \prod_{k=1}^{K} L_{LQ}(k\omega).$$

(c) The harmonic sum spectrum

Although the harmonic product spectrum was derived intuitively from the log spectrum, there is really nothing sacred about using the log spectrum. The argument for coherent summation should also be valid for frequency-compressed version of the power spectrum itself. Thus, a harmonic sum spectrum $\sigma(\omega)$ can be defined as

$$\sigma(\omega) \equiv \sum_{k=1}^{K} |F(k\omega)|^2,$$

with its low-quefrency liftered counterpart

$$\sigma_{LQ}(\omega) \equiv \sum_{k=1}^{K} \log^{-1}[L_{LQ}(k\omega)].$$

(d) Maximum likelihood estimate

Although the peaks in the harmonic product spectrum might appear exciting because of their sharpness, the only important fact is that a peak above some fixed threshold be obtained at the fundamental frequency or period. In this respect, all three methods, namely, the cepstrum, the harmonic product spectrum, and the harmonic sum spectrum, perform satisfactorily. But clearly there should be an "optimum" method for funda-mental-frequency determination, and standard signal processing techniques of analysis should give the method. This motivation resulted in a maximum likelihood estimate of the period of a periodic signal. The mathematical derivation of the maximum likelihood estimate of the periodic signal was performed by Dr. David Slepian; an intuitive justification of the technique follows.

Consider a finite length T of a periodic signal $r(t)$ with period τ_0 (see Fig. 4). Although the period τ_0 is unknown, break up the signal into intervals of length τ and add them together to give a new function

$$R(t, \tau) \equiv \begin{cases} \dfrac{1}{N+1} \displaystyle\sum_{n=0}^{N} r(t + n\tau), & 0 \leqslant t < b \\[2em] \dfrac{1}{N} \displaystyle\sum_{n=0}^{N-1} r(t + n\tau), & b \leqslant t < \tau \end{cases}$$

where $T = N\tau + b$. Next, form the integral of the square of $R(t, \tau)$ weighting the different ranges in proportion to the number in the summation

$$J(\tau) \equiv (N + 1) \int_0^b R^2(t, \tau)\, dt + N \int_b^\tau R^2(t, \tau)\, dt.$$

All of the small intervals should add coherently when τ is chosen to equal the pitch period τ_0 so that $J(\tau)$ would be maximum at $\tau = \tau_0$. This value of τ which maximizes $J(\tau)$ is the maximum likelihood estimate of the period.

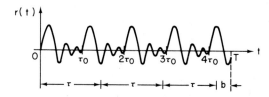

FIG. 4. Finite length T of a periodic speech signal with unknown period τ_0. The signal is broken up into N intervals of length τ so that $T = N\tau + b$.

4. Pitch Determination

Voiced-speech sounds result from the resonant action of the vocal tract on the periodic puffs of air admitted through the vocal cords. For pitch-period determination, the time periodicity of the source signal must be obtained from the observed speech signal. Also, voiced-unvoiced decisions require accurate determination of the presence or absence of such periodic puffs in the source signal. This deceptively simple problem has been the object of considerable research over the past few decades. Aside from its obvious use in analysis of speech sounds from a pure research standpoint, an accurate pitch detector must also perform adequately as an integral part of most speech-bandwidth compression schemes. The design of an accurate pitch detector that works satisfactorily with band-limited, noisy speech signals remains one of the challenging areas of speech processing research.

Determining the fundamental frequency of a periodic signal should be a relatively simple problem. Unfortunately this is not true for speech signals for a number of reasons. The pitch frequencies can cover a range of from 80 Hz for male speech to 400 Hz for female speech. These frequencies can vary considerably during a single speech utterance for a single person; in fact, sudden doubling of the pitch period is not at all uncommon. If the pitch detector is to work with telephone-quality speech, then it must work with signals for which the fundamental frequency might be absent. Also, phase shifts, poor signal-to-noise ratios, and nonlinear distortions must be expected. The patently simple problem of pitch determination becomes the very difficult and challenging problem of determining the pitch of a signal about which virtually no *a priori* statistics are known and which can be immersed in noise after being passed through nonlinear communication channels.

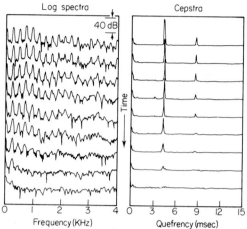

FIG. 5. Computer-calculated short-time log spectra (left) and corresponding short-time cepstra (right) for a segment of female speech recorded with a high-quality condenser microphone. For the short-time log spectra, frequency is plotted along the abscissa, log amplitude along the ordinate, and each successive plot is the log spectrum for a 40 ms speech segment. These speech segments overlap so that time progresses downwards in jumps of 10 ms. A 40 ms long hamming time window was used.

A realistic pitch detector must perform with signals for which the fundamental might be missing, for which the harmonics might be decreasing with frequency, and for which only a few harmonics might be available. All of these variations could not be evaluated using speech, so artificial computer-generated signals consisting of a sum of sinusoids immersed in computer-calculated white noise were generated. The same segment of white noise was used so that any fluctuations because of different random noise segments would be eliminated. The amplitudes and frequencies of the sinusoids were varied to produce a series of permutations with missing harmonics and decaying amplitudes in the spectra.

Some of the advantages claimed for cepstrum pitch detection and confirmed by computer simulation are, first, that the fundamental frequency component need not be present in the time signal, since the spectral ripples or fine structure caused by the harmonics give rise to the cepstral peak. For this reason, cepstrum pitch detection is particularly well suited to such bandpass-filtered signals as telephone speech. Since only the power spectrum is used, phase is completely ignored. Additive white noise is not too degrading if it does not destroy the spectral ripples.

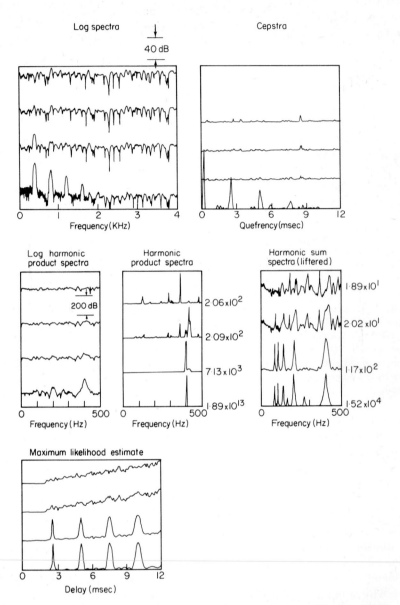

FIG. 6. Comparison of different period-determination techniques. The four signals analysed by each technique consist of four different levels of the same segment of white noise added to five sinusoids with amplitudes decreasing at 6 dB/octave. The four noise levels are decreasing in steps of 12 dB, and the average spectral level of the noise for the top curve is approximately equal to the spectral peak of the fundamental.

It seems that one is never satisfied; it was observed that cepstrum pitch determination failed for very noisy speech signals, and the quest for the "perfect" pitch detector started anew. Such hunts resulted in the clipstrum as an attempt to simplify hardware, and in the harmonic sum spectrum and harmonic product spectrum in an attempt to improve performance with noisy speech signals.

The general conclusion is that the harmonic product spectrum, harmonic sum spectrum, and maximum likelihood estimate all perform about equally well with the exception that multiple peaks with increasing amplitudes are obtained for the maximum likelihood estimate. Compared to the other methods, the cepstral peak does not become clearly recognizable until the noise level has decreased 12 dB. In other words, the other methods perform better than the cepstrum in detecting and determining the fundamental frequency of a signal made up of harmonically-related sinusoids imbedded in additive white noise.

Although in the evaluation of the various methods, there is some inferiority of the cepstrum when applied to sinusoids imbedded in white noise, the cepstrum nevertheless performs "better" than the other methods for pitch determination of non-noisy speech. In fact, the cepstrum gave the pitch period of an isolated flap of the vocal cords and also correctly indicated a doubling in pitch period at the end of a nasalized speech segment; the other methods gave either no or incorrect pitch period information for the identical speech signals. Nevertheless, the other methods, and in particular the harmonic product spectrum, perform better with very noisy signals and might also find application in areas other than pitch determination.

REFERENCES

1. Bogert, B. P., Healy, M. J. R. and Tukey, J. W. (1963). The quefrency alanysis of time series for echoes: cepstrum, pseudo-autocovariance, cross-cepstrum and saphe cracking, (M. Rosenblatt, Ed.) "Proceedings of the Symposium on Time Series Analysis", 209–243, John Wiley & Sons, Inc., New York.
2. Cooley, J. W. and Tukey, J. W. (1964). An algorithm for the machine calculation of complex Fourier series. *Math Comp.* 19, 297–301.
3. Kincaid, Thomas G. and Scudder, Henry J. III, (1967). Estimation of periodic signals in noise, *J. Acoust. Soc. Am.* 42, No. 5, 1166.
4. Miller, R. L. (1968). Pitch determination by measurement of harmonics, *J. Acoust. Soc. Am.* 44, No. 1, 390.
5. Noll, A. Michael (1964). Short-time spectrum and "cepstrum" techniques for vocal-pitch detection. *J. Acoust. Soc. Am.* 36, No. 2, 296–302.
6. Noll, A. Michael (1967). Cepstrum pitch determination. *J. Acoust. Soc. Am.* 41, No. 2, 293–309.
7. Noll, A. Michael (1968). Clipstrum pitch determination. *J. Acoust. Soc. Am.* 44, No. 6, 1585–1591.
8. Noll, A. Michael. Pitch determination of human speech by the harmonic product spectrum, the harmonic sum spectrum, and a maximum likelihood estimate.

"Proceedings of the Symposium on Computer Processing in Communications",
Vol. XIX, 779–797, Polytechnic Press, Brooklyn, New York.

9. Schroeder, M. R. (1966). Vocoders: analysis and synthesis of speech. *Proc. IEEE*
54, No. 5, 720–734.

10. Schroeder, M. R. (1968). Period histogram and product spectrum: new methods
for fundamental-frequency measurement. *J. Acoust. Soc. Am.* 43, No. 4, 829–834.

11. Sondhi, M. M. (1968). New methods of pitch extraction. *IEEE Trans. Audio
Electroacoust.* **AU-16**, 262–266.

Walsh Functions —Introduction to the Theory

F. PICHLER

Hochschule Linz, Linz, Austria

1. Introduction

This paper attempts an introduction to those parts of the theory of Walsh functions which have gained importance by their application to communication, especially in signal processing. As this theory is largely analogous to the well-known theory of sinusoidal functions and the related theory of Fourier transforms it is not difficult to acquire. In general we shall not be able to give proofs of the statements referred to here but we are trying to point to specific literature for further study. Seen mathematically Walsh functions represent in the main the characters of a specific locally compact Abelian group, viz. the dyadic group. A theory of these functions should, therefore, be considered as a special case of the general theory of characters. Similar considerations apply also to the theory of Walsh–Fourier transforms; this may also be considered in principle as a special case of the general Fourier transforms on groups. For these reasons we discuss in Section 2 briefly the most important formulations of concepts of abstract harmonic analysis. This does not, however, mean that the study of the theory of Walsh functions and of the theory of Walsh–Fourier transforms can only be attempted after study of this general theory. The point is to draw the attention of the reader to the possibility of this embedding, so that he may also better appreciate the association with the classical theory of Fourier transforms and at the same time also see the possibilities of changes in the theory which may be desirable.

Section 3 deals with the most important properties of Walsh functions. We consider there both the continuous Walsh functions as stated first by Fine [4] and the continuous-time sequency ordered Walsh functions whose mathematical presentation stems in the main from the author [22] as well as the discrete-time finite Walsh functions whose mathematical theory has been mainly developed by Gibbs [7, 8, 9, 10, 11].

Section 4 concerns the Walsh–Fourier transforms. Whilst this may in the case of continuous-time Walsh functions be mathematically more demanding only elementary results of linear algebra are used for finite discrete Walsh functions and is therefore easily understood.

23

Section 5 in conclusion aims to examine certain theoretical concepts developed for the application of Walsh functions in communications. The discussion covers *inter alia* a generalization of the sampling theorem, the dyadic linear filters and their optimization, the dyadic correlation functions and related power spectra. For a complete bibliography on Walsh Functions and their applications we refer to Bramhall [2].

2. Abstract Harmonic Analysis

A locally compact Abelian group (LCA-group) is defined as an Abelian group which is at the same time also a locally compact Hausdorff space and which has the additional property that the operations $x \mapsto -x$ and $(x, y) \mapsto x + y$ are continuous. For a given LCA group G for each $x \in G$ the composition $(x, y) \mapsto x + y$ determines an homomorphism from G in G, where the zero element 0 of G is mapped to itself. It follows that the topological properties around an arbitrary point $x \in G$ are equal to those of the point 0 of G. For each LCA group G there exists an unique measure $\mu(x)$ called the Haar measure of G which is translation invariant. For a given Haar measure $\mu(x)$ on G a theory of integration of complex valued functions, defined on G, can be introduced (Haar integral). It can be shown that the set $L(G)$ of Haar-integrable functions is a Banach space. Furthermore one can introduce the Hilbert space $L_2(G)$ of square Haar-integrable functions. For a given $f \in L_2(G)$ it is often desirable to represent it by a linear combination of simpler functions which are orthogonal. This can be achieved by the Fourier representation of f with respect to the character functions of the LCA-group G (for the LCA-group given by the real numbers R the character functions are given by the functions $\exp(i\omega(.))$, $\omega \in R$). Any character function χ of G is determined by the following properties.

   (1) χ is continuous on G
   (2) $|\chi(x)| = 1$ for all $x \in G$
   (3) $\chi(x + y) = \chi(x)\chi(y)$ for all $x, y \in G$.

Let $\hat{G}$ denote the set of all character functions on G. If we introduce in $\hat{G}$ the addition $+$ by $\chi + \chi'(x) := \chi(x) + \chi'(x)$ then $\hat{G}$ becomes an Abelian group. Furthermore it can be shown that with the compact-open topology $\hat{G}$ becomes an LCA-group which is called the character group of G. The famous duality theorem of Pontryagin states that the character group of $\hat{\hat{G}}$ is given by the group G. For any function $f \in L(G)$ the Fourier transform $\hat{f}$ of f is defined by $\hat{f}: \hat{G} \to C$ (C denotes the set of complex numbers) and

$$\hat{f}(\chi) := \int_G \chi(x) f(x) \, d\mu(x) \quad \text{for all } \chi \in \hat{G}. \tag{1}$$

The definition of a Fourier transform in the sense above can be extended to the Hilbert space $L_2(G)$. Then for a given function $f \in L_2(G)$ it can be shown that the Fourier transform $\hat{f}$ of f is an element of the Hilbert space $L_2(\hat{G})$ and in addition $\| \hat{f} \| \triangleq \| f \|$ (theorem of Plancherel). Given two functions $h \in L(G)$ and $f \in L(G)$ the convolution product $h * f$ of h and f is defined by the integral

$$(h * f)(x) := \int_{G} h(x - y)f(y) \, d\mu(y) \quad \text{for all } x \in G. \tag{2}$$

It can be shown that $h * f \in L(G)$. The convolution theorem asserts that we have

$$(h * f) (\chi) = \hat{h}(\chi)\hat{f}(\chi) \quad \text{for all } \chi \in \hat{G}. \tag{3}$$

This should conclude our introductory remarks concerning the theory of abstract harmonic analysis. For detailed study the reader is advised to consult the books of Hewitt and Ross [14] and Gelfand et al. [6].

3. Walsh Functions

(a) Walsh functions ψ_y
The Walsh functions ψ_y (Fine's definition†) are essentially the character functions of a concrete given LCA-group, i.e. the dyadic group Δ of binary numbers. Δ consists of the set of all 0–1 sequences of the form $(t_i) = (\ldots, 0, t_{-N}, \ldots, t_0.t_1, \ldots,)$ which are 0-stationary to the left. The addition $\oplus$ in Δ is given by pointwise addition modulo 2. Therefore if $t, t' \in \Delta$ the sum $t \oplus t'$ is given by the 0–1 sequence $((t \oplus t')_i)$ which is given by

$$(t \oplus t')_i := (t_i + t'_i) \bmod 2 \quad \text{for all integers } i. \tag{4}$$

The map $\lambda : \Delta \to R$ defined by $\lambda(t) := \Sigma_i t_i 2^{-i}$ associates each sequence $t \in \Delta$ with a corresponding non-negative real number $\lambda(t)$. Furthermore it is convenient to introduce the maps $\mu : R_+ \to \Delta$ and $\nu : R_+ \to \Delta$ (R_+ denotes the set of non-negative real numbers) which are defined in the following way: for $t \in R_+$ the image $\mu(t)$ is given by the 0–1 sequence $\mu(t) = (t_i)$, where the coefficients t_i are derived from the dyadic representation $t = \Sigma_i t_i 2^{-i}$. If t is dyadic rational, which means that t can be represented by a finite sum of powers of 2, then to get the coefficients t_i, we have to take the finite representation of $t \in R_+$. Similarly, to define the map ν we determine $\nu(t) = (t_i)$ with $t = \Sigma_i t_i 2^{-i}$. Now, in case that t is dyadic rational the infinite dyadic representation has to be used. Clearly for all $t \in R_+$ $\lambda(\mu(t)) = t$ and also $\lambda(\nu(t)) = t$. Furthermore $\mu(t) = \nu(t)$ if t is not a dyadic rational number. If t is dyadic rational there exists an integer M such that for the sequence $\mu(t) = (t_i)$ $t_M = 1$ and $t_i = 0$ for all $i > M$. For instance if we take the dyadic rational number $t = 7$ then $M = 0$ and $\mu(7) =$

---

† It should be mentioned that also the characters of more general LCA-groups are often called Walsh functions. Examples are given by the functions introduced by Levy [17], Vilenkin [28] and Chrestenson [3].

$(\ldots, 0, 1, 1, 1, 0, \ldots)$ and $\nu(7) = (\ldots, 0, 1, 1, 0, 1, 1, 1, \ldots)$. The maps λ and likewise μ and ν are used in the definition of the Walsh functions. We shall see that the Walsh functions ψ_y, $y \in R_+$, as introduced by Fine [5] are defined on R_+. Therefore, strictly speaking, they are not really characters since R_+ together with the dyadic addition $\oplus$, which can be defined by $t \oplus s := \lambda(\mu(t) \oplus \mu(s))$, is only a LCA-group "almost everywhere". But the exceptions are always of measure zero, so the theory differs not strongly.

For arbitrary $y \in R_+$ Walsh function ψ_y is defined by

$$\psi_y(t) := {}^{+1}_{-1} \text{ if the sum } \sum_i y_i t_{1-i} \text{ is } {}^{\text{even}}_{\text{odd}} \quad (t \in R_+), \tag{5}$$

where the coefficients y_i and t_i are given by $\mu(y) = (y_i)$ and $\mu(t) = (t_i)$ respectively. From the given definition it is easy to derive that for all $y, z \in R_+$ for which the sequence $\mu(y) \oplus \mu(z)$ is not 1-stationary to the right we have the formula

$$\psi_y(t)\psi_z(t) = \psi_{y \oplus z}(t) \quad \text{for all } t \in R_+. \tag{6}$$

Since the set of all numbers y, z for which (6) is not true is countable and therefore of measure zero, we can conclude that the set Ψ consisting of all Walsh functions ψ_y is "almost everywhere" closed by multiplication. Since in addition we have the neutral element ψ_0 and since $\psi_y \psi_y = \psi_0$ for all $y \in R_+$, we can conclude, that Ψ is "almost everywhere" a multiplicative Abelian group. A second formula which is easy to derive from our definition of ψ_y is given by

$$\psi_y(t)\psi_y(t') = \psi_y(t \oplus t') \tag{7}$$

for all $y \in R_+$ and all pairs (t, t') for which $\mu(t) \oplus \mu(t')$ is not 1-stationary to the right. Also we have

$$\psi_y(t) = \psi_t(y) \quad \text{for all } y, t \in R_+. \tag{8}$$

So far we have avoided mentioning any results about the topology of the dyadic group Δ. For our purpose it will be sufficient to state that a basis of neighbourhoods of the zero element 0 is given by the sets of the form $\{(t_i) : t_i = 0 \text{ for all } i < N\}$ where N denotes an integer. The topology generated by this basis makes Δ a locally compact group which is totally disconnected. The topology on R_+ is induced by the homeomorphism λ: $\Delta \to R_+$. With respect to this topology the Walsh functions ψ_y are continuous. On the other hand with respect to the topology on R_+ which is induced by the usual topology of the real numbers R, the Walsh functions ψ_y, $y \neq 0$, are rectangular functions and therefore discontinuous. One further property of the Walsh functions ψ_y should be mentioned: two Walsh functions ψ_y and ψ_z with y "near" to z have equal shape on a "long" common interval. The length of this interval increases as the distance between y and z decreases. More precisely: Given that $0 < \epsilon < 2^{-M}$ for almost all $y, z \in R_+$ with $z = y + \epsilon$ we have $\psi_y(t) = \psi_z(t)$ for all $t \in [0, 2M]$. This property of the Walsh functions is easy to prove by use of the given definition. For $y = 0, 1, 2, 3, \ldots$ the Walsh functions ψ_y are exactly the functions as originally introduced into mathematics by the work of J. L. Walsh [30]. We should also refer to the remarkable paper of Paley [20].

(b) Walsh functions sal(s, .) and cal(s, .)

In the work of Harmuth [13] a new method for the design of communication systems is proposed. This method is based on the representation of signals and systems by Walsh functions. For historical but also for other reasons he does not use the Walsh functions ψ_y according to Fine's definition but the "sequency ordered" Walsh functions sal(s, .) and cal(s, .). Although these Walsh functions are in essence not different from the functions ψ_y they have certain properties which make them more similar to the sinusoidal functions $\sin 2\pi f(.)$ and $\cos 2\pi f(.)$. Like the frequency parameter f of these functions, the parameter s of sal(s, .) and cal(s, .), respectively, is equal to half the number of sign changes (zero crossings) over an interval of unit length. Therefore the parameter s represents a generalized frequency, which has been called sequency.

Furthermore the functions sal(s, .) are all odd functions while the functions cal(s, .) are even, both defined on the whole real axis. For the functions sal(s, .) and cal(s, .) we introduce the following definition: for arbitrary $s > 0$ the function sal(s, .) is defined by sal(s, .):$R \to C$ which is given by

$$\text{sal}(s, t) := \begin{matrix} +1 \\ -1 \end{matrix} \text{ if the sum } \sum_k (s_k + s_{k+1})t_{1-k} \text{ is } \begin{matrix} \text{even} \\ \text{odd} \end{matrix} \qquad (9)$$

for $t \geqslant 0$ where the coefficients s_i and t_i are determined by $\nu(s) = (s_i)$ and $\mu(t) = (t_i)$, respectively. For $t < 0$ we define

$$\text{sal}(s, t) := -\text{sal}(s, -t). \qquad (10)$$

Similarly, for arbitrary $s \geqslant 0$ the function cal(s, .) is defined by cal(s, .): $R \to C$, given by

$$\text{cal}(s, t) := \begin{matrix} +1 \\ -1 \end{matrix} \text{ if the sum } \sum_k (s_k + s_{k+1})t_{1-k} \text{ is } \begin{matrix} \text{even} \\ \text{odd} \end{matrix} \qquad (11)$$

Now the coefficients s_i stem from $\mu(s) = (s_i)$ and the coefficients t_i are again given by $\mu(t) = (t_i)$. For $t < 0$ we define

$$\text{cal}(s, t) := \text{cal}(s, -t). \qquad (12)$$

As we did for the functions ψ_y we shall also mention some of the properties of the Walsh functions sal$(s, .)$ and cal$(s, .)$. The following formulae are not difficult to derive: For any integer $\bar{k}$ we have

$$sal(s, 2^k t) = sal(2^k s, t) \quad \text{for all } s > 0 \text{ and all } t \in R, \tag{13}$$

$$cal(s, 2^k t) = cal(2^k s, t) \quad \text{for all } s \geqslant 0 \text{ and all } t \in R. \tag{14}$$

Formulae (13) and (14) describe dilatations of the Walsh functions sal$(s, .)$ and cal$(s, .)$, respectively. In the case of sinusoidal functions $\sin 2\pi f(.)$ and $\cos 2\pi f(.)$ we know that a shift of $\pi/4f$ causes them to coincide. For the Walsh functions sal$(s, .)$ and cal$(s, .)$ the following formula is valid: Let s be a dyadic rational number and let the integer M be given by $s_M = 1$ and $s_i = 0$ for all $i > M$, where $\mu(s) = (s_i)$. Then we have

$$sal(s, t) = cal(s, t - 2^{M-2}) \quad \text{if } s_{M-1} = 0, \tag{15}$$

$$sal(s, t) = cal(s, t + 2^{M-2}) \quad \text{if } s_{M-1} = 1. \tag{16}$$

For $s = 2^k$, where k is an integer, the Walsh functions sal$(s, .)$ and cal$(s, .)$ have the shape of regular rectangular waves. In mathematics these special Walsh functions are known as Rademacher functions. It is easy to observe that each Walsh function can be represented by products of Rademacher functions. The following formula concerning Rademacher functions is sometimes of practical use. We have

$$cal(2^k, t) = sal(2^k, t) \, sal(2^{k+1}, t) \quad \text{for all } t \in R. \tag{17}$$

Of special interest are the Walsh functions sal$(s, .)$ and cal$(s, .)$ where s is a non-negative integer. The set $\{cal(0, .), cal(1, .), \ldots, sal(1, .), sal(2, .), \ldots\}$ forms, as one can prove, a complete set of orthogonal functions of the Hilbert space $L_2(-1/2, +1/2)$. We will discuss this set in Section 4. Sometimes it is convenient to use a slightly different notation in case of this function set, namely the Walsh functions wal$(i, .)$, which are defined by

$$wal(2i, .) := cal(i, .) \qquad \text{for } i = 0, 1, 2, \ldots, \quad \text{and}$$
$$wal(2i - 1, .) := sal(i, .) \qquad \text{for } i = 1, 2, 3, \ldots \tag{18}$$

In practice with modulation or filtering problems the following formulae are often useful: For $i, k = 0, 1, 2, \ldots$

$$cal(i, .) \, cal(k, .) = cal(i \oplus k, .),$$
$$cal(i, .) \, sal(k + 1, .) = sal((i \oplus k) + 1, .),$$
$$sal(i + 1, .) \, sal(k + 1, .) = cal(i \oplus k, .). \tag{19}$$

Furthermore for the functions wal$(i, .)$

$$\text{wal}(i, .) \, \text{wal}(k, .) = \text{wal}(i \oplus k, .). \tag{20}$$

Finally we show the correlation between the Walsh functions sal$(i, .)$ and cal$(i, .)$ and the Walsh functions ψ_y of Fine. We have

$$\text{cal}(i, t) = \psi_{i \oplus 2i}(t) \qquad \text{for all } t \geqslant 0 \text{ and } i = 0, 1, 2, \ldots, \quad \text{and}$$

$$\text{sal}(i, t) = \psi_{(i-1) \oplus (2i-1)} (t) \quad \text{for all } t > 0 \text{ and } i = 1, 2, 3, \ldots. \tag{21}$$

Figure 1 shows the Walsh functions cal$(i, .)$, $i = 0, 1, \ldots, 7$, and sal$(i, .)$, $i = 1, 2, \ldots, 8$, for the interval $(-1, +1)$.

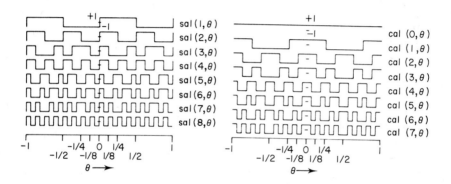

FIG. 1. Walsh functions sal$(i, .)$ and cal$(i, .)$.

(c) Discrete Walsh functions $w(i, .)$

Practical applications of the theory of Walsh functions in communications and signal processing have shown that in most of the cases it is sufficient to use a finite set of discrete Walsh functions. Then the theory becomes much easier and only elementary results of linear algebra are needed. In the following we define the discrete Walsh functions $w(i, .)$ and we discuss some of their properties. For more detailed studies the reader is advised to consult the work of Gibbs [7, 8, 11], Gibbs and Millard [12], Pearl [21] and Pichler [25].

Let $B(n)$ denote the set of all binary numbers with n digits. In $B(n)$ we define the addition $\oplus$ by the pointwise "addition modulo 2" of binary numbers. Then $B(n)$ becomes a subgroup of the dyadic group Δ. Consider the set $N(n) := \{0, 1, 2, \ldots, 2^n - 1\}$ consisting of all integers less than 2^n. If we introduce the "dyadic addition" $\oplus$ in $N(n)$ by $i \oplus k := \lambda(\mu(i) \oplus \mu(k))$ for all $i, k \in N(n)$, then $N(n)$ becomes a group which is isomorphic to $B(n)$. It will be convenient for us to identify the groups $B(n)$ and $N(n)$.

The discrete Walsh functions $w(i, .)$ associated with $B(n)$ are defined as functions $w(i, .):B(n) \to C$, $i = 0, 1, 2, \ldots, 2^n - 1$, which are given by

$$w(i, k) := \begin{matrix} +1 \\ -1 \end{matrix} \text{ if the sum } \sum_{p+q=n-1} i_p k_q \text{ is } \begin{matrix} \text{even} \\ \text{odd} \end{matrix} \quad (i, k \in B(n)). \qquad (22)$$

Let $W(n)$ denote the set of all discrete Walsh functions $w(i, .)$ which are associated with $B(n)$. For $n = 3$ the set $W(n)$ is given by Table I.

TABLE I. The set $W(3)$ of discrete Walsh functions; $+ := +1$, $- := -1$.

	000	001	010	011	100	101	110	111
$w(0, .)$	+	+	+	+	+	+	+	+
$w(1, .)$	+	+	+	+	−	−	−	−
$w(2, .)$	+	+	−	−	+	+	−	−
$w(3, .)$	+	+	−	−	−	−	+	+
$w(4, .)$	+	−	+	−	+	−	+	−
$w(5, .)$	+	−	+	−	−	+	−	+
$w(6, .)$	+	−	−	+	+	−	−	+
$w(7, .)$	+	−	−	+	−	+	+	−

In analogy to the continuous-time Walsh functions for the discrete Walsh functions the following formulae are valid:

$$w(i, k) = w(k, i) \quad \text{for all } i, k \in B(n), \qquad (23)$$

$$w(i, .)w(j, .) = w(i \oplus j, .) \quad \text{for all } i, j \in B(n), \qquad (24)$$

$$w(i, k)w(i, l) = w(i, k \oplus l) \quad \text{for all } i, k, l \in B(n). \qquad (25)$$

Sometimes, especially in coding theory, it is convenient to modify the discrete Walsh functions $w(i, .)$ into functions $b(i, .)$ which assume the values 0 and 1 by the following definition. For each $i \in B(n)$ the function $b(i, .):B(n) \to B(1)$ is defined by

$$b(i, k) := \bigoplus_{p+q=n-1} i_p k_q \quad \text{for all } k \in B(n), \qquad (26)$$

where $\oplus$ denotes modulo 2 summation. The functions $b(i, .)$ can be called the boolean Walsh functions associated with $B(n)$. We see that the boolean Walsh functions $b(i, .)$ are derived by the discrete Walsh functions by the replacements $+1 \mapsto 0$ and $-1 \mapsto 1$. In analogy to equations (24) and (25) we can write the following formulae

$$b(i, .) \oplus b(j, .) = b(i \oplus j, .) \quad \text{for all } i, j \in B(n), \qquad (27)$$

$$b(i, k) \oplus b(i, l) = b(i, k \oplus l) \text{ for all } i, k, l \in B(n). \qquad (28)$$

4. Walsh Harmonic Analysis

Since both the Walsh functions ψ_y and the Walsh functions $\text{sal}(s, .)$ and $\text{cal}(s, .)$ are essentially the character functions of a locally compact

Abelian group one can expect from the theory of abstract harmonic analysis a theory of Walsh harmonic analysis, that is a theory of Walsh-Fourier transforms. This is true and the said approach is justified. But the theory of Walsh harmonic analysis has so far been developed independently from abstract harmonic analysis. Here we are forced to restrict our considerations to some of the principal concepts of this theory. For further study the reader is advised to consult the papers of Fine [5], Vilenkin [29], Selfridge [27], Pichler [22] and Wieser [32]. In the case of the discrete Walsh harmonic analysis no special theoretical background is necessary. The mathematical tools needed are well-known results from linear algebra.

(a) Walsh–Fourier series

We have already mentioned that the set $\{\text{cal}(0, .), \text{cal}(1, .), \ldots, \text{sal}(1, .), \text{sal}(2, .), \ldots\}$ consisting of the Walsh functions $\text{sal}(s, .)$ and $\text{cal}(s, .)$ where s is an integer, represents a complete set of orthogonal functions of the Hilbert space $L_2(-1/2, +1/2)$. Therefore each function $f \in L_2(-1/2, +1/2)$ can be represented by a generalized Fourier series which is based on this set of Walsh functions (Walsh–Fourier series of f). At a point $t \in (-1/2, +1/2)$ such a series has in general the form

$$f(t) = F_c(0)\text{cal}(0, t) + \sum_{i=1}^{\infty} F_c(i)\text{cal}(i, t) + F_s(i)\text{sal}(i, t), \tag{29}$$

where the "Walsh–Fourier coefficients" $F_c(0)$, $F_c(i)$ and $F_s(i)$, respectively, are for $i = 1, 2, 3, \ldots$, given by

$$F_c(i) := \int_{-1/2}^{+1/2} f(t)\text{cal}(i, t)\, dt \quad \text{for all } i = 0, 1, 2, \ldots, \quad \text{and} \tag{30}$$

$$F_s(i) := \int_{-1/2}^{+1/2} f(t)\text{sal}(i, t)\, dt \quad \text{for all } i = 1, 2, 3, \ldots. \tag{31}$$

The theory of Walsh–Fourier series (WF-series) has a long tradition in mathematics. It was already investigated by the original paper of J. L. Walsh [30]. Later important contributions to this theory stem from Paley [20], Fine [4], Morgenthaler [19], Watari [31], Weiss [33] and others. Let us give two examples of Walsh–Fourier series.

Example 1. We consider the sawtooth function $\text{saw} \in L_2(-1/2, +1/2)$ which is given by $\text{saw}(t) := t$ for all $t \in (-1/2, +1/2)$. At a time $t \in (-1/2, +1/2)$ the Walsh–Fourier series has the form

$$\text{saw}(t) = 1/4\, \text{sal}(1, t) - \sum_{i=1}^{\infty} 2^{-i-2}\, \text{sal}(2^i, t). \tag{32}$$

Example 2. We consider the rectangular pulse $\text{rect} \in L_2(-1/2, +1/2)$ which is given by $\text{rect}(t) := 2^{k+1}$ for all $t \in [0, 2^{-k-1})$ and $\text{rect}(t) := 0$ for all $t \in [-1/2, 0) \cup [2^{-k-1}, +1/2)$. For each time point $t \in (-1/2, +1/2)$ we have for rect the Walsh–Fourier series

$$\text{rect}(t) = \sum_{i=0}^{2^k-1} \text{cal}(i, t) + \sum_{i=0}^{2^k} \text{sal}(i, t). \tag{33}$$

Each Walsh–Fourier series representation of a function $f \in L_2(-1/2, +1/2)$ as given by (29) defines the two discrete functions $F_c : \{0, 1, 2, \ldots\} \to R$ and $F_s : \{1, 2, 3, \ldots\} \to R$, respectively, which are given by the Walsh–Fourier coefficients $F_c(i)$ and $F_s(i)$, respectively. These functions can be called the Walsh–Fourier transforms of the function f. Finally we should mention the concept of the "Vielfalt" of a function $f \in L_2(-1/2, +1/2)$. Let us call Vielfalt of the Walsh function cal$(i, .)$ or sal$(i, .)$ the minimum of the numbers of Rademacher functions sal$(2^i, .)$ which are to be multiplied in order to get cal$(i, .)$ or sal$(i, .)$. Let us introduce the symbol $V(\text{cal}(i, .))$ and $V(\text{sal}(i, .))$ to denote the Vielfalt of the function cal$(i, .)$ and sal$(i, .)$, respectively. The Vielfalt $V(f)$ of a function $f \in L_2(-1/2, +1/2)$ is then defined as the supremum of the set $\{V(\text{cal}(i, .)): F_c(i) \neq 0\} \cup \{V(\text{sal}(i, .)): F_s(i) \neq 0\}$. In our examples for instance $V(\text{saw}) = 1$ and $V(\text{rect}) = 2^k$. The following theorem concerning the Vielfalt has been found by Liedl (see e.g. Weiss [33]):

(a) If $f \in L_2(-1/2, +1/2)$ is a polynomial function of degree k then we have $V(f) = k$

(b) If $f \in L_2(-1/2, +1/2)$ is continuous and if in addition $V(f) = k$ then f is a polynomial function of degree k.

It should be mentioned that the concept of the Vielfalt appears also in a paper by Polyak and Shreider [26].

(b) Walsh–Fourier transforms

Let now $f \in L(R)$. Then the Walsh–Fourier transforms F_c and F_s of the function f are defined as the functions $F_c : [0, \infty) \to R$ and

$$F_c(s) := \int_{-\infty}^{+\infty} f(t) \, \text{cal}(i, t) \, dt \quad \text{for all } s \in [0, \infty), \quad \text{and} \tag{34}$$

$$F_s(s) := \int_{-\infty}^{+\infty} f(t) \text{sal}(i, t) \, dt \quad \text{for all } s \in (0, \infty). \tag{35}$$

If we know the Walsh–Fourier transforms F_c and F_s of a function $f \in L(R)$ the following theorem shows how to compute the function f.

Theorem. If the function $f \in L(R)$ is continuous and of bounded variation then for all $t \in R$

$$f(t) = \int_0^\infty F_c(s) \text{cal}(s, t) \, ds + \int_0^\infty F_s(s) \text{sal}(s, t) \, ds. \tag{36}$$

Let ch_I denote the characteristic function $ch_I : R \to R$ of the interval $I \subset R$; that is $ch_I(t) := 1$ for all $t \in I$ and $ch_I(t) := 0$ for all $t \in R \backslash I$. Using ch_I we give the following examples of Walsh–Fourier transforms:

Example 1. Let f be given by $f := \text{cal}(i, .)ch_I$ with $I := (-1/2, +1/2)$. Then $F_c = ch_{I'}$ with $I' := [i, i + 1)$, and $F_s = 0$

Example 2. Let f be given by $f := \text{sal}(i, .)ch_I$ with $I := (-1/2, +1/2)$. Then $F_c = 0$ and $F_s = ch_{I'}$ with $I' := (i - 1, i]$

Next we discuss the Walsh–Fourier transforms of a function $f \in L_2(R)$. For each such function and for each $r \in R$ we define the window f_r by $f_r := fch_{(-r, +r)}$. Clearly $f_r \in L_2(R) \cap L(R)$. Therefore there exists the Walsh–Fourier transforms $F_{r,c}$ and $F_{r,s}$ of f_r. It can be shown that both functions $F_{r,c}$ and $F_{r,s}$ converge in the space $L_2(R)$ to functions F_c and F_s, respectively, as r goes to infinity. These limits are by definition the Walsh–Fourier transforms of the function $f \in L_2(R)$. The following "theorem of Plancherel" is true:

Theorem. For $f \in L_2(R)$ we have $F_c \in L_2(0, \infty)$ and $F_s \in L_2(0, \infty)$ and for almost all $t \in R$ we have

$$f(t) = \int_0^\infty F_c(s)\text{cal}(s, t)\, ds + \int_0^\infty F_s(s)\text{sal}(s, t)\, ds. \tag{37}$$

Furthermore we have

$$\|f\|^2 = \|F_c\|^2 + \|F_s\|^2. \tag{38}$$

Of special interest in applications of Walsh functions in communications is the following generalization of the convolution integral. Let $f, g \in L(R)$. The dyadic convolution product $f*g$ of f and g is then defined by the function $f*g: R \to R$ which is given by

$$(f*g)(t) := \int_{-\infty}^{+\infty} f(t \oplus \tau) g(\tau)\, d\tau, \tag{39}$$

where in the case that $t < 0$ the dyadic sum $t \oplus \tau$ is defined by $t \oplus \tau := -(-t \oplus \tau)$. It can be shown that the dyadic convolution has the following properties: For each $f, g, h \in L(R)$ we have

$$f*g \in L(R), \tag{40}$$

$$f*g = g*f, \tag{41}$$

$$f*(g + h) = f*g + f*h, \tag{42}$$

$$f*(g*h) = (f*g)*h. \tag{43}$$

Furthermore we have the following convolution theorem:

Theorem. Let $f, g \in L(R)$ and $h = f*g$. Let $F_c, F_s, G_c, G_s, H_c, H_s$ denote the Walsh–Fourier transforms of f, g and h, respectively. Then we have $H_c = F_c G_c$ and $H_s = F_s G_s$.

Finally we would like to mention the following shifting theorem:

Theorem. Let $f \in L(R) \cup L_2(R)$. For each $\lambda \in R$ let f_λ denote the λ-shifted function which is defined by $f_\lambda(t) := f(t \oplus \lambda)$. Let F_c, F_s and $F_{\lambda,c}$, $F_{\lambda,s}$ denote the Walsh–Fourier transforms of f and f_λ, respectively. Then we have $F_{\lambda,c}(s) = \mathrm{cal}(s, \lambda)F_c(s)$ and $F_{\lambda,s}(s) = \mathrm{sal}(s, \lambda)F_s(s)$ for all $s > 0$.

In our discussion of the theory of Walsh–Fourier transforms we have preferred to use the Walsh functions sal(s, .) and cal(s, .). An equivalent theory can be derived for the Walsh functions ψ_y.

(c) *Discrete finite Walsh transforms*

Let us now consider the case of a harmonic analysis when the discrete Walsh functions $w(i, .)$ are used. We have already mentioned that in this case the theory turns out to be part of the methods which are well-known from elementary linear algebra. Therefore it seems reasonable only to sketch the results. For more detailed studies the reader is again advised to consult the work of Gibbs [7] and Pichler [25].

Let $R(n)$ denote the linear space consisting of all functions f of the form $f:N(n) \to R$. Each function $f \in R(n)$ may be interpreted as a signal. The set $W(n)$ of discrete Walsh functions which was introduced in Section 3(c) forms a complete set of orthogonal functions of the space $R(n)$. For all $i, j, t \in N(n)$

$$\sum_{t \in N(n)} w(i, t)w(j, t) = \begin{array}{ll} 2^n & \text{if } i = j \\ 0 & \text{if } i \neq j \end{array} \tag{44}$$

With respect to the basis $W(n)$ each function $f \in R(n)$ has a representation of the form

$$f(t) = 2^{-n} \sum_{i \in N(n)} \hat{f}(i)w(i, t) \quad (t \in N(n)), \tag{45}$$

where $\hat{f}(i)$ denotes the ith "Walsh–Fourier coefficient" which is given by the sum

$$\hat{f}(i) = \sum_{\tau \in N(n)} f(t)w(i, t) \quad (i \in N(n)). \tag{46}$$

It is quite clear that equation (45) can be interpreted as a Walsh–Fourier representation of the function $f \in R(n)$. In this case equation (46) defines the Walsh–Fourier transform $\hat{f}$ of f by $\hat{f}:N(n) \to R$ with $i \mapsto \hat{f}(i)$ for all $i \in N(n)$).

The dyadic convolution product $f*g$ of the two functions $f \in R(n)$ and $g \in R(n)$ can be defined by

$$(f*g)(t) := \sum_{\tau \in N(n)} f(t \oplus \tau)g(\tau) \quad (i \in N(n)). \tag{47}$$

Analogous to the continuous case we have the following convolution theorem that for all $f, g \in R(n)$

$$(f*g)(i) = \hat{f}(i)\hat{g}(i) \quad (i \in N(n)). \tag{48}$$

For arbitrary $\tau \in N(n)$ let T_τ denote the linear operator $T_\tau : R(n) \to R(n)$ which is given by $f \mapsto T_\tau f$ with $(T_\tau f)\,(t) := f(t \oplus \tau)$ for all $t \in N(n)$ and all $f \in R(n)$. Each operator T_τ can be called a dyadic translation operator. In addition for each function $g \in R(n)$ we define the dyadic convolution operator C_g associated with it by $C_g : R(n) \to R(n)$ and $f \mapsto C_g f := f * g$. It is not difficult to see that dyadic convolution operators and dyadic translation operators commute, that is, that for all $g \in R(n)$ and for all $\tau \in N(n)$ we have

$$C_g T_\tau = T_\tau C_g. \tag{49}$$

In addition it is easy to show that all eigenfunctions of a dyadic convolution operator C_g are discrete Walsh functions. If $\hat{g}(i) \neq 0$ then $w(i, .)$ is an eigenvector of C_g which is associated with the eigenvalue $\hat{g}(i)$.

For convenience let us consider the previous concepts in matrix notations. We identify the space $R(n)$ with the 2^n-dimensional Euclidean space of real 2^n-tupels and choose in it the standard basis of unit vectors. Then each function $f \in R(n)$ corresponds to a column vector $\mathbf{f}$ of the form $\tilde{\mathbf{f}} = (f(0), f(1), \ldots, f(2^n - 1))$, where $\sim$ denotes transposition. For the operation of Walsh–Fourier transformation we have then a representation $\mathbf{W}$ given by the matrix

$$\mathbf{W} := [\mathbf{w}(0, .) \,|\, \ldots \,|\, \mathbf{w}(2^n - 1, .)]. \tag{50}$$

Therefore the Walsh–Fourier transform $\hat{f}$ of $f \in R(n)$ is represented by

$$\hat{\mathbf{f}} = \mathbf{W}\mathbf{f}. \tag{51}$$

A dyadic translation operator T_τ has a matrix representation of the form

$$\mathbf{T}_\tau := [\mathbf{e}_\tau \,|\, \mathbf{e}_{\tau \oplus 1} \,|\, \ldots \,|\, \mathbf{e}_{\tau \oplus 2^n - 1}] \tag{52}$$

where for $i \in N(n)$ the column vectors $\mathbf{e}_i$ denotes the unit vectors of the standard basis which are given by $e_i(j) = 1$ for $i = j$ and $e_i(j) = 0$ for $i \neq j$. Each dyadic convolution operator C_g is represented by a matrix $\mathbf{C}_g$ of the form

$$\mathbf{C}_g := [\mathbf{g} \,|\, \mathbf{T}_1 \mathbf{g} \,|\, \ldots \,|\, \mathbf{T}_{2^n - 1} \mathbf{g}]. \tag{53}$$

By definition $f * g = C_g f$ and applying the convolution theorem we get $(\mathbf{f} * \mathbf{g})\hat{} = (\mathbf{C}_g \mathbf{f})\hat{} = \mathbf{W}\mathbf{C}_g \mathbf{f} = \mathbf{W}\mathbf{C}_g \mathbf{W}^{-1} \hat{\mathbf{f}} = \hat{\mathbf{g}}\hat{\mathbf{f}}$. From this result we observe that each dyadic convolution operator C_g is represented with respect to the basis as given by the "Walsh–vectors" $\mathbf{w}(0, .), \mathbf{w}(1, .), \ldots, \mathbf{w}(2^n - 1, .)$, by a diagonal matrix $\mathbf{W}\mathbf{C}_g \mathbf{W}^{-1}$ which is given by

$$\mathbf{W}\mathbf{C}_g \mathbf{W}^{-1} = Diag(\hat{g}(0), \ldots, \hat{g}(2^n - 1)). \tag{54}$$

5. Concepts for Applications

In this last section we discuss some concepts which have been proved to be of some use in communication engineering. The choice depends strongly on the author's individual interest and should not reflect the state of the art. The reader who is interested to learn more about the applications of Walsh functions is advised to consult the proceedings of the annual symposium on applications of Walsh functions [1, 18, 34, 35].

(a) Sampling theorem in the case of Walsh harmonic analysis

Let $L_2(R)$ be interpreted as signal space. Then for each signal $f \in L_2(R)$ the following statements are valid: Under the assumption that there exists a "cut off" sequency s_0 such that for all $s > s_0$ we have for the Walsh–Fourier transforms F_c and F_s the property $F_c(s) = 0$ and $F_s(s) = 0$, then $f \in L_2(R)$ is uniquely determined by the sequence

$$(. . ., f(-2T), f(-T), f(0), f(T), f(2T), . . .) \tag{55}$$

of sampled values of f, where T is given by $T := 1/2s_0$.

The signal $f \in L_2(R)$ can be represented by

$$f(t) = \sum_{m=-\infty}^{\infty} f(mT)ch_{[0,T)}(t - mT). \tag{56}$$

Comparing this theorem with the classical result the analogy is obvious. But since signals which are band–limited in the sequency sense are step functions our theorem looks rather trivial. This theorem was first presented in this form by Pichler [23]. A more general sampling theorem has been proved by Kluvanek [15].

(b) Dyadic convolution systems

For our purpose let us call a relation $S \subset L_2(R) \times L_2(R)$ a time system. The theory of Walsh–Fourier transforms (similarly also the theory of Fourier or Laplace transforms) enables us to associate such a system S with the "sequency"-representations S_c and S_s which are given by

$$S_c := \{(X_c, Y_c):(x, y) \in S\} \quad \text{and} \quad S_s := \{(X_s, Y_s):(x, y) \in S\}, \tag{57}$$

where X_c, X_s and Y_c, Y_s denote the Walsh–Fourier transforms of the signals x and y, respectively. For certain classes of such time systems the sequency representations have the advantage of an easier description. One such class is that of dyadic convolution systems, which will be described in the following.

For arbitrary $h \in L(R)$ we define the dyadic convolution system $S(h)$ by

$$S(h) := \{(x, y):x \in L_2(R) \quad \text{and} \quad y = h*x\}, \tag{58}$$

where $*$ denotes dyadic convolution. We see that the output signal y of $S(h)$ is performed by dyadic convolution of the input signal x by the function h, which is called the impulse response of $S(h)$. The sequency representations $S_c(h)$ and $S_s(h)$ of $S(h)$ are given in the following form

$$S_c(h) = \{(X_c, Y_c) : (x, y) \in S \quad \text{and} \quad Y_c = H_c X_c\}, \quad \text{and} \tag{59}$$
$$S_s(h) = \{(X_s, Y_s) : (x, y) \in S \quad \text{and} \quad Y_s = H_s X_s\},$$

where H_c and H_s denote the Walsh–Fourier transforms of the impulse response h. These functions are also called the transfer functions of the dyadic convolution system $S(h)$. As an advantage of these sequency representations we have transfer functions which operate pointwise in the sequency domain. Let us mention some other properties of dyadic convolution systems. Since dyadic convolution is a linear operation each dyadic convolution system turns out to be a linear system. A second property of a dyadic convolution system $S(h)$ is given by its invariance with respect to dyadic translations of the input signals. If we have $(x, y) \in S(h)$, then for all real shifts $\lambda \in R$ we also have $(x_\lambda, y_\lambda) \in S(h)$. In other words, if $y = h*x$ then for all real numbers $\lambda \in R$ it follows that $y_\lambda = h*x_\lambda$. Therefore if we have knowledge that $(x, y) \in S(h)$ it is no problem to compute the output signal which is related to a λ-shifted input signal.

Let $S(h_1)$ and $S(h_2)$ denote dyadic convolution systems. The parallel connection of $S(h_1)$ and $S(h_2)$ can be defined by the dyadic convolution system $S(h_1 + h_2)$. It is obvious that the transfer functions of $S(h_1 + h_2)$ are given by $H_{1c} + H_{2c}$ and $H_{1s} + H_{2s}$. The serial connection of $S(h_1)$ and $S(h_2)$ is defined by the dyadic convolution system $S(h_1*h_2)$. The transfer functions of this composite system are given by $H_{1c}H_{2c}$ and $H_{1s}H_{2s}$. The set of all dyadic convolution systems forms a commutative algebra with respect to parallel–and serial connection. It should be obvious that the concept of dyadic convolution systems can also be introduced in case that the signal spaces are given by $L_2(-1/2, +1/2)$ or by $R(n)$.

(c) *Sequency band-pass filter of Harmuth*
A dyadic convolution system $S(h)$ with transfer functions H_c and H_s given by

$$H_c := ch_{[n2^k, (n+1)2^k)} \quad \text{and} \quad H_s := ch_{[n2^k, (n+1)2^k]}, \tag{60}$$

where $n \in \{0, 1, 2, \ldots\}$ and k denotes an integer, is called a sequency band-pass filter with bandwidth 2^k and cut-off sequency $n2^k$. For $n = 0$ a sequency band-pass filter is also called a sequency low-pass filter. It can be shown, that the impulse response h associated with the transfer functions H_c and H_s of equations (60) is given by

$$h(t) = 2^{k+1} \operatorname{cal}(n2^k, t) ch_{[0, 2^k)}(t) \quad \text{for all } t \in R. \tag{61}$$

Due to the special form of h it is possible to express the output signal y as given by $y = h*x$ in the form

$$y(t) = \operatorname{cal}(n2^k, t) \sum_{m=-\infty}^{\infty} a_m \, ch_{[0, 2^k)}\left(t - m2^{-(k+1)}\right) \tag{62}$$

where the coefficients a_m are given by

$$a_m = 2^{k+1} \int_{m/2^{k+1}}^{(m+1)/2^{k+1}} x(\tau)\mathrm{cal}(n2^k, \tau)\, d\tau \qquad (63)$$

Equation (62) can directly be realized by electronic devices [13, 23].

(d) Dyadic correlation functions and optimal filtering
In this section we present some results which have been achieved in the development of a theory of dyadic convolution systems which are optimal in the mean square sense. As in the sections before we consider the continuous case, although the discrete and finite case would be much easier to deal with (see e.g. [16, 21]). For more detailed studies see [24].

Let us consider the following problem: assume that $x \in L_2(R)$ is given by the sum $x = u + v$, where u is the actual input signal and v is a noise. The problem is to find a dyadic convolution system $S(h)$ such that the output signal $h*x$ (i.e. the output signal corresponding to the "noisy" input signal $x = u + v$) approximates the output signal $h*u$ (i.e. the output signal corresponding to the noiseless signal u) optimally in the mean square sense. For an equivalent formulation a function $h \in L(R)$ has to be found, such that $\|h*x - h*u\| = \|h*v\|$ is minimal. It can be shown that a causal impulse response h, that is $h(t) = 0$ for all $t < 0$, will be a solution of the given problem if and only if the following generalized Wiener–Hopf equation is fulfilled:

$$r_{ux} - r_x * h = 0, \qquad (64)$$

where the functions r_{ux} and r_x are defined by $r_{ux}(t) := (u*x)(t)$ for $t \geqslant 0$ and $r_{ux}(t) := 0$ for $t < 0$, and $r_x(t) := (x*x)(t)$ for $t \geqslant 0$ and $r_x(t) := 0$ for $t < 0$, respectively. r_{ux} is called the dyadic cross-correlation function of the signals u and x; r_x is called the dyadic autocorrelation function of x. Observe that equation (64) can also have the following interpretation: the dyadic convolution system $S(h)$ is optimal in the sense above, if and only if $(r_x, r_{ux}) \in S(h)$.

The application of the convolution theorem to equation (64) gives the following conditions for the optimal system $S(h)$

$$R_{ux,c} = H_c R_{x,c} \quad \text{and} \quad R_{ux,s} = H_s R_{x,s}, \qquad (65)$$

where $R_{ux,c}, R_{ux,s}$ and $R_{x,c}, R_{x,s}$ denotes the Walsh–Fourier transforms of the correlation functions r_{ux} and r_x, respectively. In analogy to the classical theory these functions may be regarded as "sequency power spectra". Therefore, if we know the cross-power spectra $R_{ux,c}$ and $R_{ux,s}$ of u and x and in addition also the power spectra $R_{x,c}$ and $R_{x,s}$ of x we are able to compute by means of (65) the transfer functions H_c and H_s of the optimal dyadic convolution system.

In case that the signal u and the noise v are uncorrelated in the dyadic sense $(r_{uv}(t) = 0$ for all $t \in R)$ we have $r_x = r_u + r_v$ and $r_{ux} = r_u$. Then the conditions of equation (65) may be written in the form

$$H_c = \frac{R_{u,c}}{R_{u,c} + R_{v,c}} \quad \text{and} \quad H_s = \frac{R_{u,s}}{R_{u,s} + R_{v,s}}. \tag{66}$$

Finally it should be mentioned that the given results can be extended to the wider class of signals of "finite power", that is to signals x for which the limit

$$\lim_{T \to \infty} \frac{1}{2T} \int_{-T}^{+T} x^2(t) \, dt \tag{67}$$

is finite.

REFERENCES

1. Bass, C. A. (1970). "1970 Proceedings Applications of Walsh Functions", Washington, D.C. AD 707 431.
2. Bramhall, J. N. (1972). An annotated bibliography on, and related to, Walsh Functions. Technical Memorandum TG 1198, The Johns Hopkins University, Applied Physics Laboratory, Baltimore, U.S.A.
3. Chrestenson, H. E. (1955). A class of generalized Walsh Functions, *Pacif. J. Math.* 5, 17-31.
4. Fine, N. J. (1949). On the Walsh Functions. *Trans. Am. Math. Soc.* 65, 372-414.
5. Fine, N. J. (1950). The generalized Walsh Functions. *Trans. Am. Math. Soc.* 69, 66-77.
6. Gelfand, I., Raikov, D. and Shilov, G. (1964). Commutative Normed Rings, Chelsea, Bronx, New York.
7. Gibbs, J. E. (1967). Walsh spectrometry, a form of spectral analysis well suited to binary digital computation, NPL: unpublished report.
8. Gibbs, J. E. (1969). Some properties of functions on the non-negative integers less than 2^n, NPL:DES Rept No. 3.
9. Gibbs, J. E. (1970a). Discrete complex Walsh functions in Bass, C.A. (1970), 106-122.
10. Gibbs, J. E. (1970b). Sine waves and Walsh waves in physics in Bass, C.A. (1970), 260-274.
11. Gibbs, J. E. (1970c). Functions that are solutions of a logical differential equation, NPL:DES Rept No. 4.
12. Gibbs, J. E. and Millard, M. J. (1969). Walsh functions as solutions of a logical differential equation NPL:DES Rept No. 1.
13. Harmuth, H. F. (1969). "Transmission of Information by Orthogonal Functions", Springer-Verlag, Berlin/New York.
14. Hewitt, E. and Ross, K. A. (1963, 1970). "Abstract Harmonic Analysis 1, 2", Springer-Verlag, Berlin-Heidelberg-New York.
15. Kluvanek, I. (1965). Sampling theorem in abstract harmonic analysis. *Mat. fyz. Cas.* 15, 43-48.
16. LaBarre, J. B. K. (1969). A transform technique for linear, time-varying, discrete-time systems. University of Michigan, Ann Arbor, Technical Report, SEL No. 38.
17. Levy, P. (1944). Sur une generalisation de fonctions orthogonales de M. Rademacher. *Comment. Math. Helv.* 16, 146-152.
18. Lines, P. D. (ed.) (1971). Proceedings of the Symposium Theory and Applications of Walsh Functions. The Hatfield Polytechnic, Hatfield, U.K.

19. Morgenthaler, G. W. (1957). On Walsh–Fourier Series. *Trans. Am. Math. Soc.* **84,** **24,** 350–360.

20. Paley, R. E. A. C. (1932). A remarkable series of orthogonal functions. *Proc. Lond. Math. Soc.* 34, 241-279.

21. Pearl, J. (1971). Application of Walsh transform to statistical analysis. *IEEE Trans.* **SMC-1,** 111-119.

22. Pichler, F. (1967). Dissertation, Philosophische Fakultät, Universität Innsbruck.

23. Pichler, F. (1968). Synthese linearer periodisch zeitvariabler Filter mit vorgeschriebenem Sequenzverhalten. *Arch. Elektr. Übertr.* 22, 150-161.

24. Pichler, F. (1970). Walsh–Fourier Synthese optimaler Filter. *Arch. Elektr. Übertr.* 24, 350-360.

25. Pichler, F. (1971). Dyadische Faltungsoperatoren zur Beschreibung linearer Systeme. *Sber. Ost. Akad. Wiss. Math.-naturw.Kl. II* **180.** Bd.

26. Polyak, B. T. and Shreider, Yu. A. (1962). The application of Walsh functions in approximate calculations. *Vop. Teor. Mat. Mash. Coll. II* 174-190. (translation by J. E. Gibbs).

27. Selfridge, R. G. (1955). Generalized Walsh transform. *Pacif. J. Math.* 5, 451-480.

28. Vilenkin, N. J. (1947). On a class of complete orthogonal systems. *Izv. Akad. nauk. SSSR, ser. matem.* 11, 363-400 (Russian).

29. Vilenkin, N. J. (1952). On the theory of Fourier integral on topological groups. *Mat. Sb. (N.S.)* **30**(72), 233–244 (Russian).

30. Walsh, J. L. (1923). A closed set of normal orthogonal functions. *Am. J. Math.* 45, 5-24.

31. Watari, Ch. (1958). On generalized Walsh–Fourier series. *Tohoku Math. J.* (2)10, 211-241.

32. Weiser, F. E. (1964). Walsh function analysis of instantaneous nonlinear stochastic problems. Dissertation, Polytechnic Institute Brooklyn, 1964.

33. Weiss, P. (1967). Zusammenhang von Walsh–Fourierreihen mit Polynomen. *Monats. Math.* 71(2), 165-179.

34. Zeek, R. W. and Showalter, A. E. (ed.) (1971). 1971 Proceedings Applications of Walsh Functions. Washington D.C. AD-727 000. **Also** *IEEE Trans. Electromag. Compat.* **EMC-13,** No. 3, 1-218.

35. Zeek, R. W. and Showalter, A. E. (ed.) (1972). 1972 Proceedings Applications of Walsh Functions. Washington D.C. AD-744 650.

DISCUSSION

P. M. Schultheiss: In what manner are the sal and cal functions extended if one extends the domain from (0, 1) to $(-\infty, \infty)$?

Answer: As we did one can define the sal and the cal functions for an arbitrary non negative real sequency. This set forms a complete set of kernel functions for the Hilbert space $L_2 (-\infty, \infty)$.

J. W. R. Griffiths (comment on the problem of comparing bandwidths of signals): On the subject of bandwidth I think it was Harmuth himself who, when somebody was complaining about it generating EM waves with Walsh Functions—"look at all these harmonics you are generating", complained about these people who send out sinusoidal waves, sending out all these harmonics.

T. Kooij: How large is the percentage of Rademacher Functions in a given (complete) set of Walsh functions?

Answer: Strictly speaking, only the Walsh Functions sal(2^k, .) where k is a non-negative integer are called Rademacher Functions. So a finite complete set of Walsh

Functions with 2^n members has exactly n Rademacher Functions, therefore, the percentage should be $n/2^n$.

P. M. Schultheiss: Is there a simple equivalent to the Wiener–Hopf factorization problem?

Answer: The problem of factorization does not occur in this theory, because you can always find a causal solution by solving the dyadic Wiener–Hopf equation directly by means of Walsh–Fourier transformation.

A. A. Requicha: If you have a continuous signal you sample it at equal time intervals and quantize each sample. Is this equivalent to sequency limiting the signal?

Answer: In quantizing a signal so that the result is constant over equal time intervals we get a sequency limited signal; but generally the result is not the same as sequency limiting the original signal.

R. E. Bogner (comment): True sequency limitation of the original signal is produced by setting the value during each interval equal to the average over that interval.

P. M. Schultheiss: Can you give us a simple physical example of a problem in which dyadic convolution occurs?

Answer: In looking to nature I personally have never discovered any phenomena which I would like to model taking dyadic convolution. But of course there are engineering examples where dyadic convolution can be useful.

R. E. Bogner: In filtering long sequences via DFT we need to use select/save or some other methods to avoid the circular convolution. Does a similar problem appear with DWT?

Answer: I myself never looked at such a type of problem.

Applications of Walsh Functions in Communications: State of the Art

HENNING F. HARMUTH

Department of Electrical Engineering
The Catholic University of America
Washington, D.C., U.S.A.

1. Introduction

Communications is dominated by the system of sine-cosine functions. It is often believed that nothing can be gained by using other functions except when going beyond linear circuits and processes. The argument runs as follows: Any signal has a finite energy and thus is a quadratically integrable function. Such functions can be approximated in the sense of a vanishing mean-square-error by a superposition of sine-cosine functions. Hence, anything that can be done by a nonsinusoidal function can also be done by a superposition of denumerably or nondenumerably many periodic sine-cosine functions. In other words, if we have infinite time and do not mind an infinite effort we can always use sine-cosine functions. This is a good example of a mathematically correct statement that is perfectly meaningless for the engineer. Aside from the infinities, anybody who has ever tried to generate even a few harmonics of a sine function with stable amplitude, frequency and phase will agree that this is no easy task. Indeed it is usual to synthesize such sine waves by a superposition of square waves and not vice versa.

Let us look at the problem of using general systems of orthogonal functions from the mathematical point of view and let us in particular consider the system of Walsh functions shown in Fig. 1. The important thing about these functions is the location and the height of the jumps. The constant sections between the jumps can always be filled in; they convey no information. A superposition of Walsh functions yields a step function. Again, the step function is defined by the location and height of the jumps; they are the only important part of the function from the standpoint of conveyed information. Consider now a Fourier series or transform of an individual Walsh function or of a sum of Walsh functions. The expansion converges everywhere except where we need it: The well

43

known Gibbs phenomenon occurs at the jumps as a result of the non-uniform convergence.

Let us go one step further. The Fourier expansion of functions with less radical discontinuities as those of Walsh functions may converge uniformly, but even this may not be enough. A current flowing in a Hertzian dipole produces electric and magnetic field strengths in the wave zone proportionate to its derivative. A uniformly convergent expansion of the current does not imply that one can differentiate term by term. In the case of multipole radiation, higher order derivatives are required and the problem is aggravated.

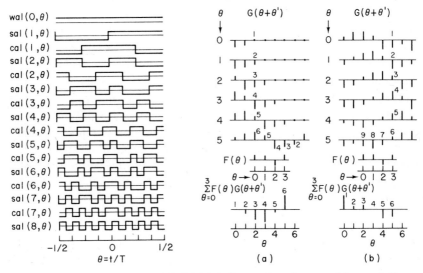

FIG. 1 (left). Walsh functions. FIG. 2 (right). Generation of a reversible correlation function (a) and a nonreversible correlation function (b).

The conclusion is that one can expect to obtain technological differences by substituting general systems of orthogonal functions for superpositions of sine–cosine functions, if mean-square convergence is the only important feature. In all other cases one can expect technological and mathematical differences.

Having clarified why and when nonsinusoidal functions can lead to useful results let us turn to two examples. In both of them the Walsh functions are only used to arrive at certain results and circuits, the functions are not actually generated. This corresponds to the role sinusoidal functions play in signal processing or the design of filters with a certain frequency response of attenuation and phase.

2. Dyadic Correlation

(a) Mathematical basis [1, 3, 4]

The presence of a known signal $F(\theta)$, $\theta = t/T$, of duration T in a received superposition $G(\theta)$ of signal plus noise can be detected by means of the correlation function $C_{FG}(\theta')$:

$$C_{FG}(\theta') = \int_{-1/2}^{1/2} F(\theta)G(\theta + \theta')\, d\theta. \tag{1}$$

In principle this operation is reversible and one can regain $G(\theta)$ from $C_{FG}(\theta')$ and $F(\theta)$, which means that no information is lost. Practically, this preservation of information may be difficult because of a requirement for storage. Consider Fig. 2a to clarify this point. It is assumed that the received superposition $G(\theta + \theta')$ is known to be zero initially. The last sample with known value zero is denoted by 1 in the line $\theta' = 0$ of Fig. 2a. The first value of the correlation function $\Sigma F(\theta)G(\theta + \theta')$ is also denoted by 1. For $\theta' = 1, 2, 3, \ldots$ additional values of $G(\theta + \theta')$ denoted by 2, 3, 4, ... line up with the four nonzero values of $F(\theta)$ and additional values of the correlation function denoted by 2, 3, 4, ... are produced. As many samples of the correlation function $\Sigma F(\theta)G(\theta + \theta')$ are generated as samples of $G(\theta + \theta')$ are shifted past $F(\theta)$. In principle, one can recompute $G(\theta + \theta')$ from $\Sigma F(\theta)G(\theta + \theta')$. Practically, this recomputation is cumbersome since the computation of sample n of $G(\theta + \theta')$ requires the knowledge of all samples $n, n - 1, \ldots, 1$ of the correlation function $\Sigma F(\theta)G(\theta + \theta')$.

Figure 2b shows what happens if $G(\theta + \theta')$ does not start with value zero or other known values. A total of nine samples of $G(\theta + \theta')$ are shifted past the nonzero values of $F(\theta)$ during the time $\theta' = 0, 1, \ldots, 5$. Only six samples of the correlation function $\Sigma F(\theta)\, G(\theta + \theta')$ are produced by these nine samples. A recomputation of $G(\theta + \theta')$ from the correlation function is impossible. This indicates that information has been lost. How can one avoid the loss of information without having to store all the previous values of the correlation function as done by the scheme of Fig. 2a?

Figure 3 shows again the received signal $G(\theta + \theta')$. The stored signal $F(\theta)$ is now denoted by $F_0(\theta)$ and it is augmented by $F_1(\theta)$, $F_2(\theta)$ and $F_3(\theta)$. These four stored signals form an orthogonal set based on $F_0(\theta)$. The received signal $G(\theta + \theta')$ is correlated with all four signals to yield the four correlation functions $\Sigma F_i(\theta) \times G(\theta + \theta')$ with $i = 0, 1, 2, 3$. At any time θ' there are now four samples of the correlation functions available from which one can recompute the four samples of $G(\theta + \theta')$ from which they were obtained.

The significant points of this process are: (a) The received signal is correlated with a linearly independent set of stored signals rather than with one signal; use of an orthogonal set rather than a linearly independent set brings the well known computational simplifications. (b) One of the signals of the set is the signal one is looking for, in order to make use of the known advantages of the autocorrelation function in the presence of noise.

There are many ways to obtain an orthogonal set from the original signal. Imposing the additional requirement that the correlation process shall be fast to perform and the equipment simple to build brings one to the concept of dyadic correlation. The ordinary correlation function $C_{FG}(\theta')$ of equation (1) is replaced by the dyadic correlation function:

$$E_{FG}(\theta', \theta_v) = \int_{-1/2}^{1/2} F(\theta \oplus \theta_v) G(\theta + \theta') \, d\theta \tag{2}$$

The sign $\oplus$ indicates addition modulo 2.

In order to derive later on a simple implementation of this equation let us substitute Walsh series for $F(\theta \oplus \theta_v)$ and $G(\theta + \theta')$:

$$F(\theta \oplus \theta_v) = \sum_{j=0}^{2^n - 1} b(j) \, \text{wal}(j, \theta \oplus \theta_v) = \sum_{j=0}^{2^n - 1} b(j) \, \text{wal}(j, \theta) \, \text{wal}(j, \theta_v), \tag{3}$$

$$b(j) = \int_{-1/2}^{1/2} F(\theta) \, \text{wal}(j, \theta) \, d\theta, \tag{4}$$

$$G(\theta + \theta') = \sum_{j=0}^{2^n - 1} a(j, \theta') \, \text{wal}(j, \theta), \tag{5}$$

$$a(j, \theta') = \int_{-1/2}^{1/2} G(\theta + \theta') \, \text{wal}(j, \theta) \, d\theta. \tag{6}$$

Let us note that $F(\theta)$ consists of 2^n samples. The shift theorem of Walsh functions,

$$\text{wal}(j, \theta \oplus \theta_v) = \text{wal}(j, \theta) \, \text{wal}(j, \theta_v), \tag{7}$$

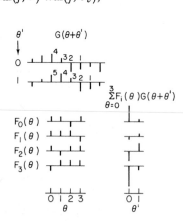

FIG. 3. Generation of a complete set of correlation functions by means of a complete set of orthogonal functions $F_i(\theta)$; $i = 0, 1, 2, 3$.

is used in equation 3. One obtains from equations (2) to (6):

$$E_{FG}(\theta', \theta_v) = \sum_{j=0}^{2^n-1} a(j, \theta')b(j)\,\mathrm{wal}(j, \theta_v). \qquad (8)$$

Let us look at a simple example. The received signal $G(\theta + \theta')$ shall have no noise superimposed and thus becomes $F(\theta + \theta')$. The coefficient $b(j)$ is identical with $a(j, 0)$ and equation (8) assumes the following form:

$$E_{FF}(\theta', \theta_v) = \sum_{j=0}^{2^n-1} a(j, \theta')a(j, 0)\,\mathrm{wal}(j, \theta_v). \qquad (9)$$

Figure 4 shows the noise free received signal $F(\theta + \theta')$ with the amplitude samples A, B, C, D shifted from left to right for successive times $\theta' = -3/8$, $-2/8, \ldots$. Note that the first arriving pulse in Fig. 4 is on the right; this

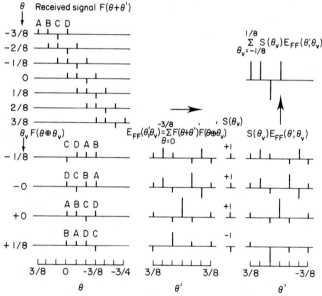

FIG. 4. Conversion of a function $F(\theta + \theta')$ into a complete set of correlation functions by a sliding correlator and reconversion into the original function.

corresponds to the way a signal is shifted through a shift register but it is just the opposite of the usual way of plotting time signals with later times being plotted toward the right.

The stored signal $F(\theta) = F(\theta \oplus + 0)$ is shown in the line $\theta_v = + 0$. The dyadically shifted signal for $\theta_v = + 1/8$ is obtained by interchanging neighbours $(A, B \to B, A$ and $C, D \to D, C)$. The dyadically shifted signal for $\theta_v = -0$ is obtained by reversing the sequence of all samples $(A, B, C, D \to D, C, B, A)$. The shifted signal for $\theta_v = -1/8$ is obtained from that for $\theta_v = -0$ again by interchanging neighbours $(D, C \to C, D$

and B, $A \rightarrow A$, B). A more detailed discussion of dyadic shifting may be found in the literature [1].

The dyadic correlation function $E_{FF}(\theta', \theta_v)$ is represented in Fig. 4 by a sum rather than an integral since $F(\theta + \theta')$ and $F(\theta \oplus \theta_v)$ are sampled signals. The ordinary autocorrelation function is obtained for $\theta_v = +0$.

Let us multiply the four functions $E_{FF}(\theta', \theta_v)$ for $\theta_v = -1/8, -0, +0,$ $+1/8$ and all times θ' by a "sidelobe suppression function" $S(\theta_v) = +1, +1,$ $+1, -1$. The products are shown on the right in Fig. 4. Summation of the products at all times θ' yields the original signal. This, of course proves nothing more than that the process is reversible and the dyadic correlation function contains at any time the same information as the samples of the received signal that line up with the nonzero samples of the stored signals $F(\theta \oplus \theta_v)$.

In order to find a first use of dyadic correlation let us observe that the multiplication by $S(\theta_v) = +1, +1, +1, -1$ and the summation may also be interpreted as a process that eliminates the sidelobes of the ordinary autocorrelation function at the times $\theta' = -3/8$ and $\theta' = -1/8$. By choosing other values of $S(\theta_v)$ one may suppress other sidelobes. Furthermore, one can reduce the larger sidelobes at the cost of enhancing the smaller ones to make all sidelobes of equal magnitude. This is a general approach to creating Barker-like codes rather than finding them. One may further choose $S(\theta_v) = 0, 0, +1, 0$. The sum then yields the ordinary autocorrelation function which is best for signal detection in the presence of thermal noise but clutters small signals by the sidelobes caused by much larger signals. Choosing $S(\theta_v) = x, y, +1, z$ one may minimize the error rate in the presence of noise and large sidelobes. This is a difficult task since the values of the sidelobe suppression function depend on the signal-to-noise ratio of the wanted as well as the cluttering signal.

(b) *Multiple decisions based on dyadic correlation* [5, 6, 12]
The discussed uses of the dyadic correlation function are based on the assumption that only one decision is made whether a signal has been received or not. One may alternately make several decisions, for instance one for each value of θ_v. Figure 5 shows an example where such multiple decisions are useful. Column "a" shows a large signal on top; its dyadic correlation function is shown in rows 1 to 4. Column "b" shows an equal signal but with one-third the magnitude and advanced in time by one sampling distance. Note that later times are plotted toward the right and not toward the left as in Fig. 4.

Let us assume the small signal alone is received and the autocorrelation function is used for detection. One may set thresholds as shown by the lines TH1 and TH2 to detect the mainlobe denoted S1. Let us now assume the sum of the large and the small signal of columns "a" and "b" is received. The resulting sum and its dyadic correlation function is shown in column "c". The ordinary correlation function in row 3 shows a large

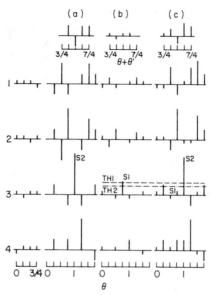

FIG. 5. Large signal (a), small signal (b) and their sum (c) with complete sets of correlation functions.

mainlobe denoted S2, and the large signal can be detected without trouble. The mainlobe of the small signal should be at the location denoted S1, but this mainlobe has been reduced to the magnitude of the smallest sidelobe; this implies a reduction of the signal-to-noise ratio by $1^2/4^2$ or -12 dB. A threshold TH1 will permit detection of the large signal, a threshold TH2 will in addition yield two ghost signals but the small signal is lost. It could be recovered if the correlation function were known for all previous times but this is not a usual way of signal detection, and it would raise problems like error accumulation if many samples were received during those previous times.

Let us consider detection by means of the dyadic correlation function. The big peak S2 in column "c" is detected by using the threshold TH1. The dyadic correlation in the rows 1, 2 and 4 would be zero if only the one signal had been received. Instead one finds the relative amplitudes $+1$, -1 and $+3$ at the time $\theta' = 1$ in column "c". The small signal can be detected from these values. Since the mainlobe S1 in column "b" has the relative amplitude 4 one obtains a reduction of the signal-to-noise ratio in the presence of the large signal by $(1^2 + 1^2 + 3^2)/4^2$ or -1.6 dB. This is more than 10 dB better than without dyadic correlation. As always one must be careful with such a statement of improvement since it applies only to the particular time shift between the two signals and an amplitude ratio of 3:1 as shown in Fig. 5. An amplitude ratio of 4:1 would yield an infinite improvement.

Let us turn from the problem of signal detection in noise and clutter to that in noise alone. It is known that the detection of the mainlobe of the autocorrelation function yields the lowest error rate in the presence of thermal noise, but how does one know there was thermal noise at the time of detection? The amplitudes of the dyadic correlation function $E_{FG}(\theta', \theta_v)$ for all values $\theta_v \neq +0$ must have a Gaussian distribution at the time $\theta' = 0$ when the ordinary correlation function $E_{FG}(\theta', +0) = C_{FG}(\theta')$ has its peak value. This is an easy check on real thermal noise. Time invariant thermal noise is the dominant source of disturbances in theoretical channels but it is not often dominant or important in real channels. Dyadic correlation provides in these practical cases information about the noise and thus the basis for improved error reduction. In a different field, the problem of detecting signals in thermal noise is sometimes compounded by the addition of artificially simulated thermal noise. Dyadic correlation is a simple way to assure that the simulated thermal noise must be as good as the one provided by nature, and expensive to produce.

(c) Circuit for dyadic correlation

Figure 6 shows on top a received and sampled signal $F(\theta)$ that is shifted through a 16 stage shift register. Four levels of adders and subtractors denoted 1 to 1000 produce at any moment the Walsh–Fourier transform according to equation (6) except that the integral is replaced by a sum and the transformation is performed according to the fast algorithm [1]:

$$a(j, \theta') = \sum_{\theta'} F(\theta + \theta')\, \text{wal}(j, \theta). \qquad (10)$$

The coefficients $a(j, \theta')$ are multiplied by the coefficients $a(j, 0)$ of the stored signal $F(\theta)$. Actually, the Walsh–Fourier coefficients $a(j, 0)$ of $F(\theta)$ are stored rather than $F(\theta)$ itself. According to equation (9) one has to perform an inverse Walsh–Fourier transform. This is again done in Fig. 6 by the fast algorithm using four levels of adders and subtractors. Note that the circuits for the fast transform and the fast inverse transform are equal [2]. The circuit of Fig. 6 will be particularly simple if all coefficients $a(j, 0)$ are either $+1$ or -1, since the multipliers would then do either nothing or reverse the sign. It follows from the definition of $a(j, 0)$,

$$a(j, 0) = \int_{-1/2}^{1/2} F(\theta)\, \text{wal}(j, \theta)\, d\theta,$$

that $F(\theta)$ should consist of a sum of Walsh functions with amplitude $+1$ or -1.

The best result obtained so far by dyadic correlation is a method that suppresses all sidelobes in the range-Doppler domain completely. This method has only been worked out for the noise-free case yet. The noise-free case is of little interest in the range domain since the main reason for

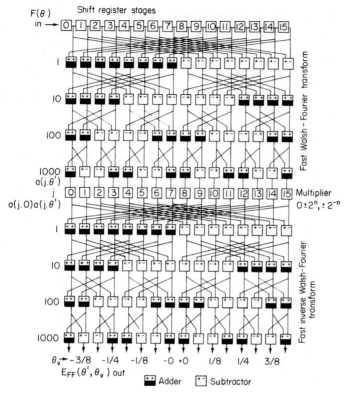

FIG. 6. Dyadic correlator for the generation of a complete set of correlation functions $E_{FF}(\theta', \theta_v)$ from a signal $F(\theta)$ with sixteen digits.

using a structural signal for a range radar is the increase in signal energy with limited peak power. In the range-Doppler domain, on the other hand, one needs long structured signals to obtain good range as well as good Doppler resolution.

3-Images of Objects Under Water by Means of Sound Waves

(*a*) *Principle of acoustic imaging using sequency filters* [7–11]
The classical method of producing an image of an object by means of waves uses lenses. Depending on the kind of waves used, they may be either optical or acoustic lenses. In essence, a lens delays waves originating at one point and striking the lense at different points in such a way as to combine the waves again in one point. A second method is based on the echo principle, particularly as used in sidelooking radar or sonar. A third method is provided by holographic techniques. A fourth method, discussed here, uses electrical filters. The change of a wave front generated by the object to that received is considered to be a linear transformation.

A filter performing the inverse transformation will then recreate an image of the object. Two major difficulties are encountered: (a) The usual electric filters work with time signals, which have time as the only variable. For image generation, one needs filters for signals with at least two space variables. Such filters have recently become known under the name "sequency filters". (b) Sequency filters cannot be implemented at the present for electromagnetic waves with frequencies as used in sidelooking radar, optical lenses or holography. Acoustic waves as used in sonar, on the other hand, are in the range from about 1 kHz to 1 MHz, and sequency filters can be built for electric voltages oscillating in this frequency range. The transformation from acoustic to electric oscillations is readily performed by microphones. The electric oscillations must eventually be transformed into light, since light is the only means that permits us to perceive two-dimensional spatial signals with any degree of resolution. This transformation is easily done by means of TV display equipment.

In most cases the generation of images by other than light waves is of academic interest only. One important exception is the production of images by sound waves in water. Light will not penetrate water very far. Sound waves will and they yield a reasonable resolution. A sinusoidal wave with frequency of 10 kHz has a wavelength of 15 cm. The attenuation in sea water is about 0.2 dB for a distance of 200 m at 10 kHz and increases to 10 dB at 100 kHz. The distance of 200 m is chosen since it is representative for the depth of the continental shelf.

It is possible to reproduce the wave fronts of sound waves by means of acoustic lenses. The reproduced acoustic image must then be transformed into an optical image. The straightforward way of doing this is to transform the acoustic image by an array of microphones into electric voltages that can be sampled and displayed as optical image on a TV screen.

There are a number of drawbacks to acoustic lenses. One may avoid them by exposing the microphone array directly to the sound waves coming from the observed object. The reproduction of the wave front has then to be done by processing the output voltages of the microphones. One may think of doing this by computer processing, but the speed required turns out to be prohibitive for resolutions of practical interest. The so-called two-dimensional sequency filters, which have become known only recently, are able to perform the processing fast enough for sinusoidal acoustic waves with frequencies up to at least 100 kHz. Although these filters for two-dimensional signals are much more expensive than the usual electric filters for one-dimensional time signals, they cost a fraction of the price of a high speed computer.

The purely electric processing of the signal deviates from what is usually done in the processing of acoustic signals. There are, however, a number of interesting features. The effect of changing from a wide angle lens used for general survey to a narrow angle lens for close-up images can be produced by changing the frequency of the sound wave used for illuminating the object. One can produce coloured images. The colour would, of course, represent different acoustic rather than optical properties of the object. One can produce three-dimensional images: If the illumination by a continuous sinusoidal wave is replaced by a pulsed wave with pulses of 1 ms duration one can observe individual layers of 1.5 m thickness by using a properly synchronized receiver. The frequency of the wave would in this case have to be above 10 kHz to have at least 10 cycles in a layer of 1.5 m thickness. This concept can be carried from naval to medical use: a sound wave with 1 MHz frequency would permit images of layers of about 1.5 cm thickness in the human body. Three-dimensional coloured images of this kind can presently not be achieved by light or X-rays since the duration of the illuminating pulse would have to be reduced by a factor 1.5/300 000.

Since sidelooking sonar has been very successful in providing images by means of sound waves, it appears indicated to discuss briefly the difference between sidelooking sonar imaging and sequency filter imaging. Sidelooking sonar requires a relative velocity between object and receiver. This is no drawback if an image of the sea bottom is to be produced by a receiver mounted aboard a ship. It would be a major handicap if a close-up view of remains of the submarine Thresher were to be obtained. One can think of many cases where relative velocity is either an advantage, a drawback or of no significance. However, it is more constructive to observe that the Doppler effect is not used in sequency filter imaging as discussed here and remains available for future developments. In other words, the principles of sidelooking sonar and sequency filter imaging may eventually be combined.

Figure 7 shows the principle of the image forming process. A source of sound waves for illumination of the object of interest is shown on top. It is assumed that a sinusoidal wave is used for illumination. This should

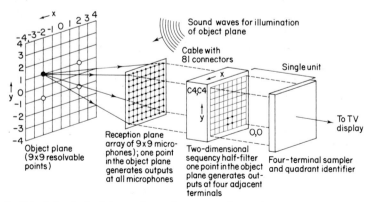

FIG. 7. Principle of obtaining an image by means of sound waves using a microphone array, a two-dimensional sequency filter and a TV display.

not be interpreted to imply that sinusoidal waves are best for this application. Sequency filters for sinusoidal waves are actually more difficult to implement than for other waves but radiators for sinusoidal waves are readily available and the properties of sea water are best known for such waves.

The sound waves are scattered in the object plane. A total of 9 x 9 points can be resolved unambiguously in the object plane since an array of 9 x 9 microphones is used in the reception plane. This reception plane may be a flat bottom of a ship in which the microphones are mounted, a microphone array suspended by a cable for a close-up view, etc.

A signal scattered by a point in the object plane will, in general, arrive at a different time at each one of the 81 microphones in the reception plane. From the arrival times one may compute the point in the object plane that scattered the signal. Signals scattered by different points will produce output voltages at the microphones according to the proportionality and the superposition law. The recomputation of the scattering points from the received waves is a filtering problem. For clarification, consider a voltage pulse of short time duration applied to one of the usual filters for time variable signals. The output can be a voltage pulse spread over a long time period. A matched filter will retransform the spread pulse into the original short pulse. In Fig. 7, a short spatial pulse is produced by one of the scattering points in the image plane; in the reception plane, a spatial pulse spread over all 81 microphones is received. The transmission medium between object plane and reception plane acts like a two-dimensional spatial filter. The output voltages of the microphones have to be fed through a two-dimensional spatial matched filter to reproduce the original short spatial pulse. The "two-dimensional sequency half filter" in Fig. 7 is such a filter.

It turns out that the sequency half filter first transforms a signal produced by one of the four symmetric points shown in the object plane of Fig. 7 into voltages at four output terminals of the sequency half filter as shown. The quadrant of the object plane in which the scattering point was located is resolved by an additional linear transformation. This transformation can be done most economically in conjunction with the sampling device needed for display on a TV screen. Hence Fig. 7 shows a block "four-terminal sampler and quadrant identifier" separated from the sequency half filter. If cost is no object, one could build the sequency half filter so that only one output terminal yields a voltage for a signal from one scattering point in the object plane.

To derive some geometrical relations consider the one-dimensional microphone array of Fig. 8. Nine microphones are spaced a distance d apart. Let the directional characteristic of the microphones have the angle α; ideally, a signal arriving within this angle will produce the same output voltage regardless of the actual angle of incidence, while a signal arriving from outside this angle will produce no output voltage. The

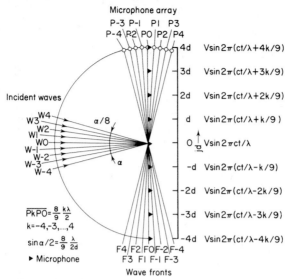

FIG. 8. Geometric relations for the reception of waves Wk by a one-dimensional array of microphones. α reception angle of the microphones; λ wavelength of the sinusoidal wave used for illumination; d distance between microphones.

angle α is subdivided into eight equal angles for the nine incident waves denoted W-4 to W4 in Fig. 8. The respective wave fronts are denoted F-4 to F4. Using the sinusoidal wave $\sin 2\pi ct/\lambda$ for illumination, one obtains from Fig. 8 the relation between reception angle α of the microphones, the wavelength λ and the distance d between the microphones:

$$\sin \alpha/2 = (8/9)\lambda/2d. \tag{11}$$

For small values of α and a large number of microphones, one may use the simpler formula $\lambda/d \doteq \alpha$.

Figure 9 shows the geometric relations in the object plane. The reception angle α defines the observable width D of the object plane at a distance L:

$$D = 2L\,tg\alpha/2. \tag{12}$$

For small angles one obtains the approximate relation $L/D \doteq d/\lambda$.

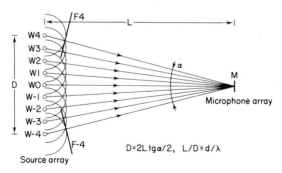

FIG. 9. Geometric relations for the object plane showing the emission of waves Wk from an array of acoustic sources.

Let us return to Fig. 8. All waves Wk produce the output voltage $V \sin 2\pi ct/\lambda$ at the microphone in the center of the array ($j = 0$). Generally, the output voltage $V \sin 2\pi(ct/\lambda + jk/9)$ is produced at microphone j, where j assumes the values $-4, -3, \ldots, +4$. One could connect nine delay lines to each microphone and connect one delay line from each microphone to one of nine summing amplifiers. One would then have nine microphones, nine summing amplifiers and 81 delay lines. Choosing the delays properly, one could produce an output voltage at only one summing amplifier for each wave Wk. This is essentially the way an optical lens produces an image. The lens provides different delays depending on location and angle of incidence in an economical way. Delay lines for time varying electric voltages are too costly to use this principle in our case.

Let us rewrite the output voltages of Fig. 8 as products:

$$V \sin 2\pi(ct/\lambda + jk/9) = V \sin 2\pi ct/\lambda \cdot \cos 2\pi jk/9 + \cos 2\pi ct/\lambda \cdot \sin 2\pi jk/9 \tag{13}$$

Let us write θ for $j/9$ and think of θ as a continuous variable. The purpose is to be able to use the continuous Fourier transform for explanation; later we will return to the discrete Fourier transform. Equation (13) is multiplied by $2 \sin 2\pi i\theta$, $2 \cos 2\pi i\theta$ or $\sqrt{2}$ and integrated over the interval $-1/2 < \theta < 1/2$ ($i = 1, 2, 3, 4$):

$$2V \int_{-1/2}^{1/2} [\sin 2\pi ct/\lambda \cdot \cos 2\pi k\theta + \cos 2\pi ct/\lambda \cdot \sin 2\pi k\theta] \sin 2\pi i\theta \, d\theta$$
$$= \delta_{ik} \frac{k}{|k|} V \cos 2\pi ct/\lambda, \tag{14}$$

$$2V \int_{-1/2}^{1/2} [\sin 2\pi ct/\lambda \cdot \cos 2\pi k\theta + \cos 2\pi ct/\lambda \cdot \sin 2\pi k\theta] \cos 2\pi i\theta \, d\theta$$
$$= \delta_{ik} V \sin 2\pi ct/\lambda, \tag{15}$$

$$\sqrt{2}V \int_{-1/2}^{1/2} [\sin 2\pi ct/\lambda \cdot \cos 2\pi k\theta + \cos 2\pi ct/\lambda \cdot \sin 2\pi k\theta] \, d\theta$$
$$= \delta_{0k} V \sin 2\pi ct/\lambda. \tag{16}$$

Let us turn from the continuous Fourier transform of equations (14) to (16) to the discrete transform. Figure 10 shows the continuous functions for $\theta > 0$ used in these equations and the samples to be taken at $\theta = j/9 = 0, \pm 1/9, \pm 2/9, \pm 3/9, \pm 4/9$ for the discrete transform in order to obtain the same results. A practical circuit for the discrete transformation is shown in Fig. 11. The microphones at the points $-4d, -3d, \ldots, 4d$ in Fig. 8 are connected to the input terminals $-4d, -3d, \ldots, 4d$ of Fig. 11. The resistors and operational amplifiers perform multiplications and summations that represent the discrete transform corresponding to equations (14) to (16). For instance, all voltages fed to the topmost operational amplifier are multiplied by 0.707 and summed. The factors

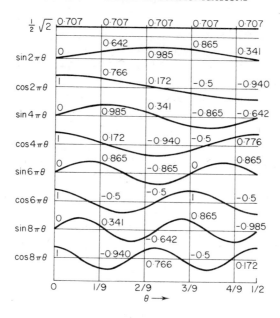

FIG. 10. Representation of the first nine functions of the Fourier series by sampled values.

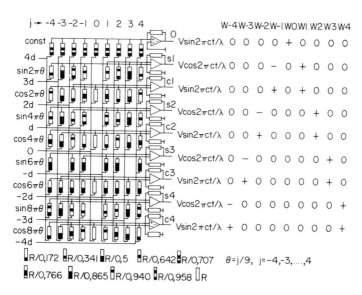

FIG. 11. Circuit for the discrete Fourier transform using nine input samples at a time. The values of the resistors correspond to the sampled values in Fig. 10. The output voltages caused by the waves W-4 to W4 of Fig. 8 are shown on the right.

0.707 represent the constant shown on top of Fig. 10. The multiplication by this function and the summation represent equation (16).

In general, the scattered sound waves will not come from discrete points W-4 to W4 as shown in Fig. 9 but also from intermediate points. The resulting change in output voltage is shown in Fig. 12. If the wave Wk comes from the point k, as assumed so far, it shall produce the relative output voltage 1. If the wave comes from some other point, the output voltage will vary according to the $(\sin x)/x$ functions shown. The worst case occurs for a wave that originates exactly in the middle between two points, for instance between k and $k - 1$. The relative voltages at the two terminals representing $k - 1$ and k will be 0.66 The voltage at the terminal representing $k + 1$ will be close to the maximum of the first sidelobe, which is about 0.21. The power ratios $0.66^2/0.21^2 = 10$ or $2 \times 0.66^2/0.21^2 = 20$ represent approximately the brightness ratio seen on the display. These ratios are about equal to the ratio of darkest to brightest parts of a printed black-and-white picture. Hence, no objectionable degradation of the image should result from the sidelobes of the $(\sin x)/x$ functions in Fig. 12.

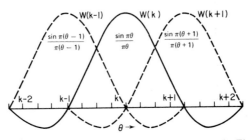

FIG. 12. Variation of the output voltages caused by the waves in Fig. 9 not originating at discrete points.

In order to obtain two-dimensional images, one has to perform a two-dimensional Fourier transform. This can be done by stacking printed circuit cards with the circuit of Fig. 11 in the way shown by Fig. 13.

(b) Two-dimensional sequency filters

The two-dimensional sequency filter shown in Fig. 13 consists of two stacks of nine printed circuit boards each. The one is stacked horizontally, the other vertically. All boards have the circuit shown by Fig. 11. Its generalization to more than 9 x 9 input terminals is straightforward. The horizontally stacked cards perform the Fourier transform for the variable x, the vertically stacked cards the transform for the variable y. The interconnection between the two sets of cards can be done by a matrix socket into which the cards are plugged from both sides. The cards can be individually removed for repair or replacement.

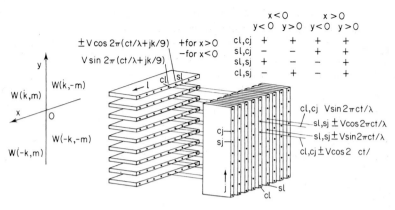

FIG. 13. Signs of the output voltages of a two-dimensional sequency filter caused by waves originating at the points $-k$, $-m$ to $+k$, $+m$ in the four quadrants of the object plane.

Let us consider the two-dimensional image transformation according to Fig. 13 in more detail. The object plane is divided into four quadrants: $x < 0, y < 0; x < 0, y > 0; x > 0, y < 0; x > 0, y > 0$. Waves $W(\pm k, m)$ and $W(\pm k, -m)$ with the same absolute value for the variable $x = |\pm m|$ produce the output voltage $V \sin 2\pi(ct/\lambda + jk/9)$ at the terminal cl of each one of the horizontally stacked cards. If the wave comes from the half-plane $x > 0$, the output voltage at the terminal sl of all cards will be equal to $V \cos 2\pi(ct/\lambda + jk/9)$, while the voltage $-V \cos 2\pi(ct/\lambda + jk/9)$ is produced for $x < 0$.

The output voltages from all terminals cl of the horizontal stack of cards are fed to the card cl of the vertical stack, the output voltages from the terminals sl to the card sl. The transformation of the voltages $V \sin 2\pi(ct/\lambda + jk/9)$ by card cl is defined by equations (13) to (16). For the transformation of $V \cos 2\pi(ct/\lambda + jk/9)$ one obtains in analogy:

$$\cos 2\pi(ct/\lambda + jk/9) = \cos 2\pi ct/\lambda \cdot \cos 2\pi jk/9 - \sin 2\pi ct/\lambda \cdot \sin 2\pi jk/9, \tag{17}$$

$$2V \int_{-1/2}^{1/2} [\cos 2\pi ct/\lambda \cos 2\pi k\theta - \sin 2\pi ct/\lambda \sin 2\pi k\theta] \sin 2\pi i\theta \, d\theta$$

$$= -\delta_{ik} \frac{k}{|k|} V \sin 2\pi ct/\lambda, \tag{18}$$

$$2V \int_{-1/2}^{1/2} [\cos 2\pi ct/\lambda \cos 2\pi k\theta - \sin 2\pi ct/\lambda \sin 2\pi k\theta] \cos 2\pi i\theta \, d\theta$$

$$= \delta_{ik} V \cos 2\pi ct/\lambda, \tag{19}$$

$$\sqrt{2}V \int_{-1/2}^{1/2} [\cos 2\pi ct/\lambda \cos 2\pi k\theta - \sin 2\pi ct/\lambda \sin 2\pi k\theta] \, d\theta$$

$$= \delta_{0k} V \cos 2\pi ct/\lambda. \tag{20}$$

TABLE I. Representative figures for acoustic imaging in water by means of two-dimensional sequency filters.

Frequency of illuminating sinusoidal wave	10 kHz		100 kHz		1 MHz	
Wavelength	15 cm		1.5 cm		1.5 mm	
Microphone array	31 × 31	255 × 255	31 × 31	255 × 255	31 × 31	255 × 255
Number of microphones	961	65025	961	65025	961	65025
α = 30°						
$d = \lambda/2 \sin(\alpha/2)$	29 cm		2.9 cm		2.9 mm	
Size of microphone array $D = 2Ltg(\alpha/2)$; $D \times D$	8.7 × 8.7 m²	74 × 74 m²	0.87 × 0.87 m²	7.4 × 7.4 m²	8.7 × 8.7 cm²	74 × 74 cm²
field of view for $L = 200$ m	106 × 106 m²					
Distance of resolvable points at $L = 200$ m	3.54 m	41.5 cm	3.54 m	41.5 cm	3.54 m	41.5 cm
Field of view for $L = 20$ m	10.6 × 10.6 m²					
Distance of resolvable points at $L = 20$ m	35 cm	—	35 cm	4.2 cm	35 cm	4.2 cm
α = 10°						
$d = \lambda/2 \sin(\alpha/2)$	86 cm		8.6 cm		8.6 mm	
Size of microphone array $D = 2Ltg(\alpha/2)$; $D \times D$	26 × 26 m²	206 × 206 m²	2.6 × 2.6 m²	20.6 × 20.6 m²	26 × 26 cm²	206 × 206 cm²
field of view for $L = 200$ m	35 × 35 m²					
Distance of resolvable points at $L = 200$ m	1.17 m	—	1.17 m	13.8 cm	1.17 m	13.8 cm
Field of view for $L = 20$ m	3.5 × 3.5 m²					
Distance of resolvable points at $L = 20$ m	—	—	11.7 cm	1.38 cm	11.7 cm	1.38 cm

The signs of the output voltages at the terminals cl, cj to sl, sj in Fig. 13 follow from these relations for waves originating in the four quadrants of the object plane. These signs are shown in Fig. 13. They represent the first four Walsh functions and form thus an orthogonal set. The quadrant ambiguity may be resolved by transforming the sinusoidal voltages $V \sin 2\pi ct/\lambda$ at the terminals cl, cj and sl, sj into cosinusoidal voltages $V \cos 2\pi ct/\lambda$ by means of integrators. The resulting voltages are multiplied by $+1$ or -1, according to the signs in the table in Fig. 13, and summed. As a result of the orthogonality of the Walsh functions, three of the four output voltages must be zero and the quadrant ambiguity is resolved.

Table I shows some representative numerical values for acoustic imaging in water by means of two-dimensional sequency filters.

REFERENCES

1. Harmuth, H. F. (1972). "Transmission of Information by Orthogonal Functions". (2nd edition), Springer-Verlag, New York/Berlin.
2. Manz, J. W. (1972). A sequency ordered fast Walsh transform. *IEEE Trans. Audio Electroacoustics* Vol. AU-2D, No. 3, 204-205.
3. Pichler, F. (1970). Some aspects of a theory of correlation with respect to Walsh harmonic analysis, AD 714-596.
4. Gibbs, J. E. (1967). Walsh spectrometry, a form of spectral analysis well suited to binary digital computation. Fundamental but unpublished manuscript.
5. Erickson, C. W. (1961). Clutter cancelling in autocorrelation functions by binary sequence pairing, AD 446-146.
6. Golay, M. J. E. (1961). Complementary series. *IRE Trans. Information Theory* IT-7, 82-87.
7. Sondhi, M. M. (1969). Reconstruction of objects from their sound diffraction patterns. *J. Acoust. Soc. Am.* 46, No. 5, Part 2, 1158-1164.
8. Smyth, C. N. and others. (1963). The ultra-sound image camera. *Proc. IEE* 110, 16-28.
9. Aoki, Y. and others. (1967). Sound wave hologram and optical reconstruction. *Proc. IEEE* 55, 1622-1623.
10. Preston, K. and Kreuzer, J. L. (1967). Ultrasonic imaging using a synthetic holographic technique. *Appl. Phys. Lett.* 10, No. 5, 150-152.
11. Kock, W. E. and Harvey, F. K. (1951). A photographic method for displaying sound wave and microwave space patterns. *Bell System Tech. J.* 20, 564-587.
12. Harmuth, H. F. Sidelobe suppression of radar signals, Proc. 1972 International Communications Conference 38-11 to 38-15. IEEE order number 72-CHO-622-1-COM.

Spectral Analysis of a Class of Random Signals

ROBERT M. FORTET

Laboratoire de Probabilites, Universite de Paris VI, Paris, France

Symboles et Notations

$\bar{a}$ nombre complexe conjugué du nombre complexe a.

$\tilde{\phi}(.)$ transformée de Fourier de la fonction $\phi(.)$.

$\tilde{f}$ transformée du Fourier de la distribution f.

$\langle f, z \rangle$ nombre que la fonctionnelle linéaire f sur un espace vectoriel $\mathscr{Z}$ fait correspondre à l'élément $z \in \mathscr{Z}$.

$$D_h(\alpha_1, \ldots, \alpha_n)f(.) = \frac{\partial^h}{\partial t_1^{\alpha_1} \ldots \partial t_n^{\alpha_n}} f(.); \quad h \text{ entier} > 0; \quad \alpha_j \geqslant 0,$$

$$\alpha_1 + \alpha_2 + \cdots + \alpha_n = h; \quad f(t) = f(t_1, \ldots, t_n) \text{ fonction de}$$

$$t = \{t_1, \ldots, t_n\} \in \mathbb{R}^n.$$

$$D_0 f(.) = f(.); f(t) = f(t_1, \ldots, t_n) \text{ fonction de } t = \{t_1, \ldots, t_n\} \in \mathbb{R}^n.$$

$$D_h f(.) = \frac{d^h}{dt^h} f(.) \quad \text{quand } n = 1 \ (t = t_1).$$

$D_h(\alpha_1, \ldots, \alpha_n)f, D_0 f, D_h f$: même signification, mais appliquée à une *distribution f*.

v.a.   abréviation pour: variable aléatoire

e.m.   abréviation pour: espérance mathematique

f.a.   abréviation pour: fonction aléatoire

$E(X)$ espérance mathématique de la v.a. X.

d.a.   abréviation pour: distribution aléatoire.

$\mathscr{B}_n$ σ-algèbre des Boréliens de $\mathbb{R}^n$.

1. Rappels sur les Distributions

L'objet de cette conférence est principalement de montrer que la Théorie des Distributions offre des moyens de calculs puissants dans l'étude des signaux aléatoires.

La Théorie des Distributions est exposée de façon très étendue dans I. M. Gelfand, G. E. Chilov, N. Y. Vilenkin [1], ou dans L. Schwartz [2]. Je me borne ici à rappeler brièvement quelques points essentiels de cette Théorie.

(a) Les fonctions de la classe S_n

Nous désignerons par S_n (n entier $\geqslant 1$) l'espace *vectoriel* des fonctions (numériques complexes) $\phi(t)$ de $t \in \mathbb{R}^n$, qui satisfont aux conditions suivantes:

(1) $\phi(.)$ possède des dérivées partielles de tous les ordres;

(2) pour tout entier $h \geqslant 0$, pour tous entiers $\alpha_j \geqslant 0$ ($j = 1, \ldots, n$) avec: $\alpha_1 + \cdots + \alpha_n = h$; pour tous entiers $q_j \geqslant 0$ ($j = 1, \ldots, n$) il existe une constante $C_h(\alpha_1, \ldots, \alpha_n; q_1, \ldots, q_n) \geqslant 0$ telle que:

$$|t_1^{q_1} \ldots t_n^{q_n} D_h(\alpha_1, \ldots, \alpha_n)\phi(t_1, \ldots, t_n)| \leqslant C_h(\alpha_1, \ldots, \alpha_n)$$

$$\forall t \in \mathbb{R}^n.$$

Soit $\{\phi_k(.); k = 1, 2, \ldots\}$ une suite de fonctions $\phi_k(.) \in S_n$; on dit qu'*elle tend vers 0 dans S_n* (lorsque $k \to +\infty$) si (avec les notations ci-dessus):

(1) pour tous h, q_j, α_j, il existe des constantes $C_h(\alpha_1, \ldots, \alpha_n; q_1, \ldots, q_n)$ indépendantes de k, telles que: pour tout k et tout $t \in \mathbb{R}^n$;

$$|t_1^{q_1} \ldots t_n^{q_n} D_h(\alpha_1, \ldots, \alpha_n)\phi_k(t_1, \ldots, t_n)| \leqslant C_h(\alpha_1, \ldots, \alpha_n;$$

$$q_1, \ldots, q_n);$$

(2) pour tous h, α_j,

$$\lim_{k \to +\infty} D_h(\alpha_1, \ldots, \alpha_n)\phi_k(t_1, \ldots, t_n) = 0$$

uniformément en t sur toute partie bornée de $\mathbb{R}^n$.

Exemple (1.1) Un exemple de fonction $\phi(.)$ appartenant à S_1, est la fonction:

$$\phi(t) = e^{-t^2} \quad (n = 1; t = t_1).$$

Transformation de Fourier. La transformée de Fourier $\tilde{\phi}(.)$ de $\phi(.) \in S_n$ est définie par:

$$\tilde{\phi}(\nu) = \int_{\mathbb{R}^n} e^{-2\pi i(\nu, t)}\phi(t) \, dt \quad (\nu \in \mathbb{R}^n), \tag{1.1}$$

où: (ν, t) désigne l'expression: $\Sigma_{j=1}^n \nu_j t_j$. On a le:

Théorème (1.1). La transformation de Fourier (1.1) est une application linéaire biunivoque et bicontinue de S_n sur lui-même, que admet la formule d'inversion:

$$\phi(t) = \int_{\mathbb{R}^n} e^{2\pi i(\nu, t)}\tilde{\phi}(\nu) \, d\nu. \ \blacksquare \tag{1.2}$$

Par la suite, nous utiliserons essentiellement:

— le cas $n = 1$. $(t = t_1, \nu = \nu_1, [\nu, t] = \nu t)$;

— le cas $n = 2$; dans ce cas $n = 2$, il nous sera commode de modifier la convention d'écriture $(1,1)$, et de poser, pour $\phi(.\,,.) \in S_2$,

$$\tilde{\phi}(\nu_1, \nu_2) = \int_{\mathbb{R}^2} e^{-2\pi i(\nu_1 t_1 - \nu_2 t_2)} \phi(t_1, t_2)\, dt_1\, dt_2. \tag{1.3}$$

(b) Les S_1-distributions

Je vais introduite la notion de distribution, en me limitant au cas $n = 1$ pour alléger l'écriture; les extensions à n quelconques sont immédiates.

On appelle S_1-distribution, toute *fonctionnelle linéaire f* sur S_1 qui est continue, c'est à dire telle que $\langle f, \phi(.) \rangle$ tend vers 0, si $\phi(.)$ tend vers 0 dans S_1.

D'une S_1-distribution f, on dit qu'elle tend vers la S_1-distribution g, si: $\langle f, \phi(.) \rangle$ tend vers $\langle g, \phi(.) \rangle$ pour toute $\phi(.) \in S_1$. $\mathscr{D}(S_1)$ désignera l'ensemble de toutes les S_1-distributions.

Exemple (1.2). Soit $m(dt)$ une fonction d'ensemble sur R, σ-additive telle qu'il existe $\alpha > 0$ tel que:

$$\int_R (1 + t^2)^{-\alpha} |m|(dt) < +\infty; \tag{1.3}$$

la formule:

$$\langle m, \phi(.) \rangle = \int_R \phi(t) \bar{m}(dt) \quad \forall m(.) \in S_1, \tag{1.4}$$

définit une S_1-distribution m, que l'on peut identifier à $m(dt)$.

Un cas particulier est celui où $m(dt) = f(t)\, dt$ admet une densité $f(t)$; alors (1.3) s'écrit:

$$\langle m, \phi(.) \rangle = \int_{-\infty}^{+\infty} \phi(t)\overline{f(t)}\, dt; \tag{1.5}$$

on peut dire de la fonction ordinaire $f(.)$ qu'elle est identifiable par (1.5) à une S_1-distribution.

Exemple (1.3). Considérons le fonction $f(t) = 1/t$; elle n'est pas directement interprétable comme une distribution selon (1.5), car l'intégrale:

$$\int_R \frac{\phi(t)}{t}\, dt$$

où $\phi(.) \in S_1$, est généralement dépourvue de sens; mais on peut définir la S_1-distribution I, ou "distribution inverse", par:

$$\langle I, \phi(.) \rangle = \lim_{\epsilon \to +0} \int_{|t| > \epsilon} \frac{\phi(t)}{t} \, dt, \quad \forall \phi(.) \in S_1; \tag{1.6}$$

équivalent à:

$$\langle I, \phi(.) \rangle = \frac{1}{2} \int_{-\infty}^{+\infty} \frac{\phi(t) - \phi(-t)}{t} \, dt,$$

et aussi à:

$$\langle I, \phi(.) \rangle = - \int_{-\infty}^{+\infty} \log|t| D_1 \phi(t) \, dt.$$

(c) *Forme générale d'une S_1-distribution.* Appelons H_m (m entier $\geqslant 0$), l'ensemble des fonctions $\phi(t)$ de $t \in \mathbb{R}$, telles que:

(1) pour $0 \leqslant h \leqslant m$, $D_h \phi(.)$ existe et est mesurable (sur $\mathcal{B}_1$);

(2) $A_m(\phi)^2 = \sum_{h=0}^{m} \int_{\mathbb{R}} |D_h \phi(t)|^2 (1 + t^2)^{2m} \, dt < +\infty.$ \tag{1.7}

H_m est un espace de Hilbert, dans lequel le produit hermitique $\phi(.)_0 \psi(.)$ de $\phi(.) \in H_m$ par $\psi(.) \in H_m$ est défini par:

$$\phi(.)_0 \psi(.) = \sum_{h=0}^{m} \int_{\mathbb{R}} D_h \phi(t) \overline{D_h \psi(t)} (1 + t^2)^{2m} \, dt; \tag{1.8}$$

$A_m(\phi)$ définie par (1.7) est la norme de $\phi(.) \in H_m$.
Si $\phi(.) \in S_1$, $\phi(.) \in H_m$ quel que soit $m \geqslant 0$.
On démontre le:

Théorème (1.2). Si $f \in \mathcal{D}(S_1)$, il existe $m \geqslant 0$ et une fonction $f(t)$ de $t \in \mathbb{R}$, appartenant à H_m, telle que pour toute $\phi(.) \in S_1$, $\langle f, \phi(.) \rangle$ est égal au produit hermitique $\phi(.)_0 f(.)$ dans H_m de $\phi(.)$ par $f(.)$. ■
En d'autre termes:

$$\langle f, \phi(.) \rangle = \sum_{h=0}^{m} \int_{\mathbb{R}} D_h \phi(t) \overline{D_h f(t)} (1 + t^2)^{2m} \, dt. \tag{1.9}$$

De cette représentation de $\langle f, \phi(.) \rangle$, on peut en déduire facilement d'autres, équivalentes, mais éventuellement plus commodes selon l'application envisagée.

(d) *Dérivée d'une S_1-distribution*

Par définition, la dérivée d'une S_1-distribution f, est la S_1-distribution $D_1 f$ définie par:

$$\langle D_1 f, \phi(.) \rangle = -\langle f, D_1 \phi(.) \rangle, \quad \forall \phi(.) \in S_1.$$

(e) *Produit d'une S_1-distribution par une fonction*

Soit $\lambda(t)$ une fonction (numérique complexe) de $t \in R$, telle que: quelle que soit $\phi(.) \in S_1$, la fonction $\overline{\lambda(.)}\phi(.) \in S_1$; ceci a lieu par exemple si $\lambda(t) = t^q$ (q entier $\geqslant 0$). On appelle produit d'une S_1-distribution f par la fonction $\lambda(.)$, la S_1-distribution $g = \lambda(.)f$ définie par:

$$\langle g, \phi(.) \rangle = \langle f, \overline{\lambda(.)}\phi(.) \rangle, \quad \forall \phi(.) \in S_1. \tag{1.10}$$

(f) *Transformée de Fourier d'une S_1-distribution*

La transformée de Fourier d'une S_1-distribution f, est la S_1-distribution $\tilde{f}$ définie par:

$$\langle \tilde{f}, \tilde{\phi}(.) \rangle = \langle f, \phi(.) \rangle, \quad \forall \tilde{\phi}(.) \in S_1.$$

La transformée de Fourier ainsi définie, est une application linéaire biunivoque et bicontinue de $\mathscr{D}(S_1)$ sur lui-même.

Théorème (1.3). La transformée de Fourier $\tilde{D}_1 f$ de la dérivée $D_1 f$ d'une S_1-distribution f, est le produit de $\tilde{f}$ par la fonction $2\pi i t$.

2. Calcul de l'Espérance Mathématique d'une Fonction non-linéaire d'une Variable Aléatoire Multidimensionnelle

Comme première application, je citerai le calcul de l'e.m. d'une fonction non-linéaire d'une v.a. multidimensionnelle. Cette application étant connue, je ne l'indiquerai que brièvement.

Soit $\{X_1, \ldots, X_n\}$ une v.a. n-dimensionnelle réelle, admettant une densité de probabilité $\rho(x_1, \ldots, x_n)$; soit $\lambda(x_1, \ldots, x_n)$ une fonction numérique, réelle ou complexe, de $\{x_1, \ldots, x_n\} \in \mathbb{R}^n$; posons nous le problème de calculer $E(\lambda(X_1, \ldots, X_n))$.

Par définition, il s'agit donc de calculer l'intégrale:

$$I = \int_{\mathbb{R}^n} \lambda(x_1, \ldots, x_n) \rho(x_1, \ldots, x_n) \, dx_1 \ldots dx_n; \tag{2.1}$$

en principe, $E[\lambda(X_1, \ldots, X_n)]$ existe si et seulement si l'intégrale I est absolument convergente. Mais faisons l'hypothèse que la fonction $\rho(.)$ appartient à la classe S_n, et que $\lambda(.)$ est interprétable comme une S_n-distribution λ; nous ferons dans la suite usage systématique des notions et notations du § 1. Alors I s'interpréte par:

$$I = \langle \lambda, \rho(.) \rangle; \tag{2.2}$$

dans ces conditions, convenons que $E[\lambda(X_1, \ldots, X_n)]$ existe et est par définition égale à $\langle \lambda, \rho(.) \rangle$.

Reste à évaluer $\langle \lambda, \rho(.) \rangle$; on peut à ce sujet noter que d'après le § 1, on a:

$$\langle \lambda, \rho(.) \rangle = \langle \tilde{\lambda}, \tilde{\rho}(.) \rangle, \tag{2.3}$$

où:

— $\tilde{\lambda}$ est la S_n-distribution, transformée de Fourier de λ;
— $\tilde{\rho}(.)$ est la transformée de Fourier de ρ; $\tilde{\rho}(\nu_1, \ldots, \nu_n)$ est donc une fonction de $\{\nu_1, \ldots, \nu_n\} \in \mathbb{R}^n$, appartenant à S_n; en fait:

$$\tilde{\rho}(\nu_1, \ldots, \nu_n) = \int_{\mathbb{R}^n} \exp\{-2\pi i \sum_{h=1}^{n} \nu_h x_h\} \rho(x_1, \ldots, x_n) \, dx_1 \ldots dx_n; \tag{2.4}$$

c'est à dire que si l'on pose:

$$\omega_h = -2\pi\nu_h \quad (h = 1, 2, \ldots, n),$$

la fonction $\tilde{\rho}[-(\omega_1/2\pi), \ldots, -(\omega_n/2\pi)]$ des ω_h n'est autre que la fonction caractéristique (n-dimensionnelle) de la v.a. n-dimensionnelle $\{X_1, \ldots, X_n\}$.

Supposons maintenant que $\tilde{\lambda}$ est identifiable à une fonction d'ensemble $\tilde{\lambda}(d\nu_1, \ldots, d\nu_n)$ sur $(\mathbb{R}^n, \mathscr{B}_n)$, σ-additive et σ-bornée, satisfaisant à la condition (1.3) (étendue à n quelconque). Alors (2.3) peut s'écrire:

$$I = \int_{\mathbb{R}^n} \tilde{\rho}(\nu_1, \ldots, \nu_n) \tilde{\lambda}(d\nu_1, \ldots, d\nu_n), \tag{2.5}$$

et le calcul de l'intégrale (2.5) peut être facile.

Par exemple, supposons dorénavant que $\{X_1, \ldots, X_n\}$ est une v.a. *Laplacienne*, avec:

$$E(X_h) = 0 \quad (h = 1, 2, \ldots, n); \tag{2.6}$$

la condition (2.6) n'est pas restrictive pour le principe de ce qui va suivre, nous la faisons simplement pour alléger l'écriture. Posons:

$$E(X_j X_k) = \gamma_{jk} \quad (j, k = 1, 2, \ldots, n); \tag{2.7}$$

les γ_{jk} sont les éléments de la matrice de covariance Γ_n de $\{X_1, \ldots, X_n\}$ (cf. R. Fortet [1], V.5); alors:

$$\tilde{\rho}(\nu_1, \ldots, \nu_n) = \exp\{-2\pi^2 \sum_{jk} \gamma_{jk} \nu_j \nu_k\}, \tag{2.8}$$

qui est de la classe S_n si et seulement si Γ_n est *définie positive, ce que nous supposerons donc.*
(2.5) s'écrit:

$$E[\lambda(X_1, \ldots, X_n)] = \int_{\mathbb{R}^n} \exp\{-2\pi^2 \sum_{jk} \gamma_{jk} \nu_j \nu_k\} \tilde{\lambda}(d\nu_1, \ldots, d\nu_n), \tag{2.9}$$

qui peut être directement calculable.
Mais on peut aussi remarquer que d'après (2.3):

$$\frac{\partial}{\partial \gamma_{jk}} E[\gamma(X_1, \ldots, X_n)] = \left\langle \tilde{\lambda}, \frac{\partial}{\partial \gamma_{jk}} \tilde{\rho}(.) \right\rangle = -2\pi^2 \langle \tilde{\lambda}, \nu_j \nu_k \tilde{\rho}(.) \rangle; \tag{2.10}$$

donc d'après le Théorème (1.3)

$$\frac{\partial}{\partial \gamma_{jk}} E[\lambda(X_1, \ldots, X_n)] = \tfrac{1}{2} \left\langle \frac{\partial^2}{\partial x_j \, \partial x_k} \lambda, \rho(.) \right\rangle =$$

$$= \tfrac{1}{2} \left\langle \lambda, \frac{\partial^2}{\partial x_j \, \partial x_k} \rho(.) \right\rangle, \tag{2.11}$$

soit:

$$\frac{\partial}{\partial \gamma_{jk}} E[\lambda(X_1, \ldots, X_n)] = \tfrac{1}{2} E\left[\frac{\partial^2}{\partial x_j \, \partial x_k} \lambda(X_1, \ldots, X_n)\right]; \tag{2.12}$$

dans (2.11) et (2.12), $(\partial^2/\partial x_j \, \partial x_k)\lambda$ désigne la dérivée seconde au sens des distributions, qui est notée au § 1 par:

$$D_2(\alpha_1, \ldots, \alpha_n)\lambda, \quad \text{avec}: \alpha_h = 0 \text{ si } h \neq j, k, \quad \alpha_j = \alpha_k = 1.$$

Si le second membre de (2.11)-(2.12) est directement calculable, on peut en déduire $E(\lambda(X_1, \ldots, X_n))$ par une intégration.

3. S_1-Distributions Aléatoires du Second Ordre

Soit $\mathscr{H}$ l'espace de Hilbert des v.a. du second ordre, c'est à dire l'ensemble des v.a. (en général complexes) Z telles que:

$$E(|Z|^2) < +\infty;$$

je rappelle que $\mathscr{H}$ est un espace de Hilbert, dans lequel le produit hermitique d'un élément $X \in \mathscr{H}$ par un élément $Y \in \mathscr{H}$ est par définition égal à:
$E(X\bar{Y})$.

Supposons qu'à chaque $\phi(.) \in S_1$ corresponde $\langle Z, \phi(.) \rangle \in \mathcal{H}$, avec les propriétés suivantes:

(1) $\forall \phi_1(.), \phi_2(.) \in S_1, \langle Z, \phi_1(.) + \phi_2(.) \rangle = \langle Z, \phi_1(.) \rangle + \langle Z, \phi_2(.) \rangle$;

(2) $\forall \phi(.) \in S_1, \rho \in \mathbb{C}, \langle Z, \rho\phi(.) \rangle = \rho \langle Z, \phi(.) \rangle$;

(3) Si lorsque $k \to +\infty$, la suite $\{\phi_k(.)\}$ de fonctions $\phi_k(.) \in S_1$, tend vers 0 dans S_1, $\langle Z, \phi_k(.) \rangle$ tend dans $\mathcal{H}$ vers l'élément nul de $\mathcal{H}$ (c'est à dire la v.a. presque-sûrement égale à 0). En d'autres termes, si $\{\phi_k(.)\}$ tends vers 0 dans S_1,

$$\lim_{k \to +\infty} E(|\langle Z, \phi_k(.) \rangle|^2) = 0.$$

Alors par définition Z est une S_1-*distribution aléatoire* du second ordre.

Comme pour toute f.a. du second ordre, l'outil de base de l'étude d'une S_1-d.a. du second ordre est sa covariance $\Gamma[\phi_1(.), \phi_2(.)]$ définie par:

$$\Gamma[\phi_1(.), \phi_2(.)] = E[\langle Z, \phi_1(.) \rangle \overline{\langle Z, \gamma_2(.) \rangle}], \phi_1(.), \phi_2(.)S_1.$$

On remarque que si $\phi_1(.), \phi_2(.) \in S_1$, alors la fonction $\phi_1(t_1)\overline{\phi_2(t_2)}$ de $\{t_1, t_2\} \in \mathbb{R}^2$, appartient à S_2. On démontre alors le:

Théorème (3.1). Si Z est une S_1-d.a., si $\Gamma[\phi_1(.), \phi_2(.)]$ est sa covariance, il existe une et une seule S_2-distribution Γ telle que:

$$\Gamma[\phi_1(.), \phi_2(.)] = \langle \Gamma, \phi_1(.)\overline{\phi_2(.)} \rangle, \phi_1(.), \phi_2(.) \in S_1. \blacksquare$$

Utilisant le Théorème (1.2) (étendu à S_2), il est alors possible de donner une expression de Γ, du type (1.9); d'où résulte une expression de $\Gamma[\phi_1(.), \phi_2(.)]$. Toutefois, pour utiliser pleinement ce résultat, il faudrait tenir compte de ce que Γ ne peut être une S_2-distribution quelconque, puisque, comme toute covariance, $\Gamma[\phi_1(.), \phi_2(.)]$ est de type non-négatif; mais la forme nécessaire et suffisante qui en résulte pour Γ, ne semble pas avoir été établie dans le littérature.

Quoiqu'il en soit, nous appellerons Γ la S_2-*distribution covariance* de Z.

(a) *Transformée de Fourier d'une S_1-d.a. du zème ordre*
Par définition, la transformée de Fourier de la S_1-d.a. du zème ordre Z est la S_1-d.a. du 2ème ordre $\tilde{Z}$ définie par:

$$\langle \tilde{Z}, \tilde{\phi}(.) \rangle = \langle Z, \phi(.) \rangle, \quad \forall \phi(.) \in S_1.$$

On obtient immédiatement le:

Théorème (3.2). La S_2-d.a. covariance de $\tilde{Z}$ est la transformée de Fourier $\tilde{\Gamma}$ de la S_2-d.a. covariance Γ de Z.

(b) *Dérivée d'une S_1-d.a. du 2ème ordre*
La formule

$$\langle D_1 Z, \phi(.) \rangle = -\langle Z, D_1\phi(.) \rangle, \quad \phi(.) \in S_1,$$

définit une S_1-d.a. du 2ème ordre D_1Z, dite dérivée de Z. On a immédiatement le:

Théorème (3.3). La S_2- distribution covariance de D_1Z, est la S_2-distribution $D_2(1, 1)\Gamma$, où Γ est la S_2-distribution covariance de Z.

4. Spectre d'un Train d'Impulsions

Pour mettre en évidence l'intérêt et l'usage de la Théorie des Distributions pour certains calculs, je vais considérer le problème particulier suivant.

A chaque instant τ, associons une *impulsion*, c'est à dire une f.a. $I(\tau, t)$ de $t \in \mathbb{R}$. Choisissons d'autre part, en général aléatoirement, des instants $\{\tau_j\}$ sur $\mathbb{R}$ $(j = 0, \pm 1, \ldots)$; ces $\{\tau_j\}$ constituent une *répartition ponctuelle aléatoire* $\mathscr{R}$ sur $\mathbb{R}$. Soit $S(t)$ le signal défini par:

$$S(t) = \sum_j I(\tau_j, t). \tag{4.1}$$

Il est commode de représenter $\mathscr{R}$ de la façon suivante; à tout $\omega \in \mathscr{B}_1$, associons le nombre (aléatoire) $M(\omega)$ des τ_j appartenant à ω; de sorte que $M(\omega)$ est une *mesure aléatoire* sur $(\mathbb{R}, \mathscr{B}_1)$. (4.1) peut s'écrire:

$$S(t) = \int_{\mathbb{R}} I(\tau, t)M(d\tau). \tag{4.2}$$

On démontre que:

$$E[M(d\tau_1)M(d\tau_2)] = m_2(d\tau_1, d\tau_2)$$

est une mesure sur $(\mathbb{R}^2, \mathscr{B}_2)$; donc (sous des conditions très peu restrictives) c'est une S_2-distribution.

Il serait intéressant d'étudier des cas où $M(.)$ est corrélée avec les $I(\tau, t)$; cependant cela ne semble pas avoir été tenté jusqu'à présent; aussi nous supposerons que la famille des v.a. $I(\tau, t)$ $(\{\tau, t\} \in \mathbb{R}^2)$ est *indépendante* de la mesure aléatoire $M(.)$.

En posant:

$$\gamma(t_1, t_2; \tau_1, \tau_2) = E(I(\tau_1, t_1)\overline{I(\tau_2, t_2)}),$$

et à partir de la formule:

$$\Gamma(t_1, t_2) = \int_{\mathbb{R}^2} \gamma(t_1, t_2; \tau_1, \tau_2)m_2(d\tau_1, d\tau_2), \tag{4.3}$$

il est facile d'étudier la covariance:

$$\Gamma(t_1, t_2) = E(S(t_1)\overline{S(t_2)})$$

du signal $S(t)$.

(a) Restriction diagonale

Dans cette étude, on voit intervenir les notions suivantes, qui ne semblent pas avoir été jusqu'ici suffisamment dégagées.

Soit D la "diagonale" $t_2 = t_1$, de $\mathbb{R}^2$; soit $m_2(dt_1, dt_2)$ une fonction d'ensemble (réelle ou complexe) σ-additive sur $(\mathbb{R}^2, \mathscr{B}_2)$; par exemple, $m_2(dt_1, dt_2)$ peut être une mesure.

Soit $_dm_2(\omega)$ la fonction de $\omega \in \mathscr{B}_2$ définie par:

$$_dm_2(\omega) = m_2(\omega \cap D), \quad \forall \omega \in \mathscr{B}_2;$$

nous appellerons $_dm_2(.)$ la *restriction diagonale* de $m_2(.)$.

Interprétons $\mathbb{R}^2$ comme un plan rapporté à deux axes $0t_1$, $0t_2$ ortho-normée. Soit π l'opération de projection de D sur l'axe $0t_1$: π est une correspondance biunivoque entre les sous-ensembles δ de D, et les sous-ensembles $e = \pi(\delta)$ de $\mathbb{R}$.

Considérons alors la fonction d'ensemble $\nu(e)$ de $e \in \mathscr{B}_1$ définie par:

$$\nu(e) = m_2[\pi^{-1}(e)];$$

disons de $\nu(e)$ qu'elle est la *restriction diagonale projetée* de $m_2(.)$; d'une façon générale, la restriction diagonale projetée d'une fonction $m_2(dt_1, dt_2)$ σ-additive sur $(\mathbb{R}^2, \mathscr{B}_2)$ sera notée $_pm_2(dt)$.

Exemple (4.1). Soient $\mu(dt)$ et $\nu(dt)$ deux *mesures* sur $(\mathbb{R}, \mathscr{B}_1)$; considérons:

— la composante discrète $\overset{d}{\mu}(dt)$ de $\mu(dt)$; si elle n'est pas nulle elle est constituée par des masses a_j placées respectivement aux points $x_j (j = 1, 2, \ldots)$;

— la composante discrète $\overset{d}{\nu}(dt)$ de $\nu(dt)$; elle est constituée par des masses b_k placées respectivement aux points $y_k (k = 1, 2, \ldots)$.

Soit z un point de $\mathbb{R}$ tel que: il existe un j et un k tels que: $z = x_j = x_k$; en ce z, plaçons la masse $a_j b_k$; nous constituons ainsi une mesure discrète, que nous désignerons par:

$$\overset{d}{\mu} \oplus \overset{d}{\nu};$$

on vérifie facilement que:

la restriction diagonale projetée $_dm_2(dt)$ de $m_2(dt_1, dt_2) = \mu(dt_1)\nu(dt_2)$, est $\overset{d}{\mu} \oplus \overset{d}{\nu}$.

Remarque (4.1). Dans l'Exemple (4.1), il est supposé que $\mu(dt)$, $\nu(dt)$ sont des mesures; la généralisation au cas où ce serait des fonctions d'ensemble σ-additives réelles ou complexes, est immédiate.

Les notions de restriction diagonale, de restriction diagonale projetée, ont été introduites pour des $m_2(dt_1, dt_2)$ supposées fonctions d'ensemble

σ-additives sur $(\mathbb{R}^2, \mathscr{B}_2)$; des extensions au cas où $m_2(dt_1, dt_2)$ serait une S_2-distribution quelconque ne semblent pas possibles.

(b) *Exemple des impulsions indépendantes entre elles*
Supposons que pour $\tau_2 \neq \tau_1$, les f.a. $I(\tau_1, .)$ et $I(\tau_2, .)$ sont indépendantes; en outre:

(1) Supposons que $E[I(\tau, t)]$ ne dépend que de $(t - \tau)$, et posons:

$$h_1(t - \tau) = E[I(\tau, t)];$$

(2) Supposons que $E\{[I(\tau, t_1) - h_1(t_1 - \tau)] \overline{[I(\tau, t_2) - h_1(t_2 - \tau)]}\}$ ne dépend que de $(t_1 - \tau)$ et $(t_2 - \tau)$, et posons:

$$h_2(t_1 - \tau; t_2 - \tau) = E\{[I(\tau, t_1) - h_1(t_1 - \tau)] \overline{[I(\tau, t_2) - h_1(t_2 - \tau)]}\}$$

(3) posons:

$$m_1(\omega) = E[M(\omega)];$$

(4) posons:

$$\mu(\omega) = E[|M(\omega) - m_1(\omega)|^2];$$

et supposons que:

$$E\{[M(\omega_1) - \mu(\omega_1)][M(\omega_2) - \mu(\omega_2)]\} = 0 \text{ si } \omega_1 \cap \omega_2 = \emptyset.$$

On trouve alors:

$$\Gamma(t_1, t_2) = \int_{\mathbb{R}} h_1(t_1 - \tau)m_1(d\tau) \times \int_{\mathbb{R}} \overline{h_1(t_2 - \tau)}m_1(d\tau)$$

$$+ \int_{\mathbb{R}} h_2(t_1 - \tau; t_2 - \tau)[\left(\overset{d}{m_1} \oplus \overset{d}{m_1}\right)(d\tau) + \mu(d\tau)]$$

$$+ \int_{\mathbb{R}} h_1(t_1 - \tau)\overline{h_1(t_2 - \tau)}\mu(d\tau). \tag{4.4}$$

(c) *Covariance harmonique*
On peut appeler *covariance harmonique* du signal $S(t)$, la S_2-distribution $\tilde{\Gamma}(d\nu_1, d\nu_2)$ transformée de Fourier de $\Gamma(t_1, t_2)$; soit symboliquement:

$$\Gamma(t_1, t_2) = \int_{\mathbb{R}^2} e^{2\pi i(\nu_1 t_1 - \nu_2 t_2)}\tilde{\Gamma}(d\nu_1, d\nu_2); \tag{4.5}$$

en général, $\tilde{\Gamma}(d\nu_1, d\nu_2)$ est une fonction d'ensemble complexe σ-additive sur $(\mathbb{R}^2, \mathscr{B}_2)$; supposons une fois pour toutes être dans ce cas.

Spectre. Alors le *spectre* du signal $S(t)$ n'est autre que la restriction diagonale projetée $_p\tilde{\Gamma}(dv)$ de $\tilde{\Gamma}(dv_1, dv_2)$.

Exemple (4.2). A titre d'exemple, partons de la formule (4.4) supposée valide. Supposons que les transformées de Fourier $\tilde{h}_1(v)$, $\tilde{h}_2(v_1, v_2)$ de $h_1(x)$, $h_2(x, y)$ sont des fonctions.

Désignons par $\tilde{m}_1$, $\tilde{\mu}$, $\tilde{n}$ les transformées de Fourier respectives de $m_1(d\tau)$, $\mu(d\tau)$,

$$\left(\overset{d}{m_1} \oplus \overset{d}{m_1}\right)(d\tau) + \mu(d\tau);$$

on trouve alors:

$$\tilde{\Gamma}(dv_1, dv_2) = \tilde{h}_1(v_1)\overline{\tilde{h}_1(v_2)}\tilde{m}_1(dv_1)\overline{\tilde{m}_1(dv_2)} +$$
$$+ \tilde{h}_2(v_1, v_2)\tilde{n}(dv_1 - v_2)\, dv_2 + \tilde{h}_1(v_1)\overline{\tilde{h}_1(v_2)}\tilde{\mu}(dv_1 - v_2)\, dv_2. \quad (4.6)$$

Voici l'interprétation de la distribution $\tilde{n}(dv_1 - v_2)\, dv_2$; dans le plan $0v_1$, v_2, traçons les axes $0x$, $0y$ définis par:

Ox est la droite $v_1 = -v_2$, orientée dans le sens des v_1 croissants;
Oy est la droite $v_1 = v_2$, orientée dans le sens des v_1 croissants.

Supposons que la distribution $\tilde{n}(dv_1)$ est une fonction σ-additive d'ensemble sur $(\mathscr{A}v_1, \mathscr{B}_1)$; désignons par $_x\tilde{n}(dx)$ sa projection sur $0x$; la distribution sur $(0v_1v_2, \mathscr{B}_2)\tilde{n}(dv_1 - v_2)\, dv_2$, est la même que la distribution sur $(0xy, \mathscr{B}_2)$ définie par:

$$\frac{1}{\sqrt{2}}\,_x\tilde{n}(dx)\, dy;$$

laquelle, naturellement, est une fonction d'ensemble σ-additive sur $(0xy, \mathscr{B}_2)$.

La distribution $\tilde{\mu}(dv_1 - v_2)\, dv_2$ est susceptible d'une interprétation analogue.

En utilisant des remarques faites plus haut, le spectre $_p\tilde{\Gamma}(dv)$ se déduit de (4.6) immédiatement:

$$_p\tilde{\Gamma}(dv) = |h_1(v)|^2 \left(\overset{d}{\tilde{m}_1} \oplus \overset{d}{\tilde{m}_1}\right)(dv) + a^2 \tilde{h}_2(v, v)\, dv + b^2 |h_1(v)|^2; \quad (4.7)$$

a^2 désigne la masse en $v = 0$, que comporte la fonction d'ensemble $\tilde{n}(dv)$;
b^2 désigne de même la masse en $v = 0$ que comporte la fonction d'ensemble $\tilde{\mu}(dv)$.

Dans (4.7), le premier terme est toujours un spectre de raies; le deuxième et le troisième termes sont des spectres continus, à densités.

Remarque (4.2). L'échantillonnage, à des instants aléatoires ou non, d'une f.a. $Z(t)$ de $t \in \mathbb{R}$, peut conduire à considérer des signaux $S(t)$ de la forme:

$$S(t) = \int_{\mathbb{R}} I(\tau, t)Z(\tau)M(d\tau);$$

ils s'étudient comme les signaux de la forme (4.2), à laquelle d'ailleurs ils se ramènent.

BIBLIOGRAPHIE

1. Gelfand, I. M., Chilov, G. E. et Vilenkin, N. Y., [1]. "Les distributions"; edit. franç., Paris 1967, Dunod edit.
2. Schwartz, L. [1]. "Théorie des distributions"; 3e ed., Paris 1966, Hermann edit.

Nonlinear Transformations of Random Processes

NORMAN ABRAMSON

University of Hawaii, Honolulu

Abstract

This paper provides a general method of calculating the mean-square bandwidth (and other spectral moments) of an arbitrary zero-memory nonlinear transformation of a stationary random process. The method is valid when the original process is an arbitrary combination of other random processes. It can be used to determine the mean-square bandwidth (or the spectral moments) of the transformed process either before or after that process is passed through a bandpass filter. Five examples of the application of this method are provided simplifying and generalizing known results, as well as providing new results. The method presented is published in the July, 1967 issue of the *IEEE Transactions on Information Theory*.

DISCUSSION

J. W. R. Griffiths: To what extent are the results presented dependant upon the Gaussian assumption?
Answer: Of the three equations presented the first two are valid for any stationary stochistic process, including the Gaussian. The third is valid for the cass of D. I. processes which includes the Gaussian is a special case but is much broader.
Question: Is the same simplicity maintained when talking about higher order moments?
Answer: For any particular higher order moment the same type of simple result can be obtained. If you try to obtain a general expression for the mth moment and use that in an expansion for the spectrum these results do not appear to be useful.

Minimum Detectable Signal for Spectrum Analyser Systems

C. NICHOLAS PRYOR

Naval Ordnance Laboratory, Silver Spring, Maryland, U.S.A

I feel almost as though I should apologize to a group such as this for discussing a topic as well known as the detection of a sine wave in noise. In principle the mathematics have been established for at least a quarter century as evidenced by Marcum's classic paper of 1948 [1], and the topic appears in most textbooks on signal processing. Yet a great deal of folklore has grown up about spectrum analysis systems and their performance, and fragmentary results are often misapplied in attempts at performance prediction. When papers were being gathered for this meeting I was in the midst of compiling a consistent method [2] of predicting performance (minimum detectable signal) of practical spectrum analysers, hopefully in a way that would remove some of the mystery from the topic. This paper is a summary of that work and a comparison of the results with one experimental system.

The classical structure of a spectrum analyser is shown in Fig. 1a, with a number of parallel processing channels each estimating the power in a different narrow frequency band. The input to each channel is passed through a narrow-band bandpass filter and then to a square-law power detector. The output of the square-law detector may be further smoothed by averaging over some time T, and this average is then compared with a threshold value K to determine whether the power in the narrow band exceeds that expected from noise alone. The threshold is set to limit the false alarm probability to an acceptable value, and one wishes to know the minimum sine wave power which can be detected with a desired probability. This is what we mean by the Minimum Detectable Signal or MDS.

Most practical spectrum analysers are not actually built as in Fig. 1a because of the large number of parallel channels involved. One common approach is the delay line time compressor (Deltic) approach shown in Fig. 1b in which a single channel is effectively time shared by repeatedly playing back time-compressed replicas of the input signal. The "Ubiquitous" spectrum analyser made by Federal Scientific is probably the best known of such systems. Fast Fourier Transform (FFT) systems such as shown in Fig. 1c are rapidly becoming commonplace as spectrum analysers. In each case the detailed implementation of the spectrum analyser affects its performance somewhat relative to the classic model in Fig. 1a. The approach used in this paper is to derive an "Ideal System MDS" based on the model of Fig. 1a and then make first-order corrections to this result to

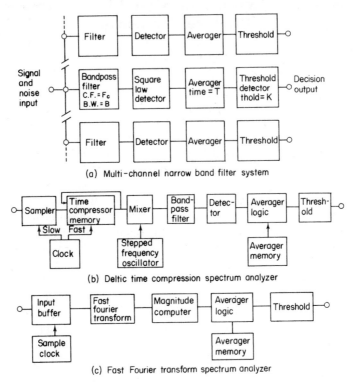

(a) Multi-channel narrow band filter system

(b) Deltic time compression spectrum analyzer

(c) Fast Fourier transform spectrum analyzer

FIG. 1.

predict the performance of a given practical system. This allows a direct appreciation of the penalties paid for any compromises in implementation of a spectrum analyser.

1. Ideal System Performance

Suppose the input to the spectrum analyser consists of a sine wave of power S and a white noise of power N watts/Hertz. If the channel containing the signal frequency in an ideal spectrum analyser is considered to have a perfect bandpass filter of bandwidth B Hertz and to integrate the square-law detector output uniformly over a time T seconds (T is assumed greater than or equal to $1/B$) the output of the averager has a modified chi-square distribution with mean $S + NB$ and variance $\sigma^2 = N^2 B/T + 2SN/T$. Suppose the threshold detector is set so that $K = NB + d\sigma$, where d is selected to give the desired false alarm probability. Since by definition the false alarm probability is determined with no signal present, the appropriate value of σ to use here is $N\sqrt{B/T}$ or the value found for $S = 0$.

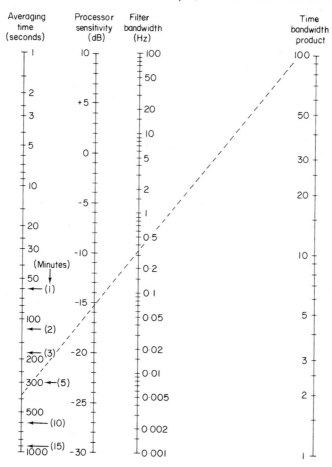

FIG. 2. Nomograph for determining processor sensitivity.

Now it happens that the mean and the median of the chi-squared distribution rapidly converge as the number of independent samples averaged by the averager (that is, the BT product) becomes reasonably larger than one. Thus if an amount of signal is added to make the mean output $S + NB$ of the averager equal to the threshold K, the probability is very near 50% that at any instant the random output will exceed K. Thus the probability of detecting the signal is very nearly 50%. Inserting the proper value for σ

and setting K equal to $S + NB$ allows this result to be written in the form $(S/N) = d\sqrt{B/T}$. This then is the minimum detectable signal (relative to the noise in a one Hertz band) for 50% probability of detection in this ideal spectrum analyser. The result can be written in decibel form as

$$(S/N)_{db} = 5 \log B - 5 \log T + 10 \log d$$
$$= \text{Basic MDS} + D,$$

where the "Basic MDS" is $5 \log B - 5 \log T$ and $D = 10 \log d$ must be determined for the desired false alarm probability.

Figure 2 is a nomograph helpful in computing the "Basic MDS" or processor sensitivity for a given B and T. In the example shown (which

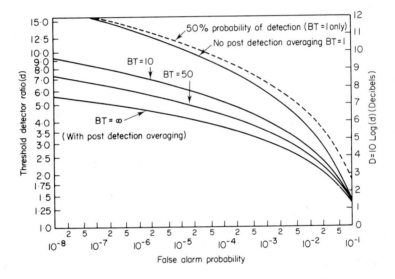

FIG. 3. Threshold detector level *vs.* false alarm probability.

will be used throughout this paper), the dotted line connects a processor bandwidth of 1/3 Hz with an integration time of 6.4 minutes and crosses the processor sensitivity scale at -15.3 dB. The time-bandwidth product BT of the averager may also be determined from the fourth scale of Fig. 2 by extending the line passing through B and T. In this example the result falls slightly off scale at $BT = 128$.

The correction D for the required false alarm probability is determined from the chi-square distribution which results at the averager output with no signal present. The solid curves of Fig. 3 give the threshold detector ratio d (from the left hand scale), which determines the threshold K, as a function of false alarm probability for several values of BT. The curve for $BT = \infty$ is based on the limiting gaussian distribution, while the curves for

$BT = 1$, 10, and 50 use the chi-square distribution with $2BT$ degrees of freedom. Note that the gaussian limit is approached rather slowly and should not be used without correction for time-bandwidth products less than several hundred. The right hand scale permits direct reading of $D = 10 \log d$ for calculation of the MDS, under the assumption that the mean and median of the averager output distribution coincide. This is sufficiently accurate except in the case of very small values of BT. For the specific case of $BT = 1$ (where the averager is performing no function except to remove double-frequency components from the detector output) the correction is significant, and D should be determined from the dashed curve to find the MDS. If a false alarm probability of 10^{-4} is desired in the current example with $BT = 128$, the threshold detector ratio d is about 4.2 and the adjustment D to the MDS is about 6.2 dB. Thus the predicted MDS of this ideal system is $-15.3 + 6.2$ or about -9.1 dB signal to noise ratio (referenced to noise in a one Hertz band).

2. Variation with Detection Probability

While the 50% probability of detection point is the one most often used in specifying a detection system, it is sometimes desirable to determine the signal required for other detection probabilities. For these cases the mean output $S + NB$ of the averager must be set above (or below) K by an amount $d'\sigma_s$, where d' is a factor determined by the desired detection probability and the standard deviation σ_s must be determined with the signal present. With some manipulation, these requirements may be written in the form

$$(S/N) = d\sqrt{B/T}[1 + (d'/d)\sqrt{1 + (2/B)(S/N)}],$$

which contains S/N only implicitly. Fortunately this expression can be solved explicitly for S/N and the result written so that the correction for arbitrary detection probability appears as a simple factor of S/N. This means that only an additive correction must be made to the MDS in decibels to account for the desired detection probability. This correction is a function of only two parameters, $p = d'/d$ and $r = d\sqrt{BT}$, and the correction is shown in terms of these two parameters in Fig. 4.

The first step of accounting for an arbitrary detection probability is to determine d' from the shape of the averager output distribution. For values of P_d between about 0.1 and 0.9 an approximate value of $d' = 3.2$ $(P_d - 0.5)$ can be used, based on a linearized form for the gaussian distribution. For more extreme values of P_d Fig. 3 may be used, with d' equal to the negative of the value of d read from the curve for P_d below 0.1 and with d' equal to the d obtained for $1 - P_d$ for P_d above 0.9. While this is not theoretically exact because of the shape of the modified chi-squared distribution, the result is sufficiently close to give a good estimate of d'.

Then given d, d', and BT for the system, the parameters p and r may be determined and the MDS correction read from Fig. 4. In our example, suppose corrections are desired for 25% and 90% detection probability. The corresponding values of d' are -0.8 and $+1.3$ respectively. For the $d = 4.2$ and $BT = 128$ values previously determined for the processor, r is about 0.37 and the values of p are -0.19 and $+0.31$ respectively. Using these parameters in Fig. 4 gives an MDS correction of about -1.2 dB for a 25% probability of detection and $+1.6$ dB for 90%.

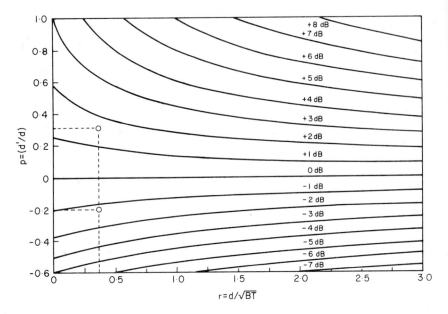

FIG. 4. MDS change for arbitrary probability of detection.

3. Effect of Finite Observation Time

One of the common implementation choices in a practical spectrum analyser is to replace the averager with a recursive filter (or exponential averager) with a time constant equal to T. It can be shown that this reduces the output variance by a factor of two, but that the output mean only approaches the steady state value exponentially after a signal is applied. Since in most cases it is important that the signal be detected within a finite time after it is first applied, a correction must be determined for the decrease in apparent signal strength within a finite observation interval. This correction is equal to $-10 \log [1 - \exp(-T_d/T)] - 1.5$ where T_d is the permitted observation time before a detection must be made and T is the averager time constant. This correction is plotted in Fig. 5. In the system being used as an example the 6.4 minute averager was in fact an exponential averager with 6.4 minute time constant. If a ten minute observation time is permitted before detection is required, T_d/T is 1.55 and an MDS correction of -0.5 dB is needed to account for the observation time. For allowed observation times of one, two, and five minutes the appropriate MDS corrections are respectively $+7$ dB, $+4.1$ dB, and $+1.1$ dB.

While the pure T second averager assumed in the ideal system builds up to its steady state output after T seconds, it too requires a correction for any required detection time less than T. This is shown as the dashed curve in Fig. 5. If the allowed detection time T_d is known in advance, an optimum choice can be made for the averager integration time. For the pure averager the optimum T is equal to T_d, while for the exponential

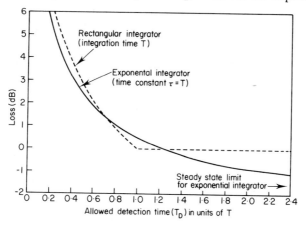

FIG. 5. MDS change due to finite detection time.

integrator the optimum time constant is equal to about 0.8 T_d. If this optimum value is chosen in each case, the net penalty for using an exponential averager in place of the pure integrator is about 0.5 dB.

4. Practical Bandpass Filter Responses

In practice it is not possible to build the ideal bandpass filter with infinitely steep skirts that was assumed in the ideal system, so some compromise must be made in any practical spectrum analyser. Choosing a realizable filter of the same nominal bandwidth but different shape causes several effects on the performance. The first of these is simply that a different total noise power may appear at the output for a given amount of white noise at the input. The noise power output divided by the noise power density at the input is defined as the noise bandwidth (NBW) of the filter. The second effect is that the spectrum of the noise operated on by the square-law detector and passed to the averager is changed, and this changes the degree of noise smoothing performed by the averager. A given filter can be characterized by a parameter F_n which is essentially the number of degrees of freedom per second in the noise appearing at the square-law detector output [3], and the performance of the averager is dependent on this F_n. The net change in system MDS due to these two effect is

$$\Delta_{\text{MDS}} = 10 \log (\text{NBW}/B) - 5 \log (F_n/B).$$

Figure 6 shows the response characteristics of six bandpass filters which are typically considered in spectrum analysers. The first is the ideal filter function, while (b) through (d) may be used or approximated in systems employing analogue or recursive filters. In each case B is the nominal bandwidth parameter of the filter. The curves shown in (e) and (f) are more typical of digital systems employing direct correlation of the input signal with a stored replica, such as FFT processors or convolutional filters. In these cases the data block size or correlation filter length is $1/B$ seconds. Characteristic bandwidths and the resultant MDS correction for each of the six filter functions are given in Table I:

TABLE I

Filter Type	3 dB Bandwidth	Noise Bandwidth	F_n	MDS Correction
Ideal	B	B	B	0
Single Tuned	B	$1.57B$	$3.14B$	-0.52 dB
3rd Order Butterworth	B	$1.05B$	$1.26B$	-0.30 dB
Gaussian	$1.67B$	$1.77B$	$2.51B$	$+0.48$ dB
Unweighted Correlator	$0.89B$	B	$1.50B$	-0.88 dB
Hanning Weighted Correlator	$1.42B$	$1.50B$	$2.08B$	$+0.17$ dB

The system under consideration in our example uses an FFT processor for spectrum analysis, so the appropriate MDS correction is either -0.88 dB or $+0.17$ dB, depending on whether Hanning weighting is applied to the data. It should be noted that this section assumes continuous spectral filtering of the input data and does not take into account the sampled nature of the data to the averager in time-shared systems such as the FFT processor.

5. Filter Scalloping Loss

Another effect of using bandpass filters with non-ideal shape is that the response to an input sine wave varies as the signal frequency deviates from the exact centre of the filter passband. For all the filters shown in Fig. 6

the response is degraded, and thus the signal power must be raised to make up for this loss. (This is known as scalloping loss or the picket-fence effect.) If the filter centre frequencies are spaced by B_0 Hertz in a spectrum analyser, any input signal will fall within $B_0/2$ of the centre of one channel and the worse case loss due to scalloping is just the minimum filter response within $B_2/2$ of the filter centre frequency. This loss is plotted in Fig. 7

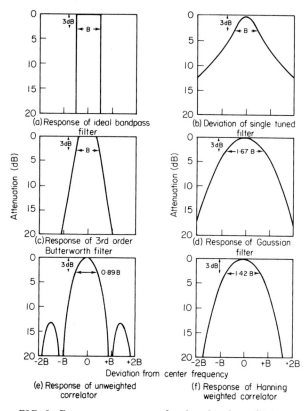

FIG. 6. Frequency response of various bandpass filters.

as a function of the frequency bin spacing B_0 for each of the six filter responses of Fig.6.

Using the worst case loss of course overestimates the seriousness of this problem for random input signal frequencies, and some sort of average must be taken over all possible frequencies.

Exact calculation of the increase in signal power required to maintain a given probability of detection is quite complex, but a reasonable estimate can be obtained by averaging the power response of the filter over frequencies within $B_0/2$ of the centre of the band. This represents the amount by which the input signal power must be raised to give the same mean output of the averager in the strongest frequency bin when the input frequency is random. These results are plotted in the dashed curves of Fig. 7.

In the usual form of FFT analyser the frequency bin spacing is equal to B, so the maximum and average scalloping losses are about 4 dB and 1.3 dB

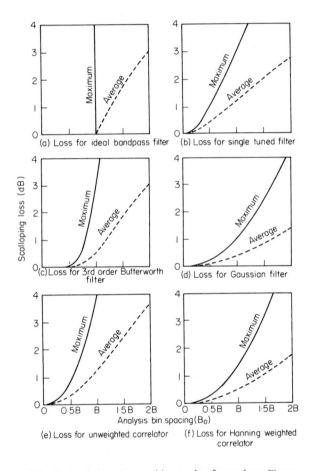

FIG. 7. Scalloping loss *vs.* bin spacing for various filters.

respectively if unweighted data is used versus 1.5 and 0.5 dB respectively if Hanning weighting is used. This is one of the reasons that Hanning weighting is often chosen. Similarly in the Butterworth filter case the maximum and average losses are 3 dB and 0.5 dB respectively if the filter bandwidth B is equal to the bin spacing B_0. If the filter bandwidth is increased by 25% without changing the bin spacing so that B_0 is equal to 0.8 B, the maximum and average scalloping losses drop to about 1 dB and 0.2 dB respectively.

Since this is in trade for an increase of about 0.5 dB in the Basic MDS of the system (5 log 1.25), the decision is often made to accept this small degradation in best-case and average MDS of the system in order to reduce the worst-case scalloping loss.

6. Filter Transient Response

Another important effect must be considered in systems of the time-compressor type (Fig. 1b) with analogue or recursive filters. This is the transient response of the bandpass filter, or the time it requires to build up to full response. The reason is that the filter is only assigned to a

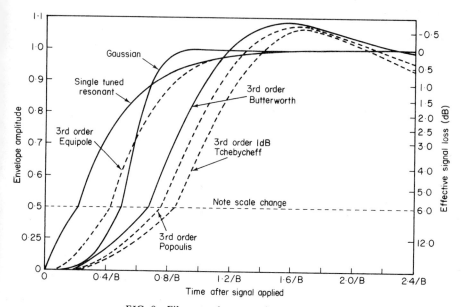

FIG. 8. Filter envelope transient response.

particular bin, on each frequency sweep, for a time T_b equal to the time compressor storage time, and its response is limited to the envelope response for the time T_b rather than its steady state response as in the parallel system of Fig. 1a.

The effect of this is shown in Fig. 8 where the envelope step response is plotted for three of the filters of Fig. 6 plus three other third order (18 dB/octave skirt response) filters. Now if it is assumed that the VFO in the Deltic system steps to a new frequency once per circulation of the time compressor loops and that the bandpass filter output is sampled at the end of the data circulation, the effective loss of signal strength is directly related to the reduction in envelope amplitude at the end of the circulation period and is shown on the right-hand scale of Fig. 8. Note that if the

filter bandwidth is the reciprocal of the data sample length, so that
$T_b = 1/B$, losses of several decibels can occur. For example the Butterworth
filter has a loss of about 1.3 dB in this case. However if the bandwidth of
the filter is increased by 25% then T_b becomes $1.25/B$ and the loss reduces
to about 0.2 dB. This is an even more compelling reason than the scalloping
loss for making the bandpass filter bandwidth somewhat larger than that
which just matches the data length.

Note that the result shown here is restricted to stepped VFO
frequencies and a specific sampling technique. If the VFO sweep is
continuous or if some other scheme is used for sampling the output of the
filter and detector, different analytical methods are required and different
results expected. However this filter transient effect remains an important
one and one of the major potential losses in Deltic type spectral analysis
systems.

7. Sampling of Averager Input

In virtually all spectrum analysis systems except the pure parallel system
of Fig. 1a, a sampling operation of some sort occurs between the square-
law detector and the input to the averager. In Deltic systems like that in
Fig. 1b a new sample is available to the averager once per full frequency
sweep of the VFO, while FFT systems provide an input to the averager
once per transform. Nominally one would expect that this point should
be sampled at a rate at least equal to the bandwidth B of the bandpass
filter. The actual rate of sampling (or generation of spectral estimates)

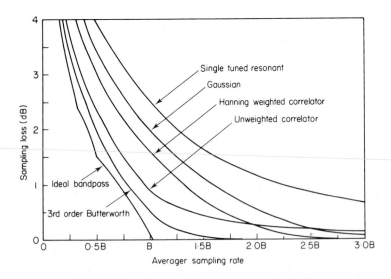

FIG. 9. Loss due to finite averager sampling rate.

relative to B is sometimes referred to as the "redundancy" of the processor. If the actual spectral content of the square-law detector output is considered, the degree of smoothing obtained from the averager for any given sampling rate can be compared to the smoothing obtained for an infinite sampling rate (continuous data) and any reduction in smoothing related to an equivalent increase in S/N required at the processor input to obtain the same detectability at the averager output. The results of this calculation [3] are shown in Fig. 9.

Note that only the ideal bandpass filter permits sampling without loss at the rate B while all other types of filters continue to improve as the processor redundancy is increased. This includes the primary digital processor forms with the unweighted correlator having a loss of 0.88 dB and the Hanning weighted system having a loss of 1.59 dB when sampling is done only once per block length of the input data. Doubling the rate at which spectra are computed (for example performing an FFT twice as often with 50% overlap of the input data each time) reduces this loss in either case to about 0.25 dB.

8. Modified Detector Characteristics

While a square-law detector is almost invariably assumed in calculations on signal processing systems because of its mathematical properties, it is often replaced in practical systems by other detector functions. One of the most common is a linear detector or envelope detector, which is often simpler to implement and reduces the dynamic range of values which must be handled by the averager and any display systems. Another approach sometimes used is to convert the detector output to a logarithmic form (effectively forming a power estimate in decibels) and average this information. This provides a "constant false alarm rate" statistic to the averager, the variance of which is independent of the actual noise power to the system. Finally in some systems a binary threshold is applied to the detector output so that the averager only has to count the number of threshold exceedances in a longer period. This is referred to in some literature as a double-threshold detector.

In systems with large time-bandwidth products the effect of a modified detector characteristic can be determined from the first and second moments of the detector output without signal and the derivative of the first moment with respect to signal power. Using this approach a performance loss (increase in required input MDS) of 0.2 dB is predicted for the linear detector and 1.08 dB for the logarithmic detector. The minimum loss for the double-threshold detector is 0.94 dB when the false alarm probability of the first detector is about 20% and this loss increases rather rapidly when the initial false alarm probability falls outside the range of about 10%

to 35%. Some care must be taken in applying these results to systems with
BT products below about 100. For example Marcum showed in one
example that the linear detector actually outperformed the square-law
detector by up to about 0.1 dB when BT products below 70 were employed.
The explanation for this is that the statistically optimum detector is the
log of a modified Bessel function, and for the high S/N values required at
the detector for useful performance with small averager BT products the
linear detector is a closer approximation to the optimum detector than is
the square-law detector.

9. Input Signal Clipping

Although it is becoming less common as digital storage devices become
more available, a commonly used technique has been hard clipping
(polarity-only sampling) of input data to the spectrum analyser so that
only one bit is required to store a sample. This obviously does terrible
things to a signal spectrum in a high S/N case. However for the detection
of a relatively weak signal in a reasonably white noise background the loss
can be shown to be tolerably small. The actual loss encountered depends
somewhat on the sampling rate used at the system input. If the spectrum
analyser operates on a signal band from nearly zero up to some cutoff
frequency F_0 the loss due to clipping is 1.96 dB when sampling is performed
at the Nyquist rate of $2F_0$. If the sampling rate is raised to $3F_0$ the loss
drops to 1.4 dB and remains fairly constant at about 0.95 dB for sampling
rates above $5F_0$. Some systems handle signals over only an octave at a time,
so the input spectrum is essentially zero outside the band from $F_0/2$ to F_0.
Nyquist rate sampling may be done at the cutoff frequency F_0 and again
results in a clipping loss of 1.96 dB. Sampling at $2F_0$ in this case produces
a sampling loss of 1.09 dB, while sampling at $5F_0$ or higher reduces the
loss to 0.65 dB.

10. Example and Experimental Verification

Figure 10 shows an example of this method of MDS calculation for a
spectrum analyser and a comparison with actual test results on the system.
The system under consideration was a laboratory FFT analyser with a 3
second data block size giving an analysis bandwidth and a frequency bin
spacing of 1/3 Hertz. Transforms were computed once per input block or
every three seconds with no data weighting. An exponential averager was
used with a time constant of 128 samples or 6.4 minutes. The threshold
detector on the output was set to produce a false alarm probability of
10^{-4} in each of the 512 output frequency channels. Calculation of the

minimum detectable signal for a five minute observation interval and for detection probabilities of 25%, 50%, and 90% is shown below:

Basic Processor MDS (1/3 Hz, 6.4 Min.)   −15.3 dB
False Alarm Probability (10^{-4}, $BT = 128$)   + 6.2
"Ideal System MDS"   − 9.1 dB

Probability of Detection $\begin{cases} 25\% \\ 50\% \\ 90\% \end{cases}$ $\left.\begin{array}{r} -1.2 \\ 0 \\ +1.6 \end{array}\right\}$

Detection Time Allowed (5 minutes)   + 1.1
Bandpass Filter Function (Unweighted Correlator)   −0.88
Scalloping Loss (Average)   + 1.3
Transient Response (not applicable)   0
Averager Sampling Rate (1:1 redundancy)   + 0.88
Detector Characteristic (Linear Detector)   + 0.2
Clipping Loss (not applicable)   0

Predicted MDS $\begin{cases} 25\% \, P_d \\ 50\% \, P_d \\ 90\% \, P_d \end{cases}$ $\left.\begin{array}{r} -7.7 \text{ dB} \\ -6.5 \text{ dB} \\ -4.9 \text{ dB} \end{array}\right\}$

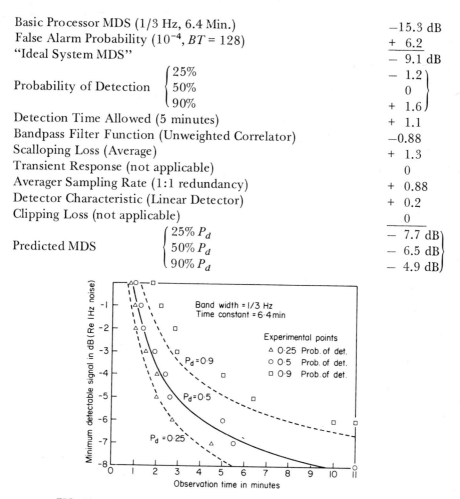

FIG. 10. Comparison of predicted performance with experiment.

The MDS is predicted in a similar manner for other allowed detection times between 1 and 10 minutes and the results plotted as the three curves on Fig. 10. Results of an MDS experiment consisting of about 25 samples at each signal-to-noise ratio are also plotted on Fig. 10. Times required to reach 25%, 50%, and 90% detection probabilities within the sample are shown for each S/N. The agreement between the predicted and experimental performance is nearly perfect for the 25% and 50% detection probabilities. However the experimental points at 90% are clearly biased away from the predicted curve in what could be interpreted either as an increased MDS or an increased required detection time. This bias may be due to the small data sample and the practice of recording the time at which the percentage of detections first *exceeded* 90% in the data reduction. There was often a considerable time lapse between this and the last previous detection, and this biases the data toward longer times.

In summary, the procedure described here is designed to separate as nearly as possible the influences of various parameters or implementation choices in spectrum analysis systems and to provide a uniform way of predicting the performance of a wide variety of such systems. Results of this prediction method seem to agree with experiment or with other analytical methods to within a fraction of a decibel.

REFERENCES

1. Marcum, J. (1948). "A Statistical Theory of Target Detection by Pulsed Radar, Mathematic Appendix", Rand Corporation Research Memo RM-753. *Also published in Transactions PGIT, I.R.E., IT-6,* April 1960.
2. Pryor, C. N. (1971). "Calculation of the Minimum Detectable Signal for Practical Spectrum Analyzers", NOLTR 71-92, Naval Ordnance Laboratory.
3. Pryor, C. N. (1971). "Effect of Finite Sampling Rates on Smoothing the Output of a Square-Law Detector with Narrow Band Input", NOLTR 71-29, Naval Ordnance Laboratory.

DISCUSSION

S. W. Autrey: Was your detection process automated in the last figure?

Answer: Yes. It was a conventional single-threshold detector where a detection occurred as soon as the average output exceeded a certain value. Incidentally, sequential detection turned out not to be any better, and one can pretty much explain why.

S. W. Autrey: If you have an observer making detections on electro-sensitive paper, do the results change according to the number of lines present?

Answer: This is another aspect of the same experiment. In manual detection experiments involving an average of four lines at a time the observer was not overloaded and no degradation was seen. However, the experiment reported here used automatic detection and should not be influenced by the number of lines.

R. E. Bogner: With regard to the loss due to recursive filters were you detecting its output at the end of the block of interest?

Answer: That is correct, and this is not necessarily the best thing to do. In fact for filters higher than third order it is essential to sample at some later instant when the filter output reaches its peak. This adds another dimension to the analysis problem, so the simpler sampling method was described here to illustrate the loss mechanism.

R. E. Bogner: Another question please. With regard to the picket fence effect, you showed for the ideal filter a difference between the average and the centre frequency performance. I could not see how the ideal filter would deteriorate as you detuned it.

Answer: The worst case loss becomes infinity as soon as the centre-to-centre spacing exceeds the filter bandwidth. The average loss then reflects the probability of the signal falling between filter bins.

G. A. van der Spek: You commented that sequential detection would not be an improvement here. Would you comment further on that?

Answer: Consider the output of the square law detector itself and the fact that you are permitting only a limited time (say five minutes) for detection. In order to detect at the end of this interval, the ideal matched filter on the detector output would be a pure integrator looking backward in time over the permitted detection interval. One should not expect the sequential detector to outperform this matched filter averager when the amount of data available is fixed.

G. A. van der Spek: But if one were to allow an average detection time of five minutes with the possibility to go over this from time to time, then what kind of performance could you get?

Answer: At the end of the five minutes allowed detection time only five minutes of signal has occurred. Thus the possibility of a longer observation time would not help the sequential detector. Now if a resource allocation problem were present, such as moving to the next frequency bin in a Deltic type system, the shorter average detection time in a sequential system could be used to some advantage. This is the scheme used in electronically beam steered radar systems.

W. S. Liggett: It seems that one correction you have not made involves the estimation of the noise mean from adjacent cells. The effect of errors in this estimation is missing from the calculation.

Answer: That is right. I assumed the use of what Don Tufts calls a "clairvoyant" threshold detector. He has done quite a bit of work on the effect of estimating the background from a small number of adjacent cells. There is a loss depending on the number of bins used to estimate the background.

P. L. Stocklin: I missed the false alarm probability used in the experiment in your last slide.

Answer: 10^{-4}.

C. van Schooneveld: You pointed out the low scalloping loss when Hanning weighting is used, but you did not mention the effect of the resulting correlation between false alarms. Did you take account of that?

Answer: No, I did not. The fact that the false alarms are correlated does not affect their probability.

C. van Schooneveld: Not per channel but in the overall system the number is less.

Answer: I did not take that into account, nor did I consider the increase in overall detection probability when the signal appears in more than one frequency bin.

P. M. Schultheiss: I noticed in some areas the time to detection is short compared to the integration time. Why does one operate in this way?

Answer: This reflects what happens when the signal-to-noise ratio is much higher than the minimum for which the system was designed. The time required for detection becomes much shorter.

P. M. Schultheiss: I understood this was the time permitted for detection.

Answer: You may consider that this curve represents the signal-to-noise required to detect in for example one minute on a system that was originally optimized for five minutes.

P. M. Schultheiss: So what you want to do if you want to detect in one minute is to shorten the averaging time.

Answer: That is right. From the matched filter argument the optimum choice would be a pure integrator with a one minute averaging time. For an exponential averager the best choice is a time constant of 0.8 times the allowed detection time.

Spectrum Analysis of Reverberations

JARL JOHNSEN

Norwegian Defence Research Establishment, Horten, Norway

1. Introduction

In the summer season the detection ranges of hull mounted sonar systems are drastically reduced due to downward refraction. Neither increase in transmitting power nor introduction of sophisticated signal processing methods will give any worthwhile improvement. The ranges can be increased, however, by moving the transducer from hull depth to a depth of 100 m and more.

Having increased the possible detection range, the environmental influence on the detection methods must be considered. In narrow and partly shallow areas such as the coastal waters of Norway, reverberations will dominate. The signal processing methods must consequently be chosen to give optimum performance under these conditions. Experience indicates that in the reverberation limited case the distorting influence of the medium is less damaging to the frequency than to the time characteristics of the signal. Frequency analysis in the signal processing should therefore be utilized.

The present limitations on frequency discrimination in operational sonars are mainly due to the use of relatively short pulses (50–100 msec). If the pulse length is increased the frequency discrimination capability is also increased. One will, however, arrive at a point when the medium itself is the limiting factor to further frequency resolution. This is due to the irregular movements of the volume and the surface scatterers.

An experimental towed active sonar system has been developed at the Norwegian Defence Research Establishment in Horten, Norway. The main objectives for the experimental Deep Sonar system are:

(a) to determine the optimum depth for sonar transducers under various conditions in coastal waters
(b) to determine the influence of the medium and the system parameters on the reverberation frequency spectrum and to develop optimum doppler detection equipment

97

(c) to study the spatial correlation characteristics of the reverberation versus target signals

(d) to establish criteria for the detectability using various types of frequency and phase modulated signals and corresponding detection methods.

This paper deals with the reverberation investigations. A short description of the sonar system and the data processing equipment is followed by a discussion of the medium and the methods used in the reverberation investigations. At the end of the paper a few results from the measurements so far carried out are presented.

2. System Description

(a) *The sonar system* [1, 2]

The operating frequency of the Deep Sonar system is in the 7–8 kHz range. The transmitting and receiving arrays, preamplifiers and the array position sensing elements are arranged in a "fish"-like platform towed at maximum depths of 600 m. Signals are transmitted between the towed body and the ship-mounted receiver and transmitter by the towing cable. In its first version, the Deep Sonar is designed to be installed on R/V "H U Sverdrup".

A block diagram of the sonar system including the transmitter, the receiver, the data registration equipment and equipment used for preliminary analysis is shown in Fig. 1. The transmitting beam, having a width of 16° vertically and 28° horizontally, can be trained mechanically in the vertical plane $(-90°-+20°)$ and electronically in the horizontal plane $(-42°-+42°)$.

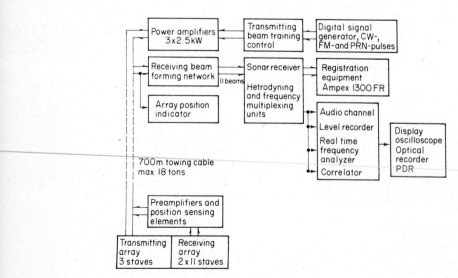

FIG. 1. Block diagram of the deep sonar system.

The digital signal generator is able to generate CW-pulses, linear FM-pulses and PRN-pulses. The pulse length may be selected in the 1 msec–10 sec range and the generator makes a very flexible choice in sweep rates, PRF, phase modulation rate, etc. possible.

The receiving array consists of 22 staves arranged in two parallel planes. 11 preformed beams are available, covering a ±45° sector in the horizontal plane. The width of the centre beam is 7° x 6° and the array may be tilted mechanically between −90° and +20°. The sonar signals are registrated on a 7 channel analogue magnetic tape recorder. A 4 bit real time spectrum analyser, a clipper correlator, adjustable audio channels, level recorder, sonar range recorder, display oscilloscope and an optical recorder are the main units used for preliminary analysis and monitoring during sea trials. The depth, the course deviation and the pitch and roll angle of the platform are monitored by the array position indicator.

A series of sea trials has been performed using the Deep Sonar system. The geometry of the reverberation experiments is shown in Fig. 2.

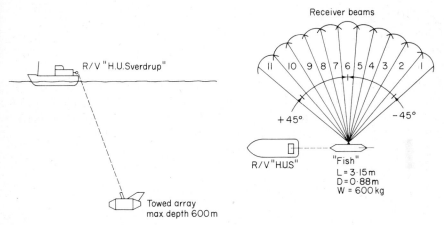

FIG. 2. Geometry of reverberation experiments.

(b) The data handling facilities
The data handling of registrated sonar signals are carried out at the NDRE computer installation on a real time digital spectrum analyser and correlator designed at NDRE [3]. The digital signal analyser has the following modes of operation:

(a) spectrum analyser
(b) general cross correlator
(c) cross correlator with "in-phase" and "quadrature" inputs
(d) correlator with stored reference
(e) correlator with built-in reference (PRN and linear FM signals).

The input signal quantization of the analyser is 8 bits and the principal specifications are:

Spectrum analysis: Max input band: $0 - \dfrac{1000}{k}$ Hz

$k = 1, 2, \ldots, 10, 20, \ldots, 100$
Frequency resolution $1/k$ Hz

Cross correlation: Max input band: $0 - \dfrac{1000}{k}$ Hz

delay sweep: $\pm k/8$ sec
delay step: $k/2$ msec

Reference correlation: Max WT product: 1000

Max input band: $0 - \dfrac{1000}{k}$ Hz

A block diagram of the signal analyser is shown in Fig. 3. In the frequency

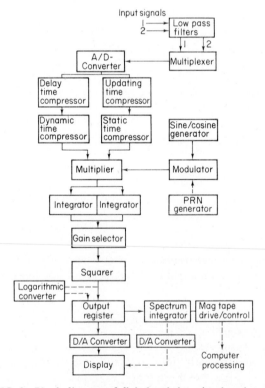

FIG. 3. Block diagram of digital real time signal analyser.

analysis mode the analyser computes the digital version of the integral:

$$P(f) = \left| \int_T x(t) \exp(-j2\pi ft) \, dt \right|^2 =$$

$$= \left[\int_T x(t) \cos(2\pi ft) \, dt \right]^2 + \left[\int_T x(t) \sin(2\pi ft) \, dt \right]^2$$

for different frequencies, i.e.:

$$P_n = \left(\sum_{i=1}^{N} x(i\Delta t) \cos(2\pi f_n i\Delta t) \right)^2 + \left(\sum_{i=1}^{N} x(i\Delta t) \sin(2\pi f_n i\Delta t) \right)^2$$

Δt is the sampling interval and $f_n = n(1/T)$ where $n = 0, 1, 2 \ldots$.

These expressions are closely related to the formula for digital computation of the square of the Fourier coefficients c_k. However, $x(t)$ is normally not a periodic function, so it is more appropriate to relate the expressions to a filtering operation. From this point of view the signal is passed through a bank of narrowband digital filters, pairwise in quadrature and with a box shaped time window function. The outputs from the quadrature filters are squared and added. The corresponding frequency responses are $(\sin x/x)^2$ functions centred at the frequencies f_n and with first zeroes at $f_n \pm 1/T$. The filtering operation is illustrated in Fig. 4.

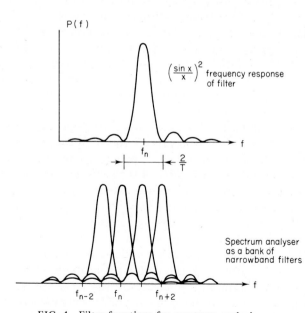

FIG. 4. Filter functions for spectrum analysis.

The different modes of operation are mainly obtained by selecting different inputs to the multiplier. By continuous spectrum analysis these inputs are the signals from the dynamic time compressor and the sine/ cosine generator. The dynamic time compressor stores T seconds of the input signal and is continuously updated with new samples. Each circulation takes 1 msec. During this 1 msec, P_n for one frequency is computed. During the next circulation the frequency from the sine/ cosine generator is different and P_n for a new frequency is computed. If it takes 1 sec to renew all information in the time compressor, computation of P_n for 1000 different frequencies is possible before a new frequency sweep is starting.

Several of the parameters of the frequency analysis may be varied. Sampling frequency, filter bandwidth, upper and lower frequency limits of the analysed band are the most important.

The 10 bits numbers stored in the output register of the analyser may be transferred to a spectrum integrator. This integrator accepts a selected number of frequency sweep widths and a maximum of 99 sweeps may be integrated. Maximum integrated value (before overflow) will be an 18 bits number. The integrator output may be displayed on a display oscilloscope or fed to a 9 tracks digital magnetic tape drive and control unit.

3. Reverberation Investigation

It is well known that acoustic signals transmitted from one point in the sea to another exhibit fluctuations causing severe degradation of the received signal. In the echo ranging situation the received signal consists of a signal reflected from a target and signals scattered and reflected back to the receiver. The reverberations are usually divided into three groups: backscattering from the inhomogeneities in the volume–volume scattering; backscattering from the bottom–bottom reverberations; backscattering from the surface–surface reverberations.

Acoustic energy travelling from the transmitting transducer to a target and reflected back to the receiving array is spread in time due to the multipath and the target extension. It is spread once by the target and twice by the medium. Due to the instability of the paths and the motions of the target, the signal is also spread in frequency, twice by the transmission through the time varying medium and once by the motions of the target. Motions of the transmitting and receiving arrays also contribute to this time and frequency spreading of the signals.

The reverberations can be regarded as reflections from independent reflecting points or scatterers distributed in the medium. The energy backscattered from this infinite number of reflecting points is composed of a summation of replicas of the original signal having various time delays and frequency shifts.

Since the signals transmitted through the underwater acoustic medium generally are subjected to time delay spread and frequency spread, the medium can be described as a linear, time-varying filter. Because the time variations of the medium very seldom can be known in detail, the filter functions must be regarded as random functions and average values must be used for characterizing the medium as a transmission channel.

Dealing with acoustic channels as linear time-variable stochastic filters a set of system functions can be defined and derived [4].

The time varying frequency response, $H(f, t)$, can be interpreted as an amplitude and phase response which is varying with time.

The inverse Fourier transform of $H(f, t)$ is a function of both absolute time, t, and delay time, τ, and is called the time-varying impulse response, $h(\tau, t)$ [i.e. the output of the filter at time t due to a dirac pulse τ (seconds) ago].

By Fourier transforming the impulse response with respect to t, a system function called the spreading function is obtained:

$$s(\tau, \phi) = \int_{-\infty}^{\infty} h(\tau, t) \exp(-j2\pi\phi t)\, dt.$$

This function gives the spectrum of the time variations of the impulse response. The output signal from the filter is formed as a sum of time and frequency shifted versions of the input signal weighted with $s(\tau, \phi)$. Thus the spreading function determines the spread in time and frequency the signal will suffer in the medium.

Another system function is obtained by Fourier transforming $H(f, t)$ with respect to t. This is the Bi-frequency function:

$$B(f, \phi) = \int_{-\infty}^{\infty} H(f, t) \exp(-j2\pi\phi t)\, dt.$$

$B(f, \phi)$ gives the spectrum of the variations of the frequency response. The meaning of this is that the output at frequency f is not dependent only upon the input at the same frequency but upon components in a frequency band around f, determined by $B(f, \phi)$.

The system functions are interconnected by Fourier transforms which can be illustrated by the well known figure, Fig. 5.

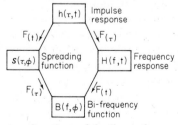

FIG. 5. The system functions of the linear time variable filter.

The combined range and doppler resolution properties inherent in a signal can be described in terms of the two-dimensional correlation function:

$$\chi(\tau, \phi) = \int_{-\infty}^{\infty} x(t)x^*(t - \tau) \exp(j2\pi\phi t) \, dt.$$

This function may be interpreted as the correlation value between $x(t)$ and a time and frequency (range and doppler) shifted version of $x(t)$. [$x(t)$ is the complex envelope of the input signal]. $\chi(\tau, \phi)$ is called the ambiguity function.

The system functions of the medium must be regarded as random functions in two variables and a reasonable average descriptor is the autocorrelation function of the system functions. Dealing with four-dimensional correlation functions is not very convenient and simplifying assumptions are often introduced. Assuming for instance that the value of the spreading function $s(\tau, \phi)$ at each time delay and frequency shift is uncorrelated with values at any other delay and shift combination, $s(\tau', \phi')$, the autocorrelation can be expressed as:

$$R_s(\tau, \phi) = \overline{|s(\tau, \phi)|^2},$$

where the bar denotes ensemble average. $R_s(\tau, \phi)$ is called the scattering function and can be regarded as a power density function. It determines the average delay and doppler spread the input signal power will be subjected to in the medium.

Due to the uncorrelation assumption, the power contributions can be summed directly, yielding a total squared output average of a matched filter receiver for delay and doppler setting τ_0 and ϕ_0:

$$P(\tau_0, \phi_0) = \int_{-\infty}^{\infty}\!\!\int |\chi(\tau_0 - \tau, \phi_0 - \phi)|^2 R_s(\tau, \phi) \, d\tau \, d\phi.$$

The scattering function of the medium must be found to design optimum active sonar systems where the signal and the signal processing methods are matched to the medium characteristics in which detection is to be performed. The scattering function will change over time as geometry and conditions change. For example in case of dominant surface reverberation the change of the scattering function with changing sea state is of great importance to the signal processor.

Measurements of the scattering function can be done using a signal with an "ideal" ambiguity function consisting of a single peak in the τ-ϕ-plane (a thumbtack-like ambiguity function). Almost ideal "probe" signals will, however, require a rather complex signal processor. More simple measurements can be done assuming the scattering function can be written as a product of a delay profile and a doppler profile:

$$R_s(\tau, \phi) = R_\tau(\tau) \cdot R_\phi(\phi).$$

The delay profile, $R_\tau(\tau)$, of the scattering function can then be measured using short CW-pulses with an ambiguity function having a fine range resolution and low doppler resolution capability. The doppler profile of the scattering function, $R_\phi(\phi)$, can be measured using long CW-pulses having an ambiguity function with a fine doppler resolution and low range resolution capability.

Due to the time dependence of the reflection field the scattering function is defined as a statistical average and consequently a larger number of measurements have to be performed to obtain estimates, i.e., a ping-to-ping integration has to be performed.

One of the first experiments with the Deep Sonar system was the measurements of the doppler profile of the scattering function for some environmental conditions in Norwegian coastal waters. One approach is to measure the energy distribution in the frequency domain for each ping and to do statistical analysis on a large quantity of such integrated reverberation spectra. Using long CW-pulses and narrow band frequency analysis, the doppler profile of the scattering function should be found.

The doppler profile of the scattering function determines the area in the τ-ϕ-plane in which doppler detection of targets will be very difficult or impossible. The reverberations must, however, be separated in three groups—volume, surface and bottom reverberations, of which the surface reverberations contribute to the broadest part of the equivalent scattering function.

Experiments show that the reverberation spectra obtained using long CW-pulses not only varies greatly as a function of the different reverberation contributions, but also as a function of numerous environmental parameters such as sea-state, wind direction, geometry of the experiments, etc. The variations of the reverberation spectra are expressed in terms of rapidly changing doppler shifts and doppler spread. Double or possibly multiple spectra are also observed. Due to these variations the reverberation spectra must also be found as a time-frequency energy distribution.

To find the time-frequency energy distribution of the reverberations, the reverberation spectra is divided and measured in a bin diagram where each bin is matched to the time and frequency resolution cell determined by the type of signal used. The frequency analyser mode used in this experiment gives spectral values each L msec where L is the number of spectral lines in the sweep. Using a pulse of 1 sec duration n_i such spectral values are summed giving a new energy value significant for the resolution cell of the signal (1 Hz × 1 sec). This integration process can be performed in the spectrum integrator described above, or on the computer which is fed with the spectral values recorded on digital magnetic tape.

The output from the real time spectrum analyser can be symbolized by a time-frequency array as shown in Fig. 6. Each line parallel to the frequency scale represents one frequency sweep, that is L numbers, one

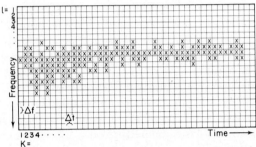

FIG. 6. Time-frequency array for one ping output from spectrum analyser.

for each frequency. During one ping the spectrum analyser computes the power content at each frequency K times. A total number of K sweeps is computed giving a total number of $L \cdot K$ numbers.

The reverberation energy distribution for each ping can then be found by an integration along the time axis for each frequency:

$$X_1 = \sum_{k=1}^{K} x_1.$$

This summation gives a frequency energy distribution for each ping. N numbers of such pings must be analysed to give a signficant average value for the reverberation energy distribution. The average value and the variance can be computed according to the following formulas:

$$\bar{X}_1 = \frac{1}{N} \sum_{n=1}^{N} X_1 \qquad \sigma_1^2 = \frac{1}{N} \sum_{n=1}^{N} X_1^2 - \bar{X}_1^2.$$

If an average time-frequency energy distribution of the reverberation is to be found, N number of time-frequency arrays must be treated according to the following formula:

$$\bar{x}_{kl} = \frac{1}{N} \sum_{n=1}^{N} x_{kl}.$$

And the variance of the content of each analyser resolution cell can be computed according to the following formula:

$$\sigma_{kl}^2 = \frac{1}{N} \sum_{n=1}^{N} x_{kl}^2 - \bar{x}_{kl}^2.$$

Using the integrator located at the output of the frequency analyser any number between 1 and 99 frequency sweeps may be integrated. The integrated reverberation spectra may be displayed on a storage oscilloscope or on an optical recorder specially designed for the reverberation analysis or recorded digitally on a 9 track magnetic tape (later the integrator output will be read directly into a NORD 1 computer installation by digital I/0 channels).

During the data handling so far the number of frequencies in each sweep is 32 or 64. The normal mode of the spectrum analyser is used giving a 1 Hz resolution and a time separation between each frequency line of 32 or 64 msec (Δf = 1 Hz, Δt = 32/64 msec).

Due to the variation of the output data as mentioned earlier, an integration along the time axis over a number of frequency sweeps is performed. In the data presented here, CW pulses of 1 sec have been used and the integrator is set up to integrate over a number of 8 sweeps. Using an 8 sec pulse repetition rate, a total number of maximum 14 integrated frequency sweeps (sweep width = 64 Hz) are stored on the magnetic tape, for each ping.

The data handling which was carried out on the NDRE computer established at Kjeller (CDC 3600/CDC Cyber 32) will later be carried out on line on the local computer installation. A number of statistical programs have been developed.

One of the most interesting results has been obtained by a program reading each integrated frequency sweep from the magnetic tape, normalizing each sweep by searching for the largest frequency value and computing the average frequency energy distribution for a number of pings. The program also computes the spectrum widths, defined at various dB levels, together with the standard deviations of the spectrum widths. The spectrum width is hard to define as the frequency energy distribution may contain several peaks. Consequently an extra facility of the program can do a search for the broadest spectrum peak, computing average widths and standard deviations of this peak, cancelling other peaks or possible noise/target signals in the time-frequency arrays.

4. Some Results from the Reverberation Investigations

The geometry of the reverberation experiments was shown in Fig. 2. The array was towed at various depths and the area of operation was mainly 30 nautical miles off the south east coast of Norway. The beams were tilted either 10 or 20 degrees upwards and examples of ray paths during the experiments are shown in Fig. 7. The array platform was towed at low speed, 2-3 knots and the "fish" appeared to be very stable during each run. The data presented in the following are recorded on the centre beam

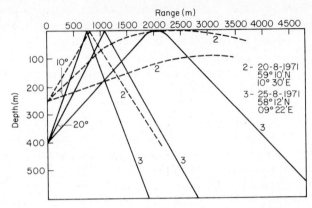

FIG. 7. Typical ray paths during reverberation experiments.

channel. The centre beam points 90° to the right of the course line of R/V "$H\,U$ Sverdrup". The influence on the reverberation spectrum due to the movements of the platform should then be very small.

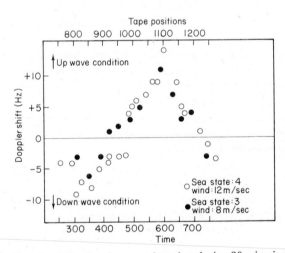

FIG. 8. Doppler shift of surface reverberation during 20 min circlings.

As expected from the ray paths, the reverberation spectra observed were rather complex, and a separation of the various reverberation contributions, volume, surface and bottom reverberations, is very difficult. A careful inspection of the reverberation spectrum as a function of time (during the ping) shows, however, that the surface contribution, lasting for a couple of seconds after the end of the transmitted pulse, can be separated. In broad terms the reverberation can be characterized by the following:

(a) the received spectra are broader than the transmitted spectrum
(b) the surface contribution of the reverberations is doppler shifted and the spectrum is much broader than the reverberations originated in the volume and at the bottom.

The doppler shifts in the surface reverberations has recently been investigated by others. The correspondence between the doppler shift and the wave direction with respect to the beam [5] direction is the same as observed by Y. Igarashi and R. Stern. Due to the distortion of the spectral shape caused by the surface reverberations, the exact doppler shift is difficult to measure. Another characteristic of the surface reverberations is the unsymmetric shape of the spectrum. For example compared to the data registered from cross wave conditions, the up wave and down wave conditions not only contribute to broader spectra, but also has its energy smeared out to higher and lower frequencies respectively. Some of the doppler shift observations made while the ship is circling are illustrated in Fig. 8.

Typical frequency analysed pings are shown in Figs 9 and 10. The transmitted pulse length is 1 sec. No amplitude modulation is used, i.e. the doppler resolution of the signal is 1 Hz. Figure 9 shows the reverberation spectra as visible on the output display of the spectrum analyser. The frequency resolution is 1 Hz and the frequency sweep length is 64 Hz. The transmitted pulse is not observed due to the time variable gain control signal suppressing the input signal during transmitting intervals and a predetermined time thereafter. Two frequency analysed pings are shown on the picture. The vertical grid lines correspond to 1 knot (i.e. 5.6 Hz at 8 kHz). A target approaching at a relative speed of 4 knots is also shown. The figure demonstrates the great variations in the reverberation spectra which can be observed from one ping to another.

Figure 10 is a computer plot of a frequency analysed ping. The time-frequency distribution of the reverberation is divided into sections, 8 sweeps in each. An integration is done for all 8 sweep sections and the result is normalized. This picture clearly demonstrates the complex nature of the surface reverberation spectrum which dominates at least the first 5 sections of the ping history. In the first two sections double and multiple spectra are observed. The two main peaks of the spectrum are approximately 4 Hz apart. The largest peak is at zero doppler (500 Hz corresponds to 8 KHz centre frequency) and the second peak corresponds to a wave velocity of 0.75 knot. The data was recorded with the array trained up wind, the sea state was 1–2 and a wind velocity of 6 m/sec was registered.

The type of data shown in Fig. 9 has been the basis for the investigations of the reverberation bandwidth discussed earlier. Two definitions

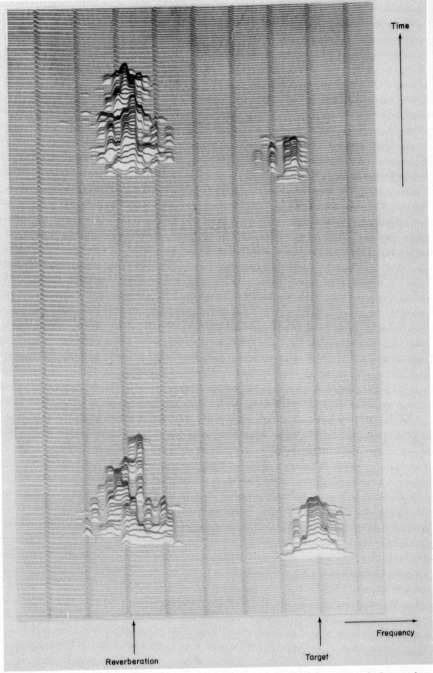

FIG. 9. Two subsequent frequency analysed pings as displayed on an optical recorder.

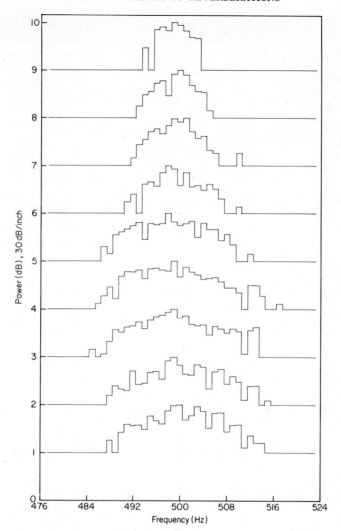

FIG. 10. Normalized reverberation spectrum (1 ping history).

of reverberation bandwidth have been used. Mode A defines the bandwidth as the frequency difference between the -6 dB points of the complete reverberation spectrum, neglecting possible zeros or multiple peaks within the spectrum. The -3 dB, -9 dB and -12 dB bandwidths points have also been calculated. In bandwidth mode B a search for the broadest spectrum peak is done if multiple peaks are present. The bandwidths found will usually be narrower than in the A mode, especially when typical surface reverberations are present.

The reverberation bandwidths express the frequency spread of the reverberation. Due to the normalizing technique used, however, a true doppler profile of the scattering function is not found. In order to plot the scattering function as a contour map the delay profile must also be measured and used as a weight function on the normalized doppler profile. This has not yet been done. The normalized doppler profile will, however, give useful information on the limits set by the medium in systems based on doppler detection.

The normalized doppler profile has been measured for various conditions such as sea state, etc. Some of the results so far computed are shown in Figs 11 and 12, where mean values for the reverberation bandwidth (−3 dB and −6 dB) are plotted as a function of time. The time scale is divided in parts equivalent to the integration intervals, one part equivalent to 8 frequency sweeps from the analyser. Each dot in the diagrams represents the average bandwidth value for 10 subsequent pings. The standard deviation for each group of 10 pings were also computed, but is not presented on the diagrams.

Even though the computed average values vary greatly, some conclusions may be drawn.

The frequency spread of the surface reverberation increases with

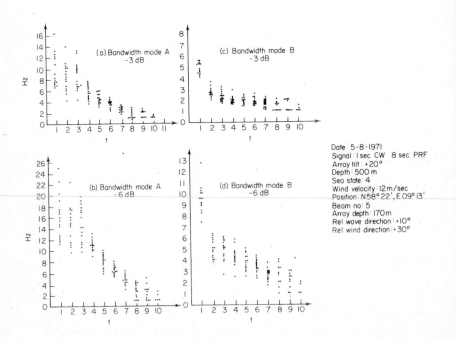

FIG. 11. Reverberation bandwidths, average values for groups of 10 pings.

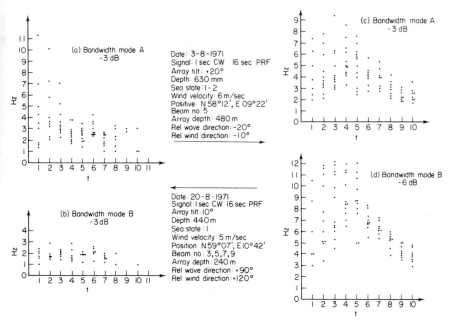

FIG. 12. Reverberation bandwidths, average values for groups of 10 pings.

increasing sea-state. Typical bandwidths of the reverberation spectra are (output pulse bandwidth: 1 Hz at 8 kHz):

TABLE I

| Bandwidth mode | Sea state | | | |
	0	1	2	4
A −3 dB	2.0 Hz	3 − 4 Hz	4 − 6 Hz	7 − 10 Hz
	± 0.18 knot	± 0.36 knot	± 0.5 knot	± 0.9 knot
A −6 dB	3.0 Hz		8 − 10 Hz	14 − 16 Hz
	± 0.27 knot		± 0.9 knot	± 1.7 knots
B −3 dB	2.0 Hz	2.0 Hz		2.5–3 Hz
	± 0.18 knot	± 0.18 knot		± 0.27 knot
B −6 dB	5.0 Hz			4.5–5.5 Hz
	± 0.45 knot			± 0.5 knot

The frequency spread of the volume and bottom reverberation is determined by local conditions such as water currents, array movements, etc. Typical volume/bottom reverberation bandwidths are (output pulse bandwidth: 1 Hz at 8 kHz):

TABLE II

bandwidth mode A:	−3 dB: 2.5 Hz (±0.22 knot)
	−6 dB: 4.0 Hz (±0.35 knot)
bandwidth mode B:	−3 dB: 1.8 Hz (±0.16 knot)
	−6 dB: 2.8 Hz (±0.25 knot)

5. Conclusions

A towed sonar has been used in reverberation experiments. The objective of the reverberation investigations was to determine the influence of the medium and the system parameters on the reverberation frequency spectrum.

Sea surface reverberations contribute to a doppler shift and a frequency smear of the reverberation spectrum. Positive doppler shifts equivalent to a wave velocity of 2.2 knots and negative doppler shifts equivalent to 1.8 knots have been observed during up wind and down wind conditions respectively.

The frequency spread of the surface reverberation spectra increase with increasing sea state. More data will be collected, but a few conclusions may already be stated. One of them is that a target with a relative speed in excess of about ± $1\frac{3}{4}$ knots may be detected if narrow band doppler analysis is used. This is valid up to sea state 4 and if the target strength exceeds −6 dB relative to the equivalent strength of the scattering field.

REFERENCES

1. Johnsen, J. (1971). "Deep Sonar—the transmitting part". TN U-268, NDRE.
2. Johnsen, J. (1971). "Deep Sonar—the receiving part, registration and monitoring equipment". TN U-270, NDRE.
3. Søstrand, K. A. (1970). "Description of a digital real time cross correlator and spectrum analyzer". IR U-247, NDRE.
4. Ellinthorpe, A. W. and Nuttal, A. H. (1965). "Theoretical and empirical results on the characterization of undersea acoustic channels". *IEEE 1st Annual Commun Conv*, June 1965.
5. Igarashi, Y. and Stern, R. (1971). "Observation of wind-wave generated doppler shifts in surface reverberation". *JASA* **49**, No. 3, p. 802.

DISCUSSION

P. H. Moose: Have you an explanation for the multiple peaks in the reverberation spectrum?

Answer: I have no specific explanation. However, the complexity of the surface reverberation spectrum must be explained by the irregular surface of the sea. We were supposed to measure the spectrum of the surface waves. However, due to the low frequency response of the present wave-buoy a measurement of the correlation between the reverberation spectrum and the surface wave spectrum has not been possible.

R. S. Thomas: How fast were you towing the transducer and how much frequency spread do you have due to the finite beam width of the transducer?

Answer: The array was towed at a speed of approximately 2 knots and the data are from the centre beam. The frequency spread due to the moving of the beam will be within the range of ± 0.6 Hz.

C. van Schooneveld: That is in the main beam but you have to do with the side lobes also.

Answer: The reverberation bandwidths shown were defined at −3, −6 and −9 dB. Contributions from the sidelobes of the centre beam should be more than 12 dB below the signals in the main lobe.

H. Mermoz: The underlining fact that the distance of the scattering function is subjected to some assumptions of the statistical properties of the spreading function, and also you have simulated the scattering function to a product of two profile versions. Do you know of any means of checking directly in the medium the validity of this assumption or do you think your results themselves are *a posteriori* controlled on this validity? As it is possible to derive from the scattering function an optimum class of signals for a time variable channel it would be important to check this assumption.

Answer: It is true that the assumptions made require that the reverberation belongs to a wide sense stationary process. This is not true for a process associated with a multipath environment, however, the surface reverberation bandwidths obtained describe from a practical point of view, the limitations due to the scattering function in doppler detection systems. The validity of the assumptions has not been the object for this work, but it would be interesting to do such a study.

R. R. Rojas: I would like to find out if you intend to make measurements with a broader or narrower beam. It seems to be that the bandwidths you are reporting certainly must be very strongly connected with the beam width of the system. This would give a good idea of what the limitations due to surface scattering are.

Answer: The present beam width is ± 3°. It would be interesting to make measurements with a narrower beam. However, the dimensions of the array will increase and this sets a limit on the beam width obtained in an experimental or operational towed sonar system.

Some Statistical Properties of Nonstationary Reverberation

TERRY D. PLEMONS, JACK A. SHOOTER and DAVID MIDDLETON

Applied Research Laboratories, The University of Texas, Austin, Texas, U.S.A.

1. Introduction

Effective signal processing of underwater acoustic signals that are corrupted by an additive reverberation process usually requires that certain statistical properties of the reverberation be known. One of the more important of these properties is the covariance of the reverberation or, equivalently, the reverberation's intensity spectrum. If the reverberation is represented by $X(t)$, then the covariance, assuming zero mean, is

$$K(t_1, t_2) \equiv \langle X(t_1)X(t_2) \rangle, \tag{1}$$

where $\langle \cdot \rangle$ denotes the infinite ensemble average. The time variable t is a measure of range and $t = 0$ is the time of transmission of each signal that generates a member of the reverberation ensemble.

If the reverberation is covariance stationary then

$$K(t_1, t_2) = K(t_2 - t_1) = K(\tau) = K(-\tau) \tag{2}$$

with $\tau = t_2 - t_1$. Since the intensity of the reverberation is generally range dependent, equation (2) is hardly ever a realistic model of the reverberation covariance. A more realistic assumption is that the reverberation is locally stationary with a covariance that can be written as

$$K(t_1, t_2) = K_1(t_1)K_2(\tau), \tag{3}$$

where $K_2(\tau)$ is a stationary covariance and $K_1(t_1)$ is a slowly varying function compared to $K_2(\tau)$. If $K_2(\tau)$ is defined such that $K_2(0) = 1$, then $K_1(t_1)$ is the intensity, $\langle X(t_1)^2 \rangle$, of the reverberation.

Another model that has been studied considerably [1] assumes that the reverberation is reducible to stationary and can be expressed as

$$X(t) = F(t)x(t), \tag{4}$$

where $F(t)$ is a deterministic function that is a measure of the non-stationarity of $X(t)$ and $x(t)$ is a random process with the stationary covariance $K_x(\tau)$. A process reducible to stationary is appealing to signal processors since it can be stationarized through division provided that $F(t)$ is known.

Equation (4) does not necessarily represent a locally stationary process for the covariance in this case is

$$K(t_1, t_2) = F(t_1)F(t_2)K_x(\tau), \tag{5}$$

and unless we can write

$$F(t_1)^2 \cong F(t_2)^2 \cong F(t_1)F(t_2), \tag{6}$$

then the process is not locally stationary.

The normalized covariance

$$K_N(t_1, t_2) = K(t_1, t_2)/[K(t_1, t_1)K(t_2, t_2)]^{1/2} \tag{7}$$

provides a test to determine if the reverberation is reducible to stationary (see Section 9 of Reference [1]). If indeed this is the case, then the normalized covariance of equation (4) is

$$K_N(t_1, t_2) = K_N(\tau) = K_x(\tau). \tag{8}$$

In a practical situation it is usually difficult to ascertain the form of the nonstationarity of the reverberation. The geometry of the experiment and the scattering mechanism must be accurately known before the behaviour of the reverberation as a function of range can be predicted with confidence. If this information is not available, then the only alternative is to estimate the covariance from the appropriate experimental data. When the data are nonstationary, the only strictly valid estimates are those derived from the computation of ensemble averages. Section 3 presents several examples of experimental nonstationary covariance functions of reverberation processes generated by scattering underwater acoustic signals from the rough, moving surface of a fresh water lake. Before proceeding, however, some theoretical aspects of the problem are discussed in Section 2.

2. A Theoretical Model of the Reverberation Covariance

Consider the case of underwater monostatic acoustic scattering from a rough surface such as that of the ocean. It is assumed that the reverberation is the sum of a large number of signals arriving from individual point scatterers located independently in space and time on the surface. This particular model has been studied extensively by several authors, notably Faure, Olshevskii, and Middleton [1, 2, 3]. Using the results of the most recent study of Plemons [4], one can show that the covariance of $X(t)$ is

$$K(t_1, t_2) = \int_{-\infty}^{\infty} G(\lambda)S(t_1 - \lambda)S(t_2 - \lambda)\, d\lambda \tag{9}$$

$$= \int_{-\infty}^{\infty} G(t_1 - \lambda)S(\lambda)S(\lambda + \tau)\, d\lambda, \tag{10}$$

where $S(\lambda)$ is the transmitted signal and $G(\lambda)$ is a nonnegative function that incorporates the experimental geometry (beam patterns, spreading loss, etc.) and the physical properties of the scattering mechanism (surface scattering strength, shadowing function, etc.). This expression for the nonstationary covariance is valid when very restrictive conditions are met. Some of these are (1) no significant scatterer motion (Doppler), (2) no platform motion, (3) no distortion of the transmitted signal $S(\lambda)$ by the scatterers, and (4) no frequency selectivities of the transducers. These restrictions can be included in the theory [3, 4], but in the discussion that follows the conclusions would not be significantly altered by their omission.

The nonstationarity of $X(t)$ is manifest through $G(\lambda)$ [1]. If $G(\lambda)$ is slowly varying in the vicinity of $\lambda = t_1$, then

$$K(t_1, t_2) \cong G(t_1) \int_{-\infty}^{\infty} S(\lambda)S(\lambda + \tau)\, d\lambda \tag{11}$$

$$= G(t_1)C(\tau) = G(t_1)C(-\tau), \tag{12}$$

where $C(\tau)$, the integral in equation (11), is the autocorrelation of the transmitted signal.

The geometry of some scattering experiments will result in a $G(\lambda)$ that varies so rapidly that it cannot be legitimately removed from the integral in equation (9). A partial removal can be obtained by expanding $G(t_1 - \lambda)$ in a Taylor series about t_1 and retaining the first two terms,

$$G(t_1 - \lambda) = G(t_1) - \lambda G'(t_1), \tag{13}$$

where $G'(t_1)$ is the derivative of $G(t)$ at $t = t_1$. Keeping only the first two terms in the series expansion puts a limit on the allowable variations of $G(t_1 - \lambda)$. More complicated variations can be included by adding higher-order terms in the series.

Using equation (13) the covariance, equation (10), can be written as

$$K(t_1, t_1 + \tau) = G(t_1)[C(\tau) + f(t_1, \tau)], \tag{14}$$

where

$$f(t_1, \tau) \equiv - \frac{G'(t_1)}{G(t_1)} \int_{-\infty}^{\infty} \lambda S(\lambda) S(\lambda + \tau) \, d\lambda. \tag{15}$$

The covariance, equation (14) in this case, is the sum [within a scale factor $G(t_1)$] of a stationary component $C(\tau)$ and a nonstationary component $f(t_1, \tau)$. If the relative variation of $G(\lambda)$ is small in the neighbourhood of $\lambda = t_1$, then

$$G'(t_1)/G(t_1) \ll 1, \tag{16}$$

and the contribution of $f(t_1, \tau)$ to $K(t_1, t_1 + \tau)$ will be small. When this is the case, the reverberation is locally stationary [see equation (3)].

If $G(t_1)$ is constant, then the intensity, $K(t_1, t_1)$ will be constant. However, whenever the intensity is a rapidly varying function over some range interval, then $G(t)$ will correspondingly have significant variations over that interval. In this case the relative variation $G'(t)/G(t)$ will be large and the contribution of $f(t_1, \tau)$ will be significant [see equation (15)].

3. The Experimental Covariance: Ensemble Averages

To obtain a set of data suitable for an experimental study of the stationarity of a reverberation process, underwater acoustic signals were scattered off the rough moving surface of a fresh water lake. The experiment was conducted at the Lake Travis Test Station of Applied Research Laboratories, Austin, Texas. The scattering was monostatic, with the transmitter and receiver co-located 8 ft beneath the scattering surface. The 3 dB beamwidth of transmitter and of the receiver was $10°$ in the horizontal and vertical planes. The acoustic axes of the transducers were parallel and intersected the scattering surface with a grazing angle of $22°$. Short pulses of centre frequency $f_0 = 110$ kHz were used to insonify the surface and generate the reverberation returns. Two ensembles of data were generated by transmitting, first, 150 pulses of length 1.0 msec and, then, 150 pulses of length 0.1 msec. Corresponding to each transmitted signal, one reverberation return $X_i(t)$ was recorded on analogue magnetic type. Thus, the two data ensembles are

$$\{X_i(t)\}_{1.0} \text{ and } \{X_i(t)\}_{0.1}, \quad i = 1, 2, \ldots, 150.$$

These narrowband functions were then sampled in quadrature [5] and stored on digital tape. Using a Kolmogorov–Smirnov test and an appropriate runs test [6], these data were tested for homogeneity and independence. Those range intervals that corresponded to homogeneous ensembles were used to construct the two dimensional covariance (the "bar" denotes the experimental estimate),

$$\bar{K}(t_1, t_2) = \frac{1}{150} \sum_{i=1}^{150} X_i(t_1)X_i(t_2). \tag{17}$$

The experimental covariance of these narrowband data can be expressed as

$$\bar{K}(t_1, t_2) = \bar{K}_0(t_1, t_2) \cos[2\pi f_0 \tau + \bar{\Phi}_0(t_1, t_2)], \tag{18}$$

where $\bar{K}_0(t_1, t_2)$ and $\bar{\Phi}_0(t_1, t_2)$ are the envelope and phase of $\bar{K}(t_1, t_2)$ respectively. In this study the envelope $\bar{K}_0(t_1, t_2)$ will be discussed.

In Fig. 1 the estimates of the intensities, $K(t_1, t_1)$ [see equation (17)], are plotted as a function of t_1 (the time of transmission is $t_1 = 0$). For

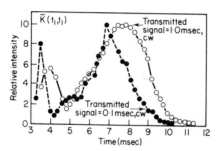

FIG. 1. Experimental intensities of short range surface reverberation.

purposes of comparison, the maxima of the 1.0 msec data and 0.1 msec data are set to 10. The main backscattered energy occurs in the interval 6 to 9 msec which corresponds to the region of intersections of the main lobes of the beam patterns with the scattering surface. It is easy to see that $\bar{K}(t_1, t_1)$ has significant variations over the duration of a pulselength of 1.0 msec, which is the correlation distance of 1.0 msec transmitted signal. There is less variation over the correlation distance of the 0.1 msec signals.

The envelope of the experimental covariance is a function of the two variables t_1 and τ, the time from transmission (range) and delay, $t_2 - t_1$, respectively. We are primarily interested in the variation with respect to τ since this provides information about the degree of stationarity of the reverberation. With t_1 set to 6.7 msec the envelope of the covariance is plotted in Fig. 2a. This estimate is derived from a 150-member ensemble. To gauge the symmetry of the covariance the values corresponding to negative τ are plotted to the right of the origin and thus they can be compared to the corresponding values of the covariance envelope when

τ is positive. The maximum values of the correlation functions are set to unity. The lack of local stationarity is exhibited as the difference in $\bar{K}(t_1, t_1 + \tau)$ and $\bar{K}(t_1, t_1 - \tau), \tau \geqslant 0$. As can be easily seen there is a considerable difference in the two plots between $\tau = 0.0$ and 1.0 msec. Considering the variation of the intensity, $\bar{K}(t_1, t_1)$, seen in Fig. 1, this lack of stationarity is to be expected considering equations (14, 15) and the discussion following equation (16).

The normalized covariance $\bar{K}_N(t_1, t_2)$ [see equation (7)] is plotted in Fig. 2b. One immediately notices that the normalization process has come close to creating a stationary covariance since $\bar{K}_N(t_1, t_1 + \tau) \approx K_N(t_1, t_1 - \tau)$

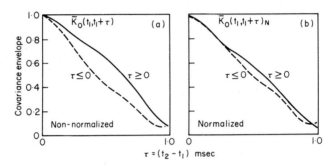

FIG. 2. Envelope structure of covariance of short range surface reverberation generated by 1.0 msec cw transmitted signals. $t_1 = 6.7$ msec reference transmission time.

although slight differences in these two functions can be observed. Thus, we conclude that at the range t_1 the reverberation is definitely not a stationary process but it is approximately a process that is reducible to stationary.

The covariance functions produced by the transmission of the shorter pulse, 0.1 msec, is shown unnormalized in Fig. 3a, and normalized in Fig. 3b. Notice that the unnormalized covariance, with less differences in the corresponding positive and negative values of τ, is more stationary than the 1.0 msec data of Fig. 2a. This is not surprising since a shorter pulse in general produces a more locally stationary process. Except for $|\tau| \geq 0.1$ msec, the normalization came close to producing a stationary covariance as Fig. 3b indicates.

Changing from the range $t_1 = 6.7$ msec to 9.0 msec the covariance of the 1.0 msec data is once again plotted in Fig. 4. The departure from stationarity is much more marked now as seen in the increased differences in $\bar{K}_0(t_1, t_1 + \tau)$ and $\bar{K}_0(t_1, t_1 - \tau)$. The normalization also failed to produce a symmetrical covariance as Fig. 4b demonstrates. Therefore in the neighbourhood of $t_1 = 9.0$ msec the reverberation produced by 1.0 msec signals is not a locally stationary process nor is it one that is reducible to stationary.

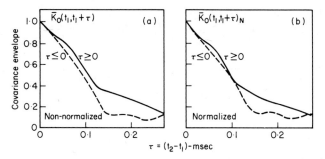

FIG. 3. Covariance envelope of surface reverberation generated by 0.1 msec cw transmitted signals. t_1 = 6.7 msec reference transmission time.

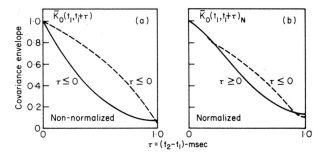

FIG. 4. Covariance envelope of surface reverberation generated by 1.0 msec cw transmitted signals. t_1 = 9.0 msec reference transmission time.

The same conclusions can be made about the reverberation process generated by the shorter 0.1 msec, transmitted pulses since neither the unnormalized or normalized covariance (Fig. 5a and b) envelopes are symmetrical.

When the experimental geometry is a less critical function of range, then a more locally stationary reverberation can be expected if the pulselength of the transmitted signal is reasonably small. An experimental set of data was available in this study to investigate this prediction. The acoustic axes of the transducers were horizontal and the significant ranges of the back-scattered energy were much greater than those of the data discussed above. The transmitted signal was an FM pulse 1.25 msec long having a bandwidth of approximately 10 kHz. In Fig. 6 the estimates of the covariance envelopes (average of 150 samples) corresponding to ranges of t_1 = 54 and 82 msec can be seen. These covariance functions are approximately symmetrical. In this example the reverberation at the extended ranges is a locally stationary process.

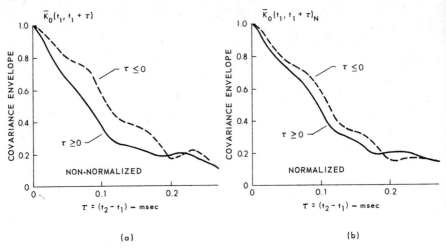

FIG. 5. Covariance envelope of surface reverberation generated by 0.1 msec cw transmitted signals. $t_1 = 9.0$ msec reference transmission time.

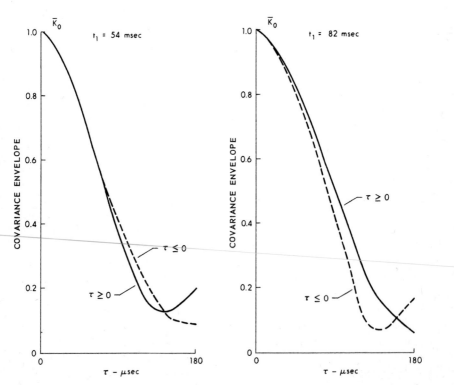

FIG. 6. Envelope of the non-normalized covariance of long range surface reverberation.

4. The Experimental Covariance: Time Averages

In many situations the estimation of the covariance by computation of ensemble averages will not be possible and the validity or invalidity of the corresponding time average of a single reverberation return must be established. If $X_j(t)$ is the available reverberation function, then the estimate of the covariance is

$$C_j(\tau)|_{T_1, T_2} = \frac{1}{T_2 - T_1} \int_{T_1}^{T_2} X_j(\tau)X_j(t + \tau) \, dt.$$

This estimate of the covariance is equal to the ensemble covariance only under very restrictive conditions. Not only must the ensemble of functions

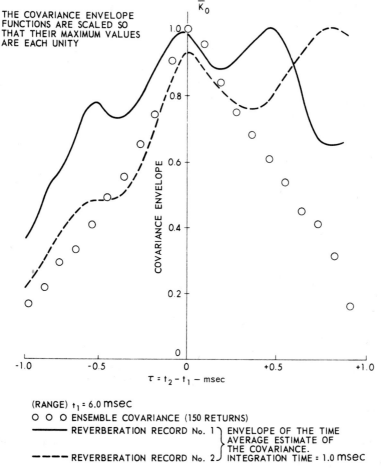

THE COVARIANCE ENVELOPE FUNCTIONS ARE SCALED SO THAT THEIR MAXIMUM VALUES ARE EACH UNITY

$\overline{K}_0$

COVARIANCE ENVELOPE

$\tau = t_2 - t_1 -$ msec

(RANGE) $t_1 = 6.0$ msec

O O O ENSEMBLE COVARIANCE (150 RETURNS)

—— REVERBERATION RECORD No. 1 ⎫ ENVELOPE OF THE TIME
⎬ AVERAGE ESTIMATE OF
THE COVARIANCE.
– – – – REVERBERATION RECORD No. 2 ⎭ INTEGRATION TIME = 1.0 msec

FIG. 7. Comparison of ensemble and time average estimates of the covariance envelopes of short range surface reverberation.

$\{X(t)\}$, of which $X_j(t)$ is a member, be stationary, but it must also be ergodic.

Providing the process is stationary, the accuracy of the time average estimate improves as the integrating time $t_2 - t_1$ increases. For locally stationary processes, however, increased integration times encompass larger intervals of nonstationarity and the interpretation of the estimate becomes difficult if not impossible. Shorter integration times bypass this

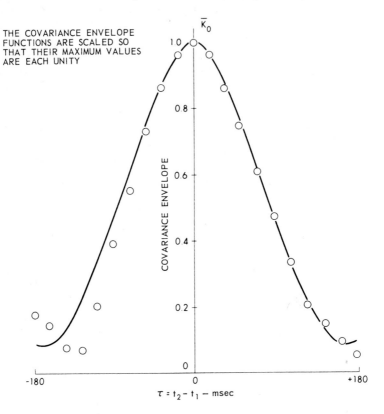

THE COVARIANCE ENVELOPE FUNCTIONS ARE SCALED SO THAT THEIR MAXIMUM VALUES ARE EACH UNITY

(RANGE) $t_1 = 82$ msec

O O O COVARIANCE ESTIMATED BY COMPUTING AN ENSEMBLE AVERAGE OVER 150 REVERBERATION RETURNS

———— COVARIANCE ESTIMATED BY COMPUTING A TIME INTEGRATION OVER AN INTERVAL EQUIVALENT TO 20 TRANSMITTED PULSELENGTHS

FIG. 8. Comparison of ensemble and time average estimates of the covariance envelopes of long range surface reverberation.

difficulty at the expense of statistical accuracy since there is a decrease in the number of independent samples in the interval of integration.

The errors that can be made with time averages are demonstrated in Fig. 7. Here the ensemble covariance can be compared with two time average estimates of the reverberation covariance. The interval of integration, at $t_1 = 6$ msec, was $T_2 - T_1 = 1.0$ msec. Obviously the time averages give meaningless results in this case.

An example of a meaningful time average estimate is illustrated in Fig. 8. The estimate is based on an integration time of 20 transmitted pulselengths. These are the same data used in Fig. 7 and are considerably more stationary than the shorter range reverberation.

5. Concluding Remarks

The accurate estimation of the reverberation covariance function requires either a detailed knowledge of the experimental geometry and scattering process or a statistically large and valid data sample. The time average estimate of the covariance function can be misleading, particularly in the range intervals where the reverberation is not locally stationary.

A reverberation process not locally stationary is unlikely to be a process that is reducible to stationary. An implication of this fact is found in sonar receivers that normalize the reverberation so that the intensity is constant with respect to range. Depending on the degree of nonstationarity, the covariance function of the normalized process will be range dependent and, as a consequence, the performance of the receiver will also be range dependent.

Departures from local stationarity will occur whenever there are abrupt changes, with respect to range, in the scattering geometry. Examples are common and easy to visualize. A projected signal intersecting a scattering surface is one example, and rapidly varying values of the projecting and receiving beam patterns in the scattering region is another. In all cases the nature of the transmitted signal is important. The chances of obtaining a locally stationary process improve as the length of the transmitted pulse is decreased.

ACKNOWLEDGMENT

This work was supported by Naval Ship Systems Command, Code 901.

REFERENCES

1. Ol'shevskii, V. V. (1967). "Characteristic of Sea Reverberation", Consultants Bureau, Plenum Press, New York.
2. Faure, P. (1964). Theoretical model of reverberation noise. *J. Acoust. Soc. Am.* **36**, 259–268.
3. Middleton, D. (1967). A statistical theory of reverberation and similar first-order scattered fields, Part I: Waveforms and the general process, Part II: Moments, spectra, and special distributions. *IEEE Trans. Information Theory* **IT-13**, 372–414.
4. Plemons, T. D. (1971). Spectra, covariance functions, and associated statistics of

underwater acoustic scattering from lake surfaces, Applied Research Laboratories Technical Report No. 71-17 (ARL-TR-71-17), Applied Research Laboratories, The University of Texas at Austin. (Also doctoral dissertation presented to the Physics Department, The University of Texas at Austin.)

5. Grace, O. D. and Pitt, S. P. (1970). Sampling and interpolation of bandlimited signals by quadrature methods. *J. Acoust. Soc. Am.* **48**, 1311–1318.

6. Middleton, D. (1969). Acoustic modeling, simulation and analysis of complex underwater targets, II. Statistical evaluation of experimental data, Applied Research Laboratories Technical Report No. 69-22 (ARL-TR-69-22), Applied Research Laboratories, The University of Texas at Austin.

DISCUSSION

J. J. Dow: To what extent is the degree of nonstationarity in the reverberation process determined by the following: (1) the type of transmitted signal? (2) acoustic propagation conditions? (3) the type of normalization?

Answer: (1) In general the asymmetry in the covariance function of reverberation can be decreased by increasing the bandwidth of the transmitted signal. This increase in bandwidth, effected by either a decrease in the pulselength of a CW signal or an increase in the swept bandwidth W of an FM slide, causes a decrease in the width of the covariance function. As the covariance function decreases in width, it becomes less susceptible to the nonstationary aspects of the experimental geometry.

(2) Whenever there is a sufficiently abrupt change, at a given range in the experimental geometry or in the scattering mechanism, then departure from local stationarities can be expected. The intersection of the propagating transmitted signal with a school of fish or the boundary of the medium (surface, botton) are examples here. Nonstationarity will result whenever the intensity of the reverberation experiences relatively rapid variations over a given range interval.

(3) If the variation in $G(\lambda)$ is large, then the conventional technique of normalization (division by the appropriate variance) will not remove the asymmetry in the covariance function. However, it may be possible, with a knowledge of $G(\lambda)$, to subject $X(t)$ to a transformation that would result in a locally stationary process.

P. H. Moose: Can you explain why the reverberation data failed the Kolmogorov–Smirnov test for homogeneity at the near ranges?

Answer: No. At these ranges, which correspond to near specular reflections, there was some instability in the scattering surface that caused the data to be invalid.

P. H. Moose: Do you think that the instabilities can be attributed to these near specular reflections?

Answer: Not at this time. That is considered only as an associated fact.

C. van Schooneveld: Would you say a few words about the required changes in a matched filter operating under these conditions?

Answer: Obviously the matched filter must be range dependent unless the reverberations were effectively normalized to a stationary process. I can say no more than that, however.

G. Vezzosi: Is it possible to define a time span of local stationarity?

Answer: Strictly speaking the process is locally stationary over a given time span during which the covariance is symmetrical and retains the same shape. One could, in practice, define a less demanding condition and require only that the total deviation, expressed perhaps by an appropriate integral, with respect to the delay τ, of the asymmetrical component of the covariance function, be within certain bounds during a time (range) interval.

B. W. Schroder: Is it possible that part of the asymmetry structure you observe in the covariance functions is due to the bias that occurs in the time integration estimate of the covariance?

Answer: In the time-average estimate this is quite possible, particularly so in the short range reverberation where the integration time had to be small. This bias, however, will not be present in the covariance estimate obtained with the ensemble averages.

Some Remarks on the Use of Auto-correlation Functions with the Analysis and Design of Signals

B. VELTMAN, A. van den BOS, R. de BRUINE, R. de RUITER and
P. VERLOREN

University of Technology, Delft, The Netherlands

1. Extrapolation of Auto-correlation Functions

In the classical method of frequency analysis of signals the power density spectrum is calculated by Fourier transforming the auto-correlation function. The numerical Fourier transformation has long been considered to be a more complicated procedure than an averaging operation. So the Fourier transformation was always carried out on truncated series of averaged products. Truncation is usually caused by a finite measuring time (improvement of statistical stability), by the allowable computation time or by a finite number of measuring pick-ups with wave length measurements. With today's numerical methods Fourier transformation is a fast operation so nowadays one calculates power density spectra directly by first transforming all data points and averaging afterwards.

There are however applications where it is desirable or even inevitable that first the correlation function is measured. E.g. with on-line measurements at very high frequencies. Parallel correlators using coarsely quantized signals provide a more rapid data reduction than the FFT procedure. Also with measurements where the argument is wavelength instead of frequency (position instead of time), correlation analysis is often preferred. In these situations one is confronted with the truncation effects of the correlation functions on the power density spectrum. Undesired truncation effects may be diminished by using appropriate windows; this is a trade-off with a deterioration in the frequency resolution.

For an increase in frequency resolution one has to extrapolate the correlation function. For the extrapolation of a particular auto-correlation function one may consider an analytic continuation based upon a well chosen series approximation of the known part of the function. Of special interest in such approximations are "non-parametric" methods, which means that the approximation requires none, or only very weak assumptions about the particular correlation function.

131

Two methods proposed in literature [1, 2] are treated in this paper. The first one may be considered as a parameter free procedure. It only assumes that the time series is bandwidth limited and that this bandwidth is known. This is usually true because a time series can only represent a bandwidth limited signal. The auto-correlation function is then approximated by a series of bandwidth limited terms: the angular prolate spheroidal wave functions.

The second method, called maximum entropy analysis, appears to be parameter free as it seems to use only an intrinsic property of auto-correlation functions. It can be shown however that there is a hidden assumption about an all-pole model for the signal.

(a) *Approximations with angular prolate spheroidal wave functions*
Usually the angular prolate spheroidal wave functions are notated as time functions; its Fourier transformed versions are thus the frequency spectra [3]. With correlation functions the argument has to be a time shift and its transforms are power density spectra. The auto-correlation function $R(\tau)$ is approximated by:

$$R(\tau) = \sum_{0}^{\infty} A_n S_{on}\left(W\tau_m, \frac{\tau}{\tau_m}\right)$$

with

$$A_n(W\tau_m) = \frac{1}{U_{on}^2} \int_{-\tau_m}^{+\tau_m} S_{on}\left(W\tau_m, \frac{\tau}{\tau_m}\right) R(\tau)\, d\tau$$

and

$$U_{on}^2 = \int_{-\tau_m}^{+\tau_m} S_{on}^2\left(W\tau_m, \frac{\tau}{\tau_m}\right)\, d\tau.$$

W is the bandwidth of the signal. τ_m the maximum delay. S_{on} are special cases of the more general wave functions S_{mn}. U_{on} is a normalization factor (with time functions, U_{on} is related to the total energy of the wave function on a time interval $-T$ to $+T$). The series developed is based on functions which are bandwidth limited; they are orthonormal over an infinite interval, and orthogonal over a finite interval. The wave functions are proportional to the eigen functions of the finite Fourier transform [3].

The numerical evaluation of the wave functions S_{on} is obtained from a series of Legendre functions. These Legendre functions and the coefficients for the calculation of wave functions are computed from standard recursive formulae.

The total calculation holds the following steps:

(a) Choose the maximum time shift of the correlation function, multiply this with the known signal bandwidth
(b) Calculate the Legendre functions and the series coefficients on the interval of interest
(c) Calculate the wave functions on the interval of interest
(d) Calculate the A_n with the given part of the correlation function
(e) Calculate the numerical approximation on the extended interval.

It will be clear from this scheme that there is an appreciable amount of computation time involved. Accumulation of rounding off errors soon tends to mask an honest evaluation of the series approximation. A careful choice of discretization steps is necessary [4].

The method is first applied to the ideal situation of a bandwidth limited correlation function:

$$R(\tau) = \frac{\sin W\tau}{W\tau}.$$

It is assumed that the extrapolation goes to twice the given interval length.

Figure 1 shows the results for $W\tau_m = 2\pi$. Figures 2 and 3 show the same functions, however with 10% deviations in the chosen bandwidths. In Fig. 4 the same calculations are made for the correlation function

$$R(\tau) = \left[\frac{\sin W\tau/2}{W\tau/2}\right]^2.$$

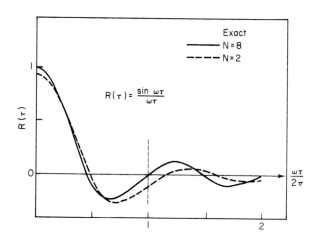

FIG. 1. $W\tau_m = 2\pi$.

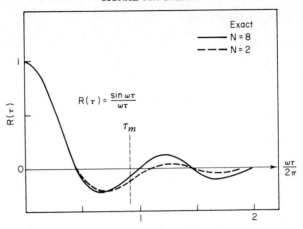

FIG. 2. $W\tau_m = 2\pi$. W 10% too large.

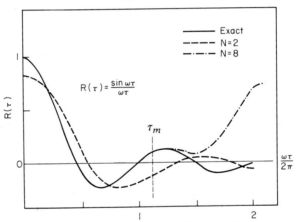

FIG. 3. $W\tau_m = 2\pi$. W 10% too small.

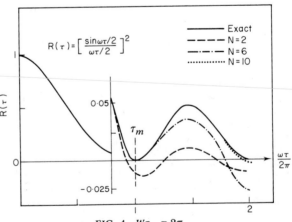

FIG. 4. $W\tau_m = 2\pi$.

The results for these essentially bandwidth limited signals are quite satisfactory provided that the signal bandwidth is estimated safely. If one however applies the same procedure to signals for which the bandwidth limitations are prohibitive the results deteriorate rapidly. Figure 5 shows the calculations for a second order Butterworth filter with an assumed bandwidth limit at the −3 dB point, Fig. 6 the same for a first order filter.

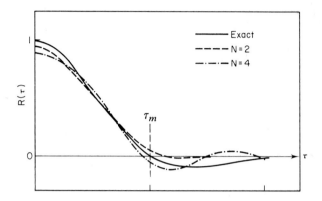

FIG. 5. Second order Butterworth filter. $W\tau_m = 2\pi$.

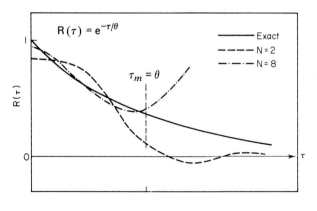

FIG. 6. $W\tau_m = 2\pi$.

One has to distinguish between the approximation on the given interval and on the extrapolated interval. If the bandwidth requirement is not fulfilled the series development clearly diverges in the extrapolated region.

(b) *Maximum entropy spectral analysis*
This method is based upon a general property of auto-correlation functions:

for every truncated auto-correlation function consisting of the lag values
0 to m the matrix

$$
C(m) = \begin{pmatrix}
c(0) & c(1) & \cdots & c(m) \\
c(1) & & & \vdots \\
\vdots & & & \vdots \\
c(m) & \cdots\cdots\cdots & c(0)
\end{pmatrix}
$$

has to be semi-positive definite. This implies that det $C(m)$ is non-negative.
Because the matrix is symmetric the addition of an unknown next lag
value $c(m+1)$ leads to an inequality of a quadratic form in the unknown
lag value. According to Parker Burg [2] the extrapolated value has to be
chosen so that the value of det $[C(m+1)]$ lies in the middle of the allowable
interval. Further inspection of the quadratic inequality shows that that
midpoint interval procedure is equivalent to maximizing the entropy. If
gaussian signals there exists an expression for the entropy of an $(m+2)$-
dimensional gaussian probability function with covariance matrix $C(m+1)$:
the entropy is [5]

$$
\ln (2\pi e)\,\frac{m+2}{2}\,\det\,[C(m+1)]^{1/2}.
$$

This implies that extrapolating the correlation function according to the
midpoint interval procedure is equivalent to maximizing the entropy. If
one lag value is extrapolated this value is used for calculating the next
point etc.

Figures 7 and 8 show some results of this extrapolation. Remarkably
accurate results are obtained over a wide extrapolation interval. It is hard to
believe that this is the result of a non-parametric method. It will be shown
now that this is indeed not the case. The calculation is in fact based upon

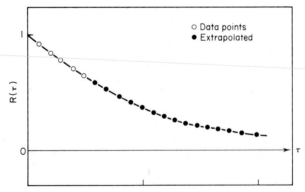

FIG. 7. $R(\tau) = e^{-\tau/\theta}$; maximum entropy extrapolation.

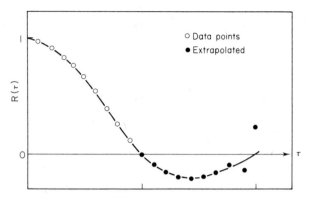

FIG. 8. $R(\tau) = \dfrac{\sin W\tau}{W\tau}$, maximum entropy extrapolation.

an all-pole model for the signal in which the allowable number of poles is given by the number of known correlation points. This result implies that signals consisting of poles only can be accurately extrapolated by the described method. However, signals containing zero's may require many more given values of the correlation functions for a desired accuracy. In these cases there are better parametric methods allowing a fitting with poles and zero's for minimum residues and which provide additional information about a good choice for the order of the model [6].

The equivalence of maximum entropy analysis and an all pole model follows from an inspection of the meaning of maximizing $\det[C(m + 1)]$ with respect to $c(m + 1)$. Taking the derivative leads to the requirement that:

$$
\det \begin{bmatrix}
c(1) & c(0) & \cdots & c(m-1) \\
c(2) & c(1) & & c(m-2) \\
\vdots & & & \vdots \\
c(m+1) & c(m) & \cdots & c(1)
\end{bmatrix} = 0.
$$

If an all-pole model is activated by a zero-mean random variable $e(n)$:

$$
x(n) + a_1 x(n-1) + \cdots + a_N(n-N) = e(n),
$$

with $E[e(i)e(j)] = 0$ for $i \neq j$; a multiplication of both sides with $x(n-k)$ and taking the mathematical expectations gives:

$$
c(k) + a_1 c(k-1) + a_2 c(k-2) + \cdots + a_N c(k-N) = 0
$$

with $c(k) = E[x(n)x(n-k)]$.

Writing this out gives:

$$c(1) + a_1 c(0) + \cdots + a_N c(N - 1) = 0$$
$$c(2) + a_1 c(1) + \cdots + a_N c(N - 2) = 0$$
$$\vdots \qquad\qquad \vdots$$
$$c(N + 1) + a_1 c(N) + \cdots + a_N c(1) = 0$$

which is equivalent to

$$
\det
\begin{bmatrix}
c(1) & c(0) & \cdots & c(N-1) \\
c(2) & c(1) & & c(N-2) \\
\vdots & & & \vdots \\
c(N-1) & c(N) & \cdots & c(1)
\end{bmatrix}
= 0.
$$

This is the same expression as the maximum entropy analysis provides. No need to say that if one has calculated the coefficients a_N, the actual extrapolation of the correlation function is unnecessary as the power density spectrum follows directly from the values of the poles in the model. For finite value time series the approximation of Nth-order pole models by a record of length m gives again equivalent expressions if the approximation is based upon least squares fitting. The true correlation functions are then replaced by their estimates from the given data points.

In conclusion: the Parker Burg method is a least squares approximation.

2. The Synthesis of Random Signals with a Prescribed Amplitude Probability Density Function and a Prescribed Power Density Spectrum

The input to the signal generator is assumed to be wide band gaussian noise with a flat spectrum, Fig. 9. The signal is first applied to a linear

FIG. 9. Linear filter and non-linear characteristic for generation signals with specified probability density function and auto-correlation function.

filter, the gaussian properties are preserved. In series with the linear filter the signal passes a non-linear amplitude characteristic. The characteristic changes the gaussian probability density function into the desired one. It also changes the power density spectrum of its input signal. However, once the non-linearity is known the influence on the power density spectrum can be calculated and used for "compensating" the linear filter so that the output of the non-linearity produces the desired frequency behaviour.

The calculation of the change in power density spectrum is done with the aid of Price's theorem [8]. This theorem applies to gaussian processes and output auto-correlations of a non-linear amplitude characteristic.

The calculations run as follows

(a) Determine the non-linear characteristic for a gaussian input and a desired output.
(b) Calculate the desired auto-correlation function from the desired spectrum.
(c) Calculate the necessary input auto-correlation function from the known output auto-correlation function and the known non-linearity.
(d) Design the linear filter from known input and output correlation functions.

The calculation of the non-linear curve is straightforward.

$$dz = \frac{p(y)}{p(z)}\,dy,$$

with $p(y)$ and $p(z)$ respectively the gaussian and the desired probability density. The non-linear gain follows from dz/dy. In most cases it is reasonable to assume that the non-linearity can be approximated by a power series

$$f(y) = \sum_{i=1}^{N} a_i y^i.$$

This simplifies the calculation of the auto-correlation function at the input of the non-linearity.

Price's theorem applied to the configuration of Fig. 9, implies [8]:

$$\frac{\partial^k \Phi_{zz}(\tau)}{\partial \Phi_{yy}^k(\tau)} = E[f^{(k)}(y_1) f^{(k)}(y_2)].$$

The output auto-correlation function can be written as a power series development of the input

$$\Phi_{zz}(\tau) = \sum_{i=1}^{N} C_i^2 \Phi_{yy}^i(\tau)$$

Thus

$$\frac{\partial^k \Phi_{zz}(\tau)}{\partial \Phi_{yy}^k(\tau)} \, k! \, C_k^2.$$

The mathematical expectation can be calculated by substituting the power series development for the non-linearity and the gaussian properties of the signal y.

As Price's theorem holds for all values of the correlation function it is advantageous to take $\Phi_{yy}(\tau) = 0$. It follows then

$$k! C_k^2 = \frac{1}{2\pi} \left[\int_0^\infty f^{(k)}(y) \exp\left(-\frac{y^2}{2}\right) dy \right]^2.$$

This allows us to compute C_k^2 by standard series approximations. If ϕ_{yy} is calculated in this way from the desired Φ_{zz} one has to design the linear filter.

$$H(z) = \frac{1}{A(z^{-1})}.$$

This is a recursive filter of infinite order which has to be approximated by a filter of finite order n.

Evaluation of the coefficients in the difference equation

$$y(t) + a_1 y(t-1) + \cdots + a_n y(t-n) = x(t)$$

goes by multiplication with $y(t-k)$ and averaging. This gives a set of linear equations in a_1 to a_n with the calculated correlation points Φ_{yy} as coefficients. Simulation on a digital computer showed that in many cases the non-linearity gives only minor changes in the power density spectrum [9].

REFERENCES

1. Slepian A. and Pollak, H. O. (1961). Prolate spheroidal wave functions. Fourier analysis and uncertainty I. *Bell System Tech. Jl.*
2. Burg, J. P. (1967). Maximum entropy spectral analysis. Presented at the 37th Annual Meeting Soc. Explor. Geophys.
3. Flammer, C. (1957). "Spheroidal Wave Functions", Stanford University Press, Calif. 1957.
4. De Bruine, R. F. (1968). "The extrapolation of truncated auto-correlation functions of bandwidth limited signals with prolate spheroidal wave functions" (in Dutch). Thesis. Department of Appl. Physics; Delft University of Technology October 1968.
5. Shannon, C. E. and Weaver, W. (1949). "The Mathematical Theory of Communication," pp. 54–57, Univ. of Illinois Press.
6. Åström, K. J. and Bohlin, T. (1966). Numerical identification of linear dynamic systems from normal operating records. Proc. IFAC Symp. on Self Adaptive Control Systems.
7. Van den Bos, A. (1971). Alternative interpretation of maximum entropy spectral analysis. *IEEE Trans. Information Theory.*
8. Price, R. (1958). A useful theorem for nonlinear devices having Gaussian input. *IRE Trans. Information Theory.*
9. Verloren, P. (1968). "The generation of stochastic signals with specified probability densities and auto-correlation functions" (in Dutch). Thesis. Department of Appl. Physics; Delft University of Technology.

An Adaptive Pre-whitening Filter for Radar Clutter Suppression

WALTER BÜHRING

Forschungsinstitut für Funk und Mathematik
Wachtberg-Werthhoven, Germany

Abstract

Surveillance radars have to cope with doppler shifted echoes of weather and sea clutter, varying in azimuth and time. MTI technique is a well known form of pre-filtering for ground clutter echoes. A time series as obtained from bins of constant range during rotation of the antenna is weighted by certain filter weights to form a tapped delay line filter. When the Toeplitz correlation matrix of the clutter process is known, the filter weights of the pre-whitening filter are given by the first column of the inverse of the correlation matrix (Wiener Theory).

In the case of slowly varying clutter the filter weights have to be estimated from the clutter data. J. P. Burg (A new analysis technique for time series data, Nato Advanced Study Institute on Signal Processing, Enschede, 1968) gives a stepwise algorithm to obtain the filter weights from a given data set in a way, that the power of the filter output process is minimized for this data set. The obtained filter corresponds to an estimate of the Toeplitz correlation matrix of a stationary clutter process.

For real-time data processing the algorithm of Burg can be approximated by a recursive algorithm with an exponential decreasing memory of the past. The time constant of the recursive solution controls the rate of convergence of the filter weights to the Wiener weights of the stationary process and controls the variances of the weights. The estimation procedure is stable even with very low time constants, but with a decreasing time constant the deterioration of a target signal superposed to the clutter increases by strongly incorporating the target into the filter estimation and by the increased variance of the filter weights.

The described algorithm is different to the filter estimation procedure given by Widrow *et al.* (Adaptive antenna systems, *Proc. IEEE* 55, No. 12, 2143–2159) which has attracted much interest in literature.

The Widrow algorithm is stable only up to a certain time constant, given by the eigenvalues of the normally unknown correlation matrix of the nonstationary clutter process. Although the Widrow algorithm is much easier to implement it fails in an environment where short time constants are required.

DISCUSSION

H. Mermoz: Am I right, that your system works continuously on clutter? So, when a target is superposed to clutter, you take account of the fact, that the target signal is much shorter than the clutter signal, in order to avoid that the spectrum of the target signal is whitened as well?

Answer: Yes, the time constant of the system, determined by the factor F, has to be chosen in a way, that the filter can adapt to a slowly varying characteristic of the clutter but not seriously affects the short target signal.

Time Domain Filter Methods

R. E. BOGNER

Electrical Engineering Department,
Imperial College, London, England

1. Introduction

Digital processing of signals is reaching the stage of maturity where it is usually considered along with other (analogue) methods when a signal processing task is at hand.

Special qualities of digital processors which are sometimes decisive include:

(a) Precision and reproducibility, permitting very complex systems to be practical, and also very low-frequency filters of precise characteristics.

(b) Flexibility, including the possibility of controlling the value of every element, and of changing function without changing hardware.

(c) Compatibility with integrated circuit technology.

(d) The speed of operation is limited, and will probably not permit processing of signals of bandwidths greater than a few tens of MHz.

This lecture aims to give a feel for the essential fundamental ideas involved in linear digital filters in which signals are processed by direct manipulation of sample values; a later lecture considers filtering by Fourier analysis and operations on the frequency components.

(a) What is a filter?
"Filtering" includes many aspects:

(a) We have filters whose purpose is to stop waveforms of one class while passing those of another class. Examples are the frequency filters used in frequency division multiplex transmitters and receivers—these operate to stop or pass sinusoids. Similarly, filters may be used to stop or pass other waveforms used for multiplexing, e.g. Walsh functions, pseudorandom sequences, chirps. Moving target indication filters have to reject signals whose characteristics do not change with time, but pass other signals;

such filters are usually described by frequency selective characteristics. Spectrum analysis filters, also, pass only sinusoids of certain frequencies.

Sampling circuits used in reception of time division multiplexed signals are formally time-selective filters, but it is not usual to consider such systems as filters.

(b) "Optimum" filters are intended to modify an input waveform, usually containing signal and noise, to enhance some desired characteristic. The formulation is often in statistical terms. Examples are:

(i) Filters for minimum mean square error (noise power) estimation of the present or future value of the input signal (Wiener and Kalman-Bucy filters).

(ii) Filters for maximizing the "visibility" of a given waveform in the presence of noise—matched filters.

(iii) Filters for maximizing the probability of correct detection of a pulse in the presence of other pulses and noise. The pulses may have been distorted by a transmission medium.

Such optimum filters may be made adaptive to accommodate changes in the characteristics of the signals or noise, perhaps due to changes in the medium. Several simultaneous input waveforms may be processed simultaneously as in processing for array steering, and diversity reception.

(c) Other filters are required to introduce specific modifications in a transmission characteristic, e.g.

(i) Equalizers to compensate for amplitude and phase characteristics of a transmission medium, before or after transmission. The characteristics of such a filter are often not spectacular, but may need to be carefully controlled. Inverse filters or deconvolvers, used for improving resolution in optical systems, or separating excitations from systems in geophysics and acoustics fall in this class.

(ii) Generation of signalling waveforms for, say, data transmission, sonar or radar.

(iii) For speech synthesis as in vocoders, helium speech processors, audio response systems.

(b) Numerical processing

The fundamental differences between analogue (or continuous) and numerical (or digital) signal processing come about from the essentially discrete nature of the signal representations in the latter, i.e.

(i) Values are represented numerically with a finite number of digits, and thus can take only discrete values, i.e. they are quantized. The numbers can be "fixed point", i.e. with fixed quantization steps (Fig. 1a), or floating point, with bigger steps for bigger values (Fig. 1b). Logarithmic representations of signals are sometimes used too (Fig. 1c). Some systems use delta modulation techniques for digitizing, but these are of limited applicability because of limitations placed on the arithmetic [1].

These effects are usually only a source of error; always they can be reduced arbitrarily by increasing the number of digits.

(ii) The signals are represented by a finite number of values (called samples), usually being the values of a continuous signal taken at equal increments of time (the sampling interval, T). This results in all sampled-data systems exhibiting periodic frequency characteristics, as we discuss later (Section 3(b)).

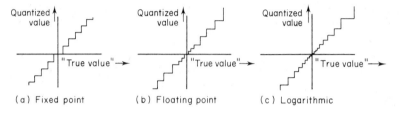

FIG. 1. Value representations.

Most signals outside of signal processors are in analogue form; for digital processing it is necessary to carry out analogue to digital (A–D) conversion; this involves sampling and conversion of each sample value to a numerical form.

2. Some Fundamental Properties and Limitations of Linear Signal Processors

All signal filters have to do something about the "shape" of the input waveform—e.g. accept components of some class while rejecting others, respond to rate of change, carry out smoothing. Thus they have to take into account values of input other than the present one—they have memory. This comes about in analogue filters through the energy storage in inductors and capacitors. The corresponding memory in digital processors is in the registers (or memory locations or "delays") used to store values of the signals during the intervals between samples. There is a close correspondence between the degrees of freedom of both types, as given by the number of storage elements, and the complexity of the task. Each storage element introduces a pole (unless it is used in a trivial sense, e.g. one capacitor in parallel with another), i.e. there are as many poles as there are storage elements.

For example, a chirp filter matched to a chirp of dispersion (bandwidth) W Hz and duration τ seconds (Fig. 2a) has to be capable of absorbing $2\tau W$ samples, and each of these eventually contributes to the peak response.

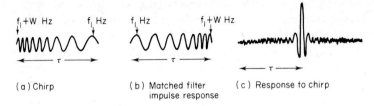

(a) Chirp   (b) Matched filter   (c) Response to chirp
                       impulse response

FIG. 2.

In this context we should notice some important economic differences between analogue and digital systems.

Storage in digital systems is very cheap and precise; it is the arithmetic operations which are expensive, in hardware and in time. Conventional analogue systems involve expensive, precise storage (C, L or C with amplifiers), but the addition and scaling are inconsequential and almost instantaneous. Hence, it is reasonable to consider very complicated (e.g. 100 pole) digital filters, but it is desirable to find architectures which economize on the amount of computation e.g. Fast Fourier Transforms, and sequential convolution [2, 3].

The absolute accuracy and losslessness of the storage in digital systems contributes to the practicality of complex systems. However, the effect of finite word length in producing roundoff errors in arithmetic operations has to be borne in mind (Section 7).

3. Elementary Characteristics of Digital Filters

The operations which we consider in digital filters are:

(a) Addition: ⊕→
(b) Multiplication by constants: →▷B→
(c) Storage for one clock period; this is frequently referred to as delay by a fixed time T: →⊔T⊔→   →⊔z^{-1}⊔→

These can be compared with the operations available in analogue systems:

Addition (e.g. $v = R_1 i_1 + R_2 i_2$)
Multiplication by constants (e.g. $v = Ri$ etc.)

Integration or differentiation $[$e.g. $v(t) = \dfrac{1}{C} \displaystyle\int_0^t i(t)\ \mathrm{d}t]$

In analogue systems the resultant equations are linear integro-differential equations, and the solutions are continuous functions of time. The solutions are always exponentially damped sinusoids characteristic of the system, plus forced responses characteristic of the inputs. Superposition applies, i.e. the system response to several simultaneous inputs is the same as the sum of the responses to the inputs applied separately.

We will find similar behaviour in digital filters, with the exception that the responses change only at discrete values of time. The first order (single delay) recursive (i.e. with feedback) system of Fig. 3, illustrates some of the behaviour.

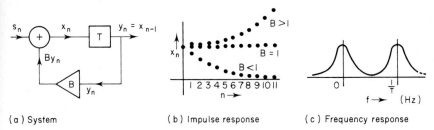

(a) System   (b) Impulse response   (c) Frequency response

FIG. 3. First order system.

The system operates as follows. Each clock period (i.e. each T second), a new input "sample" s_n is added to the existing value at the other input of the adder to produce the new x_n:

$$x_n = By_n + s_n. \tag{1}$$

The "output" of the store or delay $\boxed{T}$ is the previous value of x_n, i.e. x_{n-1} which was produced in the previous clock period. This is the output y_n, and is multiplied by B, to be added to the next input, i.e.

$$x_n = Bx_{n-1} + s_n. \tag{2a}$$

This can be arranged in the form

$$\Delta x_n = (B - 1)x_{n-1} + s_n, \tag{2b}$$

where

$$\Delta x_n = x_n - x_{n-1},$$

the first difference of x_n. Thus (2a) and (2b) are difference equations. Such difference equations have the same role for digital filters as that which differential equations have for analogue filters.

We examine the impulse response, i.e. the response to a single input sample of magnitude 1: Let the system be in the zero state, i.e. with the value 0 held in the delay. The response to the unit sample (arbitrarily applied at $n = 0$) is readily found by studying the passage of the resulting disturbance around the system:

n	-1	0	1	2	3	n
s_n	0	1	0	0	0	0
x_{n-1}	0	0	1	B	B^2	B^{n-1}
$x_n = s_n + Bx_{n-1}$	0	1	B	B^2	B^3	B^n

It is now evident that the impulse response is a *geometric progression* (Fig. 3b); this is easily shown by substitution to be the characteristic solution of (1) with the forcing function $s_n = 0$. Such a response is equivalent to the exponential response of the first order analogue system, but sampled, i.e. the values

$$1, B, B^2, B^3 \ldots B^n$$

and

$$e^0, e^T, e^{2T}, e^{3T} \ldots e^{nT}$$

are the same, if

$$e^T = B.$$

Similarly, higher order systems have characteristic functions or eigenfunctions which are geometric progressions, of the form p^n where there

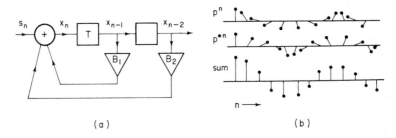

(a)   (b)

FIG. 4. Second order system.

are in general as many values of p as there are delays. The values of p for solutions of the difference equation are real, or in complex conjugate pairs if the coefficients are real. The impulse responses consist of sums of these characteristic functions. An example illustrates: The difference equation of the second order system (Fig. 4) is:

$$x_n = B_1 x_{n-1} + B_2 x_{n-2} + s_n. \tag{3}$$

We hypothesize a solution of the form $x_n = p^n$, i.e. $x_{n-1} = p^{n-1}$ etc. Hence, for zero input, we have the characteristic equation

$$x_n = p^{-1} B_1 x_n + p^{-2} B_2 x_n \tag{4}$$

and for a nontrivial solution (i.e. $x_n \neq 0$)

$$p^2 - p B_1 - B_2 = 0. \tag{5}$$

In the same pattern, higher order systems yield polynomial equations in p, whose roots are real or in complex conjugate pairs if the B's are real. For real roots, the values of x_n are, as before, equivalent to samples of exponentials. When the roots are complex the solutions are equivalent to samples of complex exponentials. Occurring in conjugate pairs, these yield damped sinusoids, e.g. a pair of roots

$$p = 0.78 \pm j0.55$$

yields the responses shown in Fig. 4b.

Notice that while all values are real in real hardware it is simple to make systems with complex values represented within them by pairs of real values. Hence digital filters with complex coefficients, and complex valued-signals are practical (References [4, 5]),

The values of p found above are eigenvalues of the system; they specify the poles of the system response, and will be discussed again later in the section on pulse transfer functions and z transforms.

(a) Linearity; superposition; convolution

The difference equations of the digital processors being discussed are linear, i.e. each term is proportional to one signal value, or is constant. As with linear analogue systems, this allows use of the principle of superposition, i.e. the system response to input $s_{1n} + s_{2n} + \cdots$ is the same as $x_{1n} + x_{2n} + \cdots$ where x_{1n} is the response to s_{1n} etc. This property follows readily from the difference equation e.g. (3). Thus if there are two inputs s_{1n} and s_{2n}, yielding solutions (responses) x_{1n} and x_{2n} respectively, we have

$$x_{1n} = B_1 x_{1,n-1} + B_2 x_{1,n-2} + s_{1n}$$

and

$$x_{2n} = B_1 x_{2,n-1} + B_2 x_{2,n-2} + s_{2n}$$

and hence

$$x_{1n} + x_{2n} = B_1 (x_{1,n-1} + x_{2,n-1}) + B_2 (x_{1,n-2} + x_{2,n-2}) + s_{1n} + s_{2n}$$

is the response corresponding to the superposition of the inputs s_{1n} and s_{2n}.

From superposition follows the convolution, the general expression for the output of a linear system in terms of the input and the impulse response.

This is readily derived for a system by reference to Fig. 5. Any impulse response may be realized by the system shown, if enough delays are allowed. A unit input sample, preceded and succeeded by zeros results in outputs $h_0, h_1, h_r \ldots$ as it moves from delay to delay. Thus, the impulse response is $h_r, r = 0, 1, \ldots$. A general input sequence, $x_n, n = \ldots -2, -1, 0, 1, 2 \ldots$ results in the delayed weighted contributions shown. The output is evidently

$$y_n = \sum_{r=0}^{\infty} h_r x_{n-r} \tag{6a}$$

$$= \sum_{r=-\infty}^{n} h_{n-r} x_r \tag{6b}$$

which is often written

$$y_n = h_n * x_n. \tag{7}$$

More generally, if the impulse response is itself dependent on the time of excitation, it has two arguments, n and r, i.e. $h_{n,r}$ can describe the output at time n to an input at time r. The resultant output at time n to a general input sequence x_n is

$$y_n = \sum_r h_{n,r} x_r. \tag{8}$$

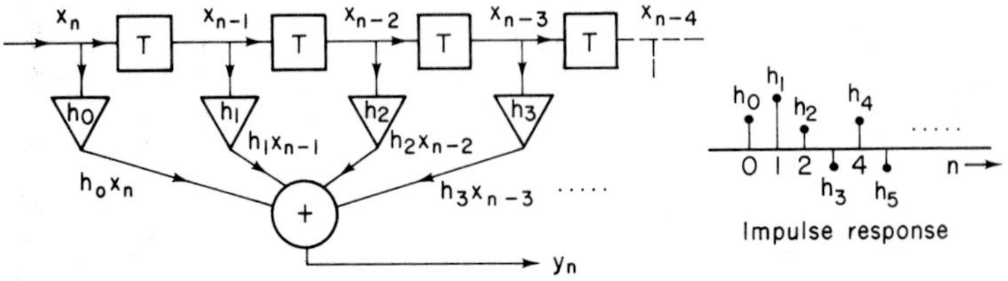

FIG. 5. Derivation of convolution.

(b) Frequency characteristics

In many applications, we are interested in the frequency responses (amplitude and phase) of filtering systems. We study this property by finding the response to sampled sinusoids of various frequencies, as is done for analogue systems with continuous inputs. It is convenient to consider the response to complex exponentials of the form $\exp(j\omega nT)$; the response to a real input $\cos \omega nT$ can then be found if desired by superposing the response to $\exp(-j\omega nT)$ since

$$2 \cos \omega nT = \exp(j\omega nT) + \exp(-j\omega nT) \tag{9}$$

but this step is usually replaced by the equivalent one of taking the real parts (such a procedure is valid if the coefficients of the system are real).

To find the frequency characteristic of the system whose impulse response h_n, $n = 1, 2 \ldots$ is as shown in Fig. 5, we apply $\exp(j\omega nT)$. Each of the samples h_n results in a delayed replica of the input, scaled by the value h_n, and we add these replicas:

h_0 results in output $h_0\, e^{j\omega nT}$

h_1 results in output $h_1\, e^{j\omega (nT-T)} = h_1\, e^{-j\omega T}\, e^{j\omega nT}$

h_2 results in output $h_2\, e^{j\omega(nT-2T)} = h_2\, e^{-j\omega 2T}\, e^{j\omega nT}$

h_r results in output $h_r\, e^{j\omega(nT-rT)} = h_N\, e^{-j\omega rT}\, e^{j\omega nT}$

i.e. the input $e^{j\omega nT}$ results in the output

$$\left(\sum_{r=0}^{\infty} h_r\, e^{-j\omega rT} \right) e^{j\omega nT}. \tag{10}$$

The input is thus modified in amplitude and phase by the complex number

$$H(j\omega) = \sum_{r=0}^{\infty} h_r \, e^{-j\omega rT}. \tag{11}$$

This is the general expression for the complex transfer function of the system characterized by h_n, $n = 0, 1, 2, \ldots$ at the frequency of ω radians per second. Notice that the expression (11) for $H(j\omega)$ is the correlation of h_r and $e^{j\omega rT}$; i.e. it is an indication of the similarity between these two, or it tells how big is the contribution of $e^{+j\omega rt}$ to h_n. A later lecture on frequency domain techniques describes the use of this resolution to effect filtering by weighting the $H(j\omega)$.

Example Consider the system whose impulse response is given by:

$$h_0 = \tfrac{1}{2}, \quad h_1 = 1, \quad h_2 = \tfrac{1}{2}.$$

The transfer function is

$$\begin{aligned}
H(\omega) &= \tfrac{1}{2} + e^{-j\omega T} + \tfrac{1}{2} \, e^{-j2\omega T} \\
&= e^{-j\omega T} \left(\tfrac{1}{2} \, e^{+j\omega T} + 1 + \tfrac{1}{2} \, e^{-j\omega T} \right) \\
&= e^{-j\omega T} \left(1 + \cos \omega T \right).
\end{aligned}$$

We note that the amplitude is modified by the periodic term $(1 + \cos \omega T)$ and the phase is modified by the periodic term $\exp(-j\omega T)$. In this case the phase, $-\omega T$, is proportional to frequency—i.e. the system is phase linear, corresponding to the impulse response being symmetrical about the central value h_1.

The *periodicity of the transfer function* in the frequency we have just noted is true in general, as we can see from (11); the period is $2\pi/T$ rad./sec or $1/T$ Hz. This property is a result of the sampled nature of the impulse response, and may be appreciated in a different way as follows. Figure 6(a) shows a continuous exponential impulse response, and the corresponding nonperiodic frequency characteristic is shown in Fig. 6(b). Figure 6(c)

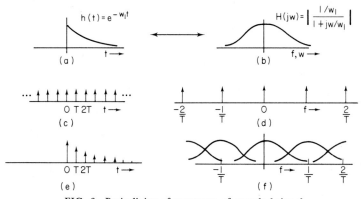

FIG. 6. Periodicity of spectrum of sampled signal.

shows a sequence of impulses $i(t)$ by which $h(t)$ can be multiplied to obtain the sampled version signal shown in Fig. 6(e). The frequency spectrum of the sequence of impulses is shown in Fig. 6(d).

When $h(t)$ is multiplied by $i(t)$, each frequency component of $i(t)$ is modulated by $h(t)$, resulting in a sideband pattern repeated at intervals of $1/T$ Hz, Fig. 6(f). Thus, any sampled signal has a periodic spectrum. For accurate representation of a continuous waveform, it is necessary to choose the sampling rate high enough to avoid significant aliassing effects, due to the overlap of the tails of the spectra Fig. 6(f), and this is achieved if the signal spectrum is negligible at frequencies greater than $1/2T$ Hz. We might note that for some special classes of signals this is not a necessary condition [6, 7].

4. z Transforms; Pulse Transfer Function

The convolution equation (6) is the general expression relating output to input and impulse response. It is cumbersome and usually inconvenient for analytical work. As with continuous systems, where we use the Laplace transform, a transform is useful; in this case it is the z transform. Use of such transforms results in multiplication instead of convolution:

	Input	System	Output	
Convolution:	x	$*$ h	$=$ y	
Transformation:	$\downarrow$	$\downarrow$	$\uparrow$	Inverse
Multiplication:	X	$\cdot$ H	$=$ Y	Transformation

As in the case of Laplace transforms, $X(s)$,

$$X(s) \underset{\text{def.}}{=} \int_0^\infty x(t)\, e^{-st}\, dt \qquad (12)$$

where the transform kernel e^{-st} is a general form of the characteristic response of the continuous linear systems, we use a z *transform* whose

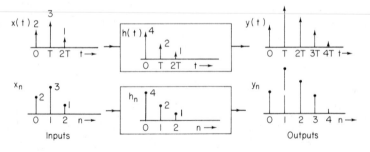

FIG. 7.

kernel is the general form the characteristic response of discrete linear systems, i.e. z^n where z is a general complex number:

$$X(z) \underset{\text{defn.}}{=} \sum_{n=0}^{\infty} x_n z^{-n}. \tag{13}$$

This defines the z transform $X(z)$ corresponding to the sequence x_n.

An example shows the parallel between Laplace and z transforms: Figure 7 shows continuous and discrete system responses and inputs. By convolution the discrete system output is found to be:

$$y_0 = 8, \quad y_1 = 16, \quad y_2 = 12, \quad y_3 = 5, \quad y_4 = 1$$

and the transform manipulations as follows:

Laplace Transform	*z Transform*

$X(s) = 2 + 3e^{-sT} + 1\ e^{-2sT}$

$H(s) = 4 + 2\ e^{-sT} + 1\ e^{-2sT}$

$Y(s) = H(s)X(s) =$

$\qquad 2.4\ e^{-0T}$

$\qquad + (4.3 + 2.2)\ e^{-sT}$

$\qquad + (4.1 + 2.3 + 1.2)\ e^{-2sT}$

$\qquad + (2.1 + 1.3)\ e^{-3sT}$

$\qquad + 1.1\ e^{-4sT}.$

$X(z) = 2 + 3z^{-1} + 1z^{-2}$

$H(z) = 4 + 2z^{-1} + 1z^{-2}$

$Y(z) = 8 + 16z^{-1} + 12z^{-2}$

$\qquad + 5z^{-3} + 1z^{-4}$

We note that z transform of a sequence is equivalent to the Laplace transform of a sequence of delta functions occurring at times nT of the same "values", with z^{-n} substituted for e^{-snT}. The validity of the z transform, which so far is no more than a method of book-keeping, depends on the fact that the convolution expression,

$$y_n = \sum_r h_r x_{n-r} \tag{14}$$

is the same as the expression for the coefficient of z^{-n} in the product of the polynomials

$$(h_0 + h_1 z^{-1} + h_2 z^{-2} + \cdots)(x_0 + x_1 z^{-1} + x_2 z^{-2} + \cdots).$$

The z transform $H(z)$ of the system impulse response,

$$H(z) = \sum_{n=0}^{\infty} h_n z^{-1} \tag{15}$$

is called the *pulse transfer function*. Notice that this expression is the same as that obtained for $H(j\omega)$, the transfer function at ω rad./sec, with $z = e^{j\omega T}$.

Example—an important building block, the unit delay. The unit delay, Fig. 8 causes an input signal to be delayed by one sampling interval, i.e. $h_0 = 0$, $h_1 = 1$, i.e. $H(z) = z^{-1}$. Then:

$$\text{input} = x_n$$
$$\text{output} = y_n = x_{n-1}$$

$$\therefore \quad Y(z) = \sum_n y_n z^{-n} = \sum_n x_{n-1} z^{-n} = z^{-1} \sum_n x_n z^{-n} = z^{-1} X(z)$$

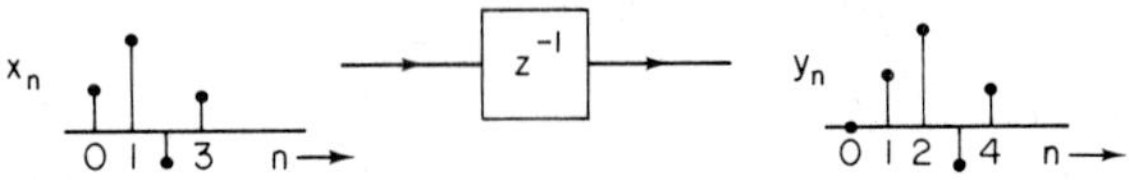

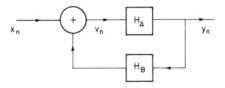

FIG. 8. Unit delay element.

(a) Use of transfer functions: a recursive filter

A general recursive (with feedback) system is shown in Fig. 9.

FIG. 9. General recursive system.

The overall pulse transfer function $H(z) = Y(z)/X(z)$ is found in a manner similar to that used for feedback in continuous systems:

$$V(z) = X(z) + H_B(z)Y(z)$$
$$Y(z) = H_A(z)V(z)$$

$$\therefore \quad H(z) = \frac{Y(z)}{X(z)} = \frac{H_A(z)}{1 - H_A(z)H_B(z)} \tag{16}$$

Example—first order recursive system (Fig. 3a). Using (16) for the first order system, in which $H_A(z) = z^{-1}$, $H_B = B$:

$$H(z) = \frac{z^{-1}}{1 - Bz^{-1}} = z^{-1}(1 + Bz^{-1} + B^2 z^{-2} + \cdots)$$

if $|Bz^{-1}| < 1$, by binomial expansion. Hence,

$$h_n = B^{n-1}, \quad n > 0 \tag{17}$$

which is seen to be the same as we obtained by physical reasoning in Section 3.

We saw earlier that the frequency characteristics of a digital filter are obtained by setting $z = e^{j\omega T}$ in the transfer function. For the present example, we find

$$H(j\omega) = \frac{e^{-j\omega T}}{1 - Be^{-j\omega T}}$$

which is shown in Fig. 3c; the system is a (repeated) low pass filter, which might be expected from the nature of its impulse response. Infinite response occurs for ω such that $Bz^{-1} = 1$, i.e. $z = B^{-1}$ and zero response at $z = \infty$; both nonreal frequencies.

Example—second order recursive system (Fig. 10a). This system is of considerable practical significance as it is frequently used as the basic building block in complicated filters. It can contribute a complex conjugate pole pair or two real poles like an inductor-capacitor-resistor combination, and two zeros.

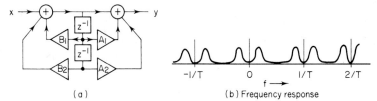

(a)

(b) Frequency response

FIG. 10.   Second order system.

For this system by using (16) with $H_A = 1$ and $H_B = B_1 z^{-1} + B_2 z^{-2}$,

$$H(z) = \frac{1 + A_1 z^{-1} + A_2 z^{-2}}{1 - B_1 z^{-1} - B_2 z^{-2}}$$

$$= \frac{1 + A_1 z^{-1} + A_2 z^{-2}}{1 - 2z^{-1} e^{-\alpha T} \cos \omega_1 T + z^{-2} e^{-2\alpha T}} \tag{18}$$

where we have substituted $e^{-2\alpha T} = -B_2$ and $e^{-\alpha T} \cos \omega_1 T = B_1$. The corresponding impulse response, h_n [found e.g. by using partial fractions and binomial expansion as in (17)] is

$$h_n = C e^{-n\alpha T} \cos (n\omega_1 T + \theta) \tag{19}$$

and the frequency response, obtained by putting $e^{j\omega T} = z$, is of the periodic form shown in Fig. 10(b). Filters of arbitrary characteristics may be produced by cascading sections with characteristics of this type.

There is the possibility that recursive systems can be unstable, i.e. with pulse responses that increase exponentially. This is not a problem in designing filters from their pole-zero descriptions [Section 4(b)], provided these are realizable, but approximations introduced by the discreteness of the coefficients (B's) may result in unstable realizations, together with modification of the designed characteristics.

(b) z plane; poles and zeros

It is convenient, mainly for conceptual purposes, to display the values of z for which pulse transfer functions become infinite or zero (poles and zeros) graphically in the complex z plane, e.g. (Fig. 11):

$$H(z) = \frac{z(z - 0.7 - j0.7)(z - 0.7 + j0.7)}{(z - 0.8)(z - 0.9 - j0.087)(z - 0.9 + j0.087)}. \tag{20}$$

Multiplication of z transforms then corresponds to superposition of their pole zero patterns. The *unit circle* $|z| = 1$ (Fig. 10a), i.e. $z = e^{j\theta}$, or $z = e^{j\omega T}$ is the locus of values of z corresponding to sampled sinusoidal excitation at frequency ω—compare equations (11) and (15). The values of $H(j\omega)$ as given by (20), with $z = ^{j\omega T}$ are simply related to the vectors indicated in the numerator and dominator factors.

The unit circle has the same role as the $j\omega$ axis in the s-plane of Laplace transforms. The substitution $z = e^{sT}$ indicated in Section 4 defines a mapping between the s and z planes. An important feature of the z plane is that the periodicity of response in frequency [Section 3(b)] corresponds to successive circuits of the unit circle.

5. Recursive and Non-Recursive

We review the essential features of two broad classes of digital filter in Table I.

TABLE I

Non-Recursive (no feedback)	Recursive (with feedback)
1. Very simple structure (Fig. 5)	•Most general structure may include non-recursive paths (Fig. 10a).
2. Design (determination of coefficients) is very simple, especially for time-domain specifications Responses and coefficients are related linearly; hence linear programming[8, 9].	•Design more sophisticated; non-linear relations exist between coefficients and response.
3. Always stable. Readily adjusted as for adaptive equalization.	•Can be unstable.
4. Impulse response is of finite duration—not greater than the number of delays, plus one.	•Impulse responses are infinite geometric (exponential) sequences of sums of these (Fig. 4b).
5. Impulse response can be arbitrarily long to achieve required resolution, but may be expensive in computations.	•Often more efficient, in the sense of frequency selectivity for a given complexity.
6. Easily made phase-linear (time-symmetrical impulse response).	•Only special classes are phase linear.

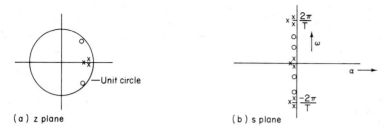

FIG. 11. *s* and *z* planes, poles and zeros.

6. Multidimensional Filters

Usually higher dimensional filters are associated with special problems, e.g. image processing. The impulse response becomes a forest of samples in N dimensions rather than a row of samples in 1 dimension, and the number of computations can be very great unless the desired response is of some special class. A general filter can be specified by its impulse response, e.g. in 2 dimensions, $h_{m,n}$. The output, obtained by convolution is

$$y_{m,n} = \sum_r \sum_s h_{r,s} x_{m-r,n-s}. \tag{21}$$

If $h_{r,s} = h_r h_s$, then (21) results in two one-dimensional convolutions which can be carried out sequentially:

$$y_{m,n} = \sum_r h_r \sum_s x_{m-r,n-s} h_s. \tag{22}$$

In such cases, the number of multiplications per output sample, is $M + N$ rather than MN, where M and N are the length of $h_{m,n}$ in the M and N directions.

Often we desire circularly symmetric impulse responses; this means that $h_{r,s}$ is a function of $\sqrt{(r^2 + s^2)}$ and if we also desire $h_{r,s} = h_r h_s$, then $h_{r,s}$ has to be of the form

$$h_{r,s} = C_1 \exp \{C_2[r^2 + s^2]\} \tag{23}$$

i.e. 2 dimensional gaussian. Approximations are often used, which are not truly symmetrical, but give reasonable subjective effects. For example, an exponential pyramid, which is easily realized recursively.

Two dimensional recursive filters are significantly more efficient than non-recursive ones, but the design and stability criteria are as yet in their infancy [10].

7. Quantization Effects [11]

Because of the finite number of digits (almost always binary, hence "bits")
in a digital filter, there are usually approximations in the representations of
the quantities, i.e. signals and coefficients. These result in the following
effects:

(i) *Quantization noise at the input*—in the case where an analogue
signal is converted to digital form (Fig. 1). Usually for the case of uniform
quantization it is a good approximation to regard this noise as being
independent of the signal, and of (two-sided) spectral density $TE^2/12$
where E is the height of a quantizing step. For floating point and logarithmic systems the effective step height depends on the actual signal value.
Thus the short term noise average power depends on the signal, and the
long term average depends on the signal probability density function.

(ii) *Quantization noise at the output of each multipler*—due to the
increase of the number of digits in a multiplication, and subsequent
truncation to the original wordlength. The effect is similar to quantization
of the input, and can be represented by an additive noise source at the
output of each multiplier, and of spectral density $TE^2/12$. For floating
point number systems the effects are more complicated, as is the case
for quantization noise at the input.

Noise transmission through digital filters can usually be treated by
spectral descriptions, i.e. the spectral density is weighted by the squared
magnitude of the transfer function of frequency, $|H(j\omega)|^2$.

Current practice favours the realization of complex filters as cascades
of second order sections, largely to reduce problems of coefficient precision. In such structures, it tends to be advantageous to place the narrow
bandwidth or lightly damped sections late in the cascade, and to realize
in the same section poles and zeros which are close together in the z plane.

(iii) *The coefficients* (A's and B's, Fig. 10) can take only values permitted
by the number system. They will usually be slightly different from the values
determined in a design procedure to realize a given specification. The
resultant responses may be very different from those for exact coefficients,
especially so for recursive systems, which may become unstable. Generally,
the lower the damping of a resonance, the more critical are the relevant
coefficients.

Example. Consider a first order recursive system (Fig. 3) in which the
coefficient B is quantized to the nearest 0.01, i.e. B can take only the
values $\pm(0, 0.01, 0.02, 0.3 \ldots 0.99, 1.0)$. Two separate specifications and
the resultant characteristics may be compared (Table II).

Generally, it is found that filters composed of cascaded second order
sections are less critical than ones with higher order recursive sections.
However, there is hope that some structures may be found where the
opposite is true (References 13, 14).

TABLE II

Case:	Value of B	Impulse Response	3dB Bandwidth, rad/s	Response at zero frequency
(a) Specification	0.997	$e^{-0.003n}$	$0.003/T$	333.3
Achieved	1.00	1.0000	0	∞
(b) Specification	0.977	$e^{-0.023n}$	$0.023/T$	43.5
Achieved	0.98	$e^{-0.02n}$	$0.02/T$	50.0

Many techniques have been described for finding coefficients to satisfy given specifications, analytically and by automatic optimization (References [8, 9, 12]. However, discrete optimization, taking into account the quantization of coefficients is as yet in its infancy and results are only just starting to appear (Reference [15]).

(iv) *Low level limit cycles* can occur because an effective multiplier changes when the multiplicand becomes small enough so that roundoff errors are systematic. Figure 12 suggests that the effective slope (equivalent to describing function) for a true product of ±0.5 units is 2 while that for a true product of ±1.5 units is 2/3. The nominal or ideal value is unity.

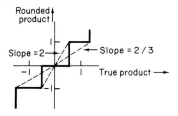

FIG. 12.

Example. A second order resonator (Fig. 4) with $B_1 = 0.6$ and $B_2 = -0.8$ has the result of each multiplication rounded to the nearest integer. The response to an isolated input sample of 3 is readily found by detailed arithmetic to result in a sustained output sequence of the form +1, +1, 0, −1, −1, 0, +1, +1, 0, ... An input of 3000 would result in a slightly noisy decaying exponential $3000\, e^{-0.89n} \cos 0.53n$ until the output decreased to be of the order of 3.

8. Modulo-Linear Systems

In this section we note some developments which might change the pattern of digital filter progress.

Arithmetic operations in digital processors are expensive, in time and hardware. Residue number systems [16] have been shown to have some advantages in this regard, allowing multiplications and additions to be performed by several simultaneous, smaller units without carry. They also offer improved reliability through simple redundancy.

Residue number systems are based on modulo arithmetic. In this, two numbers, a and b are said to be congruent, modulo M, i.e.

$$a \equiv b \pmod{M} \tag{24}$$

where

$$a = b + rM \tag{25}$$

where r is an integer. The useful property of such congruences is that they obey superposition, i.e. if

$$a \equiv a' \pmod{M} \quad \text{and} \quad b \equiv b' \pmod{M} \tag{26}$$

then

$$a + b \equiv a' + b' \pmod{M}. \tag{27}$$

Also,

$$ab \equiv a'b' \pmod{M}. \tag{28}$$

We usually need consider only numbers in the range 0 to M, and all sums are taken modulo M, which we write in the form $a \oplus b$. For example if the modulus M were 5 and $a = 4$, and $b = 3$ then

$$a \oplus b = 4 \oplus 3 = 2 \tag{29}$$

or

$$(4 + 3) \bmod 5 \equiv 2 \tag{30}$$

"Modulo Linear" systems, obeying such superposition rules have overall behaviour patterns similar to those with conventional linearity, except that unstable responses are bounded. Equalizers based on these concepts have been proposed [17] to compensate for maximum phase channels.

Using modulo linearity, it is possible to construct a theory of numerical filters with close correspondence to conventional digital filters, and in particular there is a D transform with the same role as the z transform. It is defined thus: if h_n defines a sequence of sample values, then

$$H(D) \underset{\text{defn}}{=} h_0 D^0 \oplus h_1 D^1 \oplus h_2 D^2 \oplus \cdots \oplus h_n D^n + \cdots. \tag{31}$$

Such D transforms can be manipulated just like z transforms, including their use in rational fractions, inverse filters and the expansion of denominators by the binomial theorem.

However, evaluation for particular D (which we might expect from the calculation of frequency response by setting $z = e^{j\omega T}$) does not seem to be useful. Also, the responses are bounded, and so the concept of poles does not appear.

A special class of modulo linear systems occurs for $M = 2$, i.e. each number can take only the values 0 or 1, as 1, as $0 \equiv 2$ modulo 2. Then we have binary sequence filters, and linear sequential coding theory [18, 19]. Arithmetic is easy in such systems—addition is the logical exclusive—OR, or half-add:

$$0 \oplus 0 = 0$$
$$1 \oplus 0 = 1$$
$$0 \oplus 1 = 1$$
$$1 \oplus 1 = 0$$

Multiplication is AND:

$$1.1 = 1$$
$$0.1 = 1.0 = 0.0 = 0$$

Systems of this nature are familiar in pseudorandom binary number generators and coding schemes. They may become valuable in detection systems because machines can do these operations cheaply and quickly.

REFERENCES

1. Lockhart, G. B. (1971). Binary transversal filters with quantized coefficients. *Electronics Letters*, **7**, No. 11, p. 305.
2. Ellis, J. H. (1970). A Method for greatly reducing the computation necessary for simulating certain kinds of filters by replacing the impulse response by the convolution of two simple responses, Proc. Imperial College Symposium on Digital Filtering, London.
3. Bogner, R. E. and Scott-Scott, M. (1970). Digital chirp filters, Proc. Imperial College Symposium on Digital Filtering, London.
4. Crystal, T. and Ehrman, L. (1968). The design and applications of digital filters with complex coefficients. *IEEE Trans. Audio Electroacoustics*, **AU-16**, No. 3, p. 315.
5. Bogner, R. E. (1969). Frequency sampling filters, Hilbert Transformers and Resonators, B.S.T.J., **48**, No. 3, p. 501.
6. Cain, G. D. (1971). Processors for reducing aliassing distortion in sampled signals, Imperial College Symposium on Digital Filtering.
7. Fjallbrandt, T. (1971). A network for the recovery of non-uniformly sampled signals, 1971 I.E.E.E. International Symposium on Electrical Network Theory, London, p. 32.
8. Bown, G. Optimization. *In* "Introduction to digital filtering" (editors R. E. Bogner, and A. G. Constantinides). To be published by J. Wiley and Sons.
9. Rabiner, L. R. and Hu, J. V. (1971). Applications of linear programming to design of finite duration impulse response digital filters, Imperial College Symposium on Digital Filtering.
10. Shanks, J. L. (1972) Two-dimensional digital filters, Paper read at I.E.R.E. Conference on Digital Processing of Signals in Communications, Loughborough, (Not included in Proceedings).
11. Weinstein, C. J. (1969). Quantization effects in digital filters, Technical Report 468, Lincoln Laboratory MIT, 1969.

12. Sablatash, M. (1971). The state of the art in approximation techniques for digital filter design, Proc. Imperial College Symposium on Digital Filtering.

13. Fettweiss, A. (1971). Some principles of designing digital filters imitating classical filter structures. *Trans. I.E.E.E. Circuit Theory*, **CT-18**, p. 314.

14. Crochiere, R. (1971). Digital ladder filter structures and coefficient sensitivity, *Q. Progr. Rep. Res. Lab. Electronics, M.I.T.*, **103**, 129.

15. Avenhaus, E. (1971). An optimization procedure to minimize the wordlength of digital filter coefficients, Proc. Imperial College Symposium on Digital Filtering.

16. Garner, H. L. (1959). The residue number system. *I.R.E. Trans. Electronic Computers*, **EC-8**, 140.

17. Tomlinson, M. (1971). New automatic equalizer employing modulo arithmetic. *Electronic Letters*, **7**, No. 5/7, 138.

18. Peterson, W. W. (1961). "Error-correcting codes", Wiley.

19. Huffman, D. A. (1956). A linear circuit viewpoint on error-correcting codes. *I.R.E. Trans. Infor Th.* **IT-2**, 20.

DISCUSSION

Severwright: You referred to logarithmic quantization of signals. Can you say whether it allows the processing to be implemented more conveniently?

Answer: Yes, it was intended to simplify multiplications. Additions would appear to be more complicated, but an algorithm has recently been described by N. Kingsbury which made addition very efficient—it involves a table lookup and interpolation.

F. Pichler: It is very convenient to look at these filters in a state space description, both recursive and nonrecursive. One can readily see whether two systems are isomorphic, i.e. there is a one to one correspondence in state space.

C. van Schooneveld: On terminology—Is there a term to describe systems which contain both recursive and nonrecursive paths?

Answer: Apparently not; I use the term recursive structure when the system contains feedback paths, whether nonrecursive paths are present or not.

C. van Schooneveld: Would you give some more details on two-dimensional filtering?

Answer: There are some details in the preprint which were not presented in the lecture. Briefly, we may, for example, want to filter (to blur or deblur) a two-dimensional picture by replacing the value at each point by a linear combination of weighted values at other points. This can always be done nonrecursively, and by transforms. However, these are expensive in storage, and it is often much more economical and convenient to use simple recursive systems, particularly if certain classes of response are acceptable. For example, an exponential pyramid response may be realised by using successively in the x and y directions a simple $B^{-|n|}$ response. This in turn may be realized by forward and backward passes of a simple first order filter whose impulse response is B^{-n}. Such responses (which are separable into the products of x-direction and y-direction functions) can only be circularly symmetric if they are gaussian (equation 23). Some rearrangements can allow superpositions of doubly gaussian responses to be admissible.

Comment: It strikes me that these situations are very similar to two-dimensional arrays in which you are filtering spatially in two dimensions, and you can deal with a line of line arrays to get the square pattern you are talking about. There are techniques for non-separable pattern synthesis. They may be directly applicable to filter design.

Answer: Yes, the correspondences are very close. I had imagined that the carry over might be both ways. The problem of separability does restrict the class of convenient responses, but these may be all you need. The problems of stability seem to be quite complicated; you may design a transfer function in two dimensions and find that the realization is unstable, and the criteria are not simple.

D. Creasey: Coming back to LCR systems, we often realize these by using active networks. An important problem is sensitivity. Can you comment on the situation for digital filters?

Answer: Sensitivity is relevant as the problem appears in connection with the quantization of coefficients (refers to example given on page 159). Many of the problems in architecture of digital filters relate to how the system should be arranged so that the precision (which is limited by the number of bits representing the co-efficients) will be acceptable. Large systems in which many delays appear together, i.e. many poles are generated by one difference equation, appear to be inferior. Current practice favours a system composed of a cascade of second order systems. Such systems are also easy to design. There is hope that ladder networks may have some advantages in this regard—analogue LC and active ladders do.

B. P. Th. Veltman: I would like to recommend also the use of hybrid filters where the memory elements are made in the digital way and the summation and multiplication elements in the analogue way. In today's technology of thick films, it becomes attractive to do so. The accuracy you can obtain is the same as with digital filters and is much more economical.

Answer: Yes, I have been interested in these. A later paper by K. Galpin refers to related techniques. One restriction is that they are not as flexible or adaptable as completely digital systems which you can change by software. (Communicated: the example of the second order resonator responses shown in the lectures were obtained with such a system.)

B. P. Th. Veltman: Often you don't need the adaptability or flexibility and in these cases we find these filters are much cheaper, and better than available commercial filters for laboratory use. The A–D conversion is the limiting part of the system.

The Fast Fourier Transform and its Implementation

D. BUTLER and G. HARVEY

*Plessey Company Limited, Templecombe,
Somerset, U.K.*

1. Introduction

This paper aims to discuss the background to the Fast Fourier Transform and the mathematical derivation of the Tree Graph. The properties apparent in the Tree Graph representation are used in Section 2 to point out how a novel FFT processor can be derived having advantages in speed and simplicity over existing techniques. It is thought that this technique may have application elsewhere.

Section 3 of this paper discusses current methods of implementing the FFT and contrasts their relative merits.

Some of the methods of implementing additional functions such as cross correlation are discussed in Section 4.

If the Discrete Fourier Transform (DFT) is computed without using the Fast Fourier Transform (FFT) algorithm, the number of arithmetic operations involved with N samples is proportional to N^2. With use of the FFT algorithm, the number of operations is now proportional to $N \log_2 N$.

The factor of improvement is therefore

$$N^2/N_2 \log N = N/\log_2 N.$$

With large sample numbers, the improvement can be very large.

Thus the FFT is a method of computing a discrete Fourier Transform (DFT) of N samples in $N \log_2 N$ operations when N is a power of 2.

In 1965 Cooley and Tukey published a paper outlining this method. At the time the majority of the paper's readers thought this was a new method of computing a DFT. It was thought that the DFT of N samples took something proportional to N^2 operations, the constant of proportionality depending on the symmetries of the sine and cosine in the weighting functions. Computer algorithms based on this N^2 method took up a great deal of computational time.

Soon after the publication of this paper it was realized that it was in fact a rediscovery of a method, in more general terms, used by Danielson and Lanczos in 1942 who refer to a German paper by Ronge and Konig in 1924. The paper by Ronge and Konig describes in terms of sine and cosine series a method of transforming N points where N is a power of 2, by forming $\log_2 N$ sub-series, and the algorithm doubles to form this DFT in $\log_2 N$ doublings. Thus a total number of operations of $N \log_2 N$ is necessary.

In those days of elementary computing machinery the saving of time was very small because of the small number of samples possible. Thus the algorithm was forgotten, until the advent of modern computing machinery.

2. Derivation of Tree Graphs

The lecture is intended to remind the audience of some forms of the FFT algorithm, as a background to the proposal for an FFT processor. An examination will be made of the way in which the data may be organized.

The FFT algorithm is merely an efficient way of calculating the discrete Fourier transform (DFT),

$$F(k) = \sum_{n=0}^{N-1} f(n) W^{kn}, \quad \text{for } k = 0, 1, \ldots, N-1,$$

with

$$W_N = \exp\left(-\frac{2\pi j}{N}\right)$$

This is an N-point transform, and the $f(n)$ are data points, and n is often a time suffix.

The DFT above, done directly, requires the order of $4N^2$ real multiplications and additions, whereas the FFT requires the order of $4N \log_2 N$ real multiplications and additions.

One set of FFT algorithms may be derived by separating the DFT sum above into a sum over the $N/2$ odd terms of the series, and a sum over the $N/2$ even terms of the series (here N is assumed even). This gives the "decimation in time" forms of the FFT algorithms.

It is easy to show that

$$F(k) = G(k) + W_N^k H(k),$$

where $G(k)$ and $H(k)$ are $N/2$ point transforms, which therefore repeat with period $N/2$. This is illustrated in Fig. 1, where the even points of f are labelled g, and the odd points are labelled h, and the transforms are written with capital letters. The outputs of the $N = 4$ FFT boxes are combined as

$$F_0 = G_0 + W_8^0 H_0$$

$$F_1 = G_1 + W_8^1 H_1$$

etc, in an obvious notation.

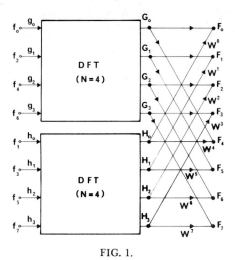

FIG. 1.

If N is a power of 2, this process may be repeated through $\log_2 N$ stages, to give a "tree graph", as shown in Fig. 2. At each node, at which two lines arrive from the left, a sum of two numbers is formed, by taking the number at the other end of each line, and multiplying it by the weight appropriate to that line. The weight is unity, unless otherwise specified on the diagram.

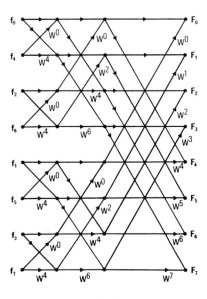

FIG. 2.

This tree graph is made up of certain basic units, which are often referred to as "butterflies", shown in Fig. 3.

If the nodes of Fig. 2 are regarded as sequential storage positions in some register, then this form of tree graph has "bit-ordered" (or BO) output,

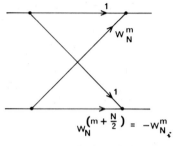

FIG. 3.

with "bit reverse ordered" (BRO) input, i.e. writing the storage location in binary, the binary number has to be reversed to give the label of the input data stored there.

A simple rearrangement leads to Fig. 4, which shows a BO input with BRO output. The algorithm is in no way altered, the stores are merely rearranged.

One property of both of these forms of the algorithm is the "in place" nature of the calculation, i.e. f_0 and f_4 are used to produce intermediate

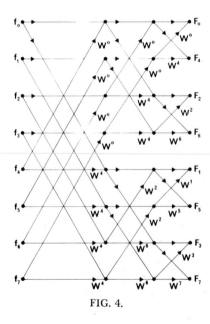

FIG. 4.

data in similar locations, and are not needed again in the calculation. The intermediate data might as well be stored in the locations of f_0 and f_4, which are then lost, so that a single store will suffice. This is not always the case.

In Fig. 5 the algorithm is rearranged to give BO input and BO output. The calculation is no longer in "place", and a single store will not now be

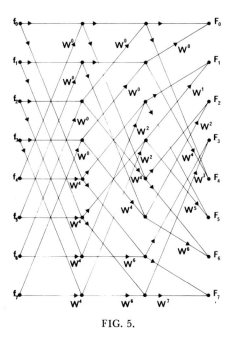

FIG. 5.

sufficient for the calculation. This particular scheme is not simple to organize, but some algorithms which are not "in place" do have other desirable properties.

In the algorithms illustrated previously in the lecture, random access to the store is required, e.g. Fig. 2, where access is required to different locations at different stages of the calculation. In the tree graph illustrated in Fig. 6, access is required sequentially at all stages of the calculation. The algorithm is not "in place", and two stores must be used. In the first stage, f_0 and f_4 in locations 1 and 2 give the first point of the next stage. The geometry of the transform is identical from one stage to the next, but the weighting functions are different. The store containing the input data for each stage is recirculated twice as fast as the store used for the output data. This particular algorithm is BRO to BO.

Figure 7 shows the version of the algorithm proposed for use in hardware. Again it requires two stores. Two access points are needed on the input store $N/2$ points apart. In Fig. 7, points, f_0, f_1, f_2, f_3 are

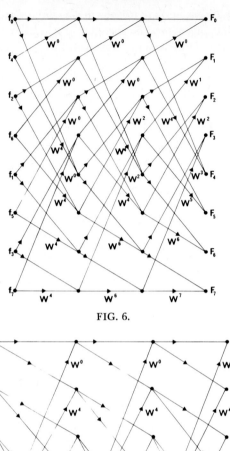

FIG. 6.

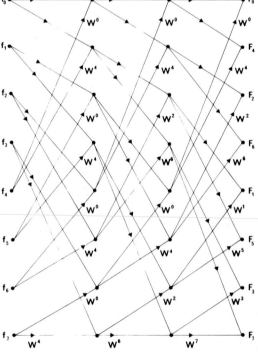

FIG. 7.

stored in one half, and f_4, f_5, f_6, f_7 are stored in the other half of the register. F_0 and f_4 are used to give the first two points of the input, etc. Weighting is applied to the output of the second half of the input store only. In this case the output store is recirculated twice as fast as the input store. The geometry of the transform is identical from stage to stage, making the hardware implementation more simple. The algorithm is BO to BRO.

A different algorithm may be derived using the so-called "decimation in frequency" approach. The initial data is not divided into a sum over even terms and a sum over odd terms, but instead the even transformed functions F_0, F_2, F_4, F_6 are each derived from the DFT of an $N/2$ point sequence, and the odd transform functions are derived from a different $N/2$ point sequence. This leads to a similar, but differently organized set of algorithms.

The one equivalent to Fig. 7 is shown in Fig. 8. Again this is BO to BRO, but the order in which the weighting functions is required is different, and is now sequential. This may be convenient in some forms of hardware. It is particularly convenient to have both algorithms, decimation in time and decimation in frequency, when convolution and autocorrelation are required. Then one algorithm may be used for the transform, and the other for the inverse transform, and no rearrangement of data is necessary, as it would be if the same algorithm were used for both tasks.

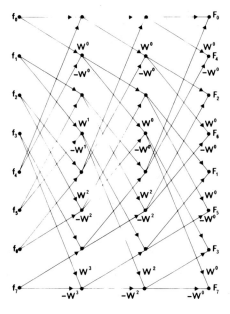

FIG. 8.

3. A Serial FFT Processing Method

As has been demonstrated above the FFT algorithm can be represented graphically by means of the Tree Graph, an example of which is shown for an eight point transform in Fig. 9. This has been drawn into a slightly different form to Figs 1–8 to facilitate the hardware discussion.

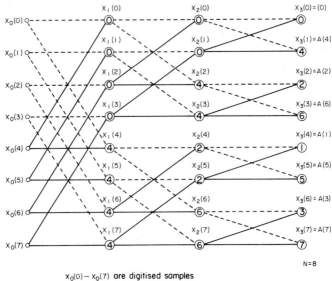

$x_0(0) - x_0(7)$ are digitised samples

$A(0) - A(7)$ are output Fourier coefficients

FIG. 9.

The description given below outlines the fundamental properties of the graph, and their relation to the design of a processor.

(i) The sampled data are represented by the left-hand vertical array of nodes: the example showing eight samples $x_0(0)$ to $x_0(7)$. (The x subscript refers to the vertical array.) The second vertical array (x_1) corresponds to intermediate coefficients calculated by operations on the first array, the third array (x_2) corresponds to coefficients calculated by operations on the first and second arrays. The last vertical array contains the final coefficients, (x_3), which are the final transform efficients, $A(n)$, ($n = 0$ to 7).

(ii) The mathematical operations required are defined by the lines joining the nodes, together with the numbers enclosed at the node. Each node represents a complex multiplication and a complex addition. The two lines entering a node define the two coefficients from the previous array which are to be added, one of which is first multiplied by a weighting function. The dotted line indicates a simple transfer of the previous coefficient. The solid line indicates that the previous coefficient is multiplied by a weighting function, the value of which is related to the number in the node circle. (The exact relationship is of the form W^z, where W^z is the value of the weighting function and z is an integer exponent, being the integer appearing in the node circle. It will normally be referred to as the weighting function exponent value.)

Thus, for example, the coefficient $x_1(0)$ is formed by operating on coefficients $x_0(0)$ and $x_0(4)$, defined by

$$x_1(0) = x_0(0) + W^0 \cdot x_0(4).$$

(iii) A second coefficient, $x_1(4)$ is also defined by operating on coefficients $x_0(0)$ and $x_0(4)$, defined by

$$x_1(4) = x_0(0) + W^4 \cdot x_0(4).$$

There is a regular relationship between such pairs of weighting functions, which simplifies organization of the processor and weighting function store. For each pair of weighting functions used to generate two coefficients from two preceding coefficients, the exponent values differ by $N/2$, where N is the number of points in the transform.

For such pairs of weighting functions, the values of the weighting functions are equal in magnitude but opposite in sign.

Thus, $W^4 = -W^0$; $W^5 = -W^1$, and so on, for $N = 8$.

This property leads to the "sum and difference" method of proceeding with the transform, i.e. from one pair of coefficients, the two associated coefficients in the next stage are formed, at the same time, by using the positive value of the weighting function in forming one coefficient and the negative value in forming the other. Thus $x_1(0)$ and $x_1(4)$ are formed by multiplying $x_0(4)$ by W_0 and $-W$, respectively, and adding the results to $x_0(0)$. In all the systems proposed, two coefficients are taken together, and two formed together.

(iv) In the case of the tree-graph used, which corresponds to one particular FFT algorithm, the input data is in time ordered form (see Fig. 7). A property inherent in the algorithm results in the output coefficients not being ordered in frequency; in fact, they are in "bit reversed order" (BRO), i.e. if the final x coefficient values are expressed in binary notation, and the binary numbers are then reversed, left to right, the resultant binary numbers are the values of the A coefficients. Thus, in the eight-point example, $x_3(0)$ corresponds to $A(0)$, $x_3(1)$ corresponds to $A(4)$, $x_3(2)$ corresponds to $A(6)$, and so on.

(a) Effects of tree-graph properties on processor design
Assuming a starting point with an input register full of the samples, inspection of the tree-graph shows that for the operations on the first

vertical array, during the first "pass" of the transform, the pairs of co-efficients required are separated by an interval of $N/2$ words. During the second pass, operating on the second array to produce the third array, the coefficient pairs are separated by an interval of $N/4$ words. This narrowing of intervals occurs progressively, until during the last pass the coefficient pairs are adjacent. The implications of this property upon the design of the processor memory are that either some kind of multi-access storage is required, or that the organization and storage of the data must be arranged in some way to compensate.

(b) Memory organization

In order to utilize the advantages of the low cost of serial MOST storage it is necessary to consider the tree-graph from Fig. 9 further.

The data samples listed in the L.H. column, i.e. $x_0(0), x_0(1) \ldots x_0(7)$ are required for computation in word pairs. These pairs change at later stages of the calculation but are always defined pairs at any particular stage.

Table I lists the pairs required, at each stage, for an eight-point transform.

TABLE I

	Stage 1	Stage 2	Stage 3
Word pairs	$X_0(0)$ & $X_0(4)$	$X_1(0)$ & $X_1(2)$	$X_2(0)$ & $X_2(1)$
	$X_0(1)$ & $X_0(5)$	$X_1(1)$ & $X_1(3)$	$X_2(2)$ & $X_2(3)$
	$X_0(2)$ & $X_0(6)$	$X_1(4)$ & $X_1(6)$	$X_2(4)$ & $X_2(5)$
	$X_0(3)$ & $X_0(7)$	$X1(5)$ & $X_1(7)$	$X_2(6)$ & $X_2(7)$

Correlating the Table and the tree-graph, the pairs required at the first Transform stage are 4 samples apart, all pairs being required, those at the second stage Transform stage are 2 samples apart, not all pairs being required; those at the third transform stage are adjacent, again not all possible pairs being required. If the initial samples $X_0(0) \ldots X_0(7)$ were stored serially then for the first pass access is required at $X_0(0)$ and $X_0(4)$. The second pass requires access at $X_0(0)$ & $X_0(2)$, and $X_0(4)$ & $X_0(6)$ at a later stage of the second pass. The third pass requires access at every word. This is shown below in Table II, where * indicates access at any stage of the transform.

Extrapolating these results on to a 2048 point transform i.e. 11 stages of transform, a total number of 23 access points would be required to implement the system.

An alternative approach is to maintain a fixed number of access points

TABLE II

Samples	$X_0(0)$	$X_0(1)$	$X_0(2)$	$X_0(3)$	$X_0(4)$	$X_0(5)$	$X_0(6)$	$X_0(7)$	
Stage at	*				*				1
which access	*		*		*		*		2
is required.	*	*	*	*	*	*	*	*	3

and reshuffle the data internally to the MOST register such that the required word pairs are always available at the access points. This method of organization, involving no redundant shifts is shown in Fig. 10. Only two access points are required at each section of the memory store, and each memory store is multiplexed to be interchangeable with its partner at subsequent transform stages.

The registers' clock frequencies are in relation f and $2f$.

By reference to Table I, and the tree-graph, the correct word pair can be seen to be accessed at each stage of the transform, and by using the "Sum and Difference" method, the corresponding pair computed are sequentially

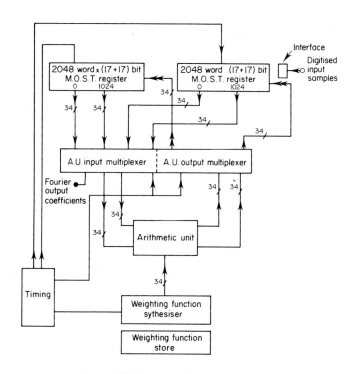

FFT block diagram

FIG. 10.

stored at the same time as the following pair are accessed from the Input
Register. This results in no redundant time procedures for re-shuffling
the data. The limits on the speed at which this process can be completed
are now set by the MOST Shift Register maximum operating frequency,
F_t. This method at present gives computation times of nanoseconds which
accord with the access times of proposed MOST stores.

The complete system block diagram for implementing the serial FFT
is shown in Fig. 11.

From the foregoing it can be seen that a serial FFT system can be implemented and it has the major advantage of high speed compared with
the traditional method of organization. 2048 complex points could be
processed in 2.5 msecs.

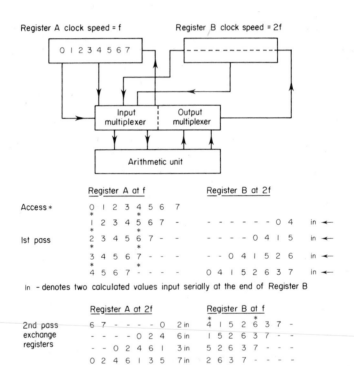

FIG. 11.

4. FFT Processing Methods

(a) Software
As explained earlier the FFT was first implemented by means of general purpose computers and software. This has meant the use of random access core store for the storage of coefficients and weighting functions. With a core cycle time of 1 μsec (typical for most GP computers) there has been a speed limit for performing the algorithm. Software programs for implementing the FFT are readily available or written for most GP computers but suffer from the disadvantage that a 1024 word transform takes several seconds to perform.

(b) Parallel processors
It will be seen by reference to the tree-graph that the processing speed can be increased by introducing parallelisms in the arithmetic units. Processors have been constructed using various degrees of parallelism, the ultimate limit being the array processor where all arithmetic operations are performed in parallel. Although extremely fast processing times are achieved with these processors the cost and size of the equipment required rules them out for any but the most specialist application.

(c) FFT peripherals
A range of processors has been developed which act as peripherals to a GP computer.

In the simplest form these take the form of a high speed arithmetic unit used in conjunction with a GP computer.

More sophisticated FFT peripherals have their own stores and arithmetic unit. A GP computer is used for organization of data, block transfer of it to and from the peripheral and system organization. These special purpose peripherals are capable of performing many more functions than just the FFT. These functions are explored in Section A. High processing speeds can be realized.

5. Incorporation of Other Products

The basic FFT processor can be used to perform other functions as explored below.

(i) *Convolution*

If the products of two frequency functions $F(f)$, $H(f)$ are multiplied to form $C(f)$, i.e.

$$C(f) = G(f), H(f) \tag{A}$$

then this multiplication can be utilized by the formation of the inverse transformation from the frequency domain to the time domain. The resultant time function $c(t)$ is termed the convolution product of the time series $g(t)$, $h(t)$, the time functions of $G(f)$ and $H(f)$ respectively. Represented mathematically this is:

$$C(\tau) = \frac{1}{N} \sum_{t=0}^{N-1} g(t)h(\tau - t). \tag{B}$$

For this sum to hold true, a provision is made on the time function $h(t)$, that is

$$h(t) = 0 \quad \text{for } t < 0 \tag{C}$$

If this process is to be utilized with the FFT processor, it would seem upon the face of it, a simple task for forming equation (A), then the inverse transform to form equation (B). In fact, this is not the case as the condition upon $h(t)$ has not been adhered to. This is because the FFT processor assumes that the time functions $h(t)$, $g(t)$ are periodically continued outside the range of sampling, i.e.

$$h(t + rN) = h(t) \quad 0 \leqslant t < N - 1$$
$$r = 0, 1, \ldots S, \ldots$$

Thus to overcome this defect in the system as it stands, zero's are added to the signal in the following way:

Let

$$
\begin{array}{llr}
g_1(t) = g(t) & 0 \leqslant t < N/2 & \text{(D)} \\
\quad\quad\quad 0 & N/2 \leqslant t < N & \text{(E)} \\
h_1(t) = h(t) & 0 \leqslant t < N/2 & \text{(F)} \\
\quad\quad\quad 0 & N/2 \leqslant t < N & \text{(G)}
\end{array}
$$

(1) Form equations (D), (E), (F), (G).
(2) Form the FFT's of $g_1(t)$, $h_1(t)$, i.e. $G_1(f)$, $H_1(f)$
(3) Form the product $C(f) = G_1(f) \cdot H_1(f)$ (H)
(4) Form the inverse transformation of $c(t)$, so defining equation (B).
The total time for this process to be completed is three FFT times together with N multiplications to form equation (H).

(ii) *Cross-correlation*
The cross-correlation sum between two signals $g(t)$, and $h(t)$ is defined by a sum similar to that of convolution above, i.e.

$$c(\tau) = \frac{1}{N} \sum_{\tau=0}^{N-1} g(t) \cdot h^*(t - \tau), \tag{α}$$

where $*$ is the complex conjugate, and $h(t) = 0$ for $t < 0$.

To achieve this product with the FFT processor a procedure is adopted similar to that described above in convolution.

When the transform from the time series described in (α) to the frequency series is performed, the equation formed is:

$$C(f) = G(f) \cdot H^*(f). \qquad (\beta)$$

Thus the only difference to the procedure described above is that equation (H) is replaced by equation (β).

(iii) *Auto-correlation*

The auto-correlation product is defined by the sum:

$$c(\tau) = \frac{1}{N} \sum_{t=0}^{N-1} g(t) \cdot g^*(t - \tau). \qquad (a)$$

The transform of this equation is

$$c(f) = g(f) \cdot g^*(f). \qquad (b)$$

Thus immediately we can see that this process is inherently quicker because of the saving in the computation of a second functions FFT. The process is described below:

(1) Form $g_1(t) = g(t) \qquad 0 \leqslant t \leqslant N/2$ (c)

           $0 \qquad N/2 \leqslant t < N$ (d)

(2) Form the FFT of $g_1(t)$ namely $G(f)$
(3) Form the product $F(g) \cdot G^*(f)$ i.e. the modulus of $G(f)$
(4) Form the inverse FFT of this modulus product to give the auto-correlation coefficients.

As in cross-correlation this process can be extended when the sample length exceeds $N/2$ samples.

(iv) *Cross-spectrum analysis*

The cross-spectrum of two time functions $x(t)$, $y(t)$ is defined in terms of the cross-correlation coefficient $C_{xy}(\tau)$.

$$R_{xy}(f) = \frac{1}{N} \sum_{\tau=0}^{N-1} C_{xy}(\tau) \, e^{-2\pi i f \tau /N}. \qquad (\alpha)$$

To form this using the FFT processor use is made of the fact that:

$$R_{xy}(f) = X_1(f) \cdot Y_1^*(f), \qquad (\beta)$$

where $X_1(f)$, and $Y_1(f)$ are defined below in terms of the input functions $x(t)$, $y(t)$.

    $x_1(t) = x(t) \qquad 0 \leqslant t < N/2$ (γ)

        $0 \qquad N/2 \leqslant t < N$ (δ)

    $y_1(t) = y(t) \qquad 0 \leqslant t < N/2$ (t)

        $0 \qquad N/2 \leqslant t < N$ (μ)

and $X_1(f)$, $Y_1(f)$ are the transforms of $x_1(t)$ and $y_1(t)$ respectively.

So the process is identical to autocorrelation analysis up to and including stage (iii) but not the inverse FFT.

6. Cepstrum

This is easily implemented using the FFT processor and the peripheral computer to perform a computation.

The process is:

(1) Form the FFT process on a time series $x(t)$
(2) Re-order the output of the processor
(3) The answer formed is in a complex form $a + ib$, thus to use the definition of cepstrum from speech analysis, the computer now takes the logarithm of this complex number.

$$a + ib = Re^{i\theta} \quad \text{where} \quad R = \sqrt{a^2 + b^2}$$

$$\text{and} \quad \theta = \tan^{-1}\left(\frac{b}{a}\right)$$

So $\log_e (a + ib) = \log_e R + i\theta$.

This is only performed on the first $N/2$ answers.

(4) Place in the real register in a frequency order the sequence $\log_e R$, and in the imaginary register θ.
(5) Perform an FFT on this frequency ordered sequence to form the quefrency series.
(6) Re-order into a quefrency ordered sequence.

The reasoning behind the use of only $N/2$ of the "samples" from the first pass of the FFT is equivalent to the Nyquist sampling criteria upon a time ordered series.

If the highest resolvable frequency is f say, then the quefrency series will be based upon the inverse of this, i.e. $1/f$ as its "frequency". The cepstrum thus forms the series.

$$\left(0, \frac{1}{f}, \frac{2}{f}, \dots \frac{f/2}{f}\right) \text{ sec. as the quefrency series:}$$

Thus the largest quefrency resolvable is $\frac{1}{2}$ sec., i.e. 2 Hz in the spectrum if $f < 1$.

DISCUSSION

Question: From my limited experience of hard wired FFT devices it seems they are often more expensive than the computer they are designed to work with, and are therefore limited to very special purpose simulation. In view of this have you considered rather more limited interpretations such as radix 4?

G. Harvey: We would agree about the price of FFT analysers. The early ones on the market have a lot of facilities because the manufacturers are uncertain just what is needed. This tends to keep the price up. We have not considered radix 4 algorithms.

Long term as the market goes up, the price will come down. The hardware implementation lends itself quite readily to doing a lot of the operations on one chip.

I. G. Liddell: Can you comment on whether there is any fundamental reason why one cannot have an FFT which has bit ordered input, and bit ordered output and which is "in place"? There is such an algorithm for the Fast Walsh Transform.

D. Butler: I know of no fundamental reason, nor do I know of an algorithm which has those properties.

Interactive Signal Processing With a Minicomputer System and its Application to Sonar Research

ROBERT SEYNAEVE, EDWARD HUG and GIANCARLO VETTORI

SACLANT ASW Research Centre,
La Spezia, Italy

1. The Saclantcen Interactive Time Series Analysis (ITSA) System

(a) Introduction

The purpose of the first part of the paper is to describe the conversational signal processing system which is evolving at SACLANTCEN. This system is extensively used in a number of projects and is providing a great variety of results. It differs considerably from our previous systems in the same field because it is essentially oriented towards interaction between the scientist and the computer, and is better adapted to be used in actual experiments.

The present system is an evolution from our experience in using a batch operated, large scale computer for sonar signal processing research.

The batch-type computer is excellent for solving pre-formulated problems, but does not satisfy some of the specifications which we think are important for a sonar research computer system:

(a) The same (or sufficiently similar) system should be available on shore as well as at sea during trials.

(b) It should include flexible inputs and outputs, and marry easily with the experimental set-up.

(c) It should be fast enough to provide real-time operation, at least for those parts of the processing which provide feedback on the experiment.

(d) It should be programmable in a simple language built from the natural operational functions of the various problems common to our work.

(e) Its control must be highly interactive.

(f) It should be very flexible, allowing complete changes in operation in a very short time.

(g) It should be as general as possible, and allow an easy communication with other systems. It should take care of all computer operations necessary in a typical experiment.

(h) It should be reliable and economical enough so that it will be used as any other part of the experimental equipment.

The need for a system of this sort has existed for a long time among us. Apart from opening up new possibilities for experiments it was expected to help the researcher considerably by removing the long delay introduced by program writing and debugging. Also most of the programming used to be lengthy and sophisticated enough to have to be written by dedicated programmers, and this often increased the psychological distance between the researcher and his tool.

The advent of rather cheap mini-computers gave an opportunity for such systems to be economically feasible. Being mostly designed for process control applications, these computers have just adequate word and memory size, speed and input/output flexibility for most signal processing applications. One of the first commercial attempts to produce a small general purpose signal processing system, based on a mini-computer, was made by Hewlett Packard under the name of "Fourier Analyser 5450". This system was designed in an intelligent and flexible way and it was very successful at SACLANTCEN. It could not however, satisfy all requirements but because of its open ended design it provided an excellent starting point for the implementation of the more specialized system we needed.

(b) Description of ITSA

ITSA is an interactive computer system designed for research in sonar signal processing and general time series analysis. Built with the commercial system as a starting point, it has kept most of its basic options:

—push-button control with immediate execution.
—continuous display of results on CRT.
—operations on blocks of data.
—on-line editing and correction.
—flexible A/D converter input.
—fully calibrated system.

ITSA operates with a larger variety of peripherals (Fig. 1), and includes many extra features:

—generalizations of language and functions by integrating the BASIC interpreter.†
—extended set of instructions.

† Being now implemented.

—multiprogramming environment.†
—interactive control by graphic terminal and joy-stick.
—direct access to all computer peripherals.
—remote control operation of external equipment under program
   control.
—data and program filing system.
—descriptive control statements.
—communication with other systems and programs.

In its present form ITSA can take care of many functions of an
experiment, including generation of waveforms, acquisition of data,
processing, display and filing. All the operations can be either entirely
automatic or be under partial or total control of the scientist via special
keyboards, or an interactive graphic console.

The program is set-up in an interactive way, and can be modified at any
moment in a very short time.

ITSA is presently communicating with the SPADA‡ system in order to
provide the latter with a signal processing capability.

It also communicates with the Oceanographic Data Acquisition System
providing Oceanographers with powerful time series analysis capabilities.

As a tool for signal processing research, ITSA allows a scientist to
investigate complex signal processing methods in a very short time.

(c) The Functional Features of ITSA

Control. In interactive systems, statements or commands can usually be
entered in one of two ways.

(a) By pressing a key or a push-button. This solution is adequate when
the language consists of a limited number of basic operations.

(b) By typing in words or codes corresponding to given functions.
Various alternatives are possible, but they all have in common an almost
infinite set of possible commands.

ITSA has initially used the push-button console concept but has
recently adopted a mixed method: the most used functions are available
from the push-button consoles, but are also addressable from a graphic
terminal. This terminal, however, provides for other functions and for
programming of BASIC modules to be inserted in the main program. In
order to minimize the typing effort, the terminal is driven by special
software which observes the letters inputted by the user until it recognizes
without ambiguity the command that the user intends to type: from then
on the computer (Fig. 2) takes over, completes the typing and asks for

† Being now implemented.

‡ See paper by A. Barbagelata (p. 205): SPADA is a sophisticated acquisition and
edition system developed at SACLANTCEN primarily for wideband multichannel
experiments.

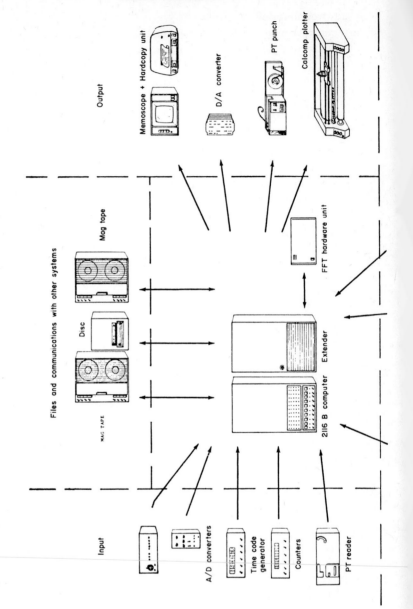

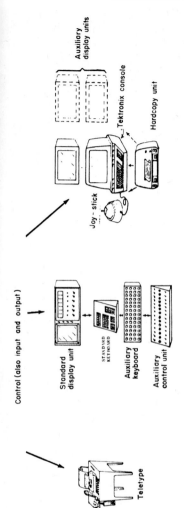

FIG. 1. Functional configuration of ITSA.

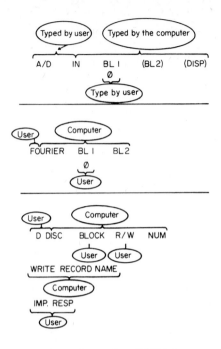

FIG. 2. Examples of ITSA input.

the required parameters. This feature allows a better use of the machine by allowing the user to concentrate more on the problem to solve, and by minimizing errors by indicating the nature of the parameters to be entered.

As a display for monitoring and output of results ITSA uses a CRT with a reasonable refreshing rate (about $\frac{1}{2}$ s for 1024 point display). This display is quite convenient, but a considerable use is also made of the memoscope screen of the graphic terminal. Its rather slow speed is compensated by the fact that it does not require refreshing and it includes a cross-hair and joy-stick option which provides a higher level of interaction:

—the design of displays of results can be made interactively.
—segments or points of numerical series can be selected by the joy-stick and their coordinates can be fed back in the subsequent steps of the program.
—a hard copy can be obtained at any time.

It is indeed our feeling that a most convenient control device would consist of a terminal with alphanumerical and special purpose keyboard, together with at least two memoscope screens (one for program listing, the other for displays), and a hard copy unit. Such a terminal is now entering in operation at SACLANTCEN.

Processing. ITSA takes advantage of the original commercial software which is essentially a set of procedures for operations on blocks of data. A large use is made of the FFT algorithm. In ITSA, this basic software has been complemented with other operations. In many cases however, there is a need to put together block-oriented time series analysis with the flexibility of other general purpose computer languages like FORTRAN. This has been achieved in ITSA using the swapping technique illustrated in Fig. 3.

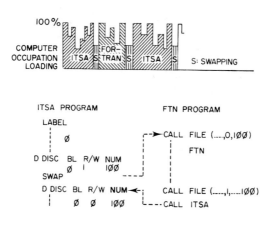

FIG. 3. ITSA Fortran operation.

At any time the computer core is occupied by only one program, i.e. either ITSA or a Fortran user program. Each program can call the other one, and thus, they can work in an interlaced mode. The data are communicated between the programs through the use of common disc files. In this way ITSA has operated extensively with SPADA and the use in this mode is attracting other users. But, although this method is simple, it has the following drawbacks:

(a) Although the user can design his program interactively when using ITSA, the Fortran part must still be compiled off-line and then loaded in the system.

(b) The computer is not used efficiently (Fig. 4a), much time can be spent in swapping software in and out of core from and to disc, and it is difficult to achieve continuous background operations like acquisition of environmental data.

The improved ITSA which is now being implemented does not suffer from these limitations. Firstly, it will run in a multi-programming† environment so that the computer can be better used (Fig. 4b). Various

† Real Time Executive.

activities can easily go on concurrently allowing, for example, simultaneous acquisition of new data and processing of already acquired signals. Secondly, ITSA is being equipped with an FFT hardware processor which should improve its overall speed of operation by an order of magnitude at least (Fig. 4c). Thirdly, ITSA will be intimately linked

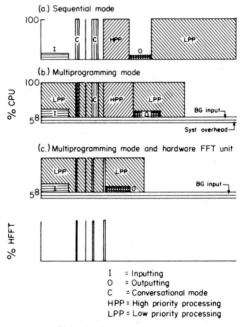

FIG. 4. Computer loading.

to a BASIC interpreter: the user will then be able to insert BASIC written subprograms in his set of instructions (Fig. 5).

Files. There are a number of reasons why a good filing system is necessary: the computer core is limited in size; there is a need for permanent storage; the user cannot work in a non-interrupted way; he likes to identify time series with information-carrying names rather than with sequential numbers, etc.

It is difficult to make the right compromise between generality, flexibility, access time, development time when a file system is designed. The file system designed at SACLANTCEN seems, however, to be working out quite well in its particular sonar research environment.

(i) *Disc file*: Each record written on disc is identified by an 8 ASCII character name. It receives a 16 word dictionary containing information about its day and time of generation, sampling frequency, type, size, etc.

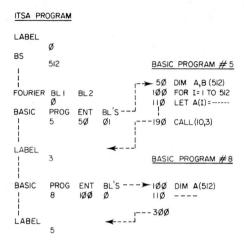

FIG. 5. Example of ITSA basic operation.

It can be read or written in random access or in automatic sequential mode.

(ii) *Magnetic file*: Records can also be written sequentially on magnetic tape together with a dictionary, just as in the disc file. No overwriting is allowed, and no mapping of the tape contents is kept in core: the search for a given record is therefore made sequentially: this simple method has proven adequate in an experimental environment where most users like to process long sequences of records, and do not access the tape randomly.

Programs can be stored on the disc or magnetic tape file as main, segment, or subroutine modules. The identification of the programs are also by name.

A user can then store, on a magnetic tape allocated to him, his data and programs, and also his results. Apart from being convenient, this way of working minimizes the set-up time of each user when starting on the computer.

Because we have standardized our filing format at SACLANTCEN, ITSA can process files of data collected with other systems like SPADA or an Oceanographic Data Acquisition System; the reverse operation is possible too.

Inputs and outputs. Although it uses a complete mini-computer system (i.e. 3 racks of equipment) ITSA tries to present itself to scientists as much as possible as a personal tool without the usual difficulties of communication inherent to the off-line computer set-up. Indeed, a great part of the attractiveness for scientists of the commercial system comes from its built-in A/D converter and well designed CRT display-digital voltmeter set-up. The same philosophy of flexibility of I/O has been kept in ITSA and the idea has been extended to other peripherals. For instance, instructions are provided to read in data from counters, voltmeters, time code generators, etc., to output computer generated analogue waveforms, to sense and to control external equipment. All the user has to do is to hook up its instruments to the right I/O slots.

(d) Future evolution of ITSA

After its integration with the Real Time Executive System and its marriage with the BASIC interpreter is completed, ITSA should be a quite general tool for sonar research, and it does not seem at this point that any major modification will be required. Still, a continuous effort will have to be provided to adapt to the constantly varying user's needs.

A re-writing of the software, however, is foreseen so that it will be able to run in a time-sharing environment on a large scale computer and be available from terminals in various locations in the building.

2. Application of ITSA to Sonar Research

(a) Introduction

The Interactive System described is useful for many fields of the experimental research, where the user's required input data rate and number of on-line operations are satisfied and where interactivity is highly desirable. Sonar is one of those fields where the use of the Interactive System may create many new users for the computer-aided conduct of experiments.

In experimental research the following sequence of actions have to be taken to verify a formulated theory: (a) establish the hypothesis; (b) design the experimental approach; (c) design and construct the experimental tool; (d) perform the experiments; (e) process and analyse the data; (f) compare the experimental and theoretical results.

In such a process the results of the comparison feed-back into the theory and into more, and possibly modified, experiments. For the process to be effective and the level of interest in the subject theory (vs. the implementation) not to decay, its duration must be contained within reasonable limits. The type of experimental tool choosen may affect considerably this duration, which consists essentially of the time required to the realization and modifications of the tool and for the time required for processing the data. Also the type of tool determines how much on-line control one has on the experiments, largely a matter of how well one can view the experimental results in real time.

Three types of experimental tool are shown in Table I: the Hardware, The Batch Processing and the Interactive System, whose principal characteristics are summarized in the Table.†

† Differences are due mainly to the signal processing part of system. Some hardware has obviously to be common to the three types (like acoustic sensors and signal conditioning, for Sonar).

TABLE I. Comparison of hardware, batch processing and interactive system type of experimental tool

| | Time required for | | Signal/data storage | Data processing | Other | |
	Design and construction of experimental tool	Modification of experimental tool			Pros	Cons
Hardware	Months to years	Weeks to months	None essentially required	By hardware, all on-line	Due to high processing speed possibility of multidimensional processing	High risk* in the cost and time invested in the hardware due to inflexibility
Batch Processing	Months	Days to weeks (For computer programs)	Large storage required—all signals are recorded for later processing	By large computer completely off-line	Minimum risk* of experimental equipment	Loss in the quality of the experimental results because of inadequate control of experiment. Long time before results are available
Interactive System	Hours to days	Minutes to hours	Medium to small required (Depending on system operating capabilities)	Partially on line line	The system is easy to modify	Only limited number of operations on line

* Risk = Risk of loss due to (a) Poorly formulated ideas, (b) Obsolescence.

The Interactive System leads to the shortest duration of the entire process, though the rate of on-line operations it can perform is inferior to that obtainable with the Hardware solution.

A trade-off between these two solutions seems however, acceptable for an experimental system. The Batch Processing, in which all the signals are recorded for later processing, leads to longest duration.

(b) Requirements of a sonar research tool

In "sonar research" one has to test certain signal processing schemes (e.g. for detection or parameter estimation) under the influence of various environmental and geometrical quantities.

Some of these quantities, or parameters, are shown in Fig. 6, where the sonar and its surrounding nature are represented schematically as a number of cascaded linear elements.

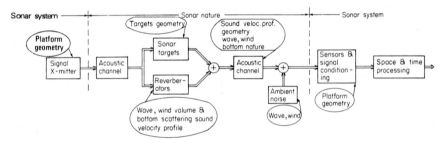

FIG. 6. Representation of sonar and sonar nature-influencing parameters.

In order to study their influence on the acoustic measurements, these parameters have to be acquired together with the acoustic data.

An ideal computerized research tool for sonar should therefore perform the following functions: acquire environmental, geometrical and acoustic signals; process each of them; display the results in graphical form. It is desirable that these functions be performed on-line. However, in practice this can only be partially done.

The signals to be acquired cover different ranges of required sampling rates, as described in Table II. The acoustic signals, which require the highest sampling rates, must usually be limited to a specified range-gate; this is possible thanks to the available means of tracking events occurring in a limited region of the range scale.

The Interactive System (ITSA) use is being developed in two phases. In Phase I (currently in operation) the computer performs only one operation at the time. With the choice of the operations to be performed on-line based on appropriate priorities. The various data channels not sampled by the computer require a separate means of recording.

In Phase II (being worked on) the computer will be able to perform

TABLE II. Principal data channels to be acquired, corresponding sampling rates

Order of magnitude of sampling rate values	Nature of Data Channels		
	Environmental	Geometrical	Acoustic
10 Hz	Wave motion, wind speed, sound velocity profile	Sonar platform and Targets geometry	
100 Hz	Bottom profile Volume reverberation profile		Background levels (both reverberation and noise)
1 kHz			Main acoustic channels— (both amplitude and and phase information preserved)

several functions in parallel, thanks to the multi-programming "Real Time Executive" mode. Furthermore with the FFT Hardware added, many more processing operations will take place per ping-period. Acquisition of many parallel channels will be possible, together with high priority on-line processing and low-priority activities in the background.

The different capabilities of these phases as applied to Sonar can be seen through the illustration shown in Fig. 7.

A typical example of sonar acquisition/processing display is shown as it occupies the computer under the different modes of operation.

A pinging interval of 45 s, a duration of the input gate of 2 s, and a block size of 4096 samples are chosen for the example.

Under the Phase I (Fig. 7a) operating mode, the processing computations take place at the computer's maximum operating speed. Under Phase II (Fig. 7b) (multi-programming) operation, the processing takes place somewhat more slowly but continuous environmental monitoring becomes possible and display driving becomes continuous and on-line interactive. Some on-line (between ping) operations are sacrificed and have to be performed off line or quasi-on-line (i.e. between runs). Under Phase II [multi-programming with hardware FFT (Fig. 7c)], considerable time is saved in all ITSA operations.† As a result even lower priority on-line operations are possible and excess computer power becomes available for background operations such as ray tracing, propagation modelling, navigation, etc.

† Overall time savings are conservatively estimated to be at least of a ratio 8:1.

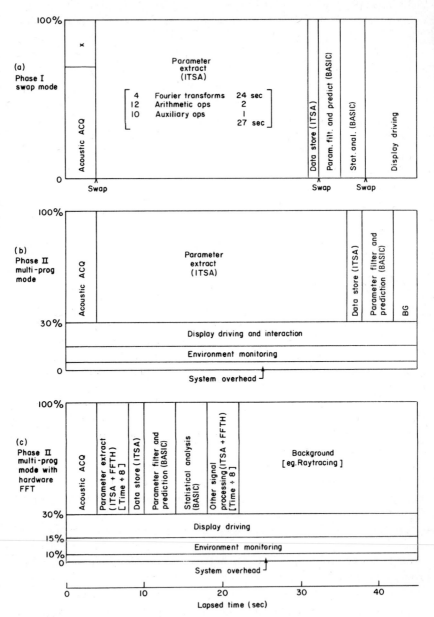

FIG. 7. Typical % rate/time diagram for experimental sonar use of real-time minicomputer.

(c) Two examples

Two examples of application of ITSA to measure some characteristics of
the acoustic channel are described next, a repetitive-ping technique and a
temporal interferometer.

Repetitive-ping technique. This is used to obtain a time-spread description
of the non-stationary acoustic channel. The characteristics considered are
those obtainable from the propagation through the channel of broadband
sonar pulses capable of resolving the various sound paths. For each of
them, mean and standard deviation of level and relative time of arrival
can be derived.

The sequence of responses is obtained by transmitting through the
channel a sequence of sonar pulses, whose time spacing is chosen shorter
than the period of the fastest variations of the channel. Linearly frequency
modulated (LFM) pulses are used to enhance the signal-to-noise ratio after
matched filter processing (see Fig. 8a). A time resolution corresponding to
the reciprocal of the signal bandwidth is available after matched filtering.
The processing chain is shown in Fig. 8b. For acquisition the received
sequence is periodically gated into the computer by a gate having the same
period as the transmitted sequence. After matched filtering, phase
information is removed by envelope detecting each processed signal. The
detected signals are presented in Fig. 8d where the successive blocks have
been automatically aligned one under the other by correlating them with
the average of the previous arrivals to obtain the necessary correction.
The misalignment of the successive responses before correction (Fig. 8c) is
mainly due to relative movements of the transmitting and the receiving
platforms.

Temporal interferometer. One is also frequently interested in obtaining
the frequency spreading characteristics of the channel. A variety of
techniques exist for doing this. This example presents one technique
which is capable of determining the rates of change of the lengths of each
of a number of propagation paths. The result is a rather fine-grain picture
of the path length/path length rate structure of the propagating medium.

The processing chain, shown in Fig. 9a, accepts a pair of LFM transmissions through the propagating channel and performs the following
basic operations:

1. Delay the first arrival to overlap the second, the delay being equal to
    the transmitted time interval. [(See Fig. 9c(1).]
2. Demodulate the LFM pulses by a frequency-shifted LFM replica to
    obtain a constant difference frequency. Each frequency component
    now has a frequency and phase corresponding to the path delay. (See
    Fig. 9b).

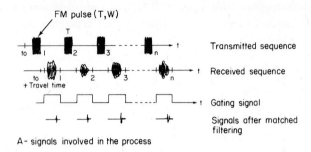

A- signals involved in the process

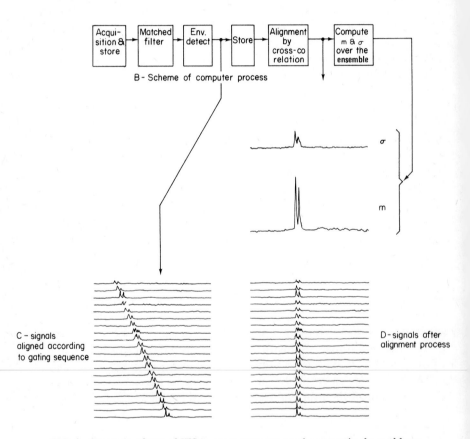

B- Scheme of computer process

C - signals aligned according to gating sequence

D- signals after alignment process

FIG. 8. Example of use of ITSA; measurements on the acoustic channel by a repetitive-ping technique.

3. Fourier transform to separate the component paths and complex-conjugate multiply (CCM). The resulting spectrum has components whose magnitude for each path is the product of the path magnitudes before and after the time interval, and whose phase for each path is the phase difference between the path before and after the time interval. From this phase difference, the path length rate can be determined as explained later. The real and imaginary parts of the CCM are shown in Fig. 9c(2).

4. Interpolate the frequency spectrum by decimating in the time domain, being careful not to alias. The result is Fig. 9c(3), showing real and imaginary part against frequency and against each other ("complex display mode"). Already the phase differences for the two paths can be seen.

5. Transform from cartesian to polar co-ordinates, giving Fig. 9c(4), and cancel phase values having too low a magnitude, giving Fig. 9c(5). In Figs 9c(4, 5) the magnitude and phase are plotted against frequency (equivalent to path delay).

The phase measurements which result are determined by the path length rate ($\dot{r}$) of change according to

$$\varphi = \left(\frac{2\pi}{\lambda}\right) T_{pp}\dot{r}$$

where T_{pp} is the time between transmitted pulses. Supposing a transmit wavelength (λ) of 2.25 m and transmit interval (T_{pp}) of 435 ms,

$$\varphi = 2\pi(0.193)\dot{r} = \pm\pi \quad \text{when} \quad \dot{r} = \pm 2.6 \text{ m/s}$$

This allows the scale of path rate which was shown in Fig. 9c(5).

Requirements for good operation of the technique, apart from obvious instrumental requirements, are:

1. Paths resolved.
2. High Signal/Interference.
3. Time base small enough that path lengths change by less than $\pm\pi$ in phase.
4. Time base large enough that paths do not interfere inter-ping.

The operations required to realize each block of Fig. 9 normally consisted of one ITSA word with three or less parameters. The block operations are normally carried to completion automatically under control of the stored program, but were stopped at illustrative points to form the plots for the figure.

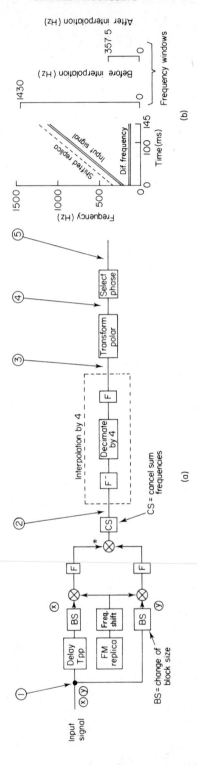

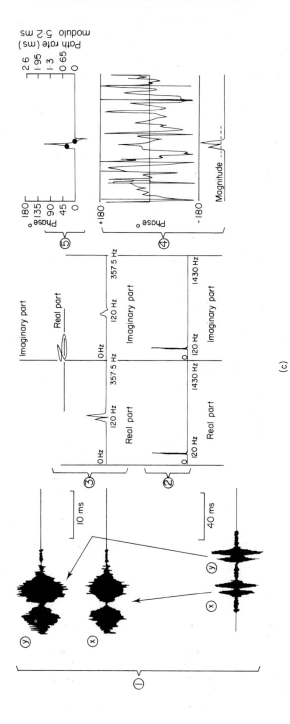

FIG. 9. Example of temporal interferometer using ITSA.

(c)

ACKNOWLEDGMENT

The authors are indebted to Mr. P. Hopford for his very significant contribution to the ITSA implementation.

DISCUSSION

T. Kooij: Did you generate those FM signals with the systems, or did you use a separate generation?

G. Vettori: In the example shown, we were using a separate generator, but the system can generate the waveforms, if required.

B. W. Schroder: In designing the systems interface did you go along the lines of the Karnak system which is used in the field of nuclear reactor control to minimize the hardware changes when different computers are considered?

R. Seynaeve: No.

C. van Schoonerveld: Could you give an idea about the number of functions you can call right away?

R. Seynaeve: Right now, about 110. We keep on implementing new ones whenever they are of general interest. The more specialized functions can be generated as small user FORTRAN programs or BASIC modules as explained earlier, and they usually do not need to be permanently part of the system. A part of our effort is dedicated to a "fast response service" to the users so that new requests can be satisfied quickly (most often between 3 days and 2 weeks).

R. R. Rojas: What are the system's capabilities for multichannel processing like array beam forming. So far you have mentioned only two channels.

R. Seynaeve: Dr. Requicha has been using this system for his work in wide band beam forming. Maybe he will like to give his impressions.

A. A. Requicha: The thing is that if you want to input analogue data from many channels, you cannot apply them to ITSA in the way it is now because it provides only a 2 channel analogue input. You need another system for the data acquisition which is in fact the SPADA system, and then you put, for example, all the data on disc. You then call ITSA to do the processing, taking the data from the disc.

R. Seynaeve: I think this example illustrates the complementarity of our systems. Although in its most recent developments ITSA has provision for more analogue input channels, it should be used with the SPADA system when continuous wide band multichannel recording is required.

R. R. Rojas: Are you able to do parallel processing or do you have to sequence?

R. Seynaeve: You have to sequence.

T. G. Birdsall: What are the cost, core and disc size?

R. Seynaeve: The core size is 24 K being now upgraded to 32 K to improve the performance when multiprogramming is used. The disc size is about 700.000 words.

The complete cost has been of about $175,000, but a smaller configuration is possible and would cost now much less.

R. Bogner: We are developing a somewhat similar installation although not nearly as sophisticated. We found that a box with knobs where one could twiddle various parameters would be very useful.

R. Seynaeve: We have exactly the same feeling. At first we thought that joy-stick would be sufficient, but now we think that such a box would be very useful.

B. Veltman: Who has written the RTx you are using and how much program effort does it take for such a computer?

R. Seynaeve: We use the Real Time Executive System of Hewlett–Packard.

B. Veltman: Do you use the computer for background operations at the same time as ITSA?

R. Seynaeve: The computer is usually in the hands of one team or one user only. In the future it could have for example ITSA running while OC data are being acquired and processed in the background.

G. Vettori: I want to insist on the fact that a great advantage of the minicomputers is that they are "personal" tools, under the complete control of the user. You should not share them between several users at the same time.

B. Veltman: The real time multiprogramming system should give you the feeling that you are the only user.

G. Vettori: I do not believe it until I see it.

R. Seynaeve: I think that in this application where you need a lot of I/O flexibility and you want to take the system to sea, a small "personal" system is often more adequate than a larger one shared by several users.

Sound Propagation Digital Analysis System

A. BARBAGELATA

SACLANT ASW Research Centre
La Spezia, Italy

1. Introduction

Some years ago, the Sound Propagation Group at SACLANTCEN started a research and development program in digital analysis. After passing through various stages we have arrived at a system which can be considered in two parts:

(a) A high density multichannel digital recording system for the direct recording in digital form of the broadband signals collected during sound propagation trials.

(b) A system for transfer of data to a digital computer and consequent data handling, editing, display and analysis.

For a better understanding of the second part, which I think is of greater interest at this Institute, a short description of the acquisition system is necessary.

Data collected during sound propagation trials at sea were normally stored on magnetic tape before analysis. Analogue instrumentation tape-recorders in FM mode were used for this purpose but the technique had many disadvantages:

(a) A frequency band ranging from a few Hertz up to 15 kHz to 20 kHz is normally needed in our investigations. For such a frequency band, a maximum dynamic range of 45 dB can be achieved, so that a precise adjustment of the gain of the receiving system becomes critical. Furthermore, our sea trials are mainly based on the use of explosive charges, i.e. on the study of the impulse response of the medium. Thus, the signals being dealt with require the highest possible dynamic range.

(b) The tape-recorder's time base is not precise because tape speed is not constant. This introduces errors in the measurement of power-spectra when high frequency resolution is required. Angular movement of the tape (skew) introduces relative time fluctuations between the various channels and prevents precise cross-correlation measurements.

In our system which digitizes during the experiment and records on to magnetic tape in digital form, all these difficulties are overcome, because the dynamic range and the accuracy of the time base no longer depend on the tape recorder, but on the preceding electronics, and the time delay fluctuations between signals can be completely compensated.

For more than three years, this equipment, which was designed and built in our laboratory (except the Analogue to Digital Converters and the Tape Recorder), has been the principal working tool of our Group.

2. Short Description of the Digital Recording System

The signals coming from the hydrophones are preamplified and are via the cable transmitted to the recording system [1]. A block diagram of the five-channel digital recording system [4] is shown in Fig. 1. The system is composed of 5 identical data channels, plus a parity-track generator and one counter-track generator. The analogue signals, 1 to 4 for each channel, are fed to a set of amplifiers with gain variable from 0 to 72 dB in 6 dB steps [2] and the outputs are filtered by a set of linear phase filters [3]. After filtering, the analogue signals are converted to digital form in a 15-bit A/D converter with multiplexer. Each 15-bit word, together with the relative 4-bit gain information from the amplifier is transformed into 12-bit floating-point format. Parity and synchronization words are added and the data are then transformed into a special format (NRZ modified) and recorded on ten tracks (two for each channel) of an Ampex AR 1600 wideband analogue tape recorder in direct mode. Track 11 of the tape records parity track which, together with the parity words of each track forms an orthogonal system of parity which allows correction and detection of most of the errors occurring during the reproduce phase. Track 12 records a 7-digit number increasing by one each millisecond, which labels the recorded signals.

Figure 2 shows the system used for reconstitution of the recorded data and transfer to the Elliott 503 computer. Owing to the angular movement of the tape, each track has a different timing, so ten equal circuits are used (one for each track), for bit-restoring, synchronizing, and for discriminating between information-carrying words and parity or synchronization words. Readers for the parity and counter tracks are also provided.

Data are transferred to the Elliott Computer through a high-capacity core memory, capable of storing 315 milliseconds of five 16 kHz bandwidth signals. The transfer is made automatically with remote control of the tape recorder from the computer, which reads the recorded counter

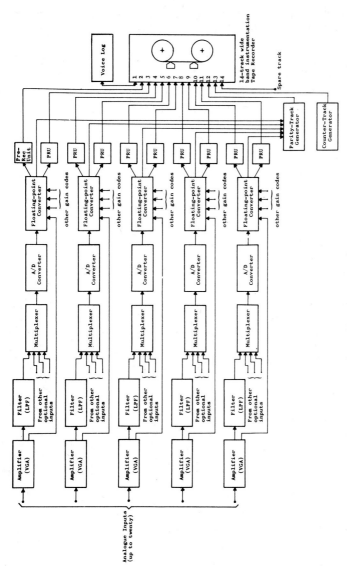

FIG. 1. Digital recording system: Record section.

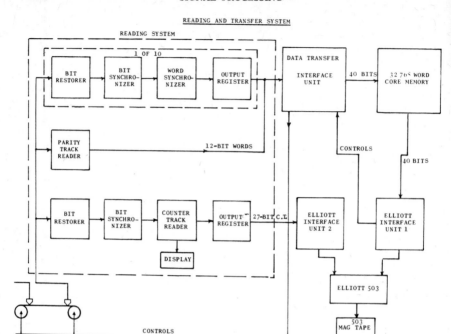

FIG. 2.

track for recognition of the portion of the signal to be transferred. When the indicated counter track number is read, data transfer is started and data flow from the tape through the reading system to the high-capacity core memory, until this is completely full. At this moment data are read block by block from the computer and, after error detection, they are stored for further analysis on the computer magnetic tape.

The system performances are as follows:

TABLE I

Input Signals	5	10	20	Combination
Samples per Second	48 000	24 000	12 000	Combination
Equivalent Bandwidth	16 kHz	8 kHz	4 kHz	Combination
Input Range	2 μV $-$ 7 V rms			
Accuracy of Samples	1%			
Dynamic Range of each Gain Setting	90 dB			
Tape Speed	30 in/s			
Bits per Inch per Track	12 133			
Bits per Second per Track	364 000			
Information-carrying Bits per Second on Tape	2 880 000			
Continuous Recording	1 hour			
Bit Error Rate	~1 x 10^{-7}			

Features of particular relevance are the dynamic range for each gain setting of the amplifier and the high recording density achieved with an error rate of 1×10^{-7}.

Figures 3, 4, 5 and 6 show an example of plots of one of the signals recorded during a sea cruise as drawn by a CALCOMP plotter connected to the Elliott computer. The scale increases from Fig. 3 to Fig. 6 in order to show the very wide dynamic range of the equipment. In order to make the system easy to use and to check, some auxiliary devices have been built. The Digital Recording Test Unit allows an error rate check and analogue display of any selected channel to be carried out.

The MTR Control and Edition Aid is capable of recording and reading the counter track, these functions being coupled with the relative controls of the tape recorder. The system has special provision in order to avoid the over-lapping of two consecutive data blocks on the tape, which is very useful if a recorded event is to be played back before the next is recorded. Registers are also provided for storing and subsequently printing any interesting counter track number, and for exact time location of the portions of signal to be analysed or displayed.

3. The Analysis System

The acquisition system once completed, has remained almost unchanged, apart from some local improvements, while the analysis system has been greatly modified.

At first, data recorded at sea were edited in the laboratory by means of the above auxiliary devices. The scientist could select the portion of signal in which he was interested. This was later transferred to our Elliott 503 computer and stored on magnetic tape. Then, the analysis programs were executed.

This can be considered the first generation of our analysis system. It has greatly improved the capabilities of the group and has made the development of experimental studies possible, which, otherwise, could not have been carried out.

On the other hand, with the increase of our needs in digital analysis, we began to find some shortcomings in this system. Apart from those connected with the use of the Elliott 503 computer, which have been presented in the previous paper, we found that a great deal of time was spent in the tedious operation of edition and increasing troubles were coming from the data transfer to a computer which, due to its age, was no longer in perfect condition.

So we decided to pass on to a second generation. With regard to the equipment, this second generation would have to be based on a modern computer, with small weight and size, in order to facilitate the transport

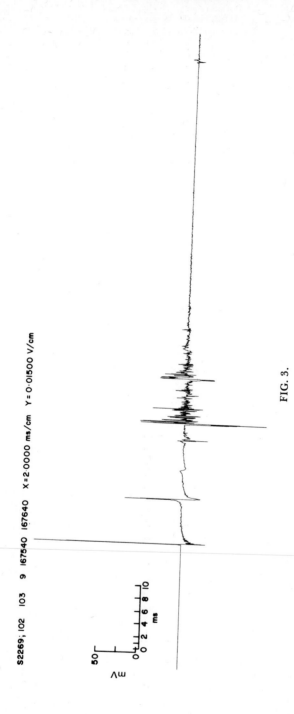

FIG. 3.

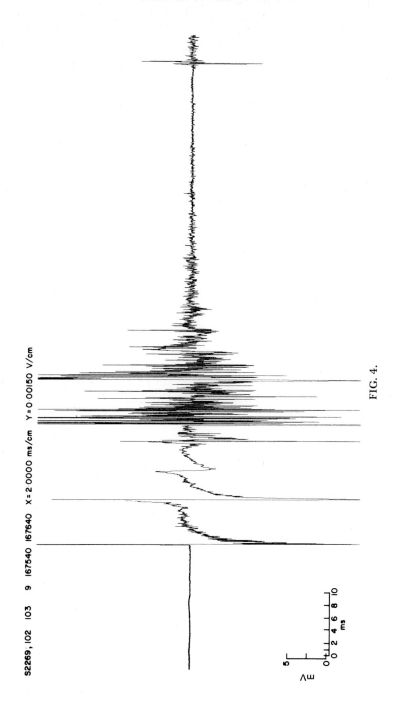

FIG. 4.

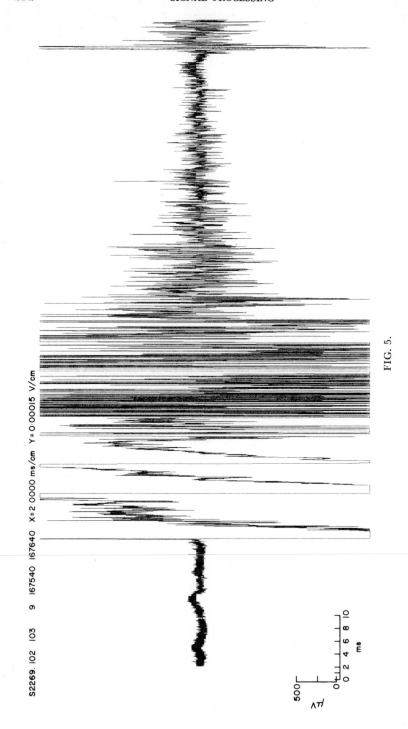

FIG. 5.

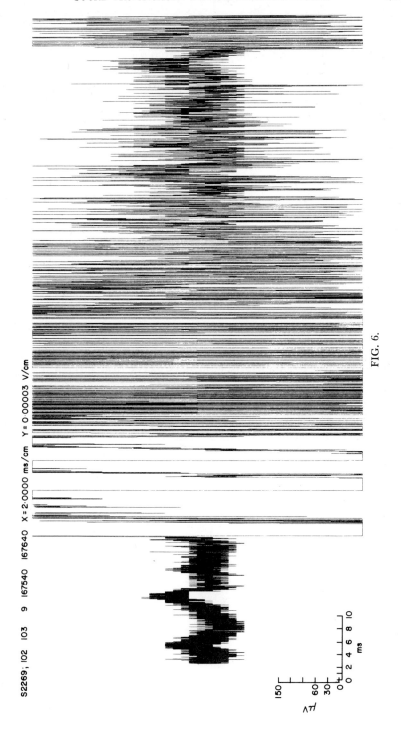

FIG. 6.

between shore and ship, and a great variety of input/output interfaces for peripherals; this meant a "minicomputer".

As regards the operation mode, we wanted a system where the scientist could follow the manipulation of data, know the partial results as they were being produced and take decisions on the further analysis or on the conditions of the experiment.

This meant that the system had to be "conversational" in the sense that the connection between computer and operator had to be easy and not relying on a good knowledge of programming procedures. We chose, therefore, a graphic terminal composed of a storage tube with joystick interactive unit and a keyboard as the central means of controlling the operations. The system also had to be suited to our requirements for use on board during sea trials, which meant slow speed data logging of environmental conditions, broadband data acquisition, ray tracing, etc. We adopted, therefore, a Real Time Executive System which permits the concurrent running of several programs and where each of these can be scheduled on a priority basis under both time and event control.

Furthermore, since in a research centre, ideas are continuously developing, we did not want to have a closed system. The improvement of existing routines, the insertion of new functions, the performing of new operations would have to be as easy as possible: this lead to a modular system.

Following these concepts we built the SPADA (which stands for Sound Propagation Automatic Data Analyser) which with its connection to ITSA (shown in the previous paper) constitutes a complete system of data acquisition, display, edition and analysis.

It is a "conversational", modular system based on the Hewlett–Packard 2116B computer and its Real Time Executive System.

4. The SPADA Hardware

The configuration chosen for the realization of SPADA is shown in Fig. 7

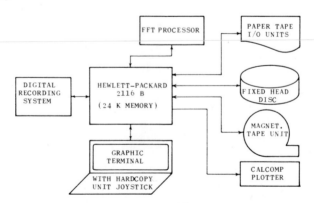

FIG. 7. SPADA configuration.

TABLE II

Unit	HP Number	Specifications
Computer HP 2116B		Memory: 24 K (extendable up to to 32 K)
		Wordsize: 16 bits
		Cycle Time: 1.6 μs
		Options:
		Two Direct Memory Access Channels with a total transfer rate of 625 Kwords per second.
		Extended Arithmetic Unit Power Failure Restart Option
		Memory Protect Option
Paper Tape I/O Units		
Punch	HP 8100 A	Speed: 75 characters/s
Reader	HP 2758	Speed: 500 characters/s
Fixed-Head Disc	HP 2771A	Maximum Size: 737 Kwords
		Transfer Rate: 176 Kwords/s
		Average Access Time: 17.4 ms
Magnetic Tape Unit	HP 7940	Format: NRZI
		Density: 800 bpi
		Speed: 37.5 ips
		Rack mountable
Calcomp Plotter Model 565		Speed: 300 steps/s
		Resolution: 0.01 in
		Plotting Width: 11 in
Graphical Terminal Tektronix T4002		Interfaced via HP TTY-Interface
Display Unit with Terminal Control		Bi-stable Storage Tube
Character Generator		Screen Size: 15.2 cm x 21 cm
		Contrast Ratio: 6:1
		Display Capabilities:
		Graphics:
		1024 x 1024 address points
		1024 x 760 viewable points
		Modes:
		Incremental
		Point
		Linear Interpolation
		Aphanumerics:
		39 lines of 85 characters
		2 size characters
		Speed: 5000 char/s
Interactive Graphic Unit (Joystick) T4901		Crosshair Cursor
Hardcopy Unit T4601		Paper: Dry Silver Paper
		Time per copy: 11 to 18 s
		Copy size: 8.5 x 6 inch up to 8.5 x 14 inch.

(the peripherals used for the oceanographic data acquisition, line printer, etc. are not considered here).

Table II describes the specifications of all the units shown in Fig. 7.

5. General Concept of the SPADA Software

The software of the SPADA consists of computer programs, that:

(i) control specially implemented peripheral devices.
(ii) enable and monitor data collection, data display and data editing.
(iii) initiate data analysis in ITSA.

These programs run automatically or under operator control by means of the Real Time Executive System, which is the standard operating system in the SACLANTCEN.

Figure 8 shows the relationship between the user's elements (I/O drivers and user's programs) and the basic modules of the RTE.

The program routines, which connect the peripheral devices to the RTE system are called DRIVERS. The type and the number of drivers depends on the user's need. The block diagram shows the drivers configured into the RTE for the SPADA. Information flows from devices to the I/O control module of the RTE. The interrupts caused by the peripherals are processed in the Interrupt Control Module.

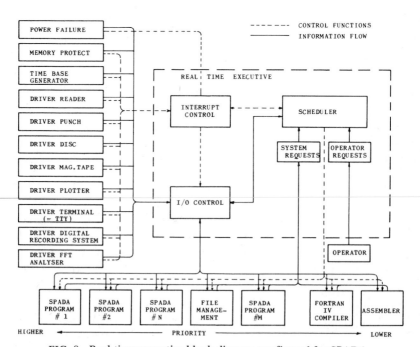

FIG. 8. Real time executive block diagram configured for SPADA.

The number of programs to be incorporated into the RTE is only restricted by the capacity of the disc unit. In our system the central program is called SPADA. It displays the available main functions and initiates them upon the operator's request. After an error-free completion of any function, the program SPADA is rescheduled again and awaits the operator's decision as to what has to follow.

A second basic program in the SPADA system is FILE [5], which is supposed to give the RTE system properties of a Filing System. The access to data files via names is made possible. Compilers for Fortran, ALGOL, and Assembler are permanently incorporated in the RTE, together with program debugging and editing procedures. These programs are background programs with the lowest priority.

The user's programs have direct access to peripheral devices by means of system-requests in the form of read/write statements. These requests are processed in the Scheduler and in the I/O control modules.

In general, the connection of the various devices to the computer is based on HP delivered software drivers and standard "Interface Cards". The connection with the Digital Recording System had, on the other hand, to be developed in our Centre.

This driver takes care of the following operations:

(a) High-speed data transfer from the DRS to the disc.
(b) Input of counter-track numbers.
(c) Remote control of the tape recorder.

The counter-track input uses one DMA channel, the tape movements are interrupt controlled.

Two important factors of the data transfer are that the transfer rate should be 286 Kwords/s and the required storage medium should have the capacity for 600 Kwords. Let us consider the available computer storage possibilities where data can be transferred during an experiment. We can have core memory, disc, and computer magnetic tape. Now, considering our above-mentioned requirements, it can be seen from Table III, referring to 16-bit words, that only the last system will completely satisfy our needs.

TABLE III

	Word Rate w/s	Word Capacity K
Core	600 000	32
Disc	150 000	700
Mag Tape	30 000	12 000
Five-channel Recording System	286 000	1 000 000

A fixed-disc had to be implemented to enable the processing of the data by computer. However, the disc has the disadvantage in that the input transfer rate has to be reduced by a factor of two as the disc is too slow. This means that using a disc memory we can perform on-line data acquisition on only two of five channels. However, all five channels are recorded on our tape recorder, and if we want to have the complete set of signals on the disc we must play back the tape at half speed thus dividing the word rate by two.

Figure 9 shows the two systems of operation, i.e. on line—off line.

The start of the data transfer is either determined by its position on

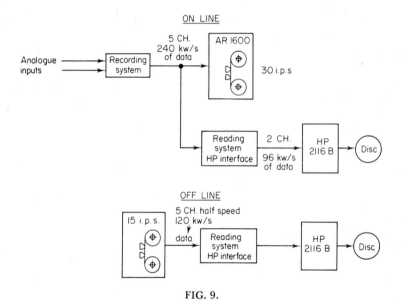

FIG. 9.

the tape (counter track) or it has to be found by means of a level detector. In this case, the data transfer continuously cycles on the disc: when the voltage level is reached, then the last cycle is initiated.

6. The Operation of "SPADA"

When the operator types ON, SPADA the system is initiated. The available function list is displayed (Fig. 10). Due to the modular constitution of SPADA, the number of available functions can easily be extended. In this case, the operator has selected "Data Acquisition" by typing "DA".

Afterwards, the operator is requested if he wants either something new or to repeat the previous sequence. The next question regards the selection of the data source which can be the High Density Digital Recording System or the WORKBOAT DIGITAL RECORDING SYSTEM, a unit capable of

```
                              SPADA

         SOUND PROPAGATION DATA ACQUISITION AND ANALYSIS SYSTEM
                    SELECT BY TYPING TWO CHARACTERS

        AN  =  ANALYSIS
        DA  =  DATA ACQUISITION
        DS  =  DISPLAY FILES
        LA  =  LABEL MAGNETIC TAPE
        MT  =  TRANSFER TO COMPUTER MAGNETIC TAPE
        PL  =  PLOT FILES
        RE  =  RELEASE DISC FILES
        ST  =  STOP SPADA

   DA
```

FIG. 10.

recording acoustical signals with a total sampling frequency of 48 kHz, a length of 4.096 16-bit words, on an IBM compatible 9-track incremental tape recorder, in standard Centre format. It has been developed especially for use on the assist ship.

Having selected "DF" the operator has now to give, if not already given, the configuration of the tape. This means the names given to the various hydrophones on the various channels of the DRS (up to twenty).

Next operation is the selection of the hydrophones to be transferred and the Transfer Start Mode.

An event code is assigned, to which the operator will always refer every time he needs to perform operations on the data to be acquired.

After having specified the acquisition, the data transfer to the computer is initiated.

If the transfer is controlled by the counter track, the tape is positioned in front of the given counter track number and only one disc transfer is executed.

If the transfer and the necessary unscrambling procedure are completed, the digital tape is stopped and program SPADA is rescheduled.

The operator can now proceed to another acquisition or select any other function, for example Display Files.

The operator is requested to give the name of the event he wants to display, then he is informed of the hydrophones available and requested to select some of them.

The selected hydrophones are displayed on the storage tube (Fig. 11). By the side of each curve appears the name given to the hydrophone. At the top, a time-scale in milliseconds is given. In this example we have the explosion of a charge as detected by an array of twenty hydrophones put in parallel two by two. The portion of signal of interest can now be saved on Magnetic Tape defining, by means of the vertical cursor of the joystick,

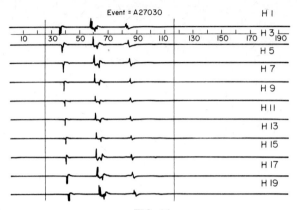

FIG. 11.

the time interval to be saved. No other parameters must be given in order to execute this transfer to Magnetic Tape.

    If a more detailed display of a signal is desired, a single hydrophone can be requested (Fig. 12) and various operations can be executed on the displayed data. The time scale can be changed at will by means of the joystick.

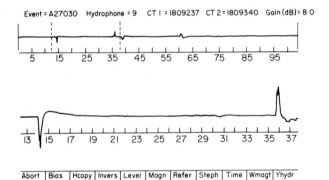

FIG. 12.

Other functions are shown in the lower part of the screen:

A = ABORT	Terminates the display operation.	
B = BIAS	Changes the 0-line.	
H = HCOPY	Copies the screen content on to the Hard Copy Unit.	
I = INVERS	Inverts the polarity.	
L = LEVEL	For measurement of level at a point selected by the joystick (Fig. 13).	
M = MAGN	Magnifies the y-scale up to a certain level given by the joystick.	
S = STEPM	As above but in dB steps.	

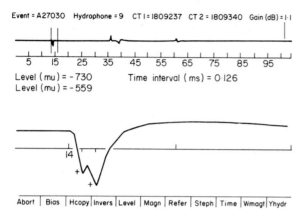

FIG. 13.

R = REFER   Displays the reference curve.
T = TIME   For measurement of time intervals.
W = WMAGT   Transfers to Magnetic Tape the selected portion of
                    displayed hydrophone.
Y = YHYDR   Transfers to Magnetic Tape the selected portion of
                    all the available hydrophones.

A similar conversational procedure is followed for the functions available in the SPADA. In particular calling AF, the ITSA program is scheduled and FFT analysis can be performed on the data acquired. Due to the compatible format, exchange of data files is possible in both directions between SPADA and ITSA.

7. Conclusions

The SPADA system is now extensively used in the Sound Propagation Group where it has met the favour of the scientists involved in signal processing. After an initial test period in laboratory in the "off-line" configuration, it is now used in "on-line" on all our cruises.

Continuous improvements are necessary in order to accommodate the particular needs of the various users, but, due to the modular constitution, these changes are not giving any particular trouble.

As regards the near future, particular emphasis will be given to the use "on-line".

A software controlled multidisplay unit has just been built which gives the possibility of displaying data, results of analysis and environmental conditions on up to four different storage units. In the design stage are the changes of the data format, the increment of the number of the available data channels, the speeding up of the unscrambling operations.

New functions are going to be added in order to broaden the range of application of SPADA.

REFERENCES

1. Figoli A. and Pazzini, M. (1970). "A Low-Noise Hydrophone Preamplifier for Experiments with Explosive Sound Sources", SACLANTCEN Technical Memorandum No. 158, NATO Unclassified.
2. Pazzini, M. (1967). "A Multichannel Receiving System for Experiments in Sound Propagation", SACLANTCEN Technical Report No. 106, NATO Unclassified.
3. Pazzini, M. (1969). "Linear-Phase Filters for a Digital Data-Acquisition System Used in Underwater Sound Propagation Experiments", SACLANTCEN Technical Memorandum No. 148.
4. Barbagelata, A., Castanet, A., Laval, R. and Pazzini, M. (1970). "A High-Density Digital Recording System for Underwater Sound Studies", SACLANTCEN Technical Report No. 170, NATO Unclassified.
5. Diess, B. "Reference Manual for a FILE MANAGEMENT SYSTEM in Connection with HP Disc and Magnetic Tape Units", SACLANTCEN Special Report M-78, NATO Unclassified.
6. Barbagelata, A., Boni, P. and Diess, B. "SPADA—a Conversational Computer System for Acoustical Data Processing", SACLANTCEN Technical Report in preparation.

DISCUSSION

Mr. E. R. Thomas asked about the compensation of the flutter and skew of the tape in the Digital Recording System.

The answer was that the flutter and dynamic skew are compensated by means of digital phase locked loops (one for each track of the tape). Therefore, due to the use of synchronization words in the format and of buffer registers in the reading circuit, it is possible to transfer the words to the computer in the right sequence.

Dr. M. H. Ackroyd asked about eventual troubles due to the use of the disc at sea.

The author replied that no mechanical faults have been found in this long period of use of the disc at sea. On the other hand, care has been taken to locate the disc where pitch and roll were minimum.

Mr. R. S. Thomas asked the author about other commerically available high density digital recording systems.

The answer was that AMPEX is selling a system based on the use of the Miller Code but limited to the data record-reproduce section. The author had no definite information about other existing systems.

Dr. M. H. Ackroyd asked how many times the tape used in the presented recording system could be played back.

The opinion of the author was that the number of errors start increasing after something like one hundred passages of the tape on the heads. This does not become a problem, since the tape is played back only a few times in order to check and edit the data and transfer them to the computer. Then, tapes are normally stored and not utilized for recording. This is not so expensive when one considers that a fifteen-day cruise implies the use of no more than four or five tapes (about 150 events for each tape).

Sound Propagation Effects on Signal Processing

R. LAVAL

SACLANT ASW Research Centre, La Spezia, Italy

1. Introduction

Classical studies of sound propagation are mainly concerned with energy transmission factors. A sufficient amount of energy available at the receiver is of course necessary for the operation of any system in the presence of background noise. Energy transmission factors are however insufficient to evaluate or optimize a system as soon as it includes relatively advanced signal processing techniques. For this, the propagation has to be studied from the point of view of Communications, i.e. from the capability of the medium to preserve the information contained in the propagating signal.

The underwater acoustics communication channel is known to be an extremely complex and imperfect one. Multipath effects, inhomogeneities in the volume of the water, and roughness of the boundaries, surface and bottom introduce various types of distortions, which tend at the same time to complicate the operations involved in optimum signal processing and to degrade the overall system performance. It is obvious that a thorough knowledge of these communication channel properties is essential for the evaluation and optimization of any detection, classification, estimation or communication system using underwater sound as the physical support of information.

It is also essential to have a great knowledge of the nature and quality of the information-carrying "messages" expected to be transmitted through this channel. In passive sonars, the message is contained in the radiated noise of the target. In active sonar, it is represented by the scattering properties of the target. These scattering properties may be expected to be rather complex in the case of a large and complicated structure such as a submarine.

A considerable amount of theoretical work has been devoted to optimum detection and estimation processes from the point of view of the transmitted signals, the array configuration and the noise and reverberation properties, on which may be based the design of sophisticated systems for detection, classification, tracking or communication. Most of these theories, however, have been applied assuming rather simple mathematical models with regard to the propagation effects and the target properties. It is the author's feeling that the realism of these models has to be substantially improved in order to take full advantage of technological progress in designing future systems.

The problem is not simple, since one of the main characteristics of the underwater acoustics communication channel is its considerable variability with the environmental conditions (location, season, weather, etc.) and the geometry.

However, the possibility of using digital computers opens a new era in the field of signal processing. It is now possible to implement by software a sequence of rather sophisticated operations that would have required the construction of monstrous hardware systems a few years ago. The quasi-unlimited flexibility being offered by the software solutions gives the possibility of adapting the processing in order to follow the variations of environmental conditions that may affect the system optimization.

In this presentation we shall try to explain in a rather condensed form the principles of statistical analysis of time–variant space–variant channels, such that they can be applied to define the underwater acoustics communication channel in a form that is convenient for applications to signal processing.

2. Characterization of a Time–variant Channel

The propagation medium between a fixed point source and a receiver may be considered as a filter that operates a transformation on the transmitted signal [1, 2]. Except for signals of very high amplitude, this transformation may be assumed to be linear. It can be defined by its transfer function $H(f)$, which is the frequency response of the equivalent filter, or by its impulse response $h(\tau)$, which is the Fourier transform of $H(f)$. In a time-varying channel, the transfer function (and impulse response) have to be considered as time dependent functions. We then have:

The time-dependent transfer function: $H(f, t)$
The time-dependent impulse response: $h(\tau, t)$

By Fourier transforming H and h with respect to t, we obtain two new functions [2] of a new variable Φ:

The bifrequency function: $B(f, \Phi)$,
The spreading function: $s(\tau, \Phi)$.

Any one of the four system functions, $H(f, t)$, $h(\tau, t)$, $B(f, \Phi)$, $s(\tau, \Phi)$, may be used to define completely the time–varying channel between source and receiver. If it were exactly known, the medium's transformation could be completely compensated for in the receiver by applying the inverse transformation to the received signals. In practice, however, this is generally not possible. The filtering functions of the propagation medium are rather complicated, as they represent the global effect of a large number of different physical factors. The signal travels via different paths between source and receiver, each of these paths corresponding to a different time of arrival, and a different distortion. The global transformation defined by $H(f, t)$ or any of its Fourier transforms may be considered as the combination of a number of elementary transformations associated with the different physical effects that may distort the signal along each path.

These elementary transformations can be divided into two categories:

(i) *Transformations of a deterministic nature*: Multipath effect, reflection from a flat-layered bottom, dispersive effects in a homogeneous wave guide (shallow water, or surface duct), etc.

As we have already mentioned, these transformations, if they are known (from theoretical models or local measurements for instance), do not reduce the information content of the signal, as they can be compensated for, in principle, at the price of an additional complexity in the structure of the optimum receiver.

(ii) *Transformations of a random nature*: They are mostly due to the scattering effects associated with the small-scale inhomogeneities that characterize the volume of the water (temperature microstructure, turbulence, interval waves) and the boundaries (surface and bottom roughness). They can only be described in a statistical way, and their effect cannot be compensated for in the receiver. They are therefore associated with an information degradation.

3. Characterization of a Random Time–variant Channel

When the transformations operated by the medium are of random nature, the four system functions defined earlier become random functions.

Following References [2] and [3] each can be described at the second order by a four variable auto-correlation function:

$$R_H(f, f', t, t') = \overline{H(f, t)H^*(f', t')}$$
$$R_h(\tau, \tau', t, t') = \overline{h(\tau, t)h(\tau', t')}$$
$$R_B(f, f', \Phi, \Phi') = \overline{B(f, \Phi)B^*(f', \Phi')}$$
$$R_S(\tau, \tau', \Phi, \Phi') = \overline{S(\tau, \Phi)S^*(\tau', \Phi')}$$

where the bar denotes ensemble average.

Introducing some restrictive assumptions as to the nature of the random process, they may be reduced to functions of two variables only.

These restrictions concern two things [3]:

(i) *Wide sense stationarity in time*: The channel correlation functions are invariant under a translation in time, thus the correlation functions R_H and R_h depend only on the time difference Δt, and not on the absolute time t and t'. As a consequence, the correlation functions, R_B and R_S are zero when $\Phi \neq \Phi'$.

(ii) *Uncorrelated scattering*: The auto-correlations R_h and R_S are zero when $\tau \neq \tau'$. As a consequence, the correlation functions R_H and R_B will only depend on the frequency difference Δf and not on the absolute frequencies f and f'. This may be expressed by saying that the system has Wide Sense Stationarity in frequency.

The condition of uncorrelated (time) scattering is equivalent to a condition of Wide Sense Stationarity in frequency.

Similarly, the condition of Wide Sense Time Stationarity is equivalent to a condition of uncorrelated frequency scattering.

When these two conditions occur together we have a Wide Sense Stationary Uncorrelated Scattering Channel (WSSUS channel). The four auto-correlation functions are reduced to functions of two variables only:

$$R_H(\Delta f, \Delta t)$$

$$R_h(\tau, \Delta t)$$

$$R_B(\Delta f, \Phi)$$

$$R_S(\tau, \Phi)$$

These four functions are Fourier transforms of each other. The function $R_H(\Delta f, \Delta t)$ is called the Time–Frequency correlation function. The function $R_S(\tau, \Phi)$ is called the scattering function. It has received some particular attention in radar and sonar for the following reason:

Under ideal propagation conditions the range and doppler resolution of an echo-locating system is given by the time frequency ambiguity function $G(\tau, \Phi)$ of the transmitted signal $x(t)$. $G(\tau, \Phi)$ is the squared output of a filter matched to $x(t)$ in the time shift τ and frequency shift Φ domain when $x(t)$ is applied to the input.

The range and doppler resolution of the same system in a random inhomogeneous medium will be expressed by an ambiguity function $E(\tau, \Phi)$ expressing the combined effect of the signal and the medium. $E(\tau, \Phi)$ may be considered as the mean squared output of the filter matched on $x(t)$ when the signal returned from a point target is applied to the input.

For narrow-band signals, and on the assumption that the channel is WSSUS it can be shown that the combined ambiguity function of the signal and the medium is obtained by convolving the ambiguity function with the scattering function of the medium:

$$E(\tau, \Phi) = \iint G(\tau - \zeta, \Phi - \xi) R_S(\tau, \Phi) \, d\zeta \, d\xi.$$

$R_S(\tau, \Phi)$ may then be considered as another ambiguity function, which describes the inherent resolution limitation of the medium in the range–doppler space.

A constant velocity point target will no longer be seen as a point but as a "cloud" in the range–doppler space. $R_S(\tau, \Phi)$ defines the energy density of this cloud.

Let us consider some properties of this scattering function:

The signal ambiguity function $G(\tau, \Phi)$ of a signal $x(t)$ is known to have strong restrictive properties.

It is a real non-negative function, symmetric in τ and Φ, responding to the "uncertainty principle" condition:

$$G(0, 0) = 1 \quad 0 \leqslant G(\tau, \Phi) \leqslant 1$$

$$\iint G(\tau, \Phi) \, d\tau \, d\Phi = 1$$

Even when these conditions occur only a limited class of functions of two variables may be regarded as possible ambiguity functions of a signal.

Except for being real non-negative, the scattering function does not have any of these restrictions and it can take any shape.

For ideal propagation conditions, it reduces to a Dirac pulse (no ambiguity). The channel is then said to be "totally coherent".

It frequently happens that $R(\tau, \Phi)$ is the sum of a Dirac pulse $k\delta(\tau_0, \Phi_0)$ and a spread function $r(\tau, \Phi)$.

$$R(\tau, \Phi) = k\delta(\tau - \tau_0, \Phi - \Phi_0) + r(\tau, \Phi)$$

where k is a constant.

The channel is then said to be partially coherent. This point will be considered later on.

The two-dimension scattering function $R(\tau, \Phi)$ (or any of its three Fourier transforms) is a complete statistical description at the second order of a WSSUS channel.

For reasons of simplicity (experimental or theoretical), this is often replaced by a set of two one-dimension functions:

The range (time) scattering function $R_\tau(\tau)$.

The doppler (frequency) scattering function $R_\Phi(\Phi)$ which can be defined by

$$R_\tau(\tau) = \int R(\tau, \Phi) \, d\Phi$$

$$R_\Phi(\Phi) = \int R(\tau, \Phi) \, d\tau.$$

$R_\tau(\tau)$ can be considered as the mean squared value of the impulse response $h(\tau, t)$, which is no longer a function of t:

$$R_\tau(\tau) = \overline{h^2(\tau, t)},$$

where the bar indicates ensemble average.

The averaging can be taken, for instance, on a series of squared impulse responses taken at large time intervals as compared with the time correlation scale of the channel (ping-to-ping averaging of the impulse response squared envelopes).

$R_\Phi(\Phi)$ can be considered as the mean squared value of the bifrequency function $B(f, \Phi)$, which is no longer a function of f:

$$R_\Phi(\Phi) = \overline{|B(f, \Phi)|^2}$$

When the transmitted signal is sinusoidal, of frequency f_0, the power spectrum of the received signal is

$$R_\Phi(f_0 + \Phi)$$

In theory, the two functions $R_\tau(\tau)$ and $R_\Phi(\Phi)$ are not sufficient to describe the process statistically. The combined range and doppler ambiguity $R(\tau, \Phi)$ indeed, cannot be uniquely reconstructed from the knowledge of the range ambiguity $R_\tau(\tau)$ and the doppler ambiguity $R_\Phi(\Phi)$.

In practice, however, it happens frequently that $R_s(\tau, \Phi)$ can be reasonably approximated by the product of $R_\tau(\tau)$ by $R_\Phi(\Phi)$:

$$R_S(\tau, \Phi) \approx R_\tau(\tau) R_\Phi(\Phi)$$

Or, in the case of partial coherence, by

$$R_s(\tau, \Phi) \approx k\delta(\tau_0, \Phi_0) + r_\tau(\tau) r_\Phi(\Phi).$$

Generally speaking, however, such a kind of simplification only applies when the scattering can be associated with a single, well-defined physical process, such as volume or surface scattering. It is not valid for the total process associated with a multipath channel.

4. Characterization of a Space–variant Random Channel

Forgetting time dependence for the moment, the frequency response $H(f)$ and the impulse response $h(\tau)$ are obviously functions of the positions of the source and receiver.

They can be written in the form

$$H(f, \vec{s}, \vec{r}) \quad \text{and} \quad h(\tau, \vec{s}, \vec{r}),$$

where $\vec{s}$ and $\vec{r}$ are the positioning vectors of the point source and receiver.

By analogy with the time-dependent channel, these functions may be Fourier transformed with respect to variables in space, as they have been with respect to the variable time in the preceding section [4].

Let us first reduce the problem to one-dimension in space by considering the variation of either the source or the receiver along the coordinate x (whatever direction the x-axis has).

The x-dependent transfer function and impulse response are $H(f, x)$ and $h(\tau, x)$, and their Fourier transforms with respect to x are $B(f, U)$ and $s(\tau, U)$. The variable U corresponds, for the space–variant channel, to the variable Φ (frequency-spread variable) for the time–variant channel.

We can define a WSSUS channel in space by replacing the condition of time stationarity by a condition of space stationarity along x, and we shall get a system of four anto-correlation functions that are Fourier transforms of each other and depend on two variables only. In particular, one of these functions is a scattering function

$$R_S(\tau, U).$$

The transformation can be extended to the case of a two-dimension variation in space, along the coordinates x and y.

By performing a two-dimension Fourier transform of $H(f, x, y)$ and $h(\tau, x, y)$ with respect to x and y we get a function of u and v:

$B(f, u, v)$ equivalent to the bifrequency function,
$S(\tau, u, v)$ equivalent to the spreading function.

We can also define the equivalent of a scattering function (within the WSSUS condition extended to the two dimensions x and y):

$$R_S(\tau, u, v).$$

The problem is now to find the physical significance of the new variables u and v.

Let us consider that x and y represent the horizontal and vertical coordinate of the receiver in a plane perpendicular to the average propagation direction.

The transfer function $H(f, x, y)$ will describe the perturbation of amplitude and phase of a monochromatic wave of frequency f intercepting the x, y plane. (We only consider a limited domain in this plane, where the sphericity of the wave can be neglected.)

The theory of plane wave decomposition states that $H(f, x, y)$ may be considered as the result of an infinite number of elementary plane waves intercepting the array.

Each one is defined by its direction $\vec{\theta}$, (which can be decomposed into a horizontal and a vertical direction θ_H and θ_V). The amplitude and phase of these elementary waves are given by the directionality function:
$d(f, \theta_H, \theta_V)$.

$H(f, x, y)$ and $d(f, \theta_H, \theta_V)$ are known to be related by the Fourier Transform relation:

$$d(f, \theta_H, \theta_V) = \int\int H(f, x, y) \exp\left\{-2\pi if\left[\frac{\sin\theta_H}{c}x + \frac{\sin\theta_V}{c}y\right]\right\} dx\, dy$$

which shows that the function $B(f, u, v)$ and $d(f, \theta_H, \theta_V)$ may be identified by putting:

$$u = \frac{f}{c}\sin\theta_H \quad \text{and} \quad v = \frac{f}{c}\sin\theta_V$$

Then:

$$B(f, u, v) = B\left(f, \frac{f}{c}\sin\theta_H, \frac{f}{c}\sin\theta_V\right) \equiv d(f, \theta_H, \theta_V)$$

We can define in the same way a spatial scattering function (corresponding to the doppler scattering function of the time variant channel):

$$R_{\vec{\theta}}(u\cdot v) = \overline{|B(f, u, v)|^2}$$

which no longer contains the term f as a separate variable.

$$R_{\vec{\theta}}(u, v) = R_{\vec{\theta}}\left(\frac{f}{c}\sin\theta_H, \frac{f}{c}\sin\theta_V\right) = |d(f, \theta_H, \theta_V)|^2 =$$

$$= R_{\vec{\theta}}\, \frac{f}{c}\sin\theta_H, \frac{f}{c}\sin\theta_V$$

then gives the angular power spectrum of the wave intercepting the receiving plane x, y, when the source transmits a pure frequency f.

$R_{\vec{\theta}}$ may be considered as an ambiguity function in the angular domain $\vec{\theta}$, in the same way as $R_\Phi(\Phi)$ was an ambiguity function in the frequency-spread domain Φ. The difference is the frequency dependence of the angular ambiguity function, due to the presence of a factor f in the expressions of u and v in function of θ_H and θ_V. We can define indeed a global system-medium angular ambiguity function which results from the combination of a medium ambiguity function and a system ambiguity function. Let us consider the system as being a receiving array in the plane x, y. Such an array has a directivity function $e(\theta_H, \theta_V)$ related to the array weighting function $w(x, y)$ by the Fourier transform relation:

$$e(f, \theta_H, \theta_V) = \int\int w(x, y) \exp\left\{-2\pi if\left[\frac{\sin\theta_H}{c}x + \frac{\sin\theta_V}{c}y\right]\right\} dx\, dy$$

The power directivity function of the array can be written in the form:

$$|e(f, \theta_H, \theta_V)|^2 = D\left(\frac{f}{c}\sin\theta_H, \frac{f}{c}\sin\theta_V\right) = D(u, v)$$

D represents the array ambiguity function in the angular space.

The total angular ambiguity function $E_{\hat{\theta}}$ of the array plus the medium results from the convolution between the system ambiguity function $D(u, v)$ and the medium's ambiguity function $R_{\hat{\theta}}(u, v)$

$$E_{\hat{\theta}}(u, v) = E_{\hat{\theta}}\left(\frac{f}{c}\sin\theta_H, \frac{f}{c}\sin\theta_V\right) = D(u, v) * R_{\hat{\theta}}(u, v)$$

In other terms, a point source will no longer be seen as a point by a very directive array, but a "cloud" in the angular space. $R_{\hat{\theta}}(f/c \sin\theta_H, f/c \sin\theta_V)$ gives the density distribution of this cloud, the angular dimension of which is proportional to $1/f$.

The angular scattering function $R_{\hat{\theta}}(u, v)$ is related to the space scattering function $R_S(\tau, u, v)$ by

$$R_{\hat{\theta}}(u, v) = \int R_S(\tau, u, v)\, d\tau$$

$R_S(\tau, u, v)$ is the medium ambiguity function in both range and angle.

The ambiguity function $G(\tau, u, v)$ of a sytem in the range–angle space, will just be the product of two independent ambiguity functions:

$$G(\tau, u, v) = R(\tau) \cdot D(u, v)$$

where $R(\tau)$ is the range ambiguity function of the transmitted signal and D the power directivity function of the directive array. The total range-angle ambiguity function will result from the convolution of the system and medium ambiguity functions:

$$E(\tau, u, v) = [R(\tau) \cdot D(u, v)] * R_S(\tau, u, v)$$

A point source will be seen by a high resolution range–angle system as a cloud in a three-dimension space.

Another way to visualize the random space–variant medium is the following:

The perturbations, due to the random nature of the medium, cause a corrugation of the front waves.

The complex transfer function H may be written in the form:

$$H(f, x, y) = A(f, x, y)\, e^{j\varphi(f, x, y)}.$$

In a first approximation, if the correlation distance of the fluctuations in the x, y plane are much larger than the wave length, the wavefront corresponding to the frequency f is the three-dimension surface defined by the equation:

$$\varphi(f, x, y) - \frac{zf}{c} = 0,$$

where z is the coordinate along the propagation axis. Resolving the equation in z gives:

$$z(f, x, y),$$

which gives the deviation of the front from the plane x, y as a function of the x and y coordinates and for a given frequency f.

The local direction of propagation $\vec{\alpha}(x, y)$ can be defined as the local perpendicular to the corrugated wavefront. $\vec{\alpha}(x, y)$ can be decomposed into a horizontal and a vertical direction α_H and α_V.

One can define the probability density $p(f\vec{\alpha}) = p(f, \alpha_H, \alpha_V)$ of this local direction of propagation, which is in general a function of f.

$p(f, \alpha_H, \alpha_V)$ is not identical to the angular scattering function $D(f/c \sin \theta_H, f/c \sin \theta_V)$. $p(\vec{\alpha})$ is related to $D(\vec{\theta})$ in the same way as the instantaneous frequency distribution of a modulated time signal is related to its Fourier power spectrum.

5. The Time-Variant Space-variant Random Channel

The time-variant space-variant random channel can be described by the transfer function or impulse response

$$H(f, t, \vec{s}, \vec{r}) \quad h(\tau, t, \vec{s}, \vec{t})$$

Reducing the space variation to two coordinates x, y of the receiver in a plane perpendicular to the propagation, (by analogy with the preceding sections), we can define the range, doppler, angle ambiguity function as a scattering function of four variables

$$R_S[\tau, \Phi, u, v]$$

$$\text{with} \quad \left| \begin{array}{l} u = \dfrac{f}{c} \sin \theta_H \\[2mm] v = \dfrac{f}{c} \sin \theta_V \end{array} \right.$$

which represents a point source as a cloud in a four-dimension range-doppler–angle space.

6. Case of Moving Source and/or Receiver

In the most general case of a time-variant space-variant channel, the transfer function and impulse response may be written

$$H(f, t, \vec{s}, \vec{r})$$

$$h(\tau, t, \vec{s}, \vec{r})$$

If source and receiver are moving, their respective coordinates $\vec{s}$ and $\vec{r}$ become functions of time, and the time–varying source-receiver channel can be characterized by

$$H[f, t, s(\vec{t}), r(\vec{t})]$$

or $h[\tau, t, s(\vec{t}), r(\vec{t})].$

Let us imagine, to simplify the problem, that we have a purely space variant medium, and that the receiver is moving along the horizontal coordinate x, perpendicular to the propagation, with a constant velocity v.

The impulse response $h(\tau, x)$ will be written $h(\tau, vt)$.

The Fourier transform of $h(\tau, x)$ with respect to x gives the spreading function:

$$s(\tau, U).$$

The Fourier transform of $h(\tau, vt)$ with respect to t will give:

$$s(\tau, v\Phi)$$

and the range–angle spreading function becomes a range–doppler scattering function:

$$s(\tau, v\Phi)$$

the scattering function $R_S(\tau, U)$ will be reduced in a similar way to $R_S(\tau, v\Phi)$.

In a dynamic situation, a purely space–variant effect, such as the bottom roughness, will then generate doppler spreading.

In a time–variant space–variant medium, the doppler spreading due to time variance and the doppler spreading due to the dynamic space–variance effect are combined together.

It may be interesting to give an indication of the relative order of magnitude of these two effects:

For surface scattering, and for a source and/or receiver moving at a few knots, the two effects are comparable.

For volume scattering, except for very particular oceanographic situations, time variance is associated with interval motions of the water masses which are generally much slower than the slowest motion of a drifting platform. In this case, volume time–variant effects are masked by the dynamic space–variant effects.

7. Partially Coherent Scattering

We have seen that the medium transfer function $H(f)$, whether it applies to a time–variant, space–variant or space–time–variant channel, may be expressed in the complex form:

$$H(f) = A(f)\, e^{i 2\pi \varphi(f)},$$

where $A(f)$ and $\varphi(f)$ represent the amplitude and phase fluctuation of a monochromatic sound field. It may be represented by a modulation vector, which varies in time, in space or both.

When the random phase function $\varphi(f, t, x, y)$ is equidistributed over 2π, the process is said to be totally incoherent.

If this is not the case, a mean phase value $\bar{\varphi}$ can be defined and the phase φ fluctuates around this mean value, as the amplitude A fluctuates around its mean value $\bar{A}$.

It is possible to decompose the vector $Ae^{i\varphi}$ into the sum of a fixed vector $\bar{A}e^{i\bar{\varphi}}$ and a phase equidistributed vector ρ of root mean square amplitude $\bar{\rho}$.

In other words, the random process can be divided between a totally coherent deterministic process and a totally incoherent random process. It is said to be "partially coherent".

The "coherence factor" can be defined as a coefficient:

$$\gamma = \frac{\bar{A}^2}{\bar{A}^2 + \bar{\rho}^2} \quad \text{with} \quad 0 \leqslant \gamma \leqslant 1$$

The scattering function $R(\tau, \Phi, \theta_H, \theta_V)$ will thus be composed of a Dirac pulse, plus a density spread scattering function r:

$$R(\tau, \Phi, \theta_H, \theta_V) = G[\gamma\delta(\tau_0 - \tau, \Phi_0 - \Phi, \theta_{OH} - \theta_H, \theta_{OV} - \theta_V)$$
$$+ (1 - \gamma)r(\tau, \Phi, \theta_H, \theta_V)]$$

where G is the total energetic gain of the channel and $G. \gamma$ the coherent gain.

The physical significance of a partially coherent scattering process is that a point source in the range–doppler–angle space will no longer be seen as a simple point source alone, but as a point-source surrounded by a diffusion halo. This effect is frequently observed in atmospheric light propagation.

8. Validity of the Concept of Scattering Function

We have seen that the channel ambiguity function in range–doppler bearing may be completely characterized by a scattering function that contains only one variable associated with the ambiguity in range, in doppler and in vertical and horizontal bearing.

The condition for the existence of such a scattering function is that the channel should be WSSUS, which can be formulated by saying that the random transfer function must be wide-sense stationary in frequency, in time and in space.

We can question whether these three conditions are verified for the case of underwater acoustics.

Formally speaking, the reply is negative: none of the three conditions of stationarity is strictly fulfilled.

(i)   All the scattering processes we have to deal with in underwater acoustics are strongly frequency dependent, therefore, associated with the frequency-dispersion effect and the "uncorrelated scattering" condition is never fulfilled.

(ii) The channel characteristics vary from season to season, day to day and day to night, and follow closely the weather conditions.

(iii) The propagation effects depend strongly on the respective positions of source and receiver and the space stationarity condition is not fulfilled.

In order to save the concept of the scattering function we may try to replace the strong conditions of wide-sense stationarity by weaker conditions of local stationarity.

Under this assumption, the condition of stationarity should only be considered within a limited range of variation in time, space and frequency.

The time–space–frequency space should be divided into elementary domains within which the process may be considered WSSUS. For each of these domains a scattering function should be defined.

A given channel (a given "province" of the ocean) should thus be characterized by a time–frequency–space dependent scattering function, which will define the medium ambiguity function for all seasons (and weathers), for all frequency bands, and for all source-receiver geometries.

This concept is only meaningful provided all the dimensions of the systems being considered for practical applications are small as compared with the local stationarity domains.

This is generally the case for active sonar, which is characterized by a narrow frequency bandwidth and a compact array.

The frequency stationarity condition is not fulfilled for broadband systems, such as passive or explosive echo-ranging.

The condition of space stationarity may not be fulfilled for large size arrays if the array is larger than the local stationarity domain. This may happen in particular for vertical arrays in the surface channel, or in shallow water, where the propagation conditions vary rapidly with depth.

Furthermore, the notion of scattering function, is based on the assumption that the propagation is a purely random process.

We have seen that the global underwater acoustics propagation process is, in general, the combined effect of several physical phenomena, which can be divided into deterministic and random processes.

The presence of some deterministic transformation processes between source and receiver generally invalidate the condition of uncorrelated scattering, and the concept of scattering function has thus to be used with great precaution. That is what we shall try to illustrate in the next chapter.

8. Composition of Several Scattering Processes

Let us consider a simple typical case of multipath propagation (Fig. 1). Transmitter and receiver are connected through a direct path, a surface-reflected path, and a bottom-reflected path. The direct path is subject to volume-scattering effect. The surface-reflected path is subject to both volume scattering and surface scattering, and the bottom-reflected path to both volume scattering and bottom scattering.

The complete channel may be considered as a composite filter (Fig. 2). The direct path is a "volume scattering filter". The surface-reflected path

FIG. 1. A simple case of multipath propagation.

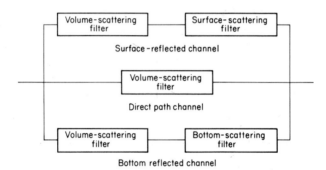

FIG. 2. Composite filter corresponding to Fig. 1.

is composed of a "volume-scattering filter" and a "surface-scattering filter" connected in series. The bottom-reflected path is composed of a volume- and a bottom-scattering filter in series. The three lines of filters are connected in parallel. Let us consider that all five filters are independent random filters (as the rays have travelled in different pieces of water). The volume and surface scattering filters are time and space-variant. The bottom filter is only space-variant.

If all the elementary filters are random and totally incoherent, a scattering function of the composite filter will be obtained by convolving the transfer function of the filters in series and adding the transfer functions of the parallel channels.

If the elementary filters are partially coherent, the composition becomes more complex. Each filter can be represented by a circuit composed of an attenuator γ representing the coherent propagation loss of the channel, placed in parallel with an incoherent random filter r representing the purely incoherent scattering (Fig. 3).

If all the filters in series in a given channel are partially coherent, the total channel will be partially coherent. The composite coherence factor of the channel will be the product of the elementary coherence factors of the filters in series. It is sufficient, however, that one of the filters in series be totally incoherent to have the complete channel incoherent.

If more than one of the parallel channels are partially (or totally) coherent, the total impulse response will contain more than one Dirac pulse at different fixed locations. This violates the condition of uncorrelated scattering and the total statistics of the process can no longer be expressed by means of a scattering function alone.

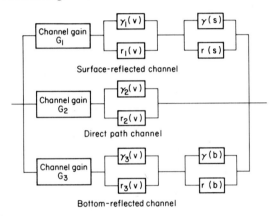

FIG. 3. Division of the filter shown in Fig. 2 into coherent and incoherent circuits.

The difficulty can be avoided by finding an equivalent composite filter to the one represented in Fig. 3. Such a filter is represented in Fig. 4.

The serial parallel combination of this equivalent filter is divided into two main channels, which are themselves in parallel:

The channel A at the top of the picture is a serial parallel combination that contains the γ terms only. Its impulse response is then reduced to a number of Dirac pulses at different locations in the range-angle space (three in the case of the picture). Channel A is then a totally deterministic filter, characterized by a multi-Dirac-pulse impulse response.

The bottom channel B is a serial parallel combination of the γ and the r circuits. It is characterized by the fact that at least one of the circuits in each parallel branch is a random incoherent filter.

Channel B is then a totally incoherent random filter, which can be statistically defined by its scattering function. This scattering function results from a rather complicated serial parallel combination of the scattering functions associated with the elementary filters. (The complexity grows very rapidly with the number of filters in series in each channel.)

Up to this point we have only considered the composition of several elementary random processes where the terms γ are constant. The composition of both random and deterministic processes can be done in a similar way. It is sufficient for this to consider the γ as deterministic functions of frequency, time, and space.

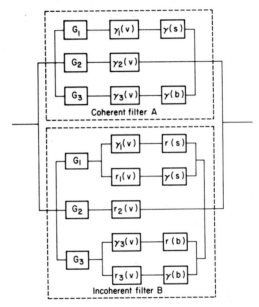

FIG. 4. Equivalent composite filter of Fig. 3 decomposed into a coherent filter A and an incoherent filter B.

The composite filter can be decomposed into a purely deterministic filter A (which may or may not include some Dirac pulses) in parallel with a totally incoherent random filter B. The deterministic filter A, in fact, represents the mean value, or first order statistics of the random process characterizing the channel. The filter B represents the deviation around the mean, or second order statistics of the process.

The distinction between the random and the deterministic processes, however, is not always obvious. The propagation in a surface duct, or in shallow water, may be considered as a super-position of a very large number of rays (or normal modes), which lead to extremely complicated impulse responses, even for the case of an extremely simple model (flat bottom and surface, and a constant sound speed gradient, for instance). Such impulse responses vary rapidly with the receiver position in depth and distance, (x and z coordinates) but is invariant in function of variations along the horizontal axis perpendicular to the propagation. It may be convenient to treat it as a random function of x and z.

9. Conclusion

As mentioned in the Introduction, sound propagation studies should no longer be limited to the energy transmission factors, but should describe the medium as a communication channel. This description can take the form of defining a composite time–space-variant filter equivalent to the channel, which can be split up into a deterministic filter in parallel with a purely incoherent random filter. Such an equivalent filter contains all the information necessary for the evaluation and the optimization of any detection, estimation or communication systems.

As the characteristics of this filter present a considerable amount of variability it cannot be precisely defined, *a priori*, for every possible location, season, weather, frequency band and geometry. The methods for improving our knowledge of the channel characteristics may be considered from three different points of view: The scientific approach; The empirical approach; The local estimation approach.

The scientific approach. This consists of conducting a research programme tending to explain the physical processes involved during propagation. The processes of a different nature, such as surface scattering, volume scattering, bottom reflection, multipath, may be first studied separately. The global properties of the channel may then be defined through a number of models that combine the influence of the elementary processes. Such studies generally associate theoretical and experimental work. Several examples of such research work carried out at SACLANTCEN will be given during this presentation at the Institute. The reader can find these examples in References [5] to [11|.

The empirical approach. This approach consists of collecting a large number of data concerning direct measurements of the global equivalent filter characteristics, with the purpose of extracting some empirical rules from a statistical analysis of all the data. Such a method may be effective when the purpose consists of evaluating the performances of an existing equipment. In general, the equipment itself constitutes the measuring tool for the evaluation of its own performance.

It suffers from an inherent defect, on the other hand, when the purpose of the operation is to define the optimum parameters and characteristics of a future system. In order to give useful information to the designer, the measurement equipment has to present a greater accuracy and a larger range of parameter variations than the final system to be designed. The measurement equipment has then to be more complex than the final system will be.

The local estimation approach (also called *"on the spot" estimation*). This consists of estimating the channel properties from measurements collected simultaneously with the operation of the system. The local estimation can be visualized at two successive stages of operation:

(a) The measurements concerning the environment (sound speed profile, wind speed, bottom depth) may be used as the input parameter of a computer model that will predict the channel properties.

(b) The channel properties can be directly evaluated from acoustic measurements. This is commonly done in communications, when the signal available at the receiver contains a certain amount of information that concerns the channel itself. This information can be used to improve the receiver, the total process of successive approximations leading to an "adaptive process". Similar examples can be found in tracking systems, or classification systems, where ping-to-ping analysis may be to the progressive separate estimation of the channel parameters and the target parameters.

With the development of digital computers, the local estimation approach will probably represent the future trend of all advanced signal-processing systems.

REFERENCES

1. Laval, R. (1964). "Transformation Aléatoire des Signaux par la Propagation", paper presented at NATO Advanced Study Institute on Signal Processing, Grenoble.
2. Sostrand, K. A. (1968). "Measurement of Coherence and Stability of Underwater Acoustic Transmissions", paper presented at NATO Advanced Study Institute on Signal Processing, Enschede 1968.
3. Bello, P. A. (1963). Characterization of randomly time-variant linear channels. *IEEE Trans. Communication Systems*, **CS-11**, No. 4, 360–393.
4. Hovem, J. M. Resolution Limitations of a Random Inhomogeneous Medium", SACLANTCEN Technical Report in preparation.
5. Laval, R., Fortuin, L., Gastmans, R., Parkes, D. J., Pazzini, M. and Castanet, A. (1967). "Coherence Problems in Underwater Acoustic Propagation", paper, presented at the NATO—Marina Italiana Advanced Study Institute on Stochastic Problems in Underwater Sound Propagation, Lerici, September 1967.
6. Fortuin, L. and Laval, R. "Wave Propagation in Random Media", SACLANTCEN Technical Report in preparation.
7. Hastrup, O. F. (1970). Digital analysis of acoustic reflectivity in the Tyrrhenian abyssal plain. *JASA*, **47**, No. 1 (Part 2) 181–190; *also* Hastrup, O. F. (1969). "A Detailed Analysis of Acoustic Reflectivity in the Tyrrhenian Abyssal Plain", SACLANTCEN Technical Report No. 145; *and* Hastrup, O. F. (1968). "The Reflectivity of the Top Layer in the Naples and Ajaccio Abyssal Plains", SACLANTCEN Technical Report No. 118.
8. Hovem, J. M. (1970). Deconvolution for removing the effects of the bubble pulses of explosive charges. *JASA*, **47**, No. 1 (Part 2) 281–284; *also* Hovem, J. M.

(1969). "Removing the Effect of the Bubble Pulses when using Explosive Charges in Underwater Acoustics Experiment", SACLANTCEN Technical Report No. 140.

9. Laval, R. (1971). "General Considerations on Reflection and Forward Scattering from a Rough Surface", paper presented at Conference on Reflection and Scattering of Sound by the Sea Surface held at SACLANTCEN on 29-31 March 1971.

10. Wijmans, W. (1971). "An Experimental Study of Sound Reflection and Forward Scattering from a Rough Surface", paper presented at Conference on Reflection and Scattering of Sound by the Sea Surface held at SACLANTCEN on 29–31 March 1971.

11. Fortuin, L. (1971). "The Sea Surface as a Random Filter for Underwater Sound", paper presented at Conference on Reflection and Scattering of Sound by the Sea Surface held at SACLANTCEN on 29–31 March 1971.

DISCUSSION

H. Mermoz: You said that a point source was seen as a cloud instead of as a point, and that this may reduce the performances of directive arrays of large dimensions. Do you think that this effect may have an influence on the optimum spacing between hydrophones? Should you still use a $\lambda/2$ spacing?

Answer: It depends upon what has to be optimized. If the problem consists of detecting this source against omnidirectional noise, the optimum hydrophone spacing is still $\lambda/2$. When the array size is larger than the spatial correlation distance of the sound field produced by the source its angular resolution is higher than the angular diameter of the "cloud". When doing beam forming the signal energy coming from the source is not concentrated in a single beam, but is divided between several adjacent beams. On the other hand, if the problem consists of estimating the direction of the source, and if we have partial coherence, i.e. if a point source looks like a point surrounded by a halo, one should try to base the bearing estimation on the direction of the point considering the halo as part of the background (forward scattering). In this case, if the total number N of hydrophones is fixed, the spacing should be larger than the spatial correlation distance of the incoherent sound field in order to get a coherent addition of the coherent energy (coming from the point) and an incoherent addition of the incoherent energy (coming from the halo). For large N the halo will then tend to be eliminated. In practice, however, this may lead to the construction of arrays of very large dimensions.

S. B. Gardner: When making the decomposition of the sound field into elementary plane waves with a direction θ, do you take into consideration the possibility that the angle θ may be imaginary?

Answer: Imaginary angles probably correspond to "lateral waves" with very high attenuation coefficients. I am not sure that this case has really to be considered in practice for propagation of underwater acoustic waves, but I have not seriously considered the case.

Space-time Relationships in Echo Wave Fields

F. WIEKHORST

FHP, Werthhoven, Germany

1. Introduction

Echo structure analysis plays an important role in active target classification, because an echo from a reflecting target may carry information about the target structure. When referring to the word "echo" one normally thinks about a time function. Therefore, echo structure analysis leads often to an analysis of a time function. However, for this paper it is important to remember that the echo from a target builds up a spatially distributed wavefield around the target which in general shows a spatial structure as well. The mathematical tool of linear system theory and related theories (such as filter theory, etc.) has been extensively applied to echo time structure analysis. The same mathematical tool can also be applied to the analysis of the spatial structure of a wavefield. Applying such a spatial filter theory to target classification problems can be quite interesting. However, if there are difficulties in the echo *time* structure analysis we cannot expect that these difficulties will not occur in an echo *space* structure analysis. Not much progress is made by studying the spatial echo structure instead of studying the temporal echo structure. However, doing both might change the situation, because there might be specific relationships between the temporal and the spatial structure of a wavefield which might be of importance. As a matter of fact, this proved to be the case. This paper attempts to demonstrate the nature of such interrelations by means of some simple examples.

2. Wavefields Around Reflecting Targets

We first describe a simple Radar or Sonar experiment using the language of linear filter theory. Let us consider that we transmit from the point S in Fig. 1 a signal $s(t)$ towards a target T. We assume that we receive an echo $\epsilon(t)$ at the point R. The point S we may consider as the input and the point R as the output of a linear filter as is shown in Fig. 2. The echo

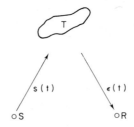

FIG. 1. Geometry of a simple radar experiment.

FIG. 2. Block diagram of a linear system.

$\epsilon(t)$ is then the convolution of $s(t)$ by a so called weighting function $h(t)$ of the filter:

$$\epsilon(t) = \int_{-\infty}^{+\infty} s(\tau) \cdot h(t - \tau) \, d\tau. \tag{1}$$

By performing a Fourier transformation for both sides of equation (1) we obtain the corresponding relationship in the frequency domain:

$$E(f) = S(f) \cdot H(f). \tag{2}$$

All information about the target that we can obtain from such an experiment, is contained in the weighting function $h(t)$ or in its corresponding Fourier transform $H(f)$.

Among other things equation (2) indicates that we might obtain no echo at all. It is quite possible that the spectrum $S(f)$ of the signal $s(t)$ and the spectrum $H(f)$ of the weighting function $h(t)$ do not overlap. In such a case $E(f)$ and consequently $\epsilon(t)$ will vanish. This fact can be used to construct a target that has no echo over a large frequency range. Unfortunately, $H(f)$ depends on the geometry in Fig. 1: it depends on the position of the point S, on the position of the point R, and on position and aspect of the target T. As can ben seen, there is—besides the problem of target classification—a strong motivation to study in what way $H(f)$ depends on the transmitter-target-receiver geometry. For simplicity, in the sequel we will consider the position of S and T as being fixed and study the dependence of $H(f)$ on the position of the receiver R.

Let us assume that a target consists of several reflecting discrete points $P_0, P_1, P_2, \ldots, P_N$. Due to the different reflectivity of these points their contributions to the total echo field strength in a certain distance from the target is given by the different amplitudes $A_0, A_1, \ldots, A_N$. In other words, we assume that each reflecting point reradiates the transmitted signal weighted by the fact A_n and that no distortion of the signal occurs at the individual points. At the receiving point the contributions of the different points in general do not arrive at the same time, depending on the location of each point. Figure 3 shows an x-y-coordinate system, the origin of which coincides with one of the reflecting points, e.g. P_0. In respect to the signal that arrives from P_0 at the receiving point R, the signals

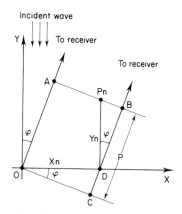

FIG. 3. Sketch for derivation of equation (3).

from the other points are delayed or advanced. For simplicity, we consider only a two-dimensional distribution of the reflecting points: they are all in the x-y plane. The incident wave is assumed to be coming opposite to the y-direction, as indicated in Fig. 3. The receiving point is assumed to be far away in the φ-direction, φ is measured as a clockwise deviation from the y-direction. We will now calculate the contribution of the point P_n to the total echo $\epsilon(t)$ at the receiver.

According to Fig. 3 we find:

$$p = \overline{AO} = \overline{DB} + \overline{DC} = y_n \cos \varphi + x_n \sin \varphi. \tag{3}$$

This means that the signal from P_n is advanced by

$$\tau_n = \frac{y_n + p}{c} = \frac{1}{c} \left[y_n (1 + \cos \varphi) + x_n \sin \varphi \right] \tag{4}$$

c being the velocity of propagation.

From this follows for the echo $\epsilon(t)$

$$\epsilon(t) = \sum_{n=0}^{N} A_n \cdot s(t + \tau_n), \quad \tau_n \text{ given by equation (4).} \tag{5}$$

If the signal is a simple sine-wave of the frequency f, we write, using complex notation:

$$s(\tau) = e^{j2\pi f t}. \tag{6}$$

Inserting (6) into equation (5) we obtain

$$\epsilon(t) = e^{j2\pi f t} \cdot \sum_{n=0}^{N} A_n \cdot \exp \{j2\pi f/c[y_n(1 + \cos \varphi) + x_n \sin \varphi]\} \tag{7}$$

We write (7) as

$$\epsilon(t) = \hat{e}(f, \varphi) \cdot e^{j2\pi f t}. \tag{8}$$

All our interest is now concentrated on the "complex amplitude" $\hat{e}(f, \varphi)$:

$$\hat{e}(f, \varphi) = \sum_{n=0}^{N} A_n \exp \left\{ j \frac{2\pi f}{c} [y_n(1 + \cos \varphi) + x_n \sin \varphi] \right\}. \tag{9}$$

If we replace A_n in equation (9) by a function $A(x, y)$ having the property that $A(x_n, y_n) = A_n$ and $A(x, y) = 0$ otherwise, we can write:

$$\hat{e}(f, \varphi) = \sum_x \sum_y A(x, y) \cdot \exp \left\{ j2\pi \left[y(1 + \cos \varphi) \cdot \frac{f}{c} + x \sin \varphi \cdot \frac{f}{c} \right] \right\} \tag{10}$$

$\hat{e}$ is now recognized as a discrete two-dimensional Fourier transform of $A(x, y)$ with the "frequencies":

$$\gamma_x = \frac{f}{c} \sin \varphi$$

and

$$\gamma_y = \frac{f}{c} (1 + \cos \varphi). \tag{11}$$

3. Restrictions on the Scatter Distribution

It seems, that we have already solved our problem by means of equation (10). If somewhere in the far field a sensor—e.g. a hydrophone—is located, the location of which is characterized by the angle φ (the range is not very interesting as we assume no structure in the range direction besides a simple $1/r$ law), equation (10) delivers the wanted field intensity as a function of frequency and the angle φ. This is all we wanted. Indeed, if the distribution $A(x, y)$ is known, then $\hat{e}(f, \varphi)$ can be calculated. However, in reality the situation is often not that simple.

Let us remember the situation in active Sonar target classification, for example. If we divide all possible targets into different classes like rocks, fish, mines, submarines, and so on, then, in order to apply equation (10), we first ask for the distribution $A(x, y)$ for rocks, fish, mines etc. It is not very realistic to hope that somebody might give us a definite answer. However, we might obtain *some* information about $A(x, y)$. For instance, the reflecting points might all be on a circle with the radius r. All we know about $A(x, y)$ is that in such a case $A(x, y)$ can have non-vanishing values only if $x^2 + y^2 = r^2$. For all other combinations of x and y $A(x, y)$ is zero. In order to demonstrate how such a problem can be attacked, we consider the simple case in which $A(x, y)$ is not zero only if $y = \mu x$. In such a case all reflecting points are located on a straight line that is determined by the equation $y = \mu x$. Note that we only assume that the reflecting points are on this line, the exact distribution we still assume to be unknown.

As x and y are not independent in this case A is a function only of x; with $y = \mu x$ equation (10) reduces to:

$$\hat{e}(f, \varphi) = \sum_x A(x) \cdot \exp \left\{ j2\pi \left[\mu(1 + \cos \varphi) \frac{f}{c} + \sin \varphi \cdot \frac{f}{c} \right] \cdot x \right\}. \qquad (12)$$

As can be seen, $\hat{e}(f, \varphi)$ is now a one-dimensional Fourier transform with the "frequency"

$$\gamma_x = \mu(1 + \cos \varphi) \cdot \frac{f}{c} + \sin \varphi \cdot \frac{f}{c}.$$

We now simplify the situation even more by assuming that $\mu = 0$. In this case all reflecting points are located on the x-Axis of the x-y coordinate system. $\hat{e}(f, \varphi)$ then reduces to

$$\hat{e}(f, \varphi) = \sum_x A(x) \cdot \exp \left[j2\pi \frac{f}{c} \sin \varphi \cdot x \right]. \qquad (13)$$

Equation (13) is well known to anyone who has ever had anything to do with directivity patterns of line-arrays. However, we emphasize again that in our situation we are unable to use equation (13) for a calculation of $\hat{e}(f, \varphi)$ because we assumed $A(x)$ as being unknown. This does not mean that we cannot say anything at all about $\hat{e}(f, \varphi)$. The fact that $\hat{e}(f, \varphi)$ has the form of (13) means that $\hat{e}(f, \varphi)$ is a function of the product $f \cdot \sin \varphi$ only:

$$\hat{e} = \hat{e}(f \cdot \sin \varphi). \qquad (14)$$

This means that we easily can find a sort of "isobars" in the f, φ-plane. These isobars are determined by

$$f \cdot \sin \varphi = \text{const.} \qquad (15)$$

If we consider f and φ as polar coordinates as is indicated in Fig. 4 it can be seen immediately that $f \cdot \sin \varphi$ is nothing but the distance from the y-axis

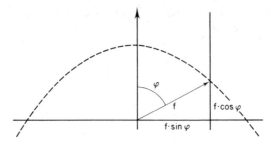

FIG. 4. f, φ-coordinate system.

(ignore the dotted line, it will be used later on). In other words, the isobars, given by (15), are straight lines parallel to the y-axis.

In order to demonstrate these isobars by means of a computer plot we instructed our computer to consider f and φ as polar coordinates, to select a frequency f and to plot with this f a full circle by changing φ from 0 to 2π. After completion of this circle the computer selects the next frequency $f_n = f_{n-1} + \Delta f$ and plots another circle. Continuing this way we obtain a set of concentric circles, a not too exciting picture. However, in addition the computer was instructed to calculate $\hat{e}(f, \varphi)$ by equation (13) for a given but arbitrarily chosen $A(x)$ and to let the plotter pen deviate in radial direction from the just mentioned concentric circles by a small amount proportional to $|\hat{e}(f, \varphi)|$. Figure 5 shows such a computer plot. The expected straight lines are clearly to be seen.

It is no problem to repeat the same experiment for $\mu \neq 0$. The computer then has to solve equation (12). The reflecting points of the target are still

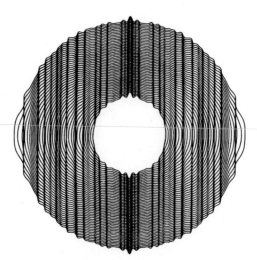

FIG. 5. f, φ-structure of a line target. Aspect angle $= 0°$ ("beam on").

located on a straight line, but the so called aspect angle has changed. Sonar operators would say the target is not in "beam-on" position any more.

The result is shown in Fig. 6. The isobars are again clearly to be seen.

FIG. 6. f, φ-structure of a line target. Aspect angle = $15°$.

They are given by the equation (16):

$$f[\mu(1 + \cos \varphi) + \sin \varphi] = \text{const.} \tag{16}$$

A lot of information can be extracted from equation (16). We shall not discuss this in detail, however, but will continue our more general considerations. We first summarize: equation (10) delivers a complex amplitude $\hat{\epsilon}$ that depends on f and φ: $\hat{\epsilon} = (f, \varphi)$. If nothing is known about the distribution $A(x, y)$ little can be said about $\hat{\epsilon}$. However, if $A(x, y) = 0$ except for $y = \mu x$ (line target), then $\hat{\epsilon} = \hat{\epsilon}[S(f, \varphi)]$, with S *not* being dependent on the actual distribution. There exist "isobars" given by $S(f, \varphi) = \text{const.}$

In order to develop a better feeling for how this fact may be used in reality, we assume that a linear FM-sweep is transmitted as a Sonar signal towards a target. We further assume that we observe such a Sonar experiment from an airplane and that we can see by some means, that we may learn from the science-fiction literature, the energy distribution of the echo-field. The picture would look similar to our computer graph in Fig. 5. This picture will, of course, change with time. There is a distinct time at which such a picture will be identical with our plots Fig. 5 or Fig. 6. This time is given by:

$$\vartheta = \frac{T}{1 - f_{\min}/f_{\max}} \tag{17}$$

ϑ = elapsed time after arrival of front end of signal at the target, T = duration of the signal.

Figure 7 shows the field distribution around a "beam on" line target that has been insonified by a linear FM-pulse. This picture has been "taken"

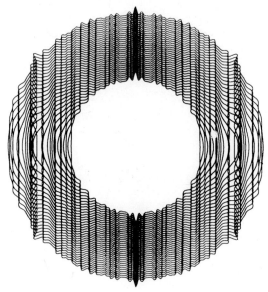

FIG. 7. FM energy distribution around a line target at the "critical" time ϑ [ϑ given by equation (17)].

exactly at the time ϑ, given by equation (17). The fact that the isobars are straight lines for such a target is clearly to be seen. If we make such a "photograph" at an instant when the elapsed time is smaller than ϑ we obtain a picture as is shown in Fig. 8. It can be seen that the isobars are now distorted. However, these distortions are fully predictable although the

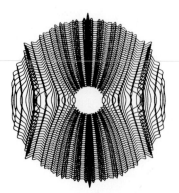

FIG. 8. FM energy distribution around a line target at a time $t \approx 0.6\vartheta$.

actual distribution $A(x, y)$ is not known. In practice, of course, it is not necessary to make the whole soundfield around the target visible. All that is needed is to invent an acoustic device that indicates the existence of isobars that have the discussed behaviour. Although it would be very interesting to investigate how this could be done we go back now to our theory and try to extend it.

We drop now the assumption that the reflecting points be located on a straight line. $A(x, y)$ is now a truly two-dimensional distribution. Our attention is now focused on equation (10) again. As we still consider the exact distribution $A(x, y)$ as being unknown we have to impose some restrictions on $A(x, y)$ in order to draw conclusion from equation (10). We consider these restrictions as being given by the assumption that $A(x, y)$ can be written in the form

$$A(x, y) = A_1(x) \cdot A_2(y). \tag{18}$$

In other words, we deal only with functions $A(x, y)$ that are separable in respect to the two coordinates x and y.

From the theory of Fourier transforms we know that the two-dimensional transform of a function $A(x, y)$ that is separable in respect to the rectangular variables x and y, is separable in respect to the two "frequencies" γ_x and γ_y. Thus

$$\hat{e}(f, \varphi) = F[A(xy)] = F_x[A_1(x)] \cdot F_y[A_2(y)] \tag{19}$$

where F denotes a Fourier transform.

The first term of the right hand of equation (19) is:

$$F_x[A_1(x)] = \sum_x A_1(x) \cdot e^{j2\pi\gamma_x \cdot x} \tag{20}$$

with $\gamma_x = f/c \cdot \sin \varphi$. As can be seen, this is identical with the complex amplitude $\hat{e}(f, \varphi)$ of a line target in "beam position". However, there is now a second "modulating factor" $F_y[A_2(y)]$. This factor is given by:

$$F_y[A_2(y)] = \sum_y A_2(y) \cdot e^{j2\pi\gamma_y \cdot y} \quad \text{with} \quad \gamma_y = f/c \cdot (1 + \cos \varphi). \tag{21}$$

In the same manner as we did for the first factor—when we treated a true line target—there can also be found isobars for the second factor. These isobars are given by:

$$f \cdot (1 + \cos \varphi) = \text{const.} \quad \text{or} \quad f + f \cos \varphi = \text{const.} \tag{22}$$

From Fig. 4 it can be seen immediately that equation (22) determines a parabola as indicated by the dotted line.

We summarize: the field $\hat{e}(f, \varphi)$ of a two-dimensional distribution $A(x, y)$ that is separable in respect to the two rectangular coordinates x and y can be written as

$$\hat{e}(f, \varphi) = \hat{e}_1(f, \varphi) \cdot \hat{e}_2(f, \varphi) \tag{23}$$

$\hat{\epsilon}_1(f, \varphi)$ having the same properties as for a "beam on" line target, i.e. there are isobars in the f, φ-polar plot consisting of straight lines parallel to the y-axis, and $\hat{\epsilon}_2(f, \varphi)$ having parabolas as isobars.

In order to demonstrate, how $\hat{\epsilon}(f, \varphi)$ looks like in such a case, we consider a special distribution: $A(x, y) = A_1(x) \cdot A_2(y)$, $A_1(x)$ being the same as we used to produce Fig. 5 and $A_2(y)$ given by:

$$A_2(y) = \begin{cases} 1 \text{ for } y = 0 \text{ and } y = 0.2 \times L & (L \text{ being the extension of} \\ -1 \text{ for } y = 0.1 \times L & \text{the target in } x\text{-direction}) \\ 0 \text{ otherwise} \end{cases}$$

Figure 9 shows the corresponding $\hat{\epsilon}(f, \varphi)$-plot. The expected modulating parabolas are clearly to be seen.

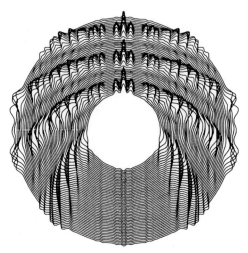

FIG. 9. f, φ-structure of a separable two-dimensional target.

This case of a separable two-dimensional distribution is of interest when we deal with a line target under multipath conditions. Under such conditions a line target acts often as a target with a separable two-dimensional distribution $A(x, y) = A_1(x) \cdot A_2(y)$.

4. Conclusion

The frequency response of a target and the field distribution around a target can be obtained by well known Fourier transform methods if the distribution of the reradiating points of a target is known. In contrast to these widely applied methods, it is assumed in this paper that the actual distribution is unknown except for some imposed restrictions.

It has been shown that certain characteristics in a suitably chosen plot—the above mentioned r, φ-plot—depend only on these restrictions. This proposes to try to classify a target by detecting those restrictions rather than by focusing on the actual scatterer distribution.

DISCUSSION

J. W. R. Griffiths: How much of the bandwidth and φ-extension do you really need when you want to apply this method?

Answer: As I did not say anything about in what way the demonstrated relations can be applied, no bandwidth considerations have been made. I only demonstrated that these relations are there and that they might be useful. It is quite another problem how to use these relations. Of course, it would be of no value if you really had to go around the whole target and measure the directivity pattern as a function of frequency in order to obtain the mentioned structures. one has to find a method that, by few measurements, indicates that these structures do exist. From this a lot of consequences could then be derived. I hope that in our next meeting I can tell you more about it.

P. M. Schultheiss: First I would like to make sure that I understand the basic assumptions. If you say that the target consists essentially of a series of scatterers which reradiate with a delay, you take in account geometry only. Does this imply that there is no coupling between scatterers?

Answer: You are right, coupling has not been taken into consideration. Assume that we measure the frequency response of the target, we will find then in general a very complicated structure. Of course, we want to know by what, say, all the maxima are caused. Are these resonances, or what are they? The assumption in my target model is that all structures in the frequency response is caused by pure interference, in other words, they are due to the superposition of all the elementary waves which come from the individual scatterers. These scatterers have not been considered as frequency-dependent nor has any coupling between them been taken into account. Geometry is here the important factor.

P. M. Schultheiss: Do you have any data on how good that assumption is?

Answer: It depends on the target. There are targets for which our assumption is just perfect. We have such targets in our laboratories. They are artificially made that way. Another question is, how good this model is if we apply it to targets which we find in nature or which are manufactured for other purposes than to behave as nice targets. There are a lot of targets you often find in acoustic laboratories, like spheres, air-filled or water-filled or solid cylinders and so on which are made to study phenomena much as creeping waves and the like. For these types of investigation our model is certainly not sufficient and in some cases even useless. However, there are strong indications that for targets such as big rocks, submarines, and so on, our model explains the majority of effects that are known about such targets.

P. M. Schultheiss: A second question is: in order to get this characteristic pattern that you observed, how large does the extent of the target have to be compared to the range for these curves to be really visible?

Answer: This is a resolution problem. Here we are back to normal physics. We have to collect information over such a distance in space or in other words, we need such an aperture that the resolution given by this aperture is such that we have at least several resolution cells on the target.

B. D. Smith: Is there any change if you include a corner reflector in your model. Is that the same as a point scatterer or is there a phase change.

Answer: It seems that a corner reflector does not create any problems within our theory. Of course, a corner reflector cannot be considered as a point scatterer but as a collection of point scatterers, which are distributed in such a way that they form a corner reflector. As our considerations do not depend on the scatterer distribution the corner reflector seems to be included. However, as the mechanism of a corner reflector is based on multiple reflection the corner reflector is excluded as we did not consider multiple reflection on the scatterers.

Question: Can you say something about the restrictions you placed on the scatterers: how many did you use and what was the approximate distance?

Answer: In the computer models I used between 5 and 15 points. I did not mention it, because the behaviour of the isobars does not depend on the number of scatterers. This behaviour can be observed as long as the f-φ-plot does show any structure at all. This means that one should take at least two points. A second condition is that the wavelength be smaller than the maximum extension of the target.

Question: Can the scatterers be complex, in other words, do you allow any phase shift?

Answer: No, I did not allow phase shift, I assumed the scatterer to be frequency independent, this excludes any phase shift. This has to do with the causability of the system. Of course, I had to assume the impulse response of each scatterer to be causal.

R. B. Gilchrist: If you also introduce a restriction on the target length can you say then something further than what you said without this restriction?

Answer: Basically nothing will change by introducing a restricted target length. Your question is related to resolution considerations. The behaviour of the mentioned structure has nothing to do with the length of the target. However, how fine the whole structure really is does depend on the extension of the target. This means for a smaller target you have to search over a bigger spatial range in order to detect any structure, at all. Once you have detected it, you will find that the behaviour of the structure is just like the one for a bigger target. In other words: all I have shown here does not depend on the size of the target. However, if we discuss the observability of the mentioned effects, resolution condition will enter the scene. Target extension related to wavelength and bandwidth as well becomes now important.

M. L. Somers: Surely the length of your target must be limited because you postulated farfield conditions.

Answer: I indeed postulated farfield conditions. If we keep the range constant, you are quite right. But if we consider the range as a free parameter I can always manage to be in the farfield no matter how big the target is.

M. L. Somers: Can you also allow negative weights on the scatterers?

Answer: Yes, As a matter of fact, in the last slide I showed you there were also negative weights in the distribution.

Detection of Underwater Sound Sources by Microwave Radiation Reflected from the Water Surface†

D. E. TREMAIN and D. J. ANGELAKOS

Department of Electrical Engineering and Computer Sciences
and *Electronics Research Laboratory, University of California,*
Berkeley, California, U.S.A.

A series of relatively simple experiments [1–3], an example of which is shown in Fig. 1, has been carried out to investigate the feasibility of detecting a low frequency, low power underwater sound source by bistatic scattering of a beam of microwave radiation. Generally, the sound source has been

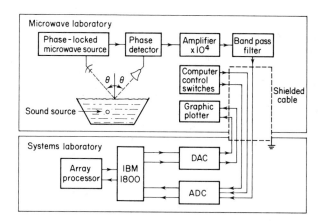

FIG. 1. Experimental Setup.

a submerged circular diaphragm set in a horizontal square plate. The diaphragm vibrates at a frequency between 45 Hz and 55 Hz with a maximum amplitude of either 0.16 cm or 0.08 cm. The water surface can be agitated by a wind blower to produce a turbulent surface on which is superimposed the low amplitude vibration produced by the submerged sound source. A small

† Research sponsored by the Office of Naval Research, Contract NOOO-14-69-A-0200-1035.

portion of the water surface is illuminated by an unmodulated beam of 8 mm wavelength microwave radiation. In one of the experiments [3], the angular positions of the transmitting and receiving antennas with respect to the surface normal to the water are independently adjustable over wide ranges.

The received microwave radiation which has been phase modulated by the turbulence is demodulated by a synchronous phase detector and passed through an amplifier and filter system to a computer. The detection system operates in the following manner. A portion of the transmitted signal is fed to a square law (crystal) detector. This component, which is assumed to be of the form $T \cos (\omega t + \phi_0)$ has a constant amplitude, T, and a phase, ϕ_0, which is set at approximately $\pi \pm \pi/4$ radians as a reference. The received signal is of the form $R(t) \cos (\omega t + \psi(t))$, where $R(t)$ is the amplitude modulation caused by the motion of the water surface, and $\psi(t)$ is the phase modulation which is also produced by the surface motion. For $\phi_0 = \pi \pm \pi/4$ the system operates off the null point and is sensitive to $\psi(t)$. The output of the detector and filter system is approximately

$$g(t) = - \frac{\sqrt{2}}{4} TR(t) + \tfrac{1}{2}R^2(t) \mp \frac{\sqrt{2}}{4} TR(t)\psi(t). \qquad (1)$$

After amplification, $g(t)$ is fed to the computer for processing. The quantity $\psi(t)$ contains components due to the submerged source as well as extraneous sources. Adjusting the magnitude of the received signal at the detector can reduce the effect of the $R^2(t)$ term in equation (1), thus increasing the sensitivity to $\psi(t)$.

The detected signal $g(t)$ is fed via a shielded cable to an IBM 1800 analogue/digital computer system for processing. The signal is sampled at a rate of about 166 points/second for a total of 2048 points. A program has been written which computes the 1024 point autocorrelation function (ACF) of the data points, and then computes the 1024 Fourier transform of the ACF (using the Cooley–Tukey FFT algorithm).[†] With this program, it is possible to average two or more transforms, if necessary, in order to suppress noise.

In one experiment [3], results are obtained for the case of specular reflection in which θ (as defined in Fig. 1) varied from 5 degrees to 50 degrees. Some of these results are shown in Fig. 2. The vertical axis in each part of Fig. 2 is a measure of the relative magnitudes of the spectral components of

† Recall that the ACF, $\Phi_{11}(\tau)$, of a periodic function $f(t)$, is a periodic function with the same period; i.e. if $f(t + T) = f(t)$, then $\Phi_{11}(\tau + T) = \Phi_{11}(\tau)$. The ACF of zero-mean noise is zero; computing the ACF of a noisy signal containing periodic components tends to suppress noise and accentuate the periodic components. Thus, if the ACF of a noisy signal is computed, and then the ACF is Fourier analysed, the periodic components of the noisy signal tend to be more readily detected than if autocorrelation were not performed.

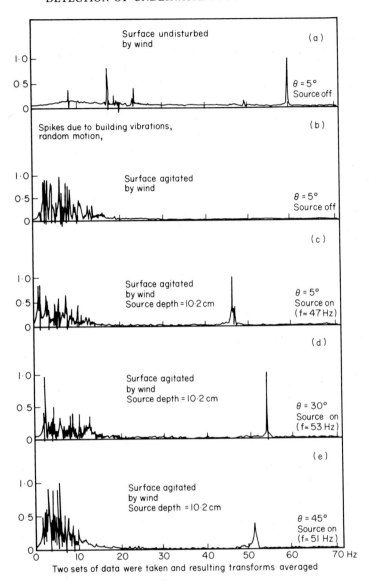

FIG. 2. Spectra of Specular Reflection experiment.

$\psi(t)$. Note that in each part of Fig. 2, the plotter output has been arbitrarily normalized.

The results indicate that a submerged source vibrating at a frequency above the water surface roughness spectrum may be detected readily for values of θ up to about 35 degrees. The source may be detected for values of θ as large as between 40 degrees and 50 degrees only if two or three transforms are averaged.

A number of cases of back-scattering have been investigated as well. However, with the simple detection system used here, detection was possible without an undue amount of signal processing only for the case of back-scattering along the normal to the water surface.

REFERENCES

1. Hutchins, H. Stuart IV and Angelakos, D. J. (1968). "Microwave Scattering from an Agitated Water Surface", University of California, ERL Report No. 68-6.
2. Comeau, J. and Angelakos, D. J. (1969). "Signal Processing of Microwaves Scattered from Water", University of California, ERL Report No. 69-6.
3. Tremain, D. E. and Angelakos, D. J. (1970). "Scattering of Microwave Radiation from a Turbulent Water Surface", University of California, ERL Report No. 70-3.

DISCUSSION

Question: What was the displacement at the water surface due to the submerged source?
Answer: About 0.02 cm—about $\frac{1}{40}$ of the incident wavelength, which is admittedly a quite large displacement.
Question: Could you see that displacement with a light?
Answer: No. That is a rough estimate of the displacement. It could have been a little less than that.
Question: How did you calculate the displacement?
Answer: As an approximation, we plugged into the formula given earlier for a spherical source.
Question: What wavelength radiation did you use in the optical experiment?
Answer: As I indicated in the talk, we used a H_e-N_e laser, which has a wavelength of roughly 7000 Å—red light.
Question: Why use red light, which will penetrate the water, when an argon laser would not?
Answer: Does not an argon laser radiate in the far infra-red?
Further comment by Questioner: I would have thought that if you were working at 10 microns or so, the depth of penetration would be quite small.
Answer: That may very well be true, but then you could not see the radiation. We preferred to have a visible beam to work with rather than have a far infra-red beam bouncing around, especially if it were high powered. That could be dangerous. But it might well be better to use an infra-red laser, from the detection standpoint.
Question: What was the displacement at the surface caused by wind? What was the amplitude of the waves?
Answer: I think I mentioned before that it was roughly 0.3 cm peak-to-trough.
Question: And the source caused a displacement of 0.02 cm?
Answer: Roughly that, yes, but we did not measure that—we just plugged into the formula I gave at the beginning, which assumed a spherical source, so the 0.02 cm figure is in error, it is probably high.
F. Wiekhorst: Do you think that it is real sound that you are detecting? You know, when you calculate the sound pressure that would cause that displacement, you must be close to cavitation. That is a tremendous displacement if it is really caused by sound.

You might detect angstroms, normally. If you have real sound, if you just pump water back and forth that is something else, but if you have real sound, the water should cavitate, due to the sound pressure involved. I do not doubt your measurements, but to relate this to detecting sound, that sound corresponds to a certain pressure, and that must be a tremendous intensity. My question is whether you have calculated this?

Answer: As I said, the diaphragm was moving up and down with an amplitude of about 0.3 cm.

F. Wiekhorst: Yes, but I understood that what you really wanted to do was detect sound, is that right?

Answer: Yes.

F. Wiekhorst: But I think what you have done is pump water. This is still useful—what you have shown is that you can detect this motion. But let us go on and relate your experiment to detecting sound, that sound corresponds to a certain pressure.

Answer: Sound will cause displacement of the surface, will it not?

F. Wiekhorst: Yes, but I think you are close to cavitation, and I wonder if you have calculated this.

Answer: No, we have not. Perhaps I should mention that the reason we used this large a displacement is that the detection system we used is quite simple, and we wanted to generate a sizeable amount of motion at the surface to see if we could actually detect that in the presence of wind noise with a simple detector.

F. Wiekhorst: Well, that is all right, but I want to go on and figure out what sound levels one can detect using this method.

Answer: If one used a more sophisticated detection scheme, for example, using product detectors to extract the phase quadrature components, then one would not have the problem of contamination by amplitude modulation. Then certainly one could detect much smaller motion on the water surface than we are talking about here.

M. L. Somers: Did you actually get an independent measure of the surface displacement?

Answer: No.

M. L. Somers: You just plugged into a free field formula?

Answer: Yes, which is definitely in error, but it should give a rough estimate of the displacement.

M. L. Somers: I would have thought that since the tank actually has smaller dimensions than a wavelength, that the interference from the boundaries would merely cancel the pressure, and the actual displacements you are measuring would be very small. You have a volume displacement in the middle, and physically you have time for the free surface to take up an equilibrium position.

Answer: That may be true, but in any case, the source was considerably closer to the water surface than it was to any of the sides.

M. L. Somers: It is a very complex situation.

R. S. Andrews: Was there any particular reason for using a 50 Hz source?

Answer: Not really. We could have set it to any frequency below 50 Hz, but it was difficult to get much above 50 Hz because the source was a mechanical vibrator.

R. S. Andrews: But you could have used a transducer.

Answer: That is true, but we chose to use this because it is easier to generate a sizeable displacement with a mechanical arrangement like this than it would be with a loudspeaker, for example. But we could have done that.

Question: Would it be possible for you to look at the very low frequency noise in this system? The natural background spectrum, say, at very low frequencies—below a cycle?

Answer: Yes, but we did not actually expand the curves at the very low end of the spectrum.

Question: Would that be feasible to do with the present set up?

Answer: It would be very easy to do. It is a matter of slowing down the sampling rate, and sampling for a longer time.

H. Cox: Did you not work back from your measurements, to see if your estimate of 0.02 cm was right?

Answer: The problem is not really that easy. To work back from these plots, especially after they have been normalized, to try to calculate the displacement, would be very difficult, or impossible. So we did not do that.

H. Cox: I think it is very important, in order for this data to be useful, to know what the displacement was, because we can argue about the sound levels that can be detected. If we knew what the exact displacement was, we would have an important piece of additional information that would be helpful in interpreting the overall results.

Answer: You have a good point. All I can tell you right now, as I mentioned during the talk, is the diameter of the diaphragm, its amplitude of vibration, and its depth, but not what the actual displacement was at the surface.

M. Blizard: What was the diameter?

Answer: The diameter was 5 cm, amplitude of vibration was 0.32 cm, and depth was 10 cm.

Additional comment by author (made on 25th October, 1972): We have measured the peak-to-peak displacement at the water surface using an optical interferometry setup, and have found it to be approximately 70 microns.

Passive Ranging Techniques

PER HEIMDAL and FINN BRYN

Norwegian Defence Research Establishment, Horten, Norway

1. Introduction

High accuracy range measurements are best obtained by using laser, radar or active sonar. However, in some situations, such as underwater warfare, such systems cannot be used since one's own position might be disclosed. Range may be found by passive acoustic means, by cross bearings. If the target moves nonlinearly, it is difficult to measure range unless simultaneous cross bearings are obtained. This can be obtained by using acoustic sensors (hydrophones) mounted on a submarine, a towed or vertical cable, or the like. Since the baselines are small for such systems, high hydrophone placing accuracy is required. A small error in measured sound wave curvature gives a large error in range estimate.

2. Geometry of Passive Ranging

A minimum of three hydrophones are required for curvature measurements. For a hydrophone system as in Fig. 1 where the target is in plane $\xi = 0$, and

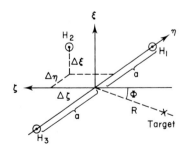

FIG. 1. Target and hydrophone system.

261

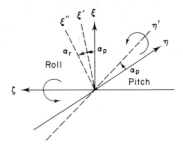

FIG. 2. Hydrophone platform rotation.

after a hydrophone platform rotation at pitch angle α_p and a roll angle α_r as in Fig. 2, the bearing Φ and range R are found from

$$\Phi = \text{arc sin} \left(\frac{c \cdot \tau}{a \cos \alpha_p} \right) \tag{2.1}$$

$$R = \frac{a^2 \cos^2 \alpha_p \cos^2 \Phi}{c \cdot \epsilon + 2l} \tag{2.2}$$

$$l = \cos \alpha_p \sin \Phi \cdot \Delta\eta + (\sin \alpha_r \sin \alpha_p \sin \Phi + \cos \alpha_r \cos \Phi)\Delta\zeta$$
$$+ (\sin \alpha_r \cos \Phi - \cos \alpha_r \sin \alpha_p \cdot \sin \Phi)\Delta\xi \tag{2.3}$$

$$\tau = \frac{\tau_{12} + \tau_{23}}{2}$$

$$\epsilon = \tau_{12} - \tau_{23}. \tag{2.4}$$

The primary measured quantities are τ_{12} (sound wave travel time difference between hydrophones H1 and H2), τ_{23} (similar meaning), sound velocity c, pitch and roll angles α_p and α_r. If $\Delta\eta = \Delta\zeta = \Delta\xi = 0$, $l = 0$, and α_r is not required. Also α_p is not required if the angle is small, since then $\cos \alpha_p$ is close to 1. This is the most simple hydrophone system. In practice, ideal hydrophone placing cannot be obtained. If the true values of $\Delta\eta$, $\Delta\zeta$ and $\Delta\xi$ are nonzero, while zero values are used in computing R, an error is made. Table I gives some examples. R is most sensitive to $\Delta\zeta$ and least to $\Delta\xi$. The sensitivity to $\Delta\eta$ decreases by decreasing Φ. An ideal, simple hydrophone system with $\Delta\eta = \Delta\xi = \Delta\zeta = \alpha_p = 0$ will be assumed in the subsequent analyses.

Reflections of the sound waves from ocean bottom and surface may disturb the wavefronts and give false range measurements. Vertical beamforming will reduce reflections, and also near surface noise. The beamwidth β of a vertical line hydrophone of height h at frequency f is

$$\beta = \text{arc sin} (c \cdot f^{-1}h^{-1}). \tag{2.5}$$

TABLE I. Errors in range measurements.

a	$\Delta\eta$	$\Delta\zeta$	$\Delta\xi$	α_p	α_r	Φ	R	% error in measured R
10 m	1 mm	0	0	$10°$	$20°$	$45°$	10 km	+29%
10 m	0	1 mm	0	$10°$	$20°$	$45°$	10 km	+45%
10 m	0	0	1 mm	$10°$	$20°$	$45°$	10 km	+5%
10 m	0	1 mm	0	$0°$	$0°$	$0°$	1 km	+2%
10 m	0	1 mm	0	$0°$	$0°$	$0°$	10 km	+20%
10 m	0	1 mm	0	$0°$	$0°$	$0°$	100 km	+200%
100 m	0	10 mm	0	$0°$	$0°$	$0°$	100 km	+20%

3. Discriminator Function

Range R and bearing Φ are based on measured delays τ_{12} and τ_{23} (and sound velocity c), by equations (2.1) to (2.4) (for an ideal, simple hydrophone system). τ_{12} and τ_{23} are treated similarly. Hence it is sufficient to consider only one of them, say τ_{12}. The first operation is to find a discriminator function $D(t, x_i - x_0)$, based on the signals from hydrophones H1 and H2, and on a delay steering value x_0, as shown in Fig. 3. $D(t, x_i - x_0)$ is is given by

$$D(t, x_i - x_0) = \int_{-(2f_0)^{-1}}^{(2f_0)^{-1}} C_{12}[(t, u - (x_i - x_0)] \cdot U(u)\, du, \qquad (3.1)$$

where the weight $U(u)$ is -1 for $u < 0$, 0 for $u = 0$, and $+1$ for $u > 0$. The filter output signal part $s(t)$ and noiseparts $n_1(t)$ and $n_2(t)$ are assumed to be

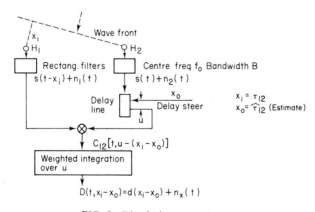

FIG. 3. Discriminator system.

statistically independent gaussian random time functions with zero means. The autocorrelation function of $s(t)$ is

$$E[s(t) \cdot s(t - u)] = P_s \cdot \frac{\sin \pi Bu}{\pi Bu} \cos 2\pi f_0 u, \qquad (3.2)$$

where P_s is signal power. The autocorrelation functions of $n_1(t)$ and $n_2(t)$ are similar, with assumed noise powers $P_{N1} = P_{N2} = P_N$. By approximating $\sin \pi Bu / \pi Bu$ with 1 in equation (3.2) the signal part $d(x_i - x_0)$ of $D(t, x_i - x_0)$ is found to be

$$d(x_i - x_0) = E[D(t, x_i - x_0)] = 4P_s(x_i - x_0) \qquad (3.3)$$

The noisepart $n_x(t)$ of $D(t, x_i - x_0)$ has an autocorrelation function

$$C_x(u) = E[n_x(t) \cdot n_x(t - u)]$$

$$= \frac{2}{\pi^2 f_0^2} (2P_s^2 + 2P_s P_N + P_N^2) \cdot \left(\frac{\sin \pi Bu}{\pi Bu} \right)^2, \qquad (3.4)$$

where $x_0 = x_i$ to have a conservative expression for $C_x(u)$ and where the approximation $\sin \pi Bu / \pi Bu = 1$ is used. The power spectrum $N_x(f = 0)$ of $n_x(t)$ is found from equation (3.4) by a Fourier transform:

$$N_x(f = 0) = \int_{-\infty}^{\infty} C_x(u) \, du = \frac{2}{\pi^2 f_0^2 B} (2P_s^2 + 2P_s P_N + P_N^2). \qquad (3.5)$$

4. Tracking Systems

(a) Simple tracking system
By inserting the discriminator system in a delay tracking loop, as shown in Fig. 4, the differential equation (4.1) is obtained

$$\dot{x}_0 = 4G_x P_S(x_i - x_0) + G_x \cdot n_x(t) + \dot{\hat{x}}_i. \qquad (4.1)$$

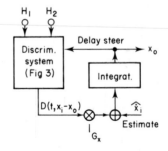

FIG. 4. Simple tracking system.

The tracking loop equivalent is shown in Fig. 5. The time constant is

$$T_x = (4P_S G_x)^{-1}.\tag{4.2}$$

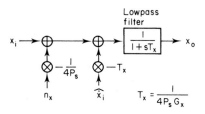

FIG. 5. Equivalent of simple tracking system.

The variance of x_0 due to $n_x(t)$ is

$$\sigma_x^2 = \frac{N_x(f = 0)}{16P_S^2} \int_{-\infty}^{\infty} \frac{df}{1 + (2\pi f T_x)^2} = \frac{K_x}{T_x},\tag{4.3}$$

where $N_x(f)$ is approximated by $N_x(f = 0)$, given by equation (3.5), and where

$$K_x = \frac{1}{16\pi^2 f_0^2 B} \left[2 + 2\left(\frac{P_N}{P_S}\right) + \left(\frac{P_N}{P_S}\right)^2 \right].\tag{4.4}$$

If x_i changes linearly by time, a lag γ will be present.

$$\gamma_x = (\dot{x}_i - \hat{\dot{x}}_i) \cdot T_x.\tag{4.5}$$

A representative measure of total error $x_i - x_0$ can be

$$\lambda_x = \gamma_x + 3\sigma_x.\tag{4.6}$$

The minimum λ_x is obtained for $\lambda_x = 3\gamma_x$

$$\lambda_{x,\min} = 3[4.5K_x \cdot (\dot{x}_i - \hat{\dot{x}}_i)]^{1/3}.\tag{4.7}$$

The corresponding time constant is

$$T_{x,\text{opt}} = [3(\dot{x}_i - \hat{\dot{x}}_i)]^{-1} \lambda_{x,\min}\tag{4.8}$$

$\lambda_{x,\min}$ depends on the error in $\hat{\dot{x}}_i$. If $\hat{\dot{x}}_i = \dot{x}_i$, $\lambda_{x,\min} = 0$ is obtained.

(b) Kalman filter application to delay tracking
A Kalman filter [1, 2] can be applied to delay tracking. A discrete filter

version with time points $t_k = k \cdot \Delta t$ will be treated here. The following vector and matrix definitions are used:

$$X_k = \begin{bmatrix} x_{ik} \\ \dot{x}_{ik} \end{bmatrix}, \quad \hat{X}_k = \begin{bmatrix} \hat{x}_{ik} \\ \hat{\dot{x}}_{ik} \end{bmatrix}, \quad W_k = \begin{bmatrix} w_k \\ \dot{w}_k \end{bmatrix}, \quad V_k = \begin{bmatrix} v_k \\ \dot{v}_k \end{bmatrix},$$

$$Z_k = \begin{bmatrix} z_k \\ \dot{z}_k \end{bmatrix}, \quad F_k = F = \begin{bmatrix} 1 & \Delta t \\ 0 & 1 \end{bmatrix},$$

where x_{ik} and $\hat{x}_{ik}$ correspond to x_i and x_0 in the system in the previous section. The "plant" and "measurement" equations (4.9) and (4.10), respectively are

$$X_k = FX_{k-1} + W_k \tag{4.9}$$

$$Z_k = X_k + V_k. \tag{4.10}$$

The plant and measurement noises W_k and V_k have covariance matrices $E[W_k \cdot W_j^T] = Q_k \cdot \delta_{kj}$ and $E[V_k \cdot V_j^T] = R_k \cdot \delta_{kj}$, where $\delta_{kk} = 1$ and $\delta_{kj} = 0$ for $j \neq k$. Also $E[W_k] = E[V_k] = E[W_k \cdot V_j^T] = 0$ for all k and j. The Kalman filter solution is

$$\hat{X}_k = F\hat{X}_{k-1} + G_k \cdot Y_k \tag{4.11}$$

where

$$Y_k = Z_k - F\hat{X}_{k-1} = \begin{bmatrix} y_k \\ \dot{y}_k \end{bmatrix} \tag{4.12}$$

Z_k is not directly available, but Y_k can be found from the discriminator output $D(t, x_i - x_0)$ by assuming constant $\ddot{x}_i$.

$$y_k = \frac{1}{4P_S} \cdot \frac{3}{\Delta t} \int_{t_{k-1}}^{t_k} D(u, x_i - x_0) \, du \tag{4.13}$$

$$\dot{y}_k = 2(\Delta t)^{-1} y_k. \tag{4.14}$$

The gain G_k is found from the following equations:

$$G_k = M_k [R_k + M_k]^{-1} \tag{4.15}$$

$$M_k = FC_{k-1} F^T + Q_{k-1} \tag{4.16}$$

$$C_k = (I - G_k) M_k, \tag{4.17}$$

where error covariance matrix C_k is

$$C_k \triangleq E[[X_k - \hat{X}_k] \cdot [X_k - \hat{X}_k]^T]. \tag{4.18}$$

If measurement noise R_k does not exist, equations (4.15) and (4.17) show that $G_k = I$ and $C_k = 0$, i.e. perfect tracking is obtained, as would be expected. If R_k is infinitely large, $G_k = 0$ and no input information is used.

X_k continues linearly according to $\hat{X}_k = F\hat{X}_{k-1}$. If x_i is known to be constant, $Q_k = 0$, F can be replaced by I, and $C_k = (I - G_k)C_{k-1}$. Both C_k and G_k decrease by time, showing that the estimate $\hat{X}_k$ is improved and less input information is required.

(c) Practical delay tracking
Without much reduction in performance, a simpler computation of loop gain can be used. A practical formula for a system as in Fig. 4 may be

$$G_x = C_G \cdot P_S \cdot P_N^{-1}\hat{\ddot{x}}_i, \tag{4.19}$$

where the constant C_G is found from selected or average values P_{S0}, P_{N0}, and $\hat{\ddot{x}}_{i0}$ and associate G_{x0} by some optimum procedure. $C_G = G_{x0} \cdot P_{N0} \cdot P_{S0}^{-1}\hat{\ddot{x}}_{i0}^{-1}$. When the signal fades out $P_S = 0$ and $G_x = 0$. Then x_0 continues linearly according to last value of $\hat{\ddot{x}}_i$. The maximum fadeout time without losing track, is approximately

$$T_{F,\text{MAX}} = [4f_0(\dot{x}_i - \hat{\dot{x}}_i)]^{-1} \cdot [1 - \ddot{x}_i(8f_0(\dot{x}_i - \hat{\dot{x}}_i)^2)]^{-1}, \tag{4.20}$$

when $|\ddot{x}_i| < 8f_0(\dot{x}_i - \hat{\dot{x}}_i)^2$ and constant $\ddot{x}_i$ are assumed. A larger $T_{F,\text{MAX}}$ could have been obtained if $\hat{\ddot{x}}_i$ were measured and applied, together with $\hat{\dot{x}}_i$, in predicting future x_0.

A change in hydrophone platform orientation acts as a virtual change in bearing to target. A gyro signal can be used for compensation. The gyro angle Φ_{GYRO} is transformed to $\dot{x}_{\text{GYRO}}$ and added to the integrator input (in Fig. 4).

$$\dot{x}_{\text{GYRO}} = -c^{-1}a(a^2 - c^2x_0^2)^{1/2}\dot{\Phi}_{\text{GYRO}}, \tag{4.21}$$

where a is distance between hydrophones and c is sound velocity.

(d) Range measurements
Range R is mainly based on $\epsilon = \tau_{12} - \tau_{23}$. The variance σ_ϵ^2 of ϵ is twice the variance of the incoherent part of the τ_{12} or τ_{23} noise. The coherent noise parts and common lag of measured τ_{12} and τ_{23} are eliminated in ϵ.

Undesired noise sources with bearings near the bearing to the target may disturb bearing and range measurements. If the indices T, U and M mean target, undesired noise source and measurement, respectively, we obtain for $\tau_T - \tau_U < (4f_0)^{-1}$ under otherwise ideal conditions

$$\tau_M = \tau_T - p(\tau_T - \tau_U) \tag{4.22}$$

$$\epsilon_M = \epsilon_T - p(\epsilon_T - \epsilon_U), \tag{4.23}$$

where $p = P_{SU}/(P_{SU} + P_{ST})$. Equal powers $P_{SU} = P_{ST}$ give, by use of equations (2.1) and (2.2), $\Phi_M \approx 2^{-1}(\Phi_T + \Phi_U)$ and $R_M \approx 2R_TR_U/(R_T + R_U)$. Target mirror pictures, occurring due to reflections in the ocean, act as undesired noise sources. Sound velocity gradients in the ocean cause sound wavefront

disturbances. Range measurements are very sensitive to wavefront disturbances. Methods of measuring wavefront perturbations are treated in
Reference [3].

5. Human Observation Methods

The performance of a human operator in a system ("man–machine system")
has been studied from various points of view for some decades [4]. In this
section, the ability of a human observer to eliminate undesired data is
discussed. This ability is demonstrated in the simple example in Fig. 6. A
machine working with a simple sample mean algorithm will not eliminate
the large noise spike.

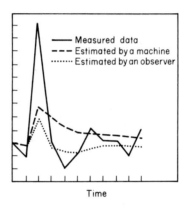

FIG. 6. Sequential estimation by a machine and by a human observer.

Tracking is easily lost due to false signals and signal fade-outs in automatic
systems as treated in Section 4. By using manual delay steering, but with a
saw tooth delay sweep sequence superimposed, a discriminator display
picture as in Fig. 7 may be obtained. Typical regions are numbered in the
figure. The discriminator curves are distinct in regions 1, 3 and 5; the signal
fades out in region 2 and false signals are present in region 4. Region 2
might be passed over in automatic tracking, but probably not region 4.

Two display pictures, one for τ_{12} and one for τ_{23} are required for a human
observer in order to estimate Φ and R.

Even if automatic delay tracking is used, a human observer is valuable for
post-averaging of range. A recording paper with a range scale as indicated in
Fig. 8 is the most useful for obtaining unbiased range estimates. The observer
eliminates regions with obvious false data and regions with heavy noise, and
draws an averaging curve through the remaining regions.

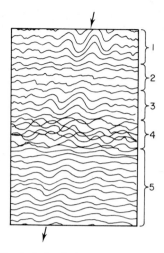

FIG. 7. Typical discriminator curve display picture.

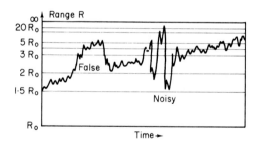

FIG. 8. Typical range recording paper.

6. Conclusion

A practical system for passive range measurements has to deal with all undesired phenomena occurring in a real ocean. A man–machine system
seems to be the best solution. The force of a human observer is to obtain an
overall survey and eliminate obvious undesired data, while the force of a
machine is high processing speed and accuracy for relevant data.

REFERENCES

1. Leondes, C. T. (Editor). (1970). "Theory and Applications of Kalman Filtering",
   NATO AGARDograph No. 139, Textbook.

2. Deutsch, R. (1969). System analysis techniques, Section 6.14 "State variable approach" pp. 232–243, Prentice-Hall Inc.
3. Lunde, E. B. (1973). Wave front stability in the ocean. *In* "Signal Processing: Proceedings of NATO Advanced Study Institute on Signal Processing, August 1972. Academic Press, London and New York.
4. Shinners, S. M. (1967). Techniques of system engineering. *In* "Man–Machine Systems" Ch. 6, pp. 178-227, McGraw-Hill Book Company.

Wave Front Stability in the Ocean

EVEN BORTEN LUNDE

Norwegian Defence Research Establishment
Horten, Norway

1. Introduction

When a wave propagates through a medium with inhomogeneities, wave front perturbation is observed. An example of this is sound wave propagation through the ocean. This perturbation is correlated both in space and time.

Both from a theoretical and practical point of view it is of interest to study this phenomenon. A very interesting practical utilization of the sound wave front is submarines' passive range meter. It measures the curvature of the wave front originated by a distant noise source in order to evaluate the distance to the source. Such a system is thoroughly treated by Heimdal [1]. The passive range meter is of course strongly sensitive to the wave front perturbations.

From a theoretical point of view it must be of interest to compare the measured parameters of the wave front with simultaneously measured parameters of the medium's microstructure in order to verify theoretical works. In this paper it is referred to the study of wave propagation in a random medium by Chernov [2].

2. Experimental Method

(a) Sound source
There are two fundamental methods to measure travel times and travel time differences, dependent on the wave source. With a continuous sound source, correlation technique is used. Correlation between emitted and received sound yields absolute travel times, while correlation between individual receiver hydrophones yields travel time differences or relative travel times.

With a sound impulse source the travel time values are directly measured. Triggering circuits on each receiver hydrophone channel are triggered by the arrival of the sound pulse, and a chronometer measures the arrival differences.

With a triggerable spark underwater sound source, the absolute travel time between source and receiver is also obtained.

The last method permits a direct recording of the quantities in question, the travel times, while the first method requires an advanced and expensive instrumentation to obtain real-time recording of travel times. Moreover, with a sound impulse source, the direct path can usually be resolved from the other paths by the differences in their times of arrival, whereas with continuous sound source the times of arrival are completely mixed. As a conclusion, the sound impulse source seems to have all advantages related to the continuous sound source.

(b) Source-receiver relation

To obtain precise measurements of travel times, it seems to require a great stability in the location of source and receiver. This can only be obtained practically by constructing a fixed bottom-mounted installation. But this creates interference problems, and makes it almost impossible to separate volume and boundary effects on the travel time fluctuations, even with use of a brief sound impulse source. And as a practical problem, this solution requires many receiver hydrophones to study spatial correlation and the dependence on range.

Another method is a system where the receiver hydrophones are fixed relative to each other, whereas the orientation and position of the receiver system relative to the source are free variables. In this case the travel time between source and receiver has only reference interest, as it will be impossible to know whether the fluctuations are due to the medium or change of geometry. Hence only relative travel times between the receiver hydrophones are of direct interest. In other words, only the relative fluctuations are directly measurable. From this statement it follows that if any fluctuations shall be measured at all, the separation between the receiver hydrophones must be of the order of the correlation distance of the fluctuations.

(c) Receiver system

When both the source and the receiver system are movable there is need of a reference system. It is of interest to find the simplest configuration of the receiver system to obtain the necessary reference system.

As earlier mentioned, with only one receiver hydrophone, it is impossible to know whether the travel time fluctuations are caused by the medium or by fluctuations in the range R. With two hydrophones, the geometry is shown in Fig. 1. The two hydrophones are separated by a distance d, and the orientation of the hydrophones are uniquely given by the angle ϕ. The sound travel time difference between hydrophone 1 and 2 is called τ_{12}. But it is impossible to know whether the fluctuations in τ_{12} are caused by the medium or by fluctuations in ϕ. When the range R is known, the curvature

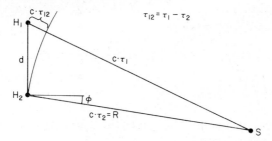

FIG. 1. Geometry with two receiver hydrophones.

of the wave front is known too, as it is the inverse of the range. But this information has been of no value so far, because it is necessary to know three points on a wave front to measure its curvature.

The simplest configuration with three hydrophones is when they are placed on a straight line, the separation between neighbouring hydrophones being d. This configuration is in fact identical to the receiver system of a passive range meter as described by Heimdal. The geometry is shown in Fig. 2, where the following symbols are used: c: sound velocity in meters per second; τ_R: range in seconds; R: range in meters; ϕ: bearing in radians; d:

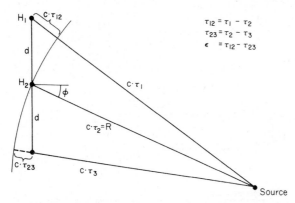

FIG. 2. Geometry with three receiver hydrophones.

hydrophone separation in meters; τ_{ij}: travel time difference between hydrophones i and j; τ_i: travel time between source and hydrophone i. A new measurable quantity ϵ is defined as:

$$\epsilon = \tau_{12} - \tau_{23} = \tau_1 - 2\tau_2 + \tau_3. \tag{2.1}$$

By simple geometric considerations and suitable manipulations, R and ϕ are found to be:

$$R = \frac{2d^2 - c^2(\tau_{12}^2 + \tau_{23}^2)}{2c \cdot \epsilon} = \frac{d^2 \cos^2 \phi}{c \cdot \epsilon} \tag{2.2}$$

$$\phi = \text{Arcsin} \frac{c(\tau_{12} + \tau_{23})}{2 \cdot d} \tag{2.3}$$

where the approximation $R \gg d$ is used. The range is also found from:

$$R = c \cdot \tau_R. \tag{2.4}$$

By combination of (2.2) and (2.4), a reference quantity ϵ_r is obtained:

$$\epsilon_r = \frac{1}{\tau_R} \cdot \left[\left(\frac{d}{c} \right)^2 - \tfrac{1}{2}(\tau_{12}^2 + \tau_{23}^2) \right]. \tag{2.5}$$

Before we discuss the quality of our reference quantity, we will sketch a practical experiment design.

(d) Design of a practical experiment

A plausible design is shown in Fig. 3. It consists of one surface vessel and one submarine. Onboard the vessel is a high-precise chronometer, which triggers a spark pulse generator at equal time intervals t_s, and a sound pulse is emitted from the spark transducer towed behind the vessel.

The submarine serves as platform for the three hydrophones. Inside the submarine is a chronometer identical to that on the vessel. They are called the source and receiver chronometers. At the same time as the source chronometer triggers the spark generator, the receiver chronometer starts a

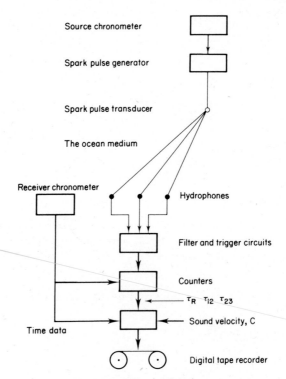

FIG. 3. Experimental design.

counter. When the sound pulse reaches hydrophone 2, via filter and trigger circuits, it stops the counter, which then contains the travel time $\tau_2 = \tau_R$.

In the same way, when the sound pulse reaches the three hydrophones, via identical filter and trigger circuits, it starts and stops two counters, which then contain the travel time differences τ_{12} and τ_{23}. τ_R, τ_{12} and τ_{23} are then recorded on magnetic tape in digital form, together with time from the chronometer and the sound velocity from the submarine's sound velocity meter. From the magnetic tape the data can now be entered directly into a general purpose computer.

Moreover, the equipment inside the submarine also ought to have the possibility of calculating ϕ and R from equation (2.3) and (2.4) in real time. It is then possible to adjust the speed and course of the submarine to maintain the desired geometry.

3. The Reference Quantity

(a) Mean deviation and calibration

In theory and with a perfect system, there should be no mean deviation between ϵ and ϵ_r. Practically it is not likely to be so. Disparities in the electrical phases of the three hydrophone channels will produce an error independent of ϕ, whereas deviation in the receiver geometry from the perfect one will produce an error dependent on the angle. However, the total mean deviation versus ϕ is found by histogram technique:

$$m_\epsilon(\phi_i) = \frac{1}{N_{i\phi}} \sum [\epsilon(\phi) - \epsilon_r(\phi)] \tag{3.1}$$

where the sum is taken over all data with angle ϕ inside a $\Delta\phi$-interval around ϕ_i, and where N_i is the number of data inside this interval.

This calibration is easily executed on a computer.

(b) Fluctuations and drift

An absolute requirement to the reference quantity ϵ_r is that fluctuations and drift are much less than the fluctuations in ϵ:

$$\Delta\epsilon_r \ll \Delta\epsilon. \tag{3.2}$$

We choose some plausible values to test this requirement: $R = 1500$ m; $d = 15$ m; $c = 1500$ m/sec; $a = 0.6$ m; $\overline{\mu^2} = 5 \cdot 10^{-9}$. The two last values are taken from Chernov, where $\overline{\mu^2}$ is the mean square fluctuation and a is the correlation distance of the refraction index of the medium. In this example $a \ll d$, and the fluctuations at the three hydrophones are uncorrelated. (2.1) gives:

$$\overline{\Delta\epsilon^2} = \overline{(\Delta\tau_1 - 2\Delta\tau_2 + \Delta\tau_3)^2} = \overline{\Delta\tau_1^2} + \overline{4\Delta\tau_2^2} + \overline{\Delta\tau_3^2} = \overline{6\Delta\tau^2}$$

$$= 3 \cdot \sqrt{\pi} \cdot \overline{\mu^2} \cdot aR/c^2 = 11 \text{ } \mu\text{sec}^2. \tag{3.3}$$

Overbar designates averaging with respect to time.

The applied formula is taken from Chernov in the case of Fraunhofer diffraction and with Gaussian medium. There are three main errors concerned with ϵ_r. As the travel time between the source and the receiver system is measured by two chronometers, a difference Δf in the frequency of the chronometer crystals makes them drift relative to each other. After a time T, this drift error is:

$$\Delta \tau_R = \frac{\Delta f}{f} \cdot T.$$

It should be rather easy to obtain $\Delta f/f = 10^{-8}$. After 5 days the drift error is: $\Delta \tau_R = 2.5$ msec.

There is one error concerned with the measurement of the sound velocity. $\Delta c_1 = 3$ m/sec.

The mean sound velocity between the source and the receiver deviates from the velocity in the neighbourhood of the receiver:

$$\Delta c_2 = 3 \text{ m/sec.}$$

It is easy to show that:

$$\Delta \epsilon_r^2 < \epsilon^2 \cdot \left[\left(\frac{\Delta c_1}{c} \right)^2 + \left(\frac{\Delta c_2}{c} \right)^2 + \left(\frac{\Delta \tau_R}{\tau_R} \right)^2 \right] + (\tau_{12}^2 + \tau_{23}^2 + \epsilon^2) \cdot \left(\frac{\Delta \tau}{\tau_R} \right)^2.$$

$$(3.4)$$

With the assumed values inserted:

$$\overline{\Delta \epsilon_r^2} < 0.12 \ \mu\text{sec}^2 \ll \overline{\Delta \epsilon^2}.$$

4. Wave Front Fluctuations

(a) Mean square fluctuation of ϵ

When the mean deviation $m_\epsilon(\phi)$ is found, the mean square fluctuation in ϵ is easily obtained:

$$\overline{\Delta \epsilon^2} = \overline{(\epsilon - \epsilon_r - m_\epsilon)^2}.$$

$$(4.1)$$

In a passive range meter, this is an important parameter. For small fluctuations, equation (2.2) yields:

$$\sqrt{\overline{\Delta R^2}}/R = \sqrt{\overline{\Delta \epsilon^2}}/\epsilon.$$

$$(4.2)$$

For a passive range meter, the relative error in range is equal to the relative error in ϵ.

(b) Transverse wave front correlation coefficient

The effect of time variation of the medium will appear at the second order in most practical applications where ships in motion are concerned. Hence when the submarine moves with a velocity v, the space variation will be almost entirely responsible for the time variation observed. Chernov has found that the longitudinal correlation extends over a distance which is considerably larger than the transverse correlation. The time correlation coefficient of ϵ is hence equal to the transverse correlation coefficient of the wave front; $R(r_T)$:

$$Q(t) = \overline{\Delta \epsilon^2} \cdot R(v \cdot \cos \phi \cdot t) \tag{4.3}$$

$Q(t)$ is the time correlation function of ϵ, and $v \cdot \cos \phi \cdot t$ is the transverse displacement of the hydrophones during the time t. If this equation shall have some practical interest, $v_T = v \cdot \cos \phi$, has to be kept as constant as possible, as transverse velocity fluctuations will produce great variations in the correlation coefficients calculated from (4.3). But if the coefficients are averaged, the averaged correlation coefficient is a fair estimate of the real coefficient for:

$$r_T \ll a \cdot v_T / \sqrt{\overline{\Delta v_T^2}}, \tag{4.4}$$

where a is the correlation distance, and $\overline{\Delta v_T^2}$ is the mean square fluctuations of v_T.

This general conclusion is drawn from the special case when both the transverse correlation coefficient and the transverse velocity are Gaussian.

(c) Mean square fluctuation of the wave front

It is now possible to calculate this fluctuation by going back to the mean square fluctuation of e:

$$
\begin{aligned}
\overline{\Delta \epsilon^2} &= \overline{(\Delta \tau_1 - 2\Delta \tau_2 + \Delta \tau_3)^2} \\
&= \overline{\Delta \tau_1^2} + \overline{4\Delta \tau_2^2} + \overline{\Delta \tau_3^2} - \overline{4\Delta \tau_1 \Delta \tau_2} - \overline{4\Delta \tau_2 \cdot \Delta \tau_3} + \overline{2\Delta \tau_1 \cdot \Delta \tau_3} \\
&= \overline{\Delta \tau^2}[6 - 8 \cdot R(d \cdot \cos \phi) + 2 \cdot R(2d \cos \phi)]
\end{aligned}
\tag{4.5}
$$

$d \cdot \cos \phi$ is the transverse separation between hydrophones 1 and 2, and between 2 and 3, while $2d \cdot \cos \phi$ is the transverse separation between hydrophones 1 and 3. The mean square fluctuation of the wave front is now found as:

$$\overline{\Delta \tau^2} = \overline{\Delta \epsilon^2} \cdot [6 - 8 \cdot R(d_T) + 2 \cdot R(2d_T)]^{-1}, \tag{4.6}$$

where $d_T = d \cdot \cos \phi$.

(d) Comparison between experimental and theoretical values

In a medium with random inhomogeneities, the fluctuations in the refractive index is a random process in space and time, described by $\mu(x, y, z, t)$. If the fluctuations are stationary in time and spatially homogeneous, the spatial correlation coefficient of the refraction index is defined by:

$$\overline{\mu^2} \cdot N(r) = \overline{\mu(x_1, y_1, z_1, t) \cdot \mu(x_2, y_2, z_2, t)}. \tag{4.7}$$

As usual, the overbar designates averaging with respect to time t. $\overline{\mu^2}$ is the mean square fluctuation of the refractive index. Chernov has found the mean square fluctuation of the wave front dependent on the fluctuations of refractive index. In the case of Fraunhofer Diffraction the formula is:

$$\overline{\Delta\tau^2} = (\overline{\mu^2} \cdot R/c^2) \cdot \int_0^\infty N(r)\, dr. \tag{4.8}$$

The spatial correlation function of refractive index can be found experimentally with a fast-acting thermometer as done by Liebermann [3], or a sound velocity meter. Then it is possible to compare the mean square fluctuation of the wave front calculated from equation (4.6) and from equation (4.8).

5. Conclusion

Methods to measure the wave front perturbation when a sound wave propagates through the ocean are discussed, and the best solution is found to be a system where both the source and the receiver are movable. The receiver consists of three hydrophones mounted on a straight line with equal separation between neighbouring hydrophones. The source is a triggerable sound pulse source. With two chronometers, the travel time between source and receiver is measured, which permits the calculation of a reference system.

The effect of the ocean's time correlation will appear at second order in most practical applications where ships in motion are concerned. This permits a calculation of the transverse correlation function of the fluctuations in the wave front from measured time correlation functions.

The micro-structure of the refractive index can be measured with a thermometer or a sound velocity meter. If the spatial correlation function of the refractive index is found, the mean square fluctuation in the wave front caused by the microstructure can be calculated. The values of the mean square fluctuation in the wave front is hence obtained both from medium measurements and from wave front curvature measurements, and an interesting comparison is possible.

REFERENCES

1. Heimdal, P. and Bryn, F. (1973). Passive ranging techniques. *In* "Signal Processing", Proceedings of NATO Advanced Study Institute on Signal Processing, August, 1972. Academic Press, London and New York.

2. Chernov, L. A. (1967). "Wave Propagation in a Random Medium", McGraw-Hill Book Co, New York.
3. Liebermann, L. J. (1951). *J. Acoust. Soc. Am.* **23**, 563.

DISCUSSION

Question: With what precision do you think you can make your time of arrival on the three hydrophones of the three pulses?

Answer: With high resolution in the counter the precision is determined by the appearance of the sound pulse and the noise level. It should be possible to make the precision better than 1 μsec for ranges less than 2 km, under normal conditions.

P. L. Stocklin: To what extent do you think the true measurable movement of your receiving hydrophones may influence the measurement?

Answer: The movement of the receiving hydrophones may cause small, but almost equal errors in τ_{12} and τ_{23}. As ϵ is the difference between τ_{12} and τ_{23}, the errors almost cancel each other, and the remaining error in ϵ is negligible. Hence the movement does not influence the measurement.

R. Laval: What is the bandwidth of the experiments?

Answer: The upper frequency is set by the sound pulse, which is determined by the spark pulse generator and transducer, and may be of the order of 5 kHz.

Shallow Water Acoustics Related to Signal Processing

R. S. THOMAS, J. C. MOLDON and J. M. ROSS

*Defence Research Establishment Atlantic,
Dartmouth, Nova Scotia, Canada*

1. Introduction

Over the past several years DREA has been investigating those statistical properties of underwater acoustic channels which most affect the structure and performance of efficient processing systems for signals used in active sonar or underwater communications. The principal considerations in describing these underwater channels are the characteristics of signal propagation, which affect the form of received signals or echoes, and the characteristics of reverberation or undesired backscatter from the surface, volume and bottom, since this is the background in which echoes must often be detected. Sufficient for many purposes have been measurements of the time spreading or frequency coherence, the frequency spreading or temporal coherence, and the directional spreading or spatial coherence of propagated sound, reverberation and ambient noise.

In deep water locations surface wave action is the dominant source of signal dispersion except sometimes in bottom bounce propagation paths. Sufficient stationarity and homogeneity thereby exist so that models providing initial estimates of these dispersion relationships, either from acoustic measurements or by inference from environmental conditions, require only a minimum of updating. In shallow water, on the other hand, multiple reflections from the surface and bottom, as well as refractive effects in the regions of strong vertical gradients in the velocity-of-sound profile, lead to a situation much more difficult to model. While time, frequency and directional dispersion of sound are still of prime interest, these characteristics may be quite different from those observed in deep water. Furthermore, in shallow water, these characteristics are much more influenced by physical features which themselves are spatially inhomogeneous and can hence lead to rapid temporal and spatial variations in the channel behaviour as a ship moves through the water. Changes in bottom type and, often more important, horizontal gradients in the

281

sound velocity are prime features giving rise to lack of stationarity and homogeneity of the acoustic channel statistics. These in turn dictate that signal processing schemes for shallow water sonar or acoustic communications systems must be capable of continuous estimation of the channel as well as continuous adaptation of the signal processing strategy.

This paper discusses first the experimental techniques employed in the measurement of signal dispersion together with the measurement of relevant environmental data. Some comments are made on the statistical behaviour of underwater propagation and reverberation in deep and shallow water where a number of relatively well controlled experiments have been carried out at frequencies between 2 and 10 kHz. Experimental results obtained under more general shallow water conditions are then presented and discussed in relation to the underlying variations in physical properties of the region. Finally some conclusions are drawn regarding the significance of these effects on processor estimation and adaptation with regard to optimal signal choice, array processing, and thresholding.

2. Propagation

In our experiments we have tried to develop models of the scattering functions of propagation and reverberation in range, frequency and direction. It is first convenient to consider propagation, which is perhaps the more fundamental of the two phenomena since the dispersion of the backscattered sound is often largely that experienced during propagation to and from the scattering elements.

Our propagation experiments were performed in a manner illustrated by Fig. 1 where two ships support the transmitting and receiving equip-

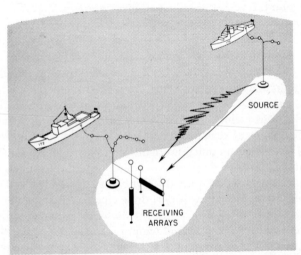

FIG. 1. Experimental arrangement for propagation measurements.

ments. In order to eliminate platform motions as a cause of signal dispersion the systems are decoupled from the surface with small float lines and large damper plates.

In the measurement of the dimensions of scattering functions in range and frequency, pseudo random signals which can achieve fine resolution in range and frequency simultaneously were not used since their undesired secondary sidelobe characteristics disqualify them for use in measurements when the scattering function is fairly smooth and extends over large range and frequency intervals. Because of this, our measurement signals have consisted of:

(1) short pulse transmissions to measure the range or multipath delay characteristics (from this can also be computed the coherent frequency bandwidth of the channel), and

(2) continuous tone transmissions to measure the spectrum spreading.

Measurement of angular dispersion was done by obtaining spatial correlations over the received wavefront in both vertical and horizontal directions with vertical and horizontal line arrays as illustrated in the diagram. This provided information suitable for assessing array performance. Environmental information collected during experiments included wind velocity and direction, wave height spectra, wave slopes and, in a few cases, directional wave spectra.

In deep water over intermediate distances sound propagates in near surface paths as a direct transmission and as a forward reflection from the surface. In the surface duct, that is, in a surface layer with an increasing sound velocity with depth, upward refraction limits the range for direct transmission and sound is conducted by repeated surface reflections. Similar situations can exist in shallow water if the velocity gradient, water depth and the vertical beamwidth of the transmitter combine to prevent significant energy reaching the bottom. We find this situation occurring only in late winter, however, when the surface of the water is very cold.

Our first studies of propagation concerned the dispersion undergone in a single forward reflection from the surface. Figure 2 shows the average envelope of received short pulses on the right while on the left are the spectra for various grazing angles on the surface as determined by the range between ships. In general, within the resolution of our measurements, we find that the direct arrival suffers no dispersion in either time, frequency or space. The surface reflection on the other hand, seen as the second arrival in the short pulse return, is spread in time due to the roughness of the surface. The time constant of this decay can be related to the source depth, surface grazing angle and mean square wave slope. At high frequencies and large grazing angles, where the surface is statistically very rough according to the Rayleigh criterion, this dependence can be deduced from simple geometric considerations and the surface wave slopes.

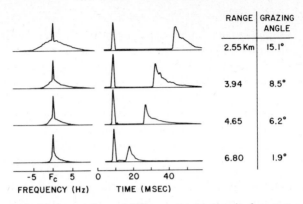

RANGE	GRAZING ANGLE
2.55 Km	15.1°
3.94	8.5°
4.65	6.2°
6.80	1.9°

FREQUENCY (Hz)   TIME (MSEC)

FIG. 2. Short pulse and CW tone propagation in deep water.

The effect of the wave motion on the spectrum of CW transmissions is evidenced here by the smooth spread of sideband energy. The spike at zero frequency is due to the direct arrival. As in the case of time spreading, the bandwidth of the surface component can be estimated quite well from simple geometric considerations and in this case it depends on the RMS vertical velocity of the surface.

The loss in spatial coherence of the scattered sound is of course due to the angular spread of the signal. Measurements made of this correlation agree with similar estimates of the angular spread used in calculating the time constant of the short pulse pressure decay. It is of course difficult to quote a single typical value because of the variety of conditions that might be encountered. Nevertheless, we find that as the range is increased, and the grazing angle correspondingly decreased, the horizontal angular spread goes from several degrees at 1 km to about one degree at 4 km. Likewise vertical spreads of 5 to 6 degrees at 4 km are observed.

Frequency spreads of at least one Hz suggest that about one half second coherent integration time is about all that one can profitably employ. Because of the very noise-like nature of the signal scattered from the surface and the nearly Rayleigh-distributed envelope, echo signals can be expected to suffer severe fading making some means of diversity desirable.

In long pulse systems which try to exploit doppler processing to discriminate a moving target from the background, the entire echo is received with the minimum two degrees of freedom and no post-detection integration can be employed as in the case of high range resolution signals which over resolve the target. Frequency diversity, however, can be achieved with most systems since the fading bandwidth of the channel is not very large. Taking as a rough rule of thumb that the coherence bandwidth is the reciprocal of the time spread we see that a hundred Hertz or so is a sufficient spacing to see uncorrelated fading in deep water surface ducts.

In shallow water, as seen in Fig. 3, the time spreading is much greater

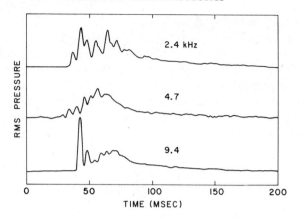

FIG. 3. Short pulse time spreading in shallow water.

with commensurate decreases in coherent bandwidths to only a few Hertz. This increase in multipath delay is caused by paths which undergo many successive bottom and surface reflections. This example was obtained with source and receiver at 20 M depth at a separation of 6 to 7 km in water of 73 M depth. It represents the average of 100 transmissions. With the 2 msec transmitted pulse it becomes impossible to distinguish arrivals beyond the first two or three peaks. Even these initial peaks are not necessarily due to single arrivals but may be the combination of two or more that are closely spaced. The difference between the 4.7 kHz and 9.4 kHz envelopes is likely due to the different interference effect of the differing wavelengths plus a slight geometry change.

Spatial coherence properties of shallow water propagation are also quite different from their deep water counterparts. As in the deep water work, we used a line hydrophone array for this measurement similar to the one in Fig. 4. Signals from the four hydrophones could be cross correlated in

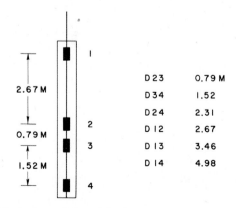

FIG. 4. Array for spatial coherence measurements.

six possible ways with the spacings indicated. Figure 5 shows two typical hydrophone signals. These were heterodyned to base band and multiplied, one by the complex conjugate of the other. This product was then averaged over 100 consecutive, independent short pulse transmissions to give the mean cross correlation as a function of multipath delay. The product was then integrated over the length of the received pulse and the absolute value normalized by the square root of the product of the individual pulse energies to define the correlation factor.

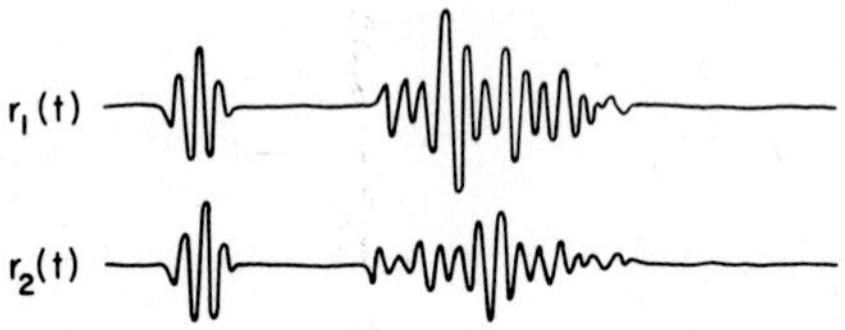

$r_1(t)$

$r_2(t)$

$$\text{CORRELATION} = \frac{\left| \int_0^T \langle r_1(t)\, r_2^*(t) \rangle\, dt \right|}{\left[\left(\int_0^T \langle r_1^2(t) \rangle\, dt \right) \left(\int_0^T \langle r_2^2(t) \rangle\, dt \right) \right]^{1/2}}$$

FIG. 5. Sample waveforms from array elements.

This correlation factor is plotted in Fig. 6 with the array horizontal. A dimensionless distance scale $2\pi d/\lambda$ is plotted logarithmically, where d is the hydrophone separation and λ is the signal wavelength. In this way the spatial correlations will be the same at all frequencies if their angular distributions of incident energy are equal. In addition to our data at discrete frequencies, results of the trend observed by Wille and Thiele [1]

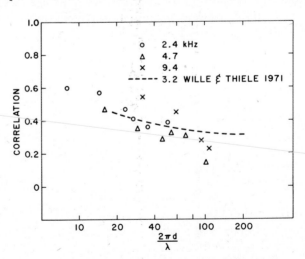

FIG. 6. Horizontal spatial correlations in shallow water.

with explosive sources in a similar geometry in the North Sea are shown to illustrate the order of consistency that may be found. Individual experimental results may depart from this trend. For example our 9.4 kHz results at this particular range show somewhat higher correlation than might be expected which is likely due to the strong contribution of early arrivals having undergone fewer reflections and hence less angular scattering. The spatial correlation functions do, however, show a smooth decay and an implied angular spread (the reciprocal of the correlation distance on this scale) of several degrees. Receiving array discrimination against either noise or reverberation from an omnidirectional source is thus limited. Likewise, there is little point in forming transmitting beams narrower than a few degrees horizontally because the forward scattering will broaden the beam in any case.

In the vertical direction the picture is somewhat worse as illustrated in Fig. 7. At all frequencies we find the vertical correlation smaller than that observed horizontally, with an implied angular spread of 10 to 15 degrees.

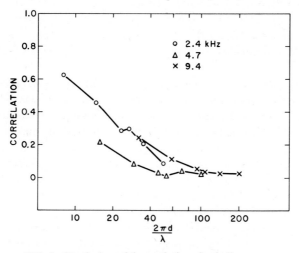

FIG. 7. Vertical spatial correlations in shallow water.

This example shows results for a propagation distance of 8 km but is typical of results observed at both shorter and longer ranges. In processing against isotropic noise, or noise with a wide vertical angular distribution, there is an available array gain until the vertical beamwidth is reduced to this range of 10 to 15 degrees. However, there can be little gain against reverberation which is generated by physical scatterers in the neighbourhood of the target and suffers much the same propagation mechanism as the echo thus showing very similar coherence properties. This statement is supported by experiments in which Urick [2] found shallow water reverberation to have vertical angular spreads similar to those observed here in propagation. However, an improvement in sonar detection performance can be achieved using spatial diversity of a vertical array. Quite modest vertical separations will yield independent observations allowing post-detection combination to significantly reduce the variance of the echo fading. This spatial diversity can be achieved with a line of receiving elements while using only a single transmitter which is usually the more bulky and expensive of the two system components. In communications, however, where propagation is in one direction only, the possibility exists for narrower vertical beams on both transmitters and receivers to take advantage of preferred propagation angles.

3. Reverberation

Reverberation, often the main background against which target echoes must be detected, also deserves considerable attention. Reverberation spectra and level variations between transmissions will now be discussed with attention first being given to frequency spread.

In the shallow water environment, as in deep water, doppler processing against reverberation is possible. The frequency spreading of the reverberation, which determines the limits of this doppler processing gain, is determined either by the motions of scatterers or by propagation mechanisms and thus shows a good deal of variation. Surface reverberation normally exhibits frequency spreading appropriate to the wind-driven surface wave motions; bottom reverberation exhibits spreads expected in propagation mechanisms; and volume reverberation falls somewhere between.

Where optimum procedures are being considered in the detection process, it is useful to study the variability of the reverberation background. The variability of mean reverberation levels will affect the accuracy of estimates of these background levels for purposes of thresholding. In shallow water, reverberation from surface, volume and bottom contribute in varying proportions with levels spatially dependent on water depth and bottom type and temporally and spatially dependent on velocity profile, sea surface conditions and volume scatterer distributions. An important feature of shallow water reverberation is also the appearance of consistent scattering highlights from the sea bottom.

A number of experiments have been performed in shallow water areas of depths less than 300 metres to study the main features of reverberation. Figure 8 illustrates the experimental procedure where a transducer with a searchlight transmit and receive beam was towed trained to the rear to keep any frequency spreading on reverberation in the main beam due to transducer motion to less than one Hz. Reverberation received in the sidelobes also has a doppler shift to one side of that in the main beam. Hence analysis can be done showing no sidelobe effects at low levels on one side

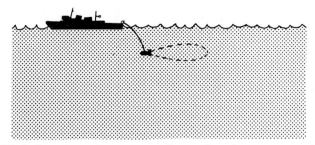

FIG. 8. Experimental arrangement for reverberation measurements.

of the reverberation spectrum. Towing speeds were generally less than 7 knots. A transmit frequency of 9.5 kHz was used where a doppler frequency shift of just under 7 Hz corresponds to one knot radial velocity. A speedometer mounted on the towed body and having an accuracy of about 0.1 knot allowed compensation to be made for doppler shifts introduced by the towed transducer itself. It should be pointed out that the water at the towing depth which ranged from 20 to 60 metres acts as the zero doppler reference.

Experiments were carried out in water depths ranging from 40 to 250 metres over a number of bottom types on the Scotian Shelf and Gulf of St. Lawrence which included mud, clay, sand, gravel, and rock. Results were obtained in winter and summer conditions and for a range of weather conditions. Normally, the ship followed fixed courses along legs forming rectangles or squares approximately 15 km to a side.

The actual experimental procedure was to transmit, at 9.5 kHz, a short CW pulse of 8 msec duration alternately with a long CW pulse with a smoothed envelope of 1.4 seconds total duration. The reverberation received from the short pulse was analysed to obtain information on the variability of levels, and that from the long pulse was analysed to obtain the reverberation doppler spectrum as a function of range.

In winter, upward refraction ensured that the dominant reverberation was scattering from the vicinity of the sea surface. An example of the form of the spectra for some winter reverberation is shown in Fig. 9 where the power spectra from about 50 transmissions were averaged and the square root taken before being plotted. The spectra at the top are for the closer ranges. The zero doppler reference is indicated by the broken vertical line and it can be seen that the reverberation spectrum has a positive doppler shift which is what would be expected from a wind driven surface current with the transducer beamed up wind. The reverberation spectrum is broader than the resolution of the signal where the spread is primarily a result of scatterer motion. This spectral width is correlated with wind velocity as illustrated in Fig. 10 where the frequency spreads between the −6 dB levels are shown as a function of wind velocity. The results obtained

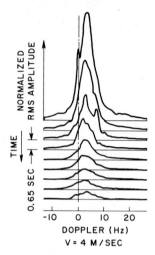

FIG. 9. Range-doppler analysis of surface reverberation sample.

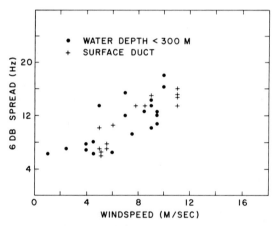

FIG. 10. Frequency spreads measured for surface reverberation.

in surface ducts in deep water are comparable to those observed in winter conditions in shallow water. In both cases the correlation with wind speed and the direction of the doppler offset (which has been found to be of the approximate magnitude mentioned by Urick [3] for wind driven surface currents) confirm the dominance of surface reverberation.

Bottom reverberation has in general much less frequency spreading as illustrated by Fig. 11, where conditions of downward refraction exist. The experimental frequency resolution is about 1.5 Hz to the −6 dB points, but we find here about 2.3 Hz spread which can be explained by propagation effects and some variability in water currents over the time in which 44 transmissions were observed.

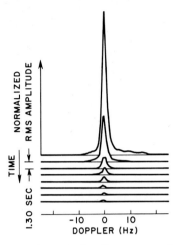

FIG. 11. Range-doppler analysis of bottom reverberation sample.

Intermediate between bottom and surface is volume reverberation where the frequency spread may be due partly to motions of individual scatterers (probably fish) and water current shears carrying scatterers along at different velocities. A typical frequency spread has been found to be about 4 Hz to the −6 dB points.

Figure 12 illustrates conditions under which volume reverberation can be dominant. The velocity profile and ray plots show a submerged sound channel which refracts energy away from the boundaries. Under these conditions the low backscattering strengths and the frequency spreads intermediate between those observed for bottom and surface reverberation clearly support volume reverberation as the dominant background. Where these sound channels are weak, conditions also exist in which two reverberation sources are of similar strength. This sometimes results in double peaks in the reverberation spectra where currents carry volume or surface scatterers

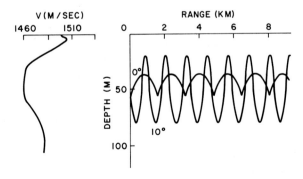

FIG. 12. Velocity profile and ray plots for a submerged channel.

over the bottom. Spectral spreads can be quite variable under these conditions.

Although no one function can be given to describe the shapes of the spectra obtained from the various reverberation backgrounds, some commen can be made on what we have observed. In most cases the spectra fell off from the main peak by 12 to 18 decibels per knot equivalent doppler shift. The spectral peaks exceeded ambient noise by approximately 40 decibels for measurements at closer ranges, thus limiting observations of lower reverberation levels.

Another topic which is being studied presently is the variation of reverberation levels from one transmission to the next. For this study the reverberation data obtained using a CW transmission of 8 msec duration was used. Before studying the variability of reverberation level from one transmission to the next it has been convenient to do some smoothing on the reverberation envelopes. The level of the reverberation received from a single transmission is regarded as consisting of a slowly decaying component combined multiplicatively with a randomly fluctuating component at higher frequencies. The slowly varying component results from geometrical spreading with propagation, and backscatter from gross features, while the high frequency component can be regarded as the result of the combination of signals with random phase and amplitude reflected from a large number of point scatterers. This description is perhaps slightly oversimplified but provides a starting point.

In order to obtain the low frequency component a low pass digital filter with a 2 Hz cut-off was applied to the logarithms of the reverberation envelopes. The results are illustrated for one sample in Fig. 13. The filter bandwidth was chosen after observing the power spectra of a number of logarithmic reverberation envelopes. The low pass version provides a reasonable estimate of level during a single return where combining a number of degrees of freedom by this filtering reduces the variance of the envelope. This makes possible the study of the variation in reverberation level through a sequence of transmissions, providing results which are also applicable to reverberation from long narrow-band CW pulses where this form of smoothing is not possible.

Figure 14 illustrates the way in which reverberation level was sampled for purposes of analysis. At a number of ranges or times after transmission the reverberation level was observed for each transmission in turn. The sequence of samples from a series of pings describes a sampled random process which in general is not stationary but may exhibit quasi stationarity for stable water conditions. For each such set of samples obtained at different ranges the mean reverberation level was obtained and also the standard deviation about the mean value.

To establish what trends, if any, the data exhibited over a number of transmissions, the difference in reverberation level between successive pings

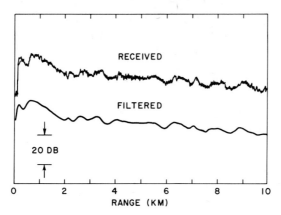

FIG. 13. Effect of low pass filtering on logarithmic reverberation envelope.

was obtained and its standard deviation also computed. Knowing the standard deviation about the mean reverberation level and the standard deviation associated with the reverberation level differences from ping to ping, the normalized autocovariance can be computed to give a measure of trends in the data from one transmission to the next.

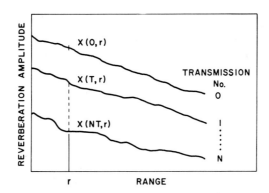

FIG. 14. Illustration of sampled process used in envelope analysis.

Values of normalized autocovariance which approach one indicate data showing strong trends, and any form of adaptive thresholding must take these trends into consideration to be effective. Where the normalized auto-covariance approaches zero, such trends do not exist, and the data can be regarded as being relatively independent from ping to ping with more of the instantaneous reverberation level attributable to the Rayleigh fading.

For the data analysed, points on the autocovariance function can be calculated only at multiples of T, the transmission period. Table I shows some of the trends observed in reverberation where the results from several hundred transmissions have been included. The normalized autocovariance, ρ, is shown for a shift of only one transmission period, 48 seconds in this case. The computations were done at three ranges, 1.6, 4 and 8 kilometers. Analysis is incomplete, but the results here are typical of the trends observed to date.

TABLE I

Reverberation	σ(dB)			ρ(T = 48 sec)		
	1.6 km	4 km	8 km	1.6 km	4 km	8 km
Volume	1.0	0.5	0.3	0.3	0	0.1
Surface	2.6	2.2	1.5	0.4	0.6	0.5
Bottom	3.0	3.9	3.3	0.7	0.8	0.7

The volume reverberation with standard deviations normally less than one decibel and low ping to ping correlation shows a relatively stationary background level and good estimates of this level for thresholding purposes can probably best be obtained by simple averaging with most weight given to recent samples. Surface reverberation, with higher standard deviations and correlations, is less stationary, exhibiting some weak trends as water conditions vary. Bottom reverberation shows the highest standard deviations and correlation coefficients, as one would expect from more consistent trends in reverberation levels as differing bottom structure passes through the transducer beam. Further analysis on the bottom reverberation, in which a form of covariance was computed diagonally across the ensemble of pings to account for the translation of the transducer over bottom features, showed correlations of up to 0.2 greater than those values obtained without taking translation into account. This indicates more strongly that bottom features show scattering highlights which are important in the detection process. Because of these larger changes in time, obtaining good estimates of reverberation level for thresholding requires that the most recent samples must be given the most weight. The choice of the most suitable approach for level estimation between the traditional automatic gain control and a more complex system will depend upon the importance of both the accuracy of these estimates and the retention of amplitude information on echoes over a number of transmissions.

To this point, there has been little discussion of the situation in which two forms of reverberation are more or less even in level. In a significant proportion of experiments a submerged sound channel existed in which volume reverberation dominated for part of the time and bottom reverberation at other times due to leakage of energy to the bottom. Since bottom backscattering strength is much higher than that for the volume, only a small percentage of the acoustic energy need reach the bottom to

give significant bottom reverberation contributions. Thus, small changes
in the velocity of sound profile can result in significant changes in the
proportion of bottom reverberation received.

A thermocouple towed with the transducer gave continuous temperature readings of the form shown in Fig. 14. Variations in the temperature
profile of this magnitude along with bottom depth variations were found

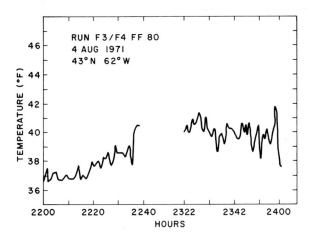

FIG. 15. Sample of output of thermocouple towed with transducer.

to be sufficient to cause large and sometimes quite rapid changes in
reverberation levels where submerged sound channels existed. This results
in perhaps the most difficult of all situations in which to accurately
estimate levels.

4. Conclusion

In conclusion, the main contrast between deep and shallow water acoustics
is that, for practical purposes, there is much more variability and less
stationarity exhibited in propagation and reverberation in shallow water.
For deep water it is often possible to develop mathematical models, either
from physical considerations or empirically, with enough accuracy to be of
practical use when related to environmental conditions. This is particularly
true for the time and frequency dispersion for signals forward scattered
from the sea surface.

In shallow water where propagation often involves multiple reflections
from boundaries, including a spatially inhomogeneous bottom, time and
frequency dispersion in both propagation and reverberation show more
variation, and it may not often be possible to obtain detailed enough
measurements of the environment to make useful predictions from

mathematical models. As a consequence it is necessary to have a knowledge of the nature of the lack of stationarity and homogeneity in shallow water areas where signal processing systems operate. It is possible to place approxi mate limits on signal dispersion as, for example, in propagation angular spreading, but optimum use of the medium for information transfer dictates that systems must be capable of continuous estimation of the channel as well as continuous adaptation of the signal processing strategy.

REFERENCES

1. Wille, P. and Thiele, R. (1971). Transverse horizontal coherence of explosive signals in shallow water. *JASA* 50, No. 1, 348–353.
2. Urick, R. J. and Lund, G. R. (1970). Vertical coherence of shallow-water reverberation. *JASA* 47, No. 1, 342–349.
3. Urick, R. J. (1967). Principles of Underwater Sound for Engineers, McGraw Hill Book Co., New York.

DISCUSSION

R. R. Rojas: In one of your slides on reverberation spectra there was a secondary peak which must be statistically significant because of the averaging. Do you have an explanation for it?
Answer: No.
J. S. Gill: Is it possible that one peak is due to gravity waves and the other due to capillary waves? Have you looked at it from this point of view to see if it would fit?
Answer: No, we have not. However, double peaks in other reverberation spectra have been considered in the past. The velocity of surface waves which give diffraction scattering at this frequency of 9.5 kHz is low enough that any doppler shifts would be only a few Hertz. In this slide (which does not appear in the paper) the secondary peak had a frequency offset of about 20 Hz and appeared at a level of −20 dB. It does not seem likely that it is due to travelling surface waves. In any case these low level secondary peaks are not typically present in results observed by us.
P. M. Schultheiss: In the spatial correlation measurements for the vertical case the values dropped to low levels at large distances, but for the horizontal case the correlation coefficient hovers around one half for a large distance. Do you have an explanation for that?
Answer: For the horizontal case the spatial scattering function is apparently partially coherent in a manner similar to that described in Mr. Laval's paper.
T. D. Plemons: In one slide you showed the calculation of the normalized correlation coefficient between two hydrophones which is not a correlation function because there is no delay between the two. Are you sure that these signals are aligned in time as they should be?
Answer: Yes. We took special care to see that they were aligned for the horizontal array. Cross correlations were done with time delays for the early part of the pulse arrivals in order to determine the orientation of the array and to compensate for any horizontal rotation of the array.
H. Mermoz: Did you calculate cross correlations as a function of time delay which might show peaks higher than those shown on the slides for the correlation coefficient at zero time shift?

Answer: Cross correlations as a function of time delay were calculated only to determine the horizontal aspect of the array. The delays chosen for the horizontal coherence should give the maximum correlation coefficients in the diagram. However, for the vertical array, no time delay was introduced because signal envelopes appeared aligned within the 2 msec resolution of the original transmitted pulse. No phase alignment was necessary since only the magnitude of the complex correlation was retained.

Question: Do you know what the structure of the water was when these large decorrelations were found in the vertical direction?

Answer: The data were collected in late spring and there was some warming at the water surface. Propagation was thus by means of multiple bottom reflections and surface reflections or downward refractions near the surface.

J. S. Gill: Have you made these measurements in deep isothermal water?

Answer: Measurements have been made in an RAP geometry with shallow source and deep receiver in deep water. Although the water was not exactly isothermal this was unimportant at close ranges. Results show a direct path, coherent in vertical and horizontal extent within our measuring capabilities followed by a surface scattered component whose decorrelation in vertical and horizontal directions was consistent with simple estimates of the angular spreading expected from the surface.

C. Van Schooneveld: I have two questions. First, about your suggestion of using frequency diversity, it is not necessarily to your advantage if your system is already working rather badly. If you must divide your energy with half to each frequency the performance will decrease.

Answer: I agree if the signal to noise ratio is already very low, but for the case where the signal to noise ratios are above a relatively modest level and Rayleigh fading exists it is better to divide the energy between independently fading bands and take advantage of the increased number of degrees of freedom to reduce the variance in the receiver output. As long as one is reverberation limited, partition of the energy in this way does not decrease signal to background ratios. Also, in some cases it may be possible to transmit pulses in sequence at maximum power levels rather than dividing the energy in a simultaneous transmission.

C. Van Schooneveld: My second question is concerning the variation of reverberation level from one transmission to the next. I assume your measurements were made from a suitably stable platform. Do you think that the exploitation of this information requires the stabilization of the platform which could affect reverberation levels if unstable?

Answer: If the platform is unstable and beamwidths are narrow there are likely to be problems. Perhaps there is a possibility of relating reverberation level changes to platform motions in a way that could be exploited.

H. O. Berktay: I would like to comment that the stability of the platform does not affect the use of diversity but may add a coherent component to the fading.

C. Van Schooneveld: Do you have measurements from unstable transducers?

Answer: No, only from a transducer in a towed body which is relatively stable.

H. Mermoz: Do you plan to do coherence measurements with fixed arrays?

Answer: There are no definite plans.

D. W. Cruse: Have you any results of the effects of fresh water outlets on reverberation?

Answer: No.

T. Kooij: Have you observed any differences in surface reverberation in developing and developed seas?

Answer: Yes. Perhaps one of the most interesting situations is the change in doppler offset as wind changes. We observed a situation where the wind changed by 180 degrees in a short time. A transducer trained upwind before the change showed reverberation with the expected positive doppler shift. About two hours after the wind changed direction and with the transducer still trained in the same direction the reverberation still showed a positive doppler shift although somewhat less in magnitude. A few hours later the reverberation showed a negative doppler shift as one would have expected for a wind driven surface current. We regard this as evidence that surface reverberation for low grazing angles may be primarily from the water just below the waves with doppler offsets being due to bulk transport of the water near the surface.

Automatic Fixed Site Propagation Tests

T. G. BIRDSALL

Cooley Electronics Laboratory
The University of Michigan, Ann Arbor, Michigan, U.S.A.

1. Introduction

The purpose of this paper is threefold. First, frankly to push John Steinberg's philosophy of underwater acoustic propagation experiments; second, to try to convince you that the small digital computers, often called minicomputers, are valuable and viable tools for propagation work; and third, to discuss the overall experimental system and to try to tie together the interactions of the many facets that go into a successful experiment.

Dr. Steinberg's philosophy on underwater acoustic propagation is that the acoustic measurements must be related, not only to time and place, but to as many observables of the ocean conditions and the atmospheric conditions as possible. Furthermore, to draw forth useful understanding of propagation, that the tests must be run over very long periods of time, in order to encounter a wide variety of oceanographic and atmospheric conditions, in order that their effects may be better understood. The philosophy of statistical decision theory is one of reasoning from observable to decision, which is a reasoning from effect to cause. Of course, in as much as is possible, this reasoning is based on well-known physical theory which relates cause to effect. However, the complexity of oceanographic and atmospheric interaction with acoustic propagation means that in addition to physical reasoning, we need to study experimentally, as a complicated random process, as much of the acoustic and environment as is possible. This study must measure long time series as its basic data. The fundamental problem that occurs when attempting to put this philosophy into practice is the enormous number of possible measurables that might be included in the experiment. In order to reduce the experiment to a viable one, certain selections must be made. In fact, these are normally made more on the basis of availability than they are based on any wise or intuitive judgment of the most important variables to attack.

This paper will use many examples from MIMI (Miami-Michigan) experiments of the last three years. I apologize for this limited viewpoint and ask the reader's indulgence in this matter.

299

This paper attempts to take an overall look at a design of an automatic fixed site propagation test and to relate both generalities and details that must be determined before the test can be begun. The paper is broken into four major topics: One, discussion of the purpose or purposes of the experiment; two, the consideration of the physical hardware that is available at the propagation test station; three, the computer hardware that is available to the experimenter; and four, the computer program itself for conducting the experiment and for the onsite reduction of the data. The final section will briefly consider some of the post-analysis that is necessary before the full scientific extraction of conclusions or information from the data can be begun.

2. Experimental Purpose

Hidden within the idea of automation is inherently the idea that the experiment will be of relatively long-term, perhaps one week, two weeks or a year, duration. Such a long time often means that the experiment should have several purposes and not just a single purpose. The effect on the style of automation of having a multi-purpose experiment is that the computer software must be set up on a multi-tasking arrangement, or time-sharing arrangement, so that all the experiments can run simultaneously. If the experimental design is based on running some experiments just part of the time, others all of the time, then the computer software must contain operator interaction so that an operator can give commands to begin and stop certain experiments without interference with other experiments.

As an example of such a multi-purpose operator interactive experiment, the MIMI experiments, beginning in July, 1971, all had a basic long-term experiment of studying time series of various signal powers. Specifically, the experiment measured the received level of a single frequency (carrier) power, the behaviour of the received phase angle of that single frequency, the surface scattered reverberation power in the near vicinity of that frequency, the total broadband signal power with the exclusion of the single carrier line, the noise power in the signal band, and perhaps the frequency shift, of this single carrier line. These five or six measurements constitute the major acoustic parameters in the long time series study. In addition to this basic experiment, the partially processed data, which is basic to the study of the surface scattered reverberation spectrum and to the carrier frequency shift, has been optionally saved on digital magnetic tape or optionally processed onsite for display of reverberation spectra; the other option is to destroy the data. In a similar fashion the partial processing necessary for development of multipath arrival pictures (or equivalently broadband signal spectrum) has been available for optional storage on digital tape, optional display on an oscilloscope, or simple destruction of the data. Another option which has been available to the operator is digital tape recording of the data demodulates, the most economical sampling of the raw data after it has been translated from bandpass to zero frequency and filtered to the band of interest.

One might question why much of the data which is potentially available in the experiment might be arbitrarily thrown away. A quick consideration of some of the numbers typical to MIMI may be helpful. The basic background experiment of powers measurement produces approximately fifteen hundred complex data values per day; the data for the reverberation studies accumulates about forty thousand complex numbers per day. On top of this, the multipath data would be about two hundred thousand samples per day and the raw demodulate dumps for a reasonable over-sampling of a 50 hertz bandwidth produce in the order of seventeen million complex numbers per day. For experiments which will last over many weeks or months, such data rates are very important. The experimenter must have in mind his ultimate analysis and processing system for handling the data back home. If he wishes to be faced with not more than a few million raw data points, then he must balance his various experiments so that the least important experiment does not produce the greatest load of data.

Very close to the choice for the purposes of the experiment is the choice of signal to be transmitted. A periodic signal is a natural signal for an unattended, automatic test. Although it is quite possible to sequence through a set of different signals, and obtain very valuable information from the differences in the measurements, the attendant problems of synchronization of the reception and overlapping of different types of signals in the water will be a major consideration if a set of signals is used.

Periodic signals have been chosen for MIMI studies for several other reasons. The first is that the ocean surface and the random scattering of the signal at the surface constitutes one of the fastest moving random processes in the signal. It is our belief that most of the time this surface scattering spectrum is limited to a small frequency band about each frequency component in the spectrum. By careful choice of the spacing of the lines in a line spectrum (periodic signal), one may separate the surface reverberation from the coherent energy of the signal. The major signal processing when dealing with a line spectrum will come from the use of a comb filter with very narrow teeth (bandpasses) which exclude both noise and scattered signal energy yet pass the coherent signal energy.

It is important in any measurement to know the quality of the measurement. When powers are being measured this means knowing the noise power at the time the signal powers were being measured. There are several techniques available for simultaneous measuring signal and noise power. For the past several years MIMI has used a spectrum separation where the noise power measurement comes from the received power passed through a narrow comb filter, where the teeth of the filter are placed midway between the signal frequencies.

Given that the signal is periodic, a choice must be made of the period for the signal. If the lines are to be spaced far enough apart to exclude surface reverberation, then the period must be at least one or two seconds (the signal lines spaced a half or one hertz apart). If multipath measurements are to be made, then the signal period should be greater than the anticipated multipath time spread. Obviously in long-range propagation the time spreads will be much longer than one or two seconds and one of these two considerations must be compromised.

The style of generation and modulation of the signal can be either discrete (digital) or continuous. The choice of signal will probably be dominated by the choice of processing that the experimenter has chosen. The MIMI signals are digital because the processing is going to be digital processing, and the interpretation of sampling is much more precise when dealing with digital signals. An equally important consideration has been the ease of maintaining very tight time and frequency control.

There are several other considerations in signal design. The power transducer, especially if it is a low-power transducer, should be operated close to its continuous peak power level at all times. This means that a good signal design criterion is constant output power, with good time resolution and good time sidelobes, if multipath measurements are going to be made. If meaningful wideband power level measurements for propagation loss are to be made, the signal spectrum should be fairly flat. This is a loose consideration since many spectra with some variation in them will work quite as well as any other signal with the same amount of variation in it. Here too, the type of processing program that is going to be used has to be taken into consideration at the same time that the signal is being designed. One must also consider whether the processing is going to be done in the time domain or in the frequency domain. Project MIMI has traditionally worked in the time domain and utilized linear maximal pseudo random sequences for its digital modulation. These always have one less than some pure power of two digits per period and hence are singularly inept for use with frequency domain processing. For frequency domain processing a modulation with some pure power of two digits per period would be ideally matched.

3. Site Hardware

Some experimenters may be blessed with having control of their own transmitting and receiving sites. The case with Project MIMI is that it utilizes transmitting and receiving sites which have been established by larger organizations, although frequently we have used a transducer which has been installed by the University of Miami. Important hardware considerations, relative to the transmitter, are the centre frequency and bandwidth (or Q) of the transducer and knowledge (or lack of knowledge) as to the exact transfer characteristic of the electrical-to-acoustic transducer.

A critical consideration in coherent propagation testing is the source of time and frequency reference available at the transmitting and at the receiving site. Technology provides us with highly stable oscillators, which are good to the order of one part in 10^{10}. Utilizing these with frequency synthesizers, a signal of great time and frequency stability is available for use. Stability of a part in 10^{10} means that at five hundred hertz, two free-running references could be expected to differ by one cycle in about five-and-a-half months. Of course, in order to insure that type of stability, frequent checks with some master frequency reference, provision for auxiliary power supply, monitoring of temperatures of the equipment, and so forth, must all be part of the test setup.

The actual signal generator for a signal designed for propagation experiments is usually not available at a site and must be constructed. Thankfully, digital technology allows us to build excellent signal generators in small boxes which still have almost perfect amplitude stability and time and frequency stability if derived from a frequency reference. One of the signal considerations is matching the Q of the signal to the Q of the transducer. It has been the MIMI practice to make the signal Q twice that of the transmitter, thereby easing the "time-sidelobe" problem considerably in processing. A much more sophisticated signal design matching could be used if the transmitter transfer were known. A compensating electrical signal is sent to the transducer, to obtain a nearly white acoustic signal. This would be accomplished at some loss of efficiency of the acoustic transducer and does require detailed knowledge of both the amplitude and phase characteristics of the transducer in its operating environment. This latter information has never been available to MIMI.

The basic reception equipment of the site often contains many surprises for the unwary experimenter. We find it convenient to carry along a dozen or so unity gain wideband buffer amplifiers so that we do not upset our host's electrical signals. The experimenter, of course, must provide his own bandpass filters and gain adjustments and a computer-controlled gain adjust ment if the experiment is going to continue for a long time and multiplexing equipment if the computer switches inputs signals. Provision should also be made for local personnel to record auxiliary data such as gain changes, time out for repairs, and catastrophic occurrences common at any experimental site.

4. Computer Hardware

The heart of any automatic propagation test is the hardware that controls the test, transmission, reception, recording and processing the data online: a computer. When one speaks of "computer" many automatically think of

IBM and large installations. Project MIMI has been utilizing minicomputers, primarily based on the PDP-8. Excluding only the teletype and the oscilloscope, the hardware we have used in the field is 55 centimeters square and 1.1 meter high. It has been carried in station wagons, a light plane, a Grumen Goose and open truck. The computer itself has 8096 twelve-bit words and an add-only arithmetic unit. In addition to the basic computer, the equipment must contain an analogue-to-digital converter for the amplified acoustic input, digital magnetic tapes (we use a pair) for storage both of program and reduced data. Additional equipment we have found extremely useful are two digital-to-analogue converters which are used for an oscilloscope display in the field and for XY plotting offline, another digital-to-analogue converter for a multipoint recorder for time series plots of the power data. The computer has also been fitted with a relay register which can be loaded by the program and which controls a set of simple relays. These relays, in turn, are used to control the gain and multiplexing of the incoming signal (changes are made during gaps in data-taking which are carefully clocked). The relays are also used for turning the teleprinter on and off, whenever the computer has a message to type out. (The teletype is the weakest and most fragile part of the whole system and must be turned off to save it from wearing out.) Relays may also be used to turn off other recording equipment, such as the multipoint recorder, which was never designed to operate day and night for months. A piece of equipment that we wish that we had had on the tests during the last three years is called, locally, a time-of-day clock. Specifically, this is simply a counter, running from the master reference which counts clock pulses and converts these to convenient readings. In the timing and development of the operating program, it should read in milliseconds and seconds. During the running of the test, it should read in seconds, minutes, hours, and days. A time-of-day clock, operating on its own battery-rechargeable supply, is important because there will undoubtedly be power failures and equipment problems. The ultimate analysis of the data depends upon the data being accurately referenced to time. If the experiment were always attended by a human operator, then he would notice and correct for changes of at least a few minutes' duration or so. Our computer is programmed to shut down for any irregularity in power supply and may be down and then bring itself back up again with only a few milliseconds gap. A computer-readable time clock would allow the program to read the time at power failure and at the return from power failure and follow programmed judgments as to what appropriate action should be taken to lose a minimum of data.

For those who might be interested, the cost of such a computer, exclusive of the frequency reference and synthesizer, is of the order of $23 000.

5. Computer Software

There is much more to programming a computer for operating an automatic test than just programming the acoustic signal processing equations. In a long-term test with various experiments being turned on and off without interruption of other experiments, and with operator interaction, the basic "support" package which runs the interrupt processor and the multi-tasking assignment must be programmed. We have utilized a three-level structure for the computer consisting of (1) those jobs which must be handled with the interrupts turned off so that they go to completion as rapidly as possible, such as the operation of the A to D converter handler and interrupt handler itself; (2) deadlined tasks which must go to completion in a relatively short time (several hundred microseconds), such as the complex demodulation task; and (3) task time where the majority of computation is done, but where the computation can be guaranteed to be completed in a fairly relaxed manner in a matter of a few seconds. Even small computers have the same basic memory cycle time as the common large ones. The PDP-8 can complete one million memory cycles in just a little over one second.

The experimenter and the computer programmer must work very closely in determining the style of computation and the accuracy desired and available so that a reasonable balance is made in the time and memory used, compared to the accuracy desired by the experimenter. In MIMI operations we have felt it necessary to have as much accuracy as the actual sampling process will permit, carried through until the data is in final form; the only exception is that processing which is carried out purely to run a momentary display for the experimenter's perusal. We have used a double-precision interpreter which holds twenty-three bits of magnitude information plus sign, a computational dynamic range of 138 dB. The reason for this will be apparent in a moment when the processing equations are displayed, showing that in many cases, two large numbers will be subtracted from each other and a reasonably accurate difference is expected. In addition to the support software for computation, there must be support software to operate the displays and to control multiplexing, gain control and the like.

Perhaps the simplest part of the entire operation is the coding of the equations for use in the processing of the acoustic signal in order to determine the basic long term powers and carrier phase. The style of processing discussed here is the MIMI processing, which works on a specified block or interval of time and processes data in nonoverlapping intervals. The real time part of the processing consists in developing a sequence of complex demodulates. Basically, this process averages the product of the bandpass real waveform, multiplied by $e^{-j\omega_c t}$, over some demodulation interval. The data block consists of an integer number of periods of the transmitted signal. The notation units for this paper are shown in equation (1).

$$r\text{th demodulate in } p\text{th period} = z(Np + r) \tag{1}$$

If a signal is averaged over one period, the result will be independent of the reception at all of the frequency lines except the carrier frequency. Specifically, the transfer function of this computation is a sinc function with zeros at all of the sideband lines and peak at the carrier. These "block" averages over single periods are the basic data from which the carrier and reverberation measurements will be obtained. The comb filtering is accomplished by averaging the period

$$B(p) = \sum_r z(Np + r) \tag{2}$$

of reception to form a representative signal of one period duration.

$$C(r) = \sum_p z(Np + r) \tag{3}$$

The tooth separation is the reciprocal of the duration of one period, and the ratio of the tooth separation to tooth width is equal to the number of periods that have been added together.

If a grand average of all of the received data is taken, the single complex number resulting represents a narrow (sinc type) filtering of the carrier line. The ratio of the bandwidth of the block average bandpass to the C bandpass is the same as the tooth spacing to tooth width ratio, the number of periods involved in the averaging.

$$C = \sum_p B(p) = \sum_r C(r) \tag{4}$$

The reverberation power measurement is obtained by subtracting the carrier power measurement from the total in the block average bandpass.

$$\sum_p |B(p)|^2 - \frac{1}{\#p} |C|^2 \tag{5}$$

This is obviously akin to the calculation of the variance of the complex numbers $B(p)$. When the reverberation power is very small compared to the carrier power, the accuracy of maintaining the numbers $B(p)$ must be very high in order to get a valid measurement from equation (5).

In a similar fashion the sideband power, with the carrier power eliminated, can be calculated by determining the variance of the $C(r)$ numbers.

$$\sum_r |C(r)|^2 - \frac{1}{\#r} |C|^2 \tag{6}$$

It is well known in classical statistics that if the z's are zero mean identically distributed independent random variables and that if there are

A times B of them, then the "variance" computation yields the same result

$$E\left[\overset{A}{\underset{}{\Sigma}} \mid \overset{B}{\underset{}{\Sigma}} z\text{'s} \mid^2\right] = ABE[\mid z \mid^2] \tag{7}$$

independent of the selection of how many terms are averaged "coherently" before being averaged "incoherently." This means that the effect of broadband white noise on the carrier, reverberation and sideband powers is virtually the same, and the slight corrections can be easily calculated. If a broadband measurement affected only by background noise were available

$$E_0[\mid C \mid^2] = E_0\left[\underset{p}{\Sigma} \mid B(p) \mid^2\right] = E_0\left[\underset{r}{\Sigma} \mid C_r \mid^2\right] \tag{8}$$

in addition to the above power measurements, then the quality of each measurement, the estimated signal-to-noise ratio involved in each measurement, would be readily apparent to the experimenter. Such a measurement can be obtained if the number of periods in the analysis block is an even number, and the $C(r)$ measurement is modified to finding a typical reception on a two-period time base. This has been dubbed a "double-length circulating average." The single-period representative reception is simply

$$D(r) = \underset{p}{\Sigma} z(2Np + r) \tag{9}$$

the sum of the first and second half of $D(r)$. If a Fourier series were to be calculated on this double period list of numbers, the eigen frequencies

$$C(r) = D(r) + D(N + r) \tag{10}$$

would be spaced one over twice the signal period. Since $C(r)$ accounts for the energy at the signal frequencies, the difference $D(r) - D(N + r)$ must account for the energy at the frequencies spaced halfway between the signal frequencies. (This can be shown rigorously to be the case.) The noise measurement utilized in MIMI's DUALCOMB experiments utilizes this difference to derive its noise measurements.

$$\underset{r}{\Sigma} \mid D(r) - D(N + r) \mid^2 \tag{11}$$

Calculation of these powers in even such a simple minicomputer as the PDP-8 with add-only arithmetic quadruple-precision calculations require less than one second.

The operations of determining the reverberation spectrum, or estimates of the carrier frequency shift, can be obtained from the complex list $B(p)$, and the multipath structure or, equivalently, the signal transfer function can be calculated from the complex list $C(r)$ and C. These can either be done onsite as background tasks and displayed to the experimenter if he is present, or the lists can be saved on magnetic tape for future processing offline. I hope the reader will be happy that I have chosen not to fill the next several pages with the equations and tricks used to calculate such functions.

6. Post-processing and Scientific Conclusions

While the test is still in the planning stages, the experimenter must keep in mind that all of the automation and all of the furor necessary in getting the test under way were only means to an end, to develop a large data base on which to draw conclusions about underwater acoustic propagation. As data becomes available from the test, the experimenter should be prepared to process this data to get a quick review of the way the test is progressing, to get some idea of what is happening, and to begin the process of reformatting the data and documentation to begin his real analysis. At this point all of his forethought in taking care of unforeseen events (equipment being inoperative for unknown periods, someone pulling his plug, or worse yet, some unknown gain changes) will come to bear. At this point we have found that with a small amount of data available one can often reach a quick overall decision about many of the uncertainties that existed before the test began.

There is no conclusion to this paper, as there is no end to the study of underwater acoustic propagation. I hope I have laid out the many considerations that must go into making a successful automated underwater acoustic fixed site propagation test. Many details have been left out, and the MIMI processing has been emphasized only by way of an example. The overall expense of running very large acoustic experiments is considerably smaller when the minicomputers are used to do the rote work. Although the computers themselves have tremendous flexibility and do not restrict the experiment that can be run, they do force much of the thinking to be done beforehand and the thinking must take care of all the interaction of the parts of a test. In contrast, a test run by a crew of intelligent human beings is far more flexible, subject to change, and much more robust in not requiring so much forethought on the experimenter's part. It is the author's opinion that the difficulties of forethought are well worth it, that the price of automation is small compared to the other expenses involved in experimentation. Automation allows the small experimenter to do the work of a very large laboratory.

DISCUSSION

R. Seynaeve: You said that the machine you used had a number of limitations. For this project did you feel that the 12 bit word length was a limitation?
Answer: We gave quite a little thought to this 12 bit compared to a 16 bit word, and I felt the 12 bit length is not, for these reasons: both 12 and 16 bits are equally inadequate. we have to go to multiple precision arithmetic. We have a 10 bit A/D convertor for 54 dB dynamic range (with computer controlled switchable 90 dB in 6 dB steps). We very quickly add together enough samples to require double precision but very seldom have to go to triple precision. After multiplications we have to go to quadruple precision. It is a limitation because in assembly language the operations are

quite limited. In fact, the signal and experimental design are closely related to the processing assembly language.

R. Seynaeve: You said you have a rather high sampling rate, four or five times Nyquist frequency. Why do you do that?

Answer: Really for ease of interpretation of the multipath analysis. Point number one: the input bandwidth is the 3 dB bandwidth of the input analogue filter, which has rather slow skirts. If we were to use Nyquist rate sampling we should use something like the 40 dB down bandwidth. Point number two: if you have a critically sampled piece of data you should use sinc ("sin X/X") type interpolation of the analysis.

R. Seynaeve: Exactly, this is what I mean. It would be interesting in this case to do the processing at close to Nyquist frequency, and then at the end interpolate the result. Would you save computer time?

Answer: No, probably not. I do use "sharp processing" which operates on just one complex value per digit, and then does several of these in parallel.

C. N. Pryor: Do you feel you are using a surface duct or submerged duct in your propagation? What percentage of the received power is typically in the sidebands compared to the carrier?

Answer: On occasion we have had good evidence of a subsurface duct. In the Straits of Florida this is an anomaly. We have had it. In November 1970 we went down to look for it. We had 19 days data encompassing the weather conditions that could cause it. There was a storm with continuing strong winds. The wind came up, a day later the noise level rose some 15 dB completely covering the biological noise, and a day later the signal rose some 20 dB. There were 9 to 12 foot seas which pretty well ruled out a surface duct. We think the propagation was close to the surface because the forward scattered reverberation rose to about equal to the carrier power. We normally measure about 1% to 10% reverberation power compared to carrier power. This anomalous behaviour happens perhaps 3-6 days out of a year in the Straits of Florida.

C. N. Pryor: You have then experienced the case when the majority of the power gets into the sidebands?

Answer: Except during deep fades, we have not in the Straits of Florida. In the open ocean experiments from Eleuthera to Bermuda we may have. I did not design that experiment to account for that situation because I expected to be deep ducted or RSR most of the time. I will have to redesign the signal for more precise measurement (place the broadband signal lines farther apart). There is some evidence that the 200 watt signal is reverberation limited after several hundred miles.

C. N. Pryor: If you had a number of propagation modes that were more or less equal amplitude, you would expect all of the power to go into the sidelobes (reverberation band about the carrier), where the multipath interferes differently.

Answer: This would be broadening of the central line. The usual measurement of the researchers at Miami is CW—one frequency. They measure the reception using 1 minute to 5 minute averages (hence 0.016 to 0.003 Hz bandwidths), and look at the resultant phase measurements plotted across months of continuous measurements. This is the way they isolate tidal and similar low frequency effects. This frequency analysis shows very small widening of the central line—certainly less than one millihertz. The changes in the multipath structure appear to give very little widening to a line.

Nonlinear Acoustics

H. ORHAN BERKTAY

Applied Research Laboratories,
The University of Texas at Austin, Austin, Texas
(On leave of absence from University of Birmingham,
Birmingham, England)

1. Introduction

At a NATO Advanced Study Institute held in Copenhagen in 1966 the author presented a paper [1] on the possibilities for the exploitation of nonlinear interactions between two high frequency acoustic waves to produce a highly directional low frequency acoustic wave. Since that time, interest in such "parametric transmitters" has grown considerably; results of a great deal of experimental and theoretical work have been published [2–20], and even sea-going systems have been produced (Walsh in Reference [14]). In a review paper of this kind it is very difficult to do full justice to all the work done to date. Therefore, the paper concentrates on presenting some thoughts on the present state of the art.

The main developments since 1966 have been a better, and more fully documented, understanding of the quasi-linear theory of parametric transmitters and of the limitations of the quasi-linear theory, and the development of parametric receivers.

The paper attempts to cover these aspects of finite-amplitude effects.

2. Parametric Transmission

(a) Quasi-linear theory and its application to parametric transmission
A quasi-linear method of evaluating the behaviour of scattered waves in nonlinear acoustic interactions [21] has been used extensively, and with good results, in studying the performance of parametric transmitters and receivers.

Basically, scattering of the interaction components from a point in the

311

field is ascribed to a volumetric source, the strength of which per unit volume is

$$q = (\beta/\rho_0^2 c_0^4) \frac{\partial}{\partial t} p_i^2, \tag{1}$$

where p_i is the instantaneous pressure at that point. It can be shown that even at high pressures, this expression does not include higher order terms in p_i. In quasi-linear (or single scattering) approach, it is assumed that the nonlinear interaction process does not draw a significant amount of energy from the primary waves and that higher order interactions can be neglected. Although these assumptions appear to be too restrictive, there is strong experimental evidence to suggest that in many applications excessive higher order effects can be avoided by suitable design.

Westervelt considered two primary waves which were confined within a column the cross-sectional dimensions of which were small compared with the wavelength at the difference frequency. Because of the absorption of the primary waves, the difference frequency source distribution then resembles a continuous end-fire array with exponential taper. Then, the farfield behaviour of the difference frequency waves can be described by the equations

$$p_-(R, \theta) = p_W(R, 0)D_R(\theta), \tag{2}$$

where

$$p_W(R, 0) = -(\omega^2 \beta \sqrt{W_1 W_2}/2\pi c_0^3 \alpha_T R) \exp\left[-(\alpha_- + jk_-)R\right], \tag{3}$$

and

$$D_R(\theta) = [1 + j(2k_-/\alpha_T) \sin^2 (\theta/2)]^{-1}. \tag{4}$$

For $\alpha_T/k_- \ll 1$ (which includes most cases), equation (4) gives 3 dB beamwidth of

$$2\theta_d = 2\sqrt{2\alpha_T/k_-}. \tag{5}$$

The value of α_T is largely controlled by the primary frequencies; hence θ_d is inversely proportional to the square root of the difference frequency, provided $f_-/f_0 \ll 1$.

For water, using $\beta \cong 3.5$, equation (3) yields a difference frequency rms source level of

$$(SL)_W \cong 37 + 20 \log F_- + 10 \log (W_1 W_2) - 40 \log \theta_d^{\circ} \text{ dB}//\mu\text{bm}, \tag{6}$$

where F_- is in kHz, W_1 and W_2 in watts and θ_d° is in degrees. A similar expression for constant α_T is

$$(SL)_W \cong -27 + 40 \log F_- + 10 \log (W_1 W_2) - 20 \log \alpha_T \text{ dB}//\mu\text{bm}, \tag{6a}$$

where α_T is in nepers per meter.

If the bulk of the interaction process takes place in the farfield of a transducer from which the two primary beams have been launched, then the primary pressures can be expressed in the form

$$p_{1,2} = (P_{1,2}/r)D_{1,2}(\gamma, \phi) \exp \left[-(\alpha_{1,2} + jk_{1,2})r\right], \tag{7}$$

where $D_{1,2}(\gamma, \phi)$ are the two-dimensional directivity functions at the two frequencies. The difference frequency source function can be evaluated at all points (r, γ, ϕ) and the pressure at a point (R, θ, η) can be expressed in the form of a scattering integral,

$$p_-(R, \theta, \eta) = -(\omega^2 P_1 P_2 \beta/4\pi\rho_0 c_0^4) \cdot$$

$$\iiint \frac{D_1(\gamma, \phi)D_2(\gamma, \phi)}{r^2 s} \exp \left[-(\alpha_1 + \alpha_2 + jk_-)r - (\alpha_- + jk_-)s\right] dv, \tag{8}$$

where s is the magnitude of the vector joining the two points and the integral is taken over the whole of the interaction volume. In practice, backscattering from regions for which $r > R$ will arrive at the observer with incorrect phasing, and is negligible. Hence, the volume integral need only be evaluated for $0 \leq r \leq R$. This method has been used in practice with great success [16, 20].

For the particular case of $f_0 \gg f_-$ and for $(\alpha_1 + \alpha_2)R \gg 1$, some normalized results can be obtained if all angles of consequence are small (i.e., $< \frac{1}{2}$ radian). Then, with the geometry of Fig. 1, equation (8) can be shown to reduce to

$$p_-(R, \theta, \eta) \cong p_W(R, 0)(S/\lambda_0^2) \int_{-\pi/2}^{\pi/2} \int \frac{D_0^2(\gamma, \phi) \cos \gamma \, d\gamma \, d\phi}{1 + j[(\theta - \gamma)^2 + (\eta - \phi)^2]/\theta_d^2} \tag{9}$$

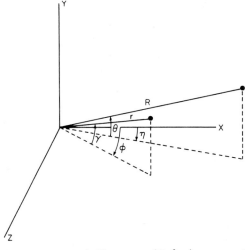

FIG. 1. The geometry used. Observer at (R, θ, η) source point (r, γ, ϕ).

This can be identified as two-dimensional convolution of the product of the primary directional patterns with $D_R(\theta, \eta)$ for small angles [19]. For θ_d much larger than the primary beamwidths, this yields

$$p_-(R, \theta, \eta) \cong p_W(R, 0)D_R(\theta, \eta) \tag{10}$$

as

$$(S/\lambda_0^2) \int\int\limits_{-\pi/2}^{\pi/2} D_0^2(\gamma, \phi) \cos \gamma \, d\gamma \, d\phi = 1. \tag{11}$$

Thus, for extremely narrow primary beams, Westervelt's result is obtained.

If, on the other hand, θ_d is much narrower than the primary beams,

$$p_-(R, \theta, \eta) \cong p_W(R, 0)D_1(\theta, \eta)D_2(\theta, \eta)(\alpha_T R_0 k_0 / k_-)$$
$$\times [(\pi/2) - j \ln (\pi/\theta_d)^2]. \tag{12}$$

In other words, if the primary beams are much wider than θ_d, the difference-frequency beam pattern is identical with the product of the primary directivity functions, and the amplitude along the axis is reduced as compared with $p_W(R, 0)$.

The directivity function of a planar transducer can be normalized with respect to its half-power point. Using this property, the difference frequency source level was evaluated [22] from equation (9), for various values of $\Psi_d = \gamma_1/\theta_d$ for a circular transducer, and of $\Psi_y = \gamma_1/\theta_d$ and $\Psi_z = \phi_1/\theta_d$ for rectangular transducers. Here subscript 1 is used to indicate half-power angles of primary beams in cylindrical geometry for a circular transducer, and in the x–y and z–x planes for the rectangular transducers. In Fig. 2 some of the results are shown. Here, V is a source level reduction factor given by

$$V = |p_-(R, 0, 0)/p_W(R, 0)|. \tag{13}$$

The results for a circular piston lie on the same curve as for a square one.

These curves, in conjunction with equation (6) or (6a), can be used for predicting the farfield performance of a parametric transmitter with some accuracy, within the limitations of the treatment already discussed. For a given set of primary frequencies, α_T is approximately known. For a required difference frequency, θ_d is determined. (For this purpose, nomographs similar to Fig. 12.2 of Reference [1] are useful.) Then, for a given transducer, Ψ_y and Ψ_z (or Ψ_d) are determined using the expressions

$$\Psi_{y,z} = 163/k_0 l_{y,z} \theta_d^\circ \tag{14}$$

for a rectangular transducer of sides l_y, l_z and

$$\Psi_d = 92.5/k_0 a \theta_d^\circ \tag{15}$$

for a circular transducer of radius a. Here θ_d° is in degrees. For the appropriate Ψ values, the pressure reduction factor V is found and the source

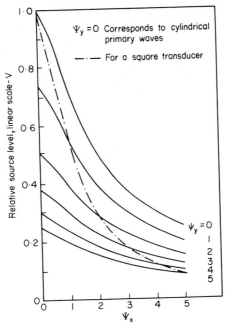

FIG. 2. Relative source level for different transducer aspect ratios.

level is determined for given acoustic powers in the primary waves.

This method has been used quite successfully in comparisons with published experimental results.

Another method [13, 17, 18] for predicting the performance of parametric transmitters uses $\alpha_T r_0 = \alpha_T R_0 k_0/k_-$ as the basic parameter. As $R_0 = S/\lambda_0$, this parameter corresponds to

$$\alpha_T r_0 = 0.650/\Psi_d^2 \tag{16}$$

for a circular transducer, and to

$$\alpha_T r_0 = 0.637/\Psi_y \Psi_z \tag{17}$$

for a rectangular transducer.

The model is developed on the basis that the difference frequency sources at ranges $r < r_0$ contribute to the difference frequency pressure at R as if these sources constituted a well-collimated end-fire array. The difference frequency sources at ranges greater than r_0 are assumed to generate spherical waves, the angular dependence of which is given by the product of the primary directivity functions. It can be shown that for $R \gg r_0$ this model yields

$$p_-(R, 0) \cong p_W(R, 0)\{[1 - \exp(-\alpha_T r_0)] + j\alpha_T r_0[E_1(\alpha_1 r_0) - E_1(\alpha_T R)]\} \tag{18}$$

where

$$E_1(x) = \int_x^\infty \frac{\exp(-t)}{t}\, dt.$$

For $\alpha_T R \gg 1$, $E_1(\alpha_T R) \to 0$, and the farfield value is obtained. For $\alpha_T r_0 \ll 1$, the farfield value reduces to

$$p_-(R, 0) \cong p_W(R, 0)\alpha_T r_0[1 + jE_1(\alpha_T r_0)]. \tag{19}$$

(This can be compared with equation (12), showing clear similarity.) As $\alpha_T r_0$ is reduced, the bulk of the contribution to $p_-(R, 0)$ comes from the region where the difference frequency waves can be assumed to be spherically spreading, with the directivity pattern given by the product of the directivity functions associated with the primary waves. This agrees with the general conclusion reached for the directivity of the difference frequency waves for large Ψ.

Comparisons made between results obtained by using Fig. 2 and from equation (18) (for $\alpha_T R \gg 1$) show good agreement, particularly when the Ψ values are greater than 1. In a particular case, the range dependence of the difference frequency source level was calculated using equation (18), and compared with those computed by the numerical evaluation of the scattering integral in equation (8), with good agreement. For $R \gg r_0$, the difference between the two sets of results did not exceed 2.5 dB.

On the basis of these manipulations, the author is of the opinion that this is a very useful model, in that it gives an insight to the propagation of the difference frequency waves, but it needs to be applied with care.

So far, the interpretation of the source function of equation (1) has been confined to the production of the difference frequency sound waves when two high frequency waves are transmitted. In general, the transmitted waves are not monochromatic. For example, if in Westervelt's model of a collimated beam the transmitted wave is a narrowband signal of the form

$$p = P_1 g(t - x/c_0) \exp(-\alpha_0 x) \cos(\omega_0 t - k_0 x), \tag{20}$$

where $g(t)$ is an envelope function the peak value of which is unity, then the low frequency component of the source function becomes

$$q = (\beta P_1^2/2\rho_0^2 c_0^4) \exp(-2\alpha_0 x) \frac{\partial}{\partial t} g^2(t - x/c_0). \tag{21}$$

If $g^2(t)$ has a Fourier transform $G_0(\omega)$, clearly Westervelt's analysis can be applied to each frequency component [23]. As the highest significant frequency component is assumed to be much smaller than ω_0, absorption of the low frequency signals may be neglected for simplicity. Then, at a point in the farfield, along the axis of the array, the pressure signal can be shown to have the waveform

$$p_-(t, R, 0) = (\beta W_0/8\pi c_0^3 R\alpha_0) \frac{\partial^2}{\partial t^2} g^2(t - R/c_0), \tag{22}$$

where W_0 is the value of the acoustic power transmitted at the peak of the pulse envelope. This result has been verified experimentally [10, 11, 15].

For one-dimensional wave propagation, the square of the envelope is differentiated only once to get the pressure waveform. For example, for transmission from a spherical source of radius r_0 (where $k_0 r_0 > 1$) the pressure waveform at range R becomes

$$p_-(t, R) = (\beta W_0/8\pi c_0^3 R)[E_1(2\alpha_0 r_0) - E_1(2\alpha_0 R)] \frac{\partial}{\partial t} g^2(t - R/c_0). \tag{23}$$

Reverting to Westervelt's model, if the same pulse is repeated at intervals of $T_0 = 2\pi/\omega_a$, say, then in the frequency domain (without the retardation term)

$$p_-(\omega, R, 0) = -(\beta W_0/8\pi c_0^3 R\alpha_0)\omega^2 G_0(\omega) \sum_{n=0}^{N-1} \exp(-jn\omega T_0) \tag{24}$$

where $N = 1/T_0$ pulses have been used in the summation, which gives

$$\sum_0^{N-1} \exp(-jn\omega T_0) = \exp[-j(N-1)\pi\omega/\omega_a]$$

$$\times \sin[\pi N\omega/\omega_a]/\sin(\pi\omega/\omega_a). \tag{25}$$

For large values of N, this term indicates that discrete frequency components are produced at $\omega = m\omega_a$. Then, equation (24) gives

$$p_-(\omega, R, 0) = -[\beta W_0/8\pi c_0^3 R\alpha_0] \sum_{m=1}^{M} (-1)^{mN}(m\omega_a)^2 NG_0(m\omega_a) \tag{26}$$

Transmitted energy per pulse is $W_0 \int g^2(t) \, dt$. For example, for square pulses with $1/2$ duty ratio, energy per pulse becomes $W_0 T_0/2$ and the average power is $W_1 = W_0/2$. Then,

$$p_-(\omega, R, 0) = -(\beta W_1/4\pi c_0^3 R\alpha_0) \sum_{m=0}^{M} (-1)^{mN} N(M\omega_a)^2 G_0(m\omega_a). \tag{27}$$

This equation is comparable with equation (4) for the two-frequency case; in fact equation (4) can be derived by using $g(t) = \sin(\omega_a t/2)$ in this analysis.

This mode of operation for a parametric transmitter has been studied extensively in Reference [15] (see, also, Merklinger in Reference [14]). It appears to have a number of advantages over the two-frequency operation. One serious drawback is that the bandwidth of the transducer used must be much wider than the low frequency to be generated, in order to be able to transmit the individual pulses without interpulse distortion. However, in applications to the generation of very low frequency signals this need not be a serious limitation.

(b) High intensity effects in parametric transmitters

As the intensity of a wave is increased, perturbation solution of the non-linear wave equation does not give a rapidly converging result. By rearranging the wave equation in the form of Burgers' equation, Blackstock [25] was able to develop an exact and complete solution for plane waves. However, only a partial solution exists for cylindrical and spherical waves. (A full treatment of the subject and a comprehensive set of references are given in Reference [26], and a historical review is given by Blackstock in Reference [9].)

Briefly, a wave which is initially sinusoidal distorts as it propagates because the propagation velocity of a particular phase associated with a particle velocity u is $c_0(1 + \beta u)$. This "phase modulation" effect continues until a shock front is produced. The waveform cannot become multivalued, a shock front develops, and the waveform becomes a sawtooth. Increased absorption at the shock front results in the rapid attenuation of the saw-tooth wave. Eventually, the amplitude of the wave is reduced to such an extent that losses through the absorption of the higher harmonics becomes greater than the rate at which they are produced as a result of continued (nonlinear) distortion, and the wave eventually reverts to a somewhat distorted sinusoid. In this, the "old age" region, the small signal propagation theory can be applied. It can be shown [25] that for the case of plane waves, the amplitude of the fundamental in the "old age" region becomes independent of the initial amplitude of the wave.

By using a coordinate transformation scheme [24, 27], it is possible to use the results obtained for plane waves for spherical and cylindrical waves as well. However, in these cases, the onset of the "old age" region cannot be determined directly.

In the sawtooth region, the amplitude of the fundamental component can be expressed in the form

$$\hat{P}_1 = 2\alpha_1 \rho_0 c_0^2/\beta k_1 \sinh(\pi\Delta/2). \tag{28}$$

For plane waves,

$$\pi\Delta/2 = (\alpha_1/\beta\epsilon k_1) + \alpha_1 x \tag{29}$$

x being distance from the source. For a spherical wave, the source-radius of which is r_0 (for $k_0 r_0 \gg 1$), at range r

$$\pi\Delta/2 = (\alpha_1 r/\beta\epsilon k_1 r_0) + \alpha_1 r \ln (r/r_0). \tag{30}$$

As the drive is increased, in either case, $\pi\Delta/2$, and hence $\hat{P}_1$, becomes independent of the initial amplitude of the wave.

$P_0/\hat{P}_1$ represents the attenuation of the fundamental component. On comparing this with the attenuation of a small signal wave to the same range, extra attenuation due to finite amplitude effects can be computed. In the "old age" region, the extra attenuation would then be expected to become independent of range. Extra attenuation curves for plane waves [25] show this effect clearly. However, those obtained for spherical waves [28] from equations (28) and (30) follow the expected trend up to a certain value of r/r_0, and then the rate of increase of extra attenuation rises rapidly. This trend cannot be justified on physical considerations. Many such curves were plotted for different values of the parameter $\beta\epsilon k_1 r_0/\alpha_1$ and in each case, the point of inflexion was found to correspond to $\pi\Delta/2 \doteq 0.6$. This is the value of Δ Blackstock proposed [25] as the limit of the sawtooth region in the plane wave case. However, in the case of plane waves, the extra attenuation increases by another 3 dB at ranges well beyond the point where this value of $\pi\Delta/2$ is reached. Arguing that in spherical waves, because of spreading loss, the nonlinear effects would tend to be diminished more rapidly, a perturbation solution was attempted to study the propagation of the spherical waves beyond the sawtooth region. As a result of a straightforward calculation, it was found that the increase in extra attenuation in this region has an upper bound of 1 dB. On this basis, it can be concluded that the "old age" region for spherical waves commences at range R_M where $\pi\Delta/2 \cong 0.6$.

From equation (28) the amplitude of the fundamental component at R_M is

$$\hat{P}_1 = 3.15\alpha_1\rho_0 c_0^2/\beta k_1. \tag{31}$$

Hence, the equivalent source level (valid for $r > R_M$) can be found.

$$SL = 20 \log (R_M\hat{P}_1) + 8.68\alpha_1(R_M - 1). \tag{32}$$

This represents the maximum source level which can be realized at ranges $r > R_M$, where R_M is given by

$$(R_M/r_0) \ln (R_M/r_0) = 0.6/\alpha_1 r_0. \tag{33}$$

The value of the maximum effective source level was computed as a function of $\alpha_1 r_0$, using the expression

$$(SL)_{max} \doteq 20 \log (R_M \hat{P}_1) + 8.68\alpha_1(R_M - r_0). \tag{32a}$$

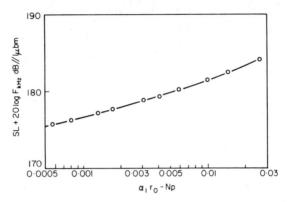

FIG. 3. Spherical waves—equivalent source level in "old age" $(r \to \infty)$

These results, which are shown in Fig. 3, need to be corrected by adding $8.68\,(\alpha_1 r_0 - \alpha_1)$.

It the source level of the transmitted wave is such that the first term in the expression for $\pi\Delta/2$ is not negligible, then equation (32) can still be used to obtain the effective source level in the "old age" region, but with R_M now obtained from

$$(R_M/r_0)[\ln\,(R_M/r_0) + (\rho_0 c_0^2/\beta k_1 \hat{P}_0 r_0)] = 0.6/\alpha_1 r_0 \qquad (34)$$

where $\hat{P}_0$ is the amplitude of the transmitted wave referred to 1 m. (A very similar study, with experimental support, is presented in Reference [29].)

Radiation of a wave from a planar transducer can be considered in two regimes. If the distortion of the wave in the nearfield of the transducer is not too great, then the farfield propagation resembles that of a spherical wave with a source radius r_0. Various experimental results suggest that r_0 lies between the Fresnel distance $(\cong S/4\lambda_1)$ and Rayleigh distance $(\cong S/\lambda_1)$ for the transducer. Good agreement is reported between experiments and theory in Reference [29].

The application of these considerations to parametric transmitters is, as yet, very much under discussion. Models are proposed for evaluating high intensity effects in parametric transmitters in References [17], [18], [30] and [31]. Some considerations of significance in such evaluations are analysed (and experimentally justified) in References [15] and [31].

Briefly, as a result of the extra-attenuation effects outlined above, the parametric array is effectively shortened, with a resultant reduction in the difference frequency source level [as compared with the values estimated as in subsection 2(a)] and a broadening of the difference frequency beam. This deterioration in the performance of the parametric transmitter becomes most significant when the shock formation occurs well within the nearfield of the transducer. For then, in the absence of spreading loss, the rate of extra attenuation in the sawtooth region is much greater.

Hence, if extra-attenuation effects in a parametric transmitter are to be minimized, as a general guide it is necessary to design the system so as to make the shock formation distance in a lossless fluid equal to or greater than the Rayleigh distance of the transducer (Berktay in Reference [7]). The ratio of these two parameters for a system, $\beta \epsilon k R_0$, is used as "the saturation parameter" χ, in the model proposed in References [17–18]. The value of χ is then used to indicate the extent of the finite amplitude effects in a parametric transmitter.

In many applications, particularly when a very narrow difference frequency beam is required, extra-attenuation effects (and hence beam broadening) can only be avoided by reducing the transmitted power. This may result in rather low difference frequency source levels. In which case, making use of the broad bandwidth of a parametric transmitter, pulse compression techniques can be used to increase detection capability of the parametric sonar (Walsh in Reference [14]).

3. Parametric Reception

In parametric transmitters, the nonlinear interaction process is used to obtain highly directional low frequency sources which use relatively small transducers. In a parametric receiver, the nonlinear interaction process is between a low frequency plane wave of low intensity (the signal wave) and a higher frequency, locally generated wave of higher intensity (the pump wave). The interaction frequency terms $\omega_\pm = \omega_1 \pm \omega_2$ are received by means of a transducer placed on the acoustic axis of the pump wave.

Earlier work [32] on parametric receivers was done using pump waves which were highly collimated in the interaction region. Very good agreement was observed between the computed performance of such parametric receivers and experimental results. Later work [33, 34] confirmed these earlier results.

Briefly, amplitudes of the interaction frequency components of the pressure at the receiving transducer is calculated to be [32]

$$\tilde{p}_\pm(L, \theta) = [L\omega_\pm P_1' P_2' \beta / 2\rho_0 c_0^3] \exp(-\alpha_\pm L) D(\theta) \tag{35}$$

where

$$D(\theta) = \sin[k_2 L(1 - \cos\theta)/2]/[k_2 L(1 - \cos\theta)/2], \tag{36}$$

L is the spacing between the pump and the receiving transducers, θ is the angle the direction of propagation of the signal wave makes with the pump axis and $P_{1,2}'$ are the pressure amplitudes of the two waves at the position of the pump transducer. The directivity pattern is the same for the sum frequency and the difference frequency, and is identical with that for a continuous line array operating in end-fire mode, receiving the low frequency signals directly.

As the length of the parametric receiver is increased, it becomes more difficult to obtain a collimated pump wave over the length of the inter-action region. Hence, more recent work [35–37] has concentrated on the use of pump waves which are spherically spreading. A numerical method of computing the performance of such a parametric receiver is described in Reference [35], and a closed form solution obtained for not too broad pump beams in Reference [36]. The closed form solution, which showed very good agreement with the results of numerical computations, is

$$\hat{p}_\pm(L, \theta) = [\omega_\pm \beta P_1 P_2'/2\rho_0 c_0^3] \exp(-\alpha_\pm L) D(\theta). \tag{37}$$

Here, P_1 is the pump pressure amplitude at 1 m from the transducer; all other quantities are the same as in equations (35) and (36). Approxima-tions made in the derivation of this expression limit its validity to pump beams which are not wider than the beamwidth of the parametric receiver which is, approximately,

$$\theta_{HP} = 1.9\sqrt{\lambda_2/L} \text{ radians.} \tag{38}$$

Two points of significance may be noted. Firstly, the pressure depends on L only through the absorption term. Secondly, the directivity function of the pump transducer does not enter the result. This is further supported by the agreement between experimental results obtained using a rectangular pump transducer, and the computed results which were for a circular pump transducer [35].

Experience with parametric receivers show that equations (35–38) can be used for design considerations of practical devices with some confidence.

As implied by equation (38), the beamwidth of the parametric receiver is completely determined by $\sqrt{\lambda_2/L}$. Hence, to reduce the beamwidth by a factor of 2, the length would have to be increased fourfold. A better engineering result can be obtained by forming a broadside array of para-metric receivers [32, 37]. Experiments have verified that the theory of linear arrays can be applied to such arrays of parametric receivers. In this manner, design of a parametric receiving system can be optimized for a given situation

The detection performance of any receiving system would be determined by its noise characteristics. The noise entering the electronics of a para-metric receiver at the frequencies $\omega_\pm$ will have various components.

(i) Water noise at the frequencies $\omega_\pm$.
    The effect of this noise can be reduced by using a receiving transducer with some directivity at these frequencies.
(ii) Water noise at the signal frequency, ω_2.
    The directivity of the parametric receiver discriminates against this component of noise, as is the case for any receiving transducer. With the directivity function of equation (36), the improvement in signal-to-noise power ratio for isotropic noise can be shown to be $4L/\lambda_2$.

(iii) Water noise at the frequencies $2\omega_1 \pm \omega_2$.

These noise components will be, in general, of much lower intensity, and will be discriminated against by the increased directivity of the parametric receiver at these frequencies.

(iv) Electronic noise in the equipment.

As the electronic circuits will be operated at frequencies much higher than the signal frequency, and with high sensitivity resonant transducers, the relative contribution of this noise component can be kept small. Special attention must be paid to the pump transmitter to reduce oscillator noise.

In general, there appears to be no reason why a carefully designed parametric receiver should not perform as well as a physical array with the same directivity function. The contribution of the noise at the signal frequency would be expected to be predominant.

The effects of inhomogeneities of the medium and of turbulence on the performance of a parametric receiver has still to be established.

4. Conclusions

A brief review of progress in the exploitation of nonlinear acoustic interactions has been presented. The developments in this field have justified the forecasts made some years ago. This "new technology", as it is sometimes referred to, provides a new variant in system design, which leads to economically more acceptable solutions to many problems.

ACKNOWLEDGMENTS

A review paper such as this draws upon the work of many people, to all of whom due credit may not be given.

Part of the work reported was supported by the Office of Naval Research.

LIST OF SYMBOLS USED

q—volume density of volumetric source

ρ_0, c_0—the density of and the velocity of propagation in the undisturbed medium

$\beta = 1 + B/2A$—the parameter of nonlinearity of a fluid (a value of 3.5 is used for water)

α—small signal absorption coefficient

ω, k, λ—frequency, wave number, and wavelength

p, P—pressure

Various subscripts are used to identify the values of these parameters at various frequencies; subscripts 1 and 2 refer to primary frequencies (ω_1, ω_2), while + or − refers to frequencies $\omega_1 \pm \omega_2$. Subscript zero refers to $\omega_0 = (\omega_1 + \omega_2)/2$. $\alpha_T = \alpha_1 + \alpha_2 - \alpha_-$

$W_{1,2}$—transmitted acoustic power at $\omega_{1,2}$

R_0—Rayleigh distance ($=S/\lambda$ for a plane transducer)

ϵ—the acoustic "Mach number", $\epsilon = P/\rho_0 c_0^2$

All units are in rationalized M.K.S. system except where otherwise stated.

REFERENCES

1. Berktay, H. O. (1967). Some finite amplitude effects in underwater acoustics. *In* "Underwater Acoustics", Vol. II, (Ed. V. M. Albers), Plenum Press.

2. Hobaek, H. (1967). Experimental investigation of an end-fire array, *J. Sound Vib.* **6**, 460–463.

3. Berktay, H. O. (1967). A study of the travelling-wave parametric amplification mechanism in nonlinear acoustics. *J. Sound Vib.* **5**, 155–163.

4. Berktay, H. O. (1967). Some proposals for underwater transmitting applications of nonlinear acoustics. *J. Sound Vib.* **6**, 244–254.

5. Zverev, V. A. and Kalachev, A. I. (1968). Measurement of the scattering of sound by sound in the superposition of parallel beams. *Sov. Phys.—Acoust.* **14**, 173–180. (English translation).

6. Berktay, H. O., Dunn, J. R. and Gazey, B. K. (1968). Constant beamwidth transducers for use in sonars with very wide frequency bandwidths. *Applied Acoustics* **1**, 81–99.

7. "Applications of Finite Amplitude Acoustics to Underwater Sound", Proceedings of a Seminar at Navy Underwater Sound Laboratory, New London, Connecticut, U.S.A. 27 May 1968, NUSL Report No. 1084.

8. Muir, T. G. and Blue, J. E. (1969). Experiments on the acoustic modulation of large amplitude waves. *J. Acoust. Soc. Am.* **46**, 227–232.

9. Muir, T. G. (Ed.). (1969). "Nonlinear Acoustics," Proceedings of a Symposium at the Applied Research Laboratories, University of Texas, Austin, Texas, U.S.A., in November 1969, (AD 719 936).

10. Moffett, M. B., Westervelt, P. J. and Beyer, R. T. (1969). Large amplitude pulse propagation, a transient effect. *J. Acoust. Soc. Am.* **47**, 1473–1474(L).

11. Moffett, M. B., Westervelt, P. J. and Beyer, R. T. (1971). Large amplitude pulse propagation—a transient effect. (II). *J. Acoust. Soc. Am.* **49**, 339–343.

12. Smith, B. V. (1971). An experimental study of a parametric end-fire array. *J. Sound Vib.* **14**, 7–21.

13. Mellen, R. H., Browning, D. G. and Konrad, W. L. (1971). Parametric sonar transmitting array measurements. *J. Acoust. Soc. Am.* **49**, 932–935(L).

14. "Proceedings of Symposium on Nonlinear Acoustics", held at University of Birmingham, Birmingham, England on 1–2 April 1971.

15. Merklinger, H. M. (1971). "High Intensity Effects in the Nonlinear Parametric Array", Ph.D. Thesis, University of Birmingham (1971).

16. Muir, T. G. (1971). "An Analysis of Parametric Acoustic Array for Spherical Wave Fields", Ph.D. Dissertation, Mechanical Engineering Department, The University of Texas at Austin, Austin, Texas.

17. Moffett, M. B. (1971). "Parametric Radiator Theory I", Naval Underwater Systems Center Technical Memorandum No. PA4-234-71.

18. Mellen, R. H. and Moffett, M. B. (1971). "A Model for Parametric Sonar Radiator Design", Naval Underwater Systems Center Technical Memorandum No. PA41-229-71.

19. Blue, J. E. (1972). "Nonlinear Acoustics in Undersea Communication", unpublished report.

20. Muir, T. G. and Willette, J. G. (1972). Parametric acoustic transmitting arrays. *J. Acoust. Soc. Am.* **52**, 1481.

21. Westervelt, P. J. (1963). Parametric acoustic array. *J. Acoust. Soc. Am.* **35**, 535–537.

22. Berktay, H. O. and Leahy, D. J. (1973). Farfield performance of parametric transmitters. *J. Acoust. Soc. Am.* (in press).

23. Berktay, H. O. (1965). Possible exploitation of nonlinear acoustics in underwater transmitting applications. *J. Sound Vib.* **2**, 435–467.

24. Smith, B. V. (1967). "Nonlinear Acoustics with Emphasis on Transmitting Applications", Ph.D. Thesis, University of Birmingham.

25. Blackstock, D. T. (1964). Thermoviscous attenuation of plane, periodic finite amplitude sound waves. *J. Acoust. Soc. Am.* **36**, 534–542.

26. Naugol'nykh, K. A. (1971). Absorption of finite amplitude waves. Part I in "High-Intensity Ultrasonic Fields" (Ed. L. D. Rozenberg), Plenum Press.

27. Blackstock, D. T. (1964). On plane, spherical and cylindrical waves of finite amplitude in lossless fluids. *J. Acoust. Soc. Am.* **36**, 217–219(L).

28. Berktay, H. O. and Smith, B. V. Unpublished work.

29. Shooter, J. A., Muir, T. G., and Blackstock, D. T. (1971). Experimental observation of acoustic saturation in water caused by nonlinear propagation effects. *J. Acoust. Soc. Am.* **49**, 119(A).

30. Bartram, J. F., and Westervelt, P. J. (1972). Nonlinear attenuation and the parametric array. *J. Acoust. Soc. Am.* **52**, 121A.

31. Merklinger, H. M. (1972). High intensity effects in the parametric transmitting array. *J. Acoust. Soc. Am.* **52**, 122A.

32. Berktay, H. O. and Al-Temimi, C. A. (1969). Virtual arrays for underwater reception. *J. Sound Vib.* **9**, 295–307.

33. Zverev, V. A. and Kalachev, A. I. (1970). Modulation of sound by sound in the intersection of sound waves. *Soviet Phys.–Acoust.* **16**, 204–208.

34. Konrad, W. L., Mellen, R. H., and Moffett, M. B. (1972). Parametric sonar receiving experiments. *J. Acoust. Soc. Am.* **51**, 82A.

35. Barnard, G. R., Willette, J. G., Truchard, J. J., and Shooter, J. A. (1972). Parametric receiving array. *J. Acoust. Soc. Am.* **52**, 1437–1441.

36. Berktay, H. O., and Shooter, J. A. (1973). Parametric receivers with spherically spreading pump waves. To be published in *J. Acoust. Soc. Am.* (in press).

37. Berktay, H. O. and Muir, T. G. (1973). Arrays of parametric receivers. *J. Acoust. Soc. Am.* **53**, 1377–1383.

DISCUSSION

C. N. Pryor: What sort of conversion efficiencies does one obtain from high frequencies to the low frequency in a parametric transmitter?

Answer: The power conversion efficiency depends upon the parameters of the system. In general, it is not greater than 1%. However, in practice the parameter of consequence is the effective source level at the difference frequency. In many applications, sufficient source level appears to be obtainable with acceptable transmitted power levels. However, the viability of each possible application should be studied in its entirety, considering the constraints on transducer size, power requirement and other considerations.

G. Pearce: Can you say anything about the bandwidth of nonlinear transmitting systems?

Answer: Two points can be made about the bandwidth of parametric transmitters.

A small fractional change in the primary frequencies is sufficient to produce a large variation in the difference frequency. Thus, the difference frequency can be varied, over a wide range while keeping the acoustic transmissions within the bandwidth of the transducer.

If the difference frequency beam pattern is mainly controlled by the primary radiation patterns (see the discussion in connection with Equation 12) then the beamwidth varies slowly with changes in the difference frequency. In various experiments, the beamwidth at the difference frequency varied by not more than 10–20% over a frequency range greater than 2 octaves.

The wide bandwidth property of parametric devices has been used to transmit signals with large time—bandwidth products, which permit the realization of large correlation gains.

Passive Sonar: Fitting Models to Multiple Time Series

W. S. LIGGETT, JR.

Raytheon, Submarine Signal Division, Portsmouth, Rhode Island, U.S.A.

1. Introduction

A sonar operator is expected to know the locations and types of all the ships in the vicinity of his array. This information is a model that the sonar system fits to the acoustic data. There are two aspects to this model fitting [4]. One is the minimization of the effects of randomness in the data. The other is the choice of the range of models to fit. This latter aspect, which is accented in this paper, is often disregarded in sonar design. A choice of range of models that is too limited or too broad can result in a system with poor performance. When the range is too limited, errors that occur because the actual situation is outside the range may be more frequent and more severe than errors due to the randomness of the data.

Passive sonars face a wide variety of data characteristics [16, 17]. Signals, which are noises from nearby ships, occur in various multiplicities and have differing spectra, nonplanar wavefronts and non-stationary behaviour. The range of ambient-noise characteristics has been partially described. Noise components with high coherence between most pairs of sensors are possible. Finally, there are unpredictable occurrences.

The demands of this range of data characteristics must be balanced against the need for meaningful estimates of model parameters. Thus, the range of models must be limited. Some assumptions, either in terms of prior probabilities or deterministic relations, are necessary. The choice of what models to fit is based on prior knowledge and on the goal of the processing.

A procedure that is designed to cope with a broad range of models, such as the one discussed in this paper, has both advantages and disadvantages. As an operational system, such processing will be robust with respect to the possibilities envisioned by the designer. As an ideal, it can be compared to processing compromised to meet cost and hardware constraints so that the benefits of increased robustness can be determined. As a laboratory procedure, it fulfills the role of exploratory data analysis: it may uncover new

327

properties of the data which, when verified, can be used in system design [4]. The disadvantage of such processing is that determining the types of incorrect results and their probabilities is difficult if not impossible. Thus, important conclusions must be verified with new data and a different statistical procedure that provides better control of error.

The first step in presenting our procedure is relating it to the state of prior knowledge for which it is appropriate. The input to our procedure is the output of several hydrophones. Using instead linear combinations of hydrophone outputs, such as beamformer outputs, is an interesting alternative which we will not discuss. The input is a multiple time series consisting of P component time series that are sampled in time with a sampling interval of Δ sec. Thus, the input is a sequence of P-dimensional (column) vectors denoted by $X(j)$, where j is the time index ($j = 0, 1, \ldots$). If sample j from hydrophone p is $x_p(j)$, then $X^*(j) = (x_1(j), \ldots, x_P(j))$, where * denotes conjugate transpose.

Our processing is based on estimates of the spectral density matrix. Since the choice of spectral estimates makes our processor less sensitive to some types of regularity in the data, we are in effect assuming that such regularity is unlikely or unimportant. Looking at second-moment properties seems reasonable for zero-mean Gaussian processes. The choice of spectral estimates seems justified when, in addition, sufficiently unbiased and stable estimates can be formed [12].

Choosing frequency-time intervals from which sufficiently unbiased and stable spectral matrix estimates can be formed is part of our processing. We assume that there is given an integer L such that over a frequency-time interval that is $(L\Delta)^{-1}$ Hz wide and $L\Delta$ sec long, the population spectral matrix does not change appreciably. Using this interval, we form the spectral-density–matrix estimates $a(f, t)$ given by [9]

$$a(f, t) = w(f, t)w^*(f, t), w(f, t) = (2\pi L)^{-1/2} \sum_{j=(t-1)L}^{tL-1} X(j) \, e^{i2\pi jf/L}, \quad (1)$$

where $f = 0, 1, \ldots$ and $t = 1, 2, \ldots$. The problem is to find regions in f and t over which to average $a(f, t)$. Our algorithm starts with a frequency-time interval that is so large that averaging over a greater range seems unreasonable. It then divides this interval into rectangular regions from which reasonably stable and unbiased spectral–matrix estimates can be obtained.

These spectral–matrix estimates must now be interpreted to yield estimates of the number of signals and the ship locations and types. Our processor estimates signal bearing, the basic location parameter, and signal spectrum, the primary basis for classification. This interpretation can be done by a variety of techniques including adaptations of classical multivariate analysis such as factor analysis [2, 8, 18] and applications of sonar techniques such as adaptive beamforming [13]. Since our processor is intended to handle multiple signals with imperfectly known wavefronts, we have chosen a modification of factor analysis.

Factor analysis requires some assumptions. First, we assume that the number of signals K is less than the number of inputs P. Second, we assume that signals are perfectly coherent between sensors, so the spectral matrix of a signal has rank one and is given by $\sigma(k)u(k)u^*(k)$, where $\sigma(k)$ is the signal spectral level, $u(k)$ describes the signal wavefront, and $u^*(k)u(k) = 1$. Third, we assume that the noise is uncorrelated between inputs and has equal spectral levels so that its spectral matrix is given by $\sigma_0 I$, where σ_0 is the noise spectral level. Thus, the model of the spectral density matrix that we fit to the spectral–matrix estimates is given by

$$F = \sum_{k=1}^{K} \sigma(k)u(k)u^*(k) + \sigma_0 I. \tag{2}$$

In this model fitting, we assume that the spectral estimates are complex–Wishart distributed [8, 18]. K and σ_0 are estimated without further assumptions. However, to allow the rest of the parameters to be determined from F when $K > 1$, we need more prior knowledge. Let $\Sigma = \text{diag}\{\sigma(1), \sigma(2), \ldots, \sigma(K)\}$, $U = (u(1), u(2), \ldots, u(K))$, and Φ be any unitary matrix $(\Phi^*\Phi = I)$. Equation (2) can be rewritten

$$F = U \sum{}^{1/2} \Phi^* \Phi \sum{}^{1/2} U^* + \sigma_0 I. \tag{3}$$

Since Φ is not determined by F, the signal levels and wavefronts cannot be uniquely determined. In order to remove this ambiguity, we maximize a criterion that measures how close the estimated signal wavefronts are to those expected on the basis of prior knowledge. This process, which produces bearing estimates, entails the assumption that the prior knowledge of the signal wavefronts has some precision.

We intend our processor to be much less dependent on assumptions about the number of signals, stationarity, spectral smoothness, and signal wavefronts than is usually true of passive sonars. There are two aspects of the processing presented here to which changes might be desirable. First, our assumption about the noise is restrictive. There are other assumptions that can be made without losing the advantages of our processor. For example, we can compensate for the differences caused by having ambient noise instead of independent noise. Coherent noise components might be eliminated if they can be identified. Second, for exploratory data analysis, data editing before spectral analysis is needed to check for extreme deviations from our assumption that the data is zero-mean Gaussian with slowly varying moments.

The sequel is divided into two sections. In Section 2, we discuss the estimation of the spectral matrix, detail the algorithm for choosing regions over which to average, and present the first part of the simulation. In

Section 3, we discuss interpretation of the spectral–matrix estimates, detail the algorithm for finding signal levels and wavefronts, and present the rest of the simulation.

2. Clustering Approach to Spectral Estimation

(a) Discussion of the algorithm

The choice of frequency and time resolution is important not only because the estimated signal spectra should have the correct resolution but also because adequate stability and lack of bias is necessary to allow processing that is robust with respect to signal multiplicity and signal wavefronts. Source–array motions that are not resolved in time and propagation delays whose contributions to the cross-spectral structure are not resolved in frequency cause signals to lose their inter-hydrophone coherence and thus their rank one property. Signal coherence is one of the essential assumptions for our interpretation of the spectral matrix. A similar concern has been exhibited in principal component analysis when the domain of averaging of the covariance matrix is limited so that slowly-varying components have rank one [6, 15]. The spectral characteristics of signals have been discussed [17]. The analogous need for time resolution is the existence of energy bursts called transients as well as other shifts in signal spectra. It is generally true that when more parameters are being estimated, more stability is needed. Thus, our processor, which estimates wavefronts as well as spectral levels, needs more stability than processors that assume the signal wavefronts are a known function of bearing.

We formulate the problem of estimating the spectral matrix at frequency-time (f_0, t_0) as that of choosing a set S of N frequency-time points over which the population spectral matrix is nearly constant and then adopting the estimate

$$\hat{F}(f_0, t_0) = N^{-1} \sum_{(f, t) \in S} a(f, t), \qquad (4)$$

where $a(f, t)$ is given in equation (1). For $N = 1$, we have assumed that this estimate is nearly unbiased (its expected value nearly equals the population value). The size of N, the time–bandwidth product, measures the stability. For $N = 1$, the estimate is too unstable to be useful. We note without further comment that by changing w in equation (1), the spectral window properties of this estimate can be improved without major modification of our processing [9].

Associating observations with similar properties is clustering. Thus, our solution to the problem of choosing the sets S is based on clustering techniques [5]. One peculiarity of our application is the requirement that each set be connected in the frequency-time plane. This requirement results from our prior knowledge that characterizes the population spectral density matrix as smoothly varying with frequency and time. Thus, we have no prior knowledge that implies similarity of the population spectral density matrix at non-adjacent frequency-time points but does not imply similarity for the points in between. Therefore, in the clustering, association of the observations is determined by both the observations $a(f, t)$ and their location in frequency-time (f, t).

From the variety of clustering techniques, we have chosen a hierarchical, divisive scheme. At the beginning of each step, there are one or more clusters that are rectangular regions in the frequency-time plane. Each cluster can be divided into two rectangular sections in different ways. For each division, a distance, which is a measure of the dissimilarity of the spectral matrices estimated from the two sections, is computed. The algorithm chooses the most dissimilar pair of sections and makes clusters out of them, thus creating a finer partition of the frequency-time plane. The algorithm produces a sequence of partitions that starts with all the data in one cluster. The choice of when to stop is discussed below.

The dissimilarity of two spectral density matrices is measured by a statistic determined by the strength of a rank one component contained in one spectral matrix but not the other. In other words, the distance between sections is based on the largest eigenvalue of $\hat{F}_1 \hat{F}_2^{-1}$, $\hat{\mu}_1$, and the largest eigenvalue of $\hat{F}_1^{-1} \hat{F}_2$, $\hat{\mu}_2$, where $\hat{F}_1$ and $\hat{F}_2$ are the spectral–matrix estimates from the two cluster sections. Since $\hat{\mu}_1 = \max_{u*u=1} u*\hat{F}_2^{-1/2}\hat{F}_1\hat{F}_2^{-1/2}u$, $\hat{\mu}_1$ (and similarly $\hat{\mu}_2$) can be thought of in terms of prewhitening $\hat{F}_1$ with $\hat{F}_2$ and searching through all wavefronts u to find the largest component [3]. There are two reasons why distances based on $\hat{\mu}_1$ and $\hat{\mu}_2$ seem more appropriate than other criteria for comparing spectral–matrix estimates [14]. First, our criterion deals with rank one components, the assumed property of signals, so that differences in signals between the two sections are accented. Second, since determining $\hat{\mu}_1$ (and similarly $\hat{\mu}_2$) does not depend on the rank of $\hat{F}_1$, the algorithm can pick out a small region containing a strong signal, whereas criteria that require both $\hat{F}_1$ and $\hat{F}_2$ to have full rank cannot.

Our algorithm provides a sequence of clusters that stops when $N < 2P + 44$ for all clusters. The best division of the frequency-time region must be chosen from this sequence. Our algorithm does not make this decision. Thus, the user must make the choice based on the values of the largest distances, on the cluster shapes including the section that contains the extra component, and on analysis of the spectral matrices generated from the clusters.

There are two alternative procedures for determining resolution. The literature on single-channel time series analysis recommends that the user compare estimates with different frequency resolutions [11]. This can be generalized to our case where both the frequency and the time resolution must be varied. This approach can have the advantage of providing clusters with equal size so that fewer assumptions are needed for inter-cluster comparisons. However, equal size clusters will usually provide neither the resolution needed at some points nor the stability available at other points. Sonar has used a variety of displays that allow the user to visually integrate in frequency or time. These displays are limited by the number of dimensions they can present. Our data, which are $P \times P$ spectral matrices indexed by frequency and time, cannot be presented all at once. Thus, some properties of the data may be lost.

(b) Details of the algorithm

Our algorithm is detailed in Fig. 1. Initially, it treats all the input data, which are $a(f, t)$ for $B1 \leqslant f \leqslant B2$ and $B3 \leqslant t \leqslant B4$, as a single cluster. For each cluster formed, the spectral matrix estimate from it is analysed as discussed in Section 3 and intra-cluster distances are computed to determine subsequent splits. These operations, which occur in Blocks 3 and 4, are applied to the cluster labelled by the integer γ. Cluster γ covers the rectangle given by $\gamma 1 \leqslant f \leqslant \gamma 2$ and $\gamma 3 \leqslant t \leqslant \gamma 4$. The function $I(f, t)$, which is updated in

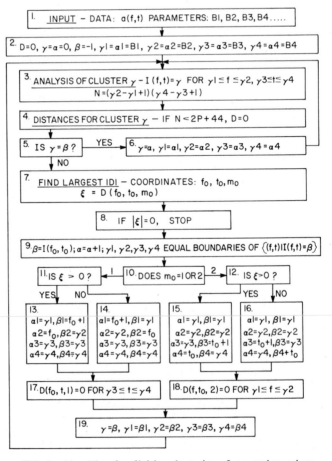

FIG. 1. Algorithm for divisive clustering of spectral matrices.

Block 3, gives the cluster number for each (f, t). The first time the algorithm performs operations 3 and 4, there is only one cluster, cluster 0. Otherwise, the operations must be repeated for clusters β and α, which cover the regions $\beta 1 \leqslant f \leqslant \beta 2$, $\beta 3 \leqslant t \leqslant \beta 4$, and $\alpha 1 \leqslant f \leqslant \alpha 2$, $\alpha 3 \leqslant t \leqslant \alpha 4$, respectively. This is accomplished through initial values assigned in Blocks 2 and 19 and the operations in Blocks 5 and 6. The intra-cluster distance $D(f, t, m)$ which is defined below contains in its magnitude a measure of the dissimilarity of two spectral matrices and in its sign a property of this dissimilarity. It is computed for every possible division of a cluster into two rectangular clusters. The first two indices of which D is a function indicate location in frequency and time, and the third differentiates between division in frequency $(m = 1)$ and division in time $(m = 2)$. $D(f, t, 1)(D(f, t, 2))$ is the value for the division between f and $f + 1$ $(t$ and $t + 1)$ unless this line is already a boundary in which case $D(f, t, 1) = 0$ $(D(f, t, 2) = 0)$. D is set equal to zero within clusters with $N < 2P + 44$ so that they are not divided further.

Blocks 7–19 choose a cluster and divide it into two sections. The cluster division chosen maximizes $|D(f, t, m)|$ over $B1 \leqslant f \leqslant B2$, $B3 \leqslant t \leqslant B4$, $1 \leqslant m \leqslant 2$. The coordinates of the maximum are (f_0, t_0, m_0); $\xi = D(f_0, t_0, m_0)$ and $\beta = I(f_0, t_0)$. When $\xi > 0$, the cluster section with lowest frequency and time becomes cluster α, where α is the next integer, and the other cluster section retains the number β. When $\xi < 0$, this is reversed. In order to find the boundaries of the clusters β and α, we first use $I(f, t)$ to find the boundaries of the cluster we are dividing. As the result of Block 9, the cluster being divided is given by $\gamma 1 \leqslant f \leqslant \gamma 2$, $\gamma 3 \leqslant t \leqslant \gamma 4$. Blocks 11–16 define the new boundaries of β and α. Blocks 17 and 18 set $D = 0$ on the new cluster boundary. Block 19 assigns γ, $\gamma 1$, $\gamma 2$, $\gamma 3$, $\gamma 4$ in preparation for the cluster analysis in Blocks 3 and 4.

In our processor, $D(f, t, m)$ is given by

$$D(f, t, m) = \begin{cases} d_1 \text{ for } d_1 \geqslant d_2 & \text{and} \quad \psi_1 > N_1/N \\ -d_2 \text{ for } d_1 < d_2 & \text{and} \quad \psi_2 > N_2/N \\ 0 \text{ for } \psi_1 \leqslant N_1/N & \text{and} \quad \psi_2 \leqslant N_2/N, \end{cases} \tag{5}$$

$$d_j = \begin{cases} \{2N_j \log[N_j/(N\psi_j)] + 2(N - N_j) \log[(N - N_j)/(N(1 - \psi_j))]\}^{1/2} \\ \qquad \text{for } N - N_j \geqslant P + 22 \\ 0 \\ \qquad \text{for } N - N_j < P + 22, \end{cases} \tag{6}$$

where N_1 is the number of frequency-time points in the cluster section with smallest frequency and time indices, N_2 is the number in the other section, $\hat{F}_1$ and $\hat{F}_2$ are the respective spectral estimates, $A_j = N_j\hat{F}_j$, and ψ_j is the largest eigenvalue of $A_j(A_1 + A_2)^{-1}$. The ψ_j can be obtained from the $\hat{\mu}_j$ since $\psi_j = N_j \hat{\mu}_j / (N + N_j(\hat{\mu}_j - 1))$.

There are two reasons why the dependence of D on N, N_1, and N_2 was chosen. First, recalling the significance of $\hat{\mu}_j$ and noting that $N - N_j$ is the size of the section without the extra component, we conclude that distances with small $N - N_j$ are unstable and not very interesting. In fact, if $N - N_j$ is less than P, $\hat{\mu}_j$ is infinite. Thus, we have taken $d_j = 0$ for $N - N_j < P + 22$ so that we do not allow divisions when the cluster section without the extra component is small (but we do allow divisions when the cluster section with the extra component is small). Second, distances should be comparable between alternatives with various N, N_1, and N_2. When the population spectral matrix does not vary, we would like the division to fall anywhere with equal probability. Our criterion does not satisfy this, but it is asymptotically independent of N, N_1, and N_2 in some limited sense.

(c) Simulation

The input to our algorithm are spectral–matrix estimates with characteristics that would be obtained by applying equation (1) to data generated by the

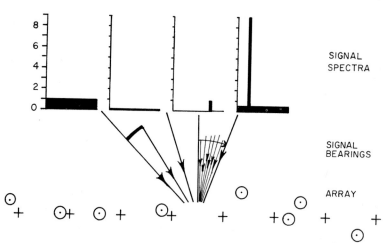

FIG. 2. Situation that generated data.

situation shown in Fig. 2. There are 20 frequencies, 60 times, and the ratio of the bandwidth of $a(f, t)$ to the centre frequency of the whole band is 0.004. The bearing angle θ is measured clockwise from broadside to the array. There are 8 hydrophones that are supposed to be separated by 0.4 wavelengths and located as indicated by + but are actually located as indicated by ⊙. There are four signals and independent noise with spectral level 1.0. The signal spectra for the frequency band analysed, which are the levels that would be received at the left-most +, and the signal bearings are shown in Fig. 2. From left to right, the first signal has a flat spectrum with spectral level 1.0 and consists of two plane waves that are in phase at the left-most + and have amplitude 0.6, bearing $\sin \theta = -0.7$ and amplitude 0.4 bearing $\sin \theta = -0.55$, respectively. The second signal is a plane wave that has a flat spectrum with spectral level 0.125 and bearing $\sin \theta = -0.3$. The third signal is a spectral line occupying the 15th frequency cell in the band analysed ($f = B1 + 14$) with spectral level 1.0 and bearing $\sin \theta = 0.0$. The fourth signal has a spectral line with spectral level 9.0 occupying the 5th frequency cell ($f = B1 + 4$) and a broadband level of 0.58. The fourth signal moves from $\sin \theta = 0.0$ to $\sin \theta = 0.36$ with $\sin \theta$ varying linearly with time.

The results of the clustering algorithm are shown in Fig. 3. The right side

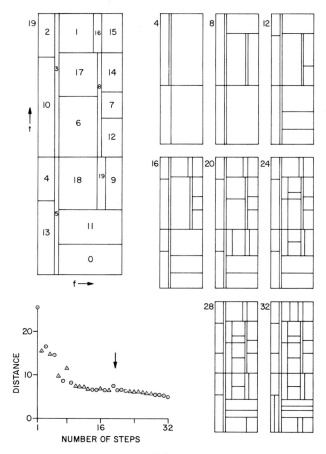

FIG. 3. Results of clustering algorithm.

shows every fourth member of the sequence of partitions. The first step, which is caused by the movement of the fourth signal, divides the data in time into nearly equal clusters. Steps 2–5 isolate the spectral line at frequency 5 ($f = B1 + 4$). After step 8, the algorithm has made another division in time and has found the spectral line at frequency 15. The algorithm stopped after 32 steps. The cluster arrangement after 19 steps along with the cluster numbers is shown on the left. At this step, the spectral line at frequency 15 has been isolated except at the beginning when the moving signal is at nearly the same bearing. In the plot of maximum distance, which is shown on the lower left, the triangles are positive values of D and the circles are negative values. The arrow is between steps 19 and 20.

For the purpose of presenting the model our algorithm created, we stopped dividing after 19 steps. This decision was biased by knowledge of the data characteristics created. However, there are some honest reasons for stopping where we did. First, the maximum distance flattens out after step 19. Second, the creation of cluster 19, which is the extra component side of the division that created it, is reasonable in light of the previous creation of clusters 8 and 16. Third, step 20, which divides cluster 18, does not indicate interesting structure since the smallest section is the one without the extra component. More reasons for stopping after step 19 might be obtained from analysis of the spectral–matrix estimates.

3. Analysis of Spectral Matrices

(a) The algorithm

In our algorithm, analysis of the spectral–matrix estimates consists of two steps, separation of the signals from the noise and separation of the signals from each other when there is more than one signal. The spectral matrix which is denoted by $\hat{F}$ is obtained from averaging over a frequency-time cluster using equation (4). Our procedure starts with orthogonal decomposition. In decreasing order, the eigenvalues of $\hat{F}$ are $\hat{\lambda}(1), \hat{\lambda}(2), \ldots, \hat{\lambda}(P)$, and the corresponding eigenvectors are $\hat{\phi}(1), \hat{\phi}(2), \ldots, \hat{\phi}(P)$ so that

$$\hat{F} = \sum_{k=1}^{P} \hat{\lambda}(k)\hat{\phi}(k)\hat{\phi}^*(k). \tag{7}$$

How we proceed differs for $N > P + 9$ and $N \leqslant P + 9$.

When $N > P + 9$, separating the signals from the noise consists of forming three estimates. As detailed below, the estimate of the number of signals $\hat{K}$ is obtained from the eigenvalues. Our estimate of the noise level is

$$\hat{\sigma}_0 = (P - \hat{K})^{-1} \sum_{k=\hat{K}+1}^{P} \hat{\lambda}(k), \tag{8}$$

and our estimate of the spectral matrix of the signals is

$$\hat{F}_s = \sum_{k=1}^{\hat{K}} (\hat{\lambda}(k) - \hat{\sigma}_0)\hat{\phi}(k)\hat{\phi}^*(k). \tag{9}$$

This decomposition of $\hat{F}_s$ into $\hat{K}$ signals is not unique when $\hat{K} > 1$ since by choosing any unitary matrix Φ and taking

$$q(k) = \sum_{j=1}^{\hat{K}} (\hat{\lambda}(j) - \hat{\sigma}_0)^{1/2} \hat{\phi}(j) \Phi_{jk}, \quad \Phi = (\Phi_{jk}), \tag{10}$$

we obtain another decomposition

$$\hat{F}_s = \sum_{k=1}^{K} q(k) q^*(k). \tag{11}$$

This change of decompositions is called rotation. To obtain the decomposition that is most satisfactory in light of prior knowledge, we choose a rotation by iteratively maximizing a criterion. Formulation of this criterion as well as bearing estimation requires prior knowledge of the signal wavefront as a function of direction. In our algorithm, this prior knowledge is a best guess at the signal spectral matrix as a function of direction and spectral level. This paper discusses the case where the hydrophones are supposed to lie on a straight line so that our guess at the signal wavefront is a function of only one parameter, θ. Our guess at the signal spectral matrix as a function of σ and θ is $\sigma v(\theta) v^*(\theta)$, where $v^*(\theta) v(\theta) = 1$. Given $q(k) q^*(k)$, a component of the chosen decomposition of $\hat{F}_s$, our spectral level estimate is $q^*(k) q(k)$ and our bearing estimate, which we denote $\hat{\theta}(k)$, is the θ that maximizes $|q^*(k) v(\theta)|$.

When $N \leqslant P + 9$, we compensate for the lack of stability by taking $\hat{K} = 1$ and taking our estimate of the signal spectral matrix to be $\hat{\lambda}(1) \hat{\phi}(1) \hat{\phi}^*(1)$. Since we ignore the noise contribution to $\hat{\lambda}(1)$, the estimate of signal power is $\hat{\lambda}(1)$. The estimate of signal bearing can be obtained from $\hat{\phi}(1)$ as above. The choice of one signal instead of none is justified since clusters with $N < P + 22$ must originate as the extra-component section. The choice of only one is based on the lack of stability.

Determination of the number of signals is based on a sequence of tests for the equality of eigenvalues. A test for the equality of all the eigenvalues, which is analogous to a covariance-matrix test [1], is based on the statistic

$$Z(N, P) = 2[N - (2P^2 + 1)/(6P)] \left[P \log \left(\sum_k \hat{\lambda}(k)/P \right) - \sum_k \log \hat{\lambda}(k) \right], \tag{12}$$

which is asymptotically χ^2-distributed with $P^2 - 1$ degrees of freedom. For testing the equality of the last $P - K$ eigenvalues, we use this statistic with $N - K$ and $P - K$ substituted for N and P, respectively, a procedure suggested for covariance matrices [10]. Let $\text{Prob}\{\chi_\nu^2 < Z\}$ be the probability that a χ^2 random variable with ν degrees of freedom is less than Z. Our algorithm selects $\hat{K} = 0$ as the estimate of the number of signals if we have

Prob $\{\chi_\nu^2 < Z(N, P)\} < 0.8$ for $\nu = P^2 - 1$. It selects $\hat{K}(1 \leqslant \hat{K} \leqslant P - 2)$ if we have Prob $\{\chi_\nu^2 < Z(N - j, P - j)\} \geqslant 0.8$ for $\nu = (P - j)^2 - 1$ and $j = 0, \ldots,$ $\hat{K} - 1$ and if we have Prob $\{\chi_\nu^2 < Z(N - \hat{K}, P - \hat{K})\} < 0.8$ for $\nu = (P - \hat{K})^2 - 1$ It selects $\hat{K} = P - 1$ otherwise.

Our criterion for determining the right rotation is a measure of how close the wavefronts in the decomposition of $\hat{F}_s$ are to the supposed signal wavefronts. Corresponding to the decomposition in equation (11), the criterion is

$$B_0 = \sum_{k=1}^{\hat{K}} \max_{\theta(k)} \frac{|q^*(k)v(\theta(k))|^2}{q^*(k)q(k)}. \tag{13}$$

The kth term in the sum is the cosine squared of the angle between the wavefront of the component $q(k)q^*(k)$ and the supposed wavefront $v(\theta)$ at the best-matched bearing. We have chosen to normalize by the component power because there is no prior knowledge that suggests a relation between signal strength and match to $v(\theta)$.

Our iterative algorithm for maximizing the criterion is shown in Fig. 4. Blocks 1–3 contain the input operation, initial bearing estimation, and computations needed to choose the first iteration. Block 1 shows that the algorithm starts with the decomposition given in equation (9) and initializes η to count iterations. For $k = 1, 2, \ldots, \hat{K}$, Block 2 finds $\theta(k)$, the θ that maximizes $|q^*(k)v(\theta)|^2/q^*(k)q(k)$, and sets $b(k) = |q^*(k)v(\theta(k))|^2/q^*(k)q(k)$. This is done by evaluating $|q^*(k)v(\theta)|^2/q^*(k)q(k)$ at values of θ covering $(-\pi/2, \pi/2)$, finding the maximum over these θ values, and then using this point as a start for Newton's method. The next step is to apply a rotation that replaces $q(j)$ with $r(j, k)q(j) - s(j, k)q(k)$ and $q(k)$ with $s^*(j, k)q(j)$ $+ r(j, k)q(k)$ and leaves the other components the same. For this transformation to be unitary, the complex number $s(j, k)$ must satisfy $|s(j, k)| \leqslant 1$ and $r(j, k) = (1 - |s(j, k)|^2)^{1/2}$. In order to choose the pair of components to rotate, we compute various rotations and compare the size of the improvements before adjusting $\theta(1), \theta(2), \ldots, \theta(\hat{K})$. Let

$$B(q(j), q(k)), = |q^*(j)v(\theta(j))|^2/q^*(j)q(j) + |q^*(k)v(\theta(k))|^2/q^*(k)q(k). \tag{14}$$

To choose $s(j, k)$, we approximate $B((1 - |s|^2)^{1/2}q(j) - sq(k), s^*q(j)$ $+ (1 - |s|^2)^{1/2}q(k))$ with the constant, linear, and quadratic terms of its Taylor series in Re(s) and Im(s) about $s = 0$. Using the technique of Reference [7], we find the s that maximizes this approximation inside the sphere $|s| \leqslant \rho$, where ρ is determined by R and $\rho \leqslant R^{-1}$. Initially, we let $R = 2$. If the resulting s produces a positive value for

$$C(j, k) = B(rq(j) - sq(k), s^*q(j) + rq(k)) - B(q(j), q(k)),$$
$$r = (1 - |s|^2)^{1/2}, \tag{15}$$

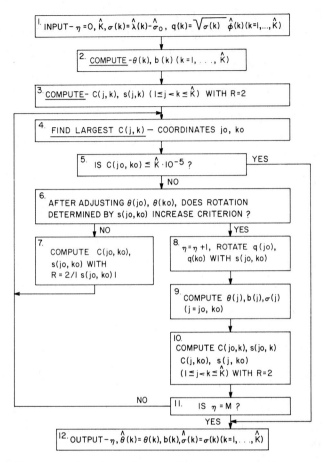

FIG. 4. Algorithm for estimation of signal spectra and bearings.

then we let $s(j, k) = s$. If it does not, then we set $R = 2/|s|$ and repeat the procedure until a positive value for C results.

Blocks 4–11 perform the iteration. It starts by finding the largest $C(j, k)$, which is denoted $C(j_0, k_0)$. If $C(j_0, k_0) \leqslant \hat{K} \cdot 10^{-5}$, the algorithm outputs and stops on the basis that it is sufficiently close to a peak. Clearly, we have no guarantee that the absolute maximum has been found. If $C(j_0, k_0) > \hat{K} \cdot 10^{-5}$, then before applying the rotation, $\theta(j_0)$ and $\theta(k_0)$ are adjusted to see whether the rotation increases B_0. If it doesn't, $C(j_0, k_0)$ is recomputed with a smaller step (by setting $R = 2/|s(j_0, k_0)|$) and the algorithm returns to Block 4. If the rotation increases B_0, the algorithm applies the rotation, increments η, and using the new values of $q(j_0)$ and $q(k_0)$, adjusts the affected values of $\sigma(k)$, $\theta(k)$, $b(k)$, $C(j, k)$, and $s(j, k)$ using the techniques described above. If the algorithm has done less than M iterations, it returns to Block 4. Otherwise, it outputs current values and stops.

(b) Simulation

We now finish our discussion of the simulation, first giving details for Clusters 11 and 4 (see Fig. 3) and then results for all clusters formed through step 19 of the clustering algorithm. We chose Cluster 11 for detailed examination because, as is typical of many clusters, the initial decomposition is reasonably good and the rotation algorithm converges quickly without producing much change. We chose Cluster 4 because the initial decomposition is poor and the rotation algorithm improves it. For Cluster 11, the algorithm finds three signals although there are four, partly because the broadside signal and the moving signal are close together during the Cluster 11 time interval. For Cluster 4, the algorithm finds three signals which is correct since the broadside signal does not appear in the Cluster 4 frequency interval. Figure 5 presents the total field and the signal components by plotting the response of a conventional beamformer to them. In other words, the top two graphs show $v^*(\theta)\hat{F}v(\theta)$ plotted for equally spaced values

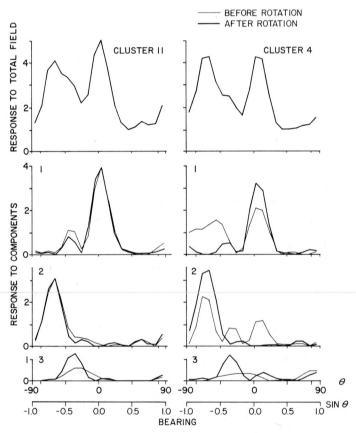

FIG. 5. Response of conventional beamforming to total field and components.

of $\sin \theta$, and the bottom six graphs show $|q^*(k)v(\theta)|^2$ before and after rotation. The signal model $v(\theta)$ is derived by assuming plane waves incident on hydrophones at their supposed locations, shown by $+$ in Fig. 2. In the response to the total field, only two signals are evident. For Cluster 11, rotation improves the match of the signal in the fourth graph and decreases its presence in the other components. Rotation increases B_0 from 1.9 to 2.3 and changes the estimates of signal spectral level from 6.0, 4.4, 1.0 to 5.5, 4.3, 1.6, a questionable improvement.[†] The result for Cluster 4 is more dramatic. The maximum response in the fourth graph is initially on the right side at $\sin \theta = 0.88$, where there are no signals. The second graph contains two signals, and the third graph contains three signals. After rotation, the signals are well separated. Rotation increases B_0 from 1.4 to 2.1 and changes the estimates of signal spectral level from 5.6, 4.3, 1.1 to 4.2, 4.7, 2.2.

Figure 6 contains a table and a display giving the results after step 19. The table lists for each cluster the estimate of the number of signals $\hat{K}$, the number of iterations η, and the estimate of noise spectral level $\hat{o}_0$. Only for Cluster 2 where $\hat{K} = 7$ does the rotation algorithm perform 50 iterations, the maximum number it is allowed. Otherwise, it stopped because the potential for more improvement was small.

The display in Fig. 6 portrays, for each signal obtained from each cluster, its intensity, bearing, and location in frequency and time. Distance along the bearing axis is proportional to $\sin \theta$. The time-frequency axis is divided according to the partition after Step 19. At each time, Fig. 3 shows the arrangement of clusters with frequency (e.g., at the earliest time, Clusters 13, 5, and 0 describe the variation with frequency). Over the total time interval, there are 10 different arrangements. The time-frequency axis is first divided to account for these time variations in the cluster arrangement, the length of each division being proportional to the time span it represents. Each such division is further divided proportional to frequency. Thus, our display is stacked bearing-frequency displays from successive time intervals. Each signal is represented by one of four line widths, the heaviest line for spectral levels $\hat{o}(k)$ greater than 16, the next for $4 < \hat{o}(k) \leq 16$, the next for $1 < \hat{o}(k) \leq 4$, and the lightest for $\hat{o}(k) \leq 1$. The line denoting a signal will be broken if the cluster results of which it is part must be divided because of the variation in time at some other frequency. The length of a line including all segments is proportional to the time-bandwidth product N of the cluster from which it was created. Thus, line width and length indicate the reliability of a particular result.

---

† Estimates of signal spectral levels at the hydrophones comparable to those shown in Fig. 2 are the diagonal elements of the component spectral matrices $q(k)q^*(k)$. Since $\hat{o}(k)$ is the sum of these diagonal elements, $\hat{o}(k)$ is eight times larger than the average level at the hydrophones.

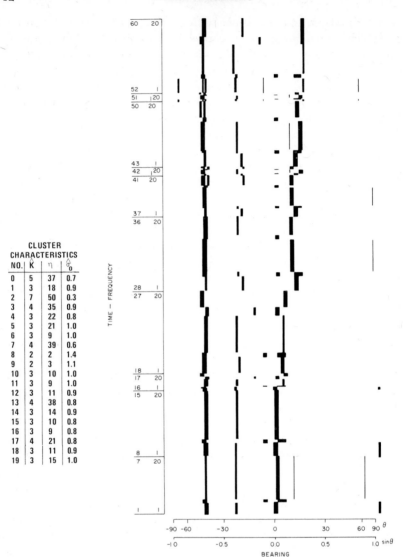

CLUSTER CHARACTERISTICS			
NO.	K	η	$\hat{\theta}_0$
0	5	37	0.7
1	3	18	0.9
2	7	50	0.3
3	4	35	0.9
4	3	22	0.8
5	3	21	1.0
6	3	9	1.0
7	4	39	0.6
8	2	2	1.4
9	2	3	1.1
10	3	10	1.0
11	3	9	1.0
12	3	11	0.9
13	4	38	0.8
14	3	14	0.9
15	3	10	0.8
16	3	9	0.8
17	4	21	0.8
18	3	11	0.9
19	3	15	1.0

FIG. 6. Results from clusters after 19 steps of spectral estimation algorithm.

On the display, three of the signals stand out. The other, the broadside signal, is evident in the results of Cluster 8. In addition, at broadside there are false alarms where the moving signal is split into two signals. There is a basis for resolving this. At frequency 5, there is a very strong line (for the strongest signal in Clusters 3 and 5, $\hat{\sigma}(1) = 67.5$ and 73.9, respectively). Since the results indicate that it is part of the moving signal, we can expect splitting. In Cluster 8, there is no response near the bearing of the moving target yet the signal near broadside is strong ($\hat{\sigma}(1) = 10.0$), thus indicating a separate signal. Other noteworthy properties of the display are: the bearing of the weak broadband target is biased towards the left; the algorithm splits the multipath target occasionally although this is not too disconcerting; there are other false alarms, most of which are weak.

Our processor and its simulation demonstrate some general properties of exploratory data analysis. The simulation shows that our processor can perform reasonably well in analysing complex, poorly-known situations. However, it does produce misleading results such as split signals and superfluous divisions in frequency and time. Interpreting such results requires sophistication, and even a sophisticated user may be fooled. Thus, if a result of our processing pointed to a conclusion of lasting importance, this conclusion would have to be verified with new data and a different statistical analysis providing better control of error. A neat analysis of our processor, such as the normal theory results relating false alarms and false dismissals, is impossible. One reason for this is the large variety of mistakes our processor can make. Another reason is the use of iterative maximization of a criterion whose properties are poorly understood and hard to visualize.

Use of our processor is economically feasible albeit expensive. In the simulation, our algorithm analysed the spectral matrix of every cluster it created, applying the iterative rotation 65 times and calculating distances for cluster division. The run, not including compilation of the Fortran program or generation of the artificial data, took 130 seconds of central processor time on a CDC 6600 computer.

REFERENCES

1. Anderson, T. W. (1958). "An Introduction to Multivariate Statistical Analysis", pp. 203–207, 247–256, Wiley, New York.
2. Brillinger, D. R. (1970). The frequency analysis of relations between stationary spatial series. *In* "Proc. of the Twelfth Biennial Seminar of the Canadian Mathematical Congress" (R. Pyke, Ed.), pp. 39–81. (Canadian Mathematical Congress, Montreal.)
3. Cox, H. (1968). "Interrelated Problems in Estimation and Detection I and II," Proceedings of the NATO Advanced Study Institute on Signal Processing with Emphasis on Underwater Acoustics (Twente Institute of Technology, Enschede, Netherlands).
4. Dempster, A. P. (1971). An overview of multivariate data analysis. *J. Multivariate Analysis* 1, 316–346.
5. Fisher, L. and Van Ness, J. W. (1971). Admissible clustering procedures. *Biometrika* 58, 91–104.
6. Fukunaga, K. and Olson, D. R. (1971). An algorithm for finding the intrinsic dimensionality of data. *IEEE Trans. on Computers* C-20, 176–183.
7. Goldfeld, S. M., Quandt, R. E. and Trotter, H. F. (1966). Maximization by quadratic hill-climbing. *Econometrika* 34, 541–551.
8. Goodman, N. R. (1963). Statistical analysis based on a certain multivariate complex Gaussian distribution (An introduction). *Ann. Math. Statist.* 34, 152–177.

9. Hannan, E. J. (1970). "Multiple Time Series", pp. 213, 265, 276, Wiley, New York.

10. James, A. T. (1969). Tests for equality of latent roots of the covariance matrix. *In* "Multivariate Analysis II" (P. R. Krishnaiah, Ed.), pp. 205–218. Academic Press, New York and London.

11. Jenkins, G. M. and Watts, D. G. (1968). "Spectral Analysis and Its Applications", pp. 280–282. Holden–Day, San Francisco.

12. Liggett, W. S., Jr. (1971). On the asymptotic optimality of spectral analysis for testing hypotheses about time series. *Ann. Math. Statist.* **42**, 1348–1358.

13. Liggett, W. S., Jr. (1972). Passive sonar processing for noise with unknown co-variance structure. *J. Acoust. Soc. Am.* **51**, 24–30

14. Pillai, K. C. S. and Jayachandran, K. (1968). Power comparisons for tests of equality of two covariance matrices based on four criteria. *Biometrika* **55**, 335–342.

15. Portnoy, S. (1971). "Local Linear Analysis". Lecture presented at Clemson University.

16. Tolstoy, I. and Clay, C. S. (1966). "Ocean Acoustics: Theory and Experiment in Underwater Sound". McGraw-Hill, New York.

17. Urick, R. J. (1967). "Principles of Underwater Sound for Engineers". McGraw-Hill, New York.

18. Wahba, G. (1968). On the distribution of some statistics useful in the analysis of jointly stationary time series. *Ann. Math. Statist.* **39**, 1849–1862.

DISCUSSION

Besides requests for clarification, the discussion concentrated on the performance of the procedure presented. Questions were asked about its ability to detect low-level signals, about ways of improving this performance, about sensitivity to the assumption of un-correlated, equi-power noise, and about sensitivity to the guess at signal amplitude-and-phase relations, $v(\theta)$. These questions cannot be answered completely since the only basis for answers is the simulation and the logic on which the procedure is based.

Dr. N. L. Owsley questioned the ability of the processor to detect low-level signals. In the simulation, the weakest signal has a signal-to-noise ratio of -9 dB at each hydro-phone. This is balanced by the array gain which is 9 dB against uncorrelated, equi-power noise. The procedure detects this signal. However, as Dr. Owsley commented, this signal is not really weak. With more hydrophones or data that allow more stable spectral–matrix estimates, the sensitivity to weak signals would improve.

Dr. Owsley asked whether the number of signals might be better determined from interrogating the eigenvectors of $\hat{F}$ instead of, or in addition to, the eigenvalues. The answer to this depends on how much prior knowledge is assumed. If there are K perfectly coherent signals where $K < P$ and uncorrelated, equi-power noise, the population spectral density matrix will have $P - K$ equal eigenvalues due to noise alone and K larger eigenvalues due to signals and noise. This fact, which is the basis for determining $\hat{K}$ in the procedure presented, does not depend on any prior knowledge of the signal amplitude-and-phase relations. Thus, the procedure presented searches for signals over all possible amplitudes and phases. If, on the basis of prior knowledge, some combinations can be eliminated or considered unlikely, better performance can be obtained. For example, conventional beamforming will perform better than the procedure presented if the proper conditions hold.

The procedure presented can be modified so that prior knowledge about signal amplitude-and-phase relations influences the choice of $\hat{K}$. Having chosen $\hat{K}$ and applied the rotation, the criterion may show that some signals do not match the guess, $v(\theta)$. If this does not indicate a poor match between the actual signal and the guess, it might indicate that $\hat{K}$ is too large or too small. Thus, new values of $\hat{K}$ might be chosen and the rotation repeated. The match produced by each $\hat{K}$ is then considered in the final selection.

Professor P. M. Schultheiss questioned the sensitivity of the procedure to the assumption of uncorrelated, equi-power noise. If this assumption is violated, can the number of signals be determined, can the signal spectral matrix be determined? Determining $\hat{K}$ and $\hat{F_s}$ is probably the weakest part of the procedure. The difficulties for the analogous case of the eigenvalues of a covariance matrix are well known. If the noise does not satisfy the uncorrelated equi-power assumption, the statistical test for $\hat{K}$ must be discarded. In some cases, the following facts allow an approximate determination of $\hat{K}$ and $\hat{F_s}$. First, strong signals will dominate the largest eigenvalues and eigenvectors regardless of the noise properties. Thus, the problem is less severe for strong signals. Second, mistakes in this determination will sometimes be apparent after the rotation either because some components do not match the guess or because the signals in one frequency-time interval do not match the signals in adjacent intervals. Finally, the assumption that the smallest eigenvalues are due to noise and are nearly the same in adjacent frequency-time intervals may be appropriate.

Mr. T. Kooij asked where the array perturbations and non-planar wavefronts manifested themselves in the procedure. The values of $q_j(q_j{}^*q_j)^{-1/2}$ after rotation are estimates of the signal amplitude-and-phase relations that actually occur. The guess, which is $v(\theta) = (8^{-1/2} \exp\{-i2\pi(0.4)(j-1)\sin\theta\})$ for the simulation, does not account for array perturbations or non-planar wavefronts.

Dr. H. Cox asked how the procedure presented compares to conventional beamforming in its sensitivity to deviations from the assumed signal amplitude-and-phase relations. He noted that the criterion to be maximized in equation (13) contains an operation equivalent to conventional beamforming. In the procedure presented, the signal spectral matrix $\hat{F_s}$ is determined without using any knowledge of signal amplitude-and-phase relations, and the sum of the signal components must equal $\hat{F_s}$. Conventional beamforming does not provide such a decomposition. In the procedure presented, the guess $v(\theta)$ is used only to determine which decomposition of $\hat{F_s}$ is chosen. Thus, the data in the form of $\hat{F_s}$ dominate the use of $v(\theta)$. In this sense, the procedure presented makes minimal use of prior knowledge of signal wavefronts and hydrophone locations.

Professor Schultheiss asked what would happen if there were a strong near-field source present. This would be a signal since it would be coherent. This case has not been simulated. The simulation provides some indication of how much discrepancy between actual and assumed signal amplitude-and-phase relations can be tolerated.

Relation of Decision and Physical Wavefield Spaces: Concept and Example

PHILIP L. STOCKLIN

Loughborough University of Technology, Loughborough, Leicestershire, England

1. Introduction

The capability of a system to process information-bearing energy fields is limited by both its physical and technological constraints. A common approach to predicting system processing capability is first to model an ideal environment together with an idealized system then predict processing performance from the combination of these two models. Differences between such predicted performance and measured (experimentally estimated) performance are then ascribed to differences between these models and reality. These differences normally bring out the effect of *technological compromises* such as amplitude quantization or limitation in processing time, or the effect of *environmental approximations* such as direct path signal propagation, in terms of processing performance. As such, they are extremely useful in providing guidance for both further technical development and further sophistication of environmental models.

With the advances in both these areas which have been realized since the late 1940's, however, improving performance has become very costly. Rather then considering, for example, the area of pure processing improvement within an otherwise unchanged system, potential processing improvement is compared with other means of improving system performance, such as changes in fundamental system—e.g., size, speed, hydrodynamic characteristics—and that technical area which offers the greatest overall improvement for a given investment tends to receive the greatest support.

It has thus become increasingly important to gauge the ultimate limit to which each technical area can contribute improved performance, as a means of judging the likelihood that investment in that area can produce desired improvements. In particular, it appears necessary to develop a useful analytic tool which gives, for specified physical constraints (e.g. sensing volume, number of sensors, bandwidth, system and signal movement and environmental conditions), limits to processing capability regardless of technological advances which may be realized.

It is equally important that the derivation of such limits does not depend upon the *ad hoc* assumption of a particular processing principle, but rather that such principles can be compared, relatively and absolutely, once limits to performance are found. As decision theory teaches us [1], prior knowledge—in terms of probability (density) functions—determines the statistically best (most efficient) processing principle together with estimates of its performance. Thus, the approach needed is one relating in the first instance to the physical reality of space-time and secondly to statements of prior knowledge on the use of space-time data [2, 3].

2. Approach

The fundamental physical quantities with which every information-processing system must deal are:

(a) finite number of space sensors or a finite sensing volume,
(b) finite period of time (total processing time),
(c) finite amplitude aperture. Amplitude is taken to be the acoustic pressure for acoustic fields, the field-amplitude vector for electromagnetic fields, etc.

Within the constraints imposed by these three quantities, a frequency range of operation is normally chosen. In fact, frequency appears to play the rôle of a connecting parameter, relating these three fundamental quantities such that their canonical properties become apparent.

Space, time and amplitude share the property of quantization. While this may be imposed by the use of discrete, point-like sensors or digital systems, more basically each suffers self-quantization through uncertainties either in measurement or ultimately by thermal noise—that is, temperature. Thus the approach proposed begins with the definition of a fundamental three-space,whose dimensions are space, time and amplitude, each quantized as shown in Fig. 1.

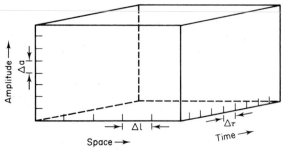

FIG. 1. Fundamental quantity space—the natural decision space (*D*-space).

The quanta Δa, Δl and $\Delta \tau$ may be determined by system design or by natural limitations; detailed consideration of their determination is a critical step in finding system limitations. Given these quanta, however, it is possible to specify the number of distinguishable amplitude patterns—called decisions—which the space is capable of yielding. If M is the number of (point) space sensors, n_T the number of (point) time sensors, and n_A the number of amplitude quanta available, then the total number of distinguishable amplitude decisions D is given by

$$D = n_A^M \cdot {}^n T$$

In terms of these decisions, conventional functional decisions like direction determination or waveform estimation may be described and so the *efficiency* of such functional decisions deduced as a fraction of D. Working with the quanta in D-space and using canonical relations among then, either a D-space can be derived for which a given set of functional decisions can be made most efficiently (minimum system) or for a given D-space, the ultimate limits in making a specified set of functional decisions can be found.

Finally, changes in environment and in target and interference fields are seen as changes in D-space and their effects on decision efficiency can be found. In particular, uncertainties among the amplitude patterns, caused by probabilistic fields such as noise, can be studied in detail and pertinent statistical tests, such as likelihood ratio, deduced.

Example: Assume a linear array of M point elements with interelement spacing d, with n_A amplitude quantization steps over amplitude aperture $(-A/2, A/2)$. Assume a plane wave periodic signal field with known fundamental wavelength λ_1, but of unknown direction as the only field in D-space (noise-free case). Find limits to the functional decision of signal direction, δ_θ.

3. Amplitude Quantization

We assume that the amplitude quantization rule—the manner in which the amplitude aperture is divided up into n_A amplitude quanta—can affect limits to δ_θ, and ask whether there is a "best" amplitude quantization rule.

First, consider the quadratic criterion of adopting a quantization rule which minimizes the mean square error E_Q between actual signal waveform and quantized signal waveform. Let A_r be the midpoint of the rth quantization interval of width Δ_r, let $s(t)$ be the signal waveform and let $p(s)$ be the probability density that the signal, if stopped randomly, will have value between s and $s + ds$. The Δ_r are constrained by the fact that *in toto* they must cover the amplitude aperture A, that is:

$$\sum_{r=1}^{n_A} \Delta_r = A \tag{1}$$

The mean square error over A, E_Q, is then:

$$E_Q = \frac{1}{A} \sum_{r=1}^{n_A} \int_{A_r - \Delta r/2}^{A_r + \Delta r/2} (A_r - s)^2 p(s)\, ds \tag{2}$$

The problem, then, is to minimize E_Q given by equation (2) subject to the constraint given by equation (1). Performing the squaring operation indicated in equation (2) and using the substitution

$$s = A_r + u \tag{3}$$

simplifies E_Q to

$$E_Q = \frac{1}{A} \sum_{r=1}^{n_A} \int_{-\Delta_r/2}^{\Delta_r/2} u^2 p(A_r + u)\, du \tag{4}$$

Using equations (4) and (1), we now define g:

$$g = \frac{1}{A} \sum_{r=1}^{n_A} \int_{-\Delta_r/2}^{\Delta_r/2} u^2 p(A_r + u)\, du - \lambda \sum_{r=1}^{n_A} \Delta_r \tag{5}$$

where λ is an undetermined multiplier (method of La Grange). We then find the extremum of g with respect to one particular Δ_r:

$$\frac{\partial g}{\partial \Delta_r} = 0 \tag{6}$$

which for the optimum Δ_r results in

$$\left(\frac{\Delta_r}{2}\right)^2 = \frac{\lambda A}{\left\{ \frac{1}{2}\left[p\left(A_r + \frac{\Delta_r}{2}\right) + p\left(A_r - \frac{\Delta_r}{2}\right)\right] \right\}} \tag{7}$$

While not necessary to the derivation, we simplify equation (7) by assuming that

$$\frac{1}{2}\left[p\left(A_r + \frac{\Delta_r}{2}\right) + p\left(A_r - \frac{\Delta_r}{2}\right)\right] = p(A_r) \tag{8}$$

so that

$$\left(\frac{\Delta_r}{2}\right) = \left(\frac{\lambda A}{p(A_r)}\right)^{1/2} \tag{9}$$

Substitution of equation (9) into equation (1) to find λ and insertion of this value into equation (9) gives

$$\Delta_r = A \frac{[p(A_r)]^{-1/2}}{\sum\limits_{r=1}^{n_A} [p(A_r)]^{-1/2}} \tag{10}$$

or

$$\Delta_r [p(A_r)]^{1/2} = \text{constant} = K_1 \tag{11}$$

where

$$K_1 = \frac{A}{\sum\limits_{r=1}^{n_A} [p(A_r)]^{-1/2}} \tag{12}$$

The basic result is given by equation (11)—the width of the rth quantization step, Δ_r, should vary inversely as the square root of the probability that the signal waveform occurs at A_r in order to minimize mean square error subject to equation (1).

In D-space, however, we are in this example interested in the amplitude rather than the squared amplitude. Thus we examine the linear criterion E_L (Absolute amplitude error):

$$E_L = \frac{1}{A} \sum\limits_{r=1}^{n_A} \left\{ \int\limits_{A_r - \Delta_r/2}^{A_r} (A_r - s)p(s)\, ds + \int\limits_{A_r}^{A_r + \Delta_r/2} (s - A_r)p(s)\, ds \right\} \tag{13}$$

Working the problem again, using the constraint equation (1), the change of variable equation (3), and the approximation equation (8), results in

$$(\Delta_r)' = A \frac{[p(A_r)]^{-1}}{\sum\limits_{r=1}^{n_A} [p(A_r)]^{-1}} \tag{14}$$

or

$$(\Delta_r)' p(A_r) = \text{constant} = K_2 \tag{15}$$

where

$$K_2 = \frac{A}{\sum\limits_{r=1}^{n_A} [p(A_r)]^{-1}} \tag{16}$$

Equation (15) states that to minimize absolute amplitude error, the rth quantization step should vary inversely as the probability that the signal waveform occurs at A_r.

Thus to minimize absolute error, the output of each of the M point sensors may be passed through the quantizer of equation (15), which effectively makes each A_r equally probable in D-space. For a periodic waveform, this means that the waveform should spend an equal time, per period, in each quantization interval.

Thus, most efficient (minimum average absolute amplitude error) quantization may be obtained for n_A quantization intervals by choosing those values of s, s_r, which divide up $p(s)$ into n_A equal areas. If $p(s) = 0$ beyond a peak value $\pm A_p$, then there is no reason for the aperture A to extend beyond $\pm A_p$ since the values of s_r for a given n_A do not change with such extension. It can be shown that $(E_L)_{\min}$ varies as n_A^{-1}, going to zero as n_A becomes infinite.

4. Time Sampling

In this example, our interest in time sampling is in terms of its effect upon the number and resolution of distinguishable signal directions. Since we are concerned with a periodic signal, a convenient parameter is n_1, the number of time samples per period. Since all values of the quantized signal waveform pass through the time sampler for each point sensor output equal time intervals between successive time samples are considered.

Consider two quantized periodic signal waveforms, from two point sensor outputs, with the signal a plane wave signal. With the signal coming from the broadside position of the array ($\theta = 0°$), the two point sensor outputs are identical. The direction of the signal is now gradually changed away from broadside until a change from broadside appears in the amplitude pattern of the two point sensor outputs. How does this minimum change in signal direction relate to n_A and n_1?

Consider first that the probability filtering, equation (15), has been done. If the signal is a sine wave of period T_1, then equation (15) transforms this sine wave into a "triangular wave", with the time necessary to change one quantization level, T_A given by

$$T_A = \frac{T_1}{2(n_A - 1)} \tag{17}$$

Now, if there were just one time sample per period T_1 ($n_1 = 1$), then the minimum change in signal direction to *ensure* a new pattern in one point sensor output versus the other would correspond to a delay $(A\tau)_r$ between signal arrivals of one T_A, that is,

$$(\Delta\tau)_1 \geqslant T_A \tag{18}$$

If there are two time samples per period $T_1(n_1 = 2)$, a new pattern again will not occur (with certainty) until $(\Delta\tau)_1$ is achieved, but in this case the new pattern will contain two different amplitudes relative to the broadside pattern, rather than just one different amplitude as in the case of $n_1 = 1$. As n_1 is increased to the point where there is one time sample per amplitude quantization bin (i.e., $n_1 = 2(n_A - 1)$ for a single sine wave signal) the minimum direction change to create a new pattern will remain at $(\Delta\tau)_1$. If n_1 is now increased to $2(n_A - 1) + 1$, however, at least one amplitude bin will contain two time samples, and so will require at the most a shift $(\Delta\tau)_2$:

$$(\Delta\tau)_2 = \frac{T_A}{2} \tag{19}$$

to create a new pattern. Thus, a plot of $(\Delta\tau)_{min}$ versus n_1 would appear as sketched in Fig. 2:

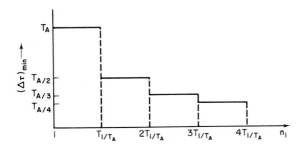

FIG. 2. Variation of delay corresponding to minimum resolvable direction with n_1 and T_A.

In Fig. 2, the effect of both n_1 and n_A can be seen. Algebraically, Fig. 2 can be expressed:

$$(\Delta\tau)_{min} = \frac{T_A}{k}, \quad (k - 1)2(n_A - 1) + 1 \leqslant n_1 \leqslant k \cdot 2(n_A - 1) \tag{20}$$

From the relation:

$$\Delta\tau = \frac{md \sin\theta}{c} \tag{21}$$

where m = number of interelement distances and separating the two point sensors, it is clear that for fixed $\Delta\tau$ minimum angle θ_1 will result for maximum separation of the two point sensors—that is, $m = (M - 1)$. Thus,

$$(\Delta\tau)_{min} = \frac{T_A}{k} = \frac{(M - 1)d}{c} \sin\theta_1 \tag{22}$$

The amplitude pattern of the outputs of the point sensors, other than the Mth, will all be identical to their pattern at broadside, for this angle.

As $\Delta\tau$ is increased (θ increased to θ_2), until the next lowest interelement separation yields to delay of $(\Delta\tau)_{min\ 1}$ then a new pattern, with the Mth and $(M-1)$th point sensor outputs now the same, will result. The condition is:

$$\frac{(M-2)\ d}{c}\sin\theta_2 = \frac{T_A}{k} \tag{23}$$

As θ is increased, distinguishable amplitude patterns will continue to occur in D-space up to angle θ_r:

$$\frac{(M-r)\ d}{c}\sin\theta_r = \frac{T_A}{k} \tag{24}$$

where r is the largest integer satisfying

$$r < \frac{M+1}{2} \tag{25}$$

As r is increased beyond the limit of equation (25), additional patterns appear. Between broadside and the first direction θ_{c1} for which conventional beam-forming can be performed—a progressive delay of $(\Delta\tau)_{min}$ between successive point sensors outputs—the total number of distinguishable amplitude patterns is given in Table I versus M, for $2 \leqslant M \leqslant 32$. This number, θ_M, appears to vary exponentially with M, for $M > 10$, as

$$\theta_M \sim K^M \tag{26}$$

TABLE I. Number of distinguishable signal directions in one conventional beam interval θ_M, versus M. See discussion following equation (25).

M	θ_M	M	θ_M	M	θ_M
1	0	11	41	21	139
2	1	12	45	22	149
3	3	13	57	23	171
4	5	14	63	24	179
5	9	15	71	25	199
6	11	16	79	26	211
7	17	17	95	27	231
8	21	18	101	28	243
9	27	19	119	29	271
10	31	20	127	30	279
				31	309
				32	325

where

$$K = \sigma(1.25)$$

As θ is increased beyond θ_{c1} to the angle θ_{c2} for which a progressive delay of $2(\Delta\tau)_{min}$ appears between successive elements, θ_M distinguishable amplitude patterns (directions) again appear in D-space. Thus, over the sector $(-\pi, \pi)$, a total of n_θ directions can be distinguished, where

$$n_\theta = 2N_d \cdot \theta_M + 1 \tag{27}$$

where N_d is the largest integer smaller than or equal to $d/[c(\Delta\tau)_{min}]$. Using equation (20) and (17), N_d is the largest integer smaller than or equal to

$$2\left(\frac{d}{\lambda_1}\right) k(n_A - 1)$$

so that equation (27) may be written:

$$n_\theta = 2N_d \cdot \theta_M + 1 \leqslant 4\left(\frac{d}{\lambda_1}\right) k(n_A - 1)\theta_M + 1. \tag{28}$$

Equations (27) and (28) illustrate the quantization of direction arising from quantization in D-space. It is also implied that for sparsely settled arrays, e.g., $M = 2$ and $d \gg \lambda_1$, a multiplicity of directions corresponding to one pattern in D-space can occur. Such occurrences can be labelled false alarms in direction.

5. Interpretation

The purpose of the example is to relate quantization in space, time and amplitude of the received acoustic field to quantization of the decision of target direction, under the conditions of a single monochromatic target, a linear array of point receiving sensors and the absence of interfering noise. From the set of quantized signal directions which result, both direction resolution and direction accuracy may be estimated, as well as the presence of false target directions when they exist. Clearly, addition of a stochastic interference field and/or additional discrete sources will alter these results on the direction decision and indeed may be anticipated to give a scale of fineness to space-time sampling and amplitude quantization beyond which no improvement in the direction decision may be possible. These are the subjects of future research.

REFERENCES

1. Peterson, W. W., Birdsall, T. G. and Fox, W. C. (1954). The theory of signal detectability. *Trans. I.R.E. (Information Theory)*, **PGIT-4**, 171.

2. Stocklin, P. L. (1963). Space-time sampling and likelihood ratio processing in acoustic pressure fields. *J. Brit. I.R.E.* **26**, No. 1, p. 79.

3. Stocklin, P. L. (1968). Space-Time Decision Theory. Relation of Wavefield and Decision Spaces. *In* Proceedings of the NATO Advanced Study Institute on Signal Processing with Emphasis on Underwater Acoustics. August 1968. Twente Institute of Technology, Enschede, The Netherlands.

Practical Limits of Time Processing

RICHARD B. GILCHRIST

Naval Ship Systems Command, Washington, D.C., U.S.A.

1. Introduction

In the development of new sonar systems, a great many compromises must
be made in order to arrive at a practical overall design. Most of these design
compromises cause an attendent degradation in the system performance,
and can have a significant effect on the proper choice of system design
parameters and the estimate of system operational performance. Unfortunately, these degradations have been frequently overlooked in the
past. The purpose of this paper is to point out, by an example, the
principal design compromises that are involved in the time processing of
data from an array of hydrophones and to cite typical values for each of
the degradations involved.

2. Sonar System Example

Figure 1 is a block diagram of a narrowband detection system that will be
the basis for the following discussion. It could represent either the receiver
portion of an active doppler detection system, or a passive line detection
system.

The array consists of 32 hydrophones of length, L, arranged in a circle
of diameter, D, with

$$L = \frac{3\lambda_0}{2}$$

$$D = 5\lambda_0$$

$$\lambda_0 = \frac{C}{f_0}$$

where C is the velocity of sound and f_0 is the upper design frequency of
the system. This produces a natural vertical beamwidth of about $34°$ and
a natural horizontal beamwidth of about $10°$.

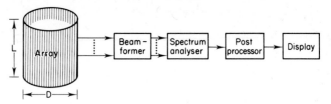

FIG. 1. Block diagram of example system.

The beamformer is a conventional time delay and sum type with multiple preformed beams. In order to reduce the susceptibility to interference, the array is shaded in the horizontal plane to lower side lobe levels. This broadens the horizontal beamwidth to 15°. There are 24 preformed beams spaced 15° apart.

The spectrum analyser is an FFT with 1024 real input points per transform and 512 frequency output bins per transform. The sampling frequency is approximately $6f_0$ and each sample is quantized to 4 bits. The FFT input is also shaded to reduce susceptibility to filter side lobe interference. The result is 170 equally spaced filter bins in the band from 0.0 Hz to f_0 Hz of width

$$B = 0.009 \, f_0$$

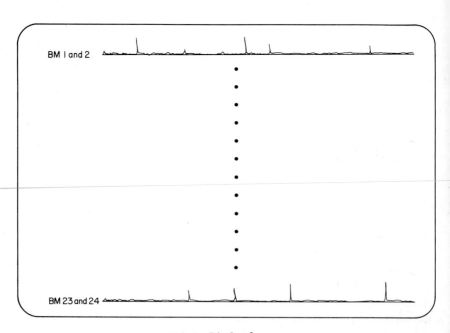

FIG. 2. Display format.

with bin cross overs at −1.5 dB. A new spectral estimate is produced every

$$\Delta T = \frac{170}{f_0} \text{ sec}$$

The post processor consists of a full wave rectifier detector, followed by an exponential integrator for each frequency bin with a time constant

$$\tau = 10 \ \Delta T.$$

At the integrator output, each frequency bin is normalized by subtracting the mean of the 16 adjacent bins and then re-quantized to 6 bits.

The display is a CRT on which 12 A-scan traces of beam power versus frequency are displayed. Each trace is 170 bins wide and 64 bins high as depicted in Fig. 2. Every two adjacent beams are "or"ed together bin-by-bin to collapse the 24 beams into 12 traces.

3. System Degradations

Table I lists the principal degradations that result from the design compromises of the example. If the equipment is improperly built or not kept operating at peak performance, the losses will be even greater. The overall degradation of 9–15 dB obviously causes a very significant loss in detection range. Moreover, if all losses are not considered, improper design choices may be made. For instance, additional (narrower) beams may be added to reduce beam scalloping and to improve the directivity index. However, most of the improvement may be lost because of increased "or"ing losses, display resolution losses, and operator losses in order to display the extra data from the additional beams. Thus very little improvement may be gained from this increase in hardware complexity and cost.

4. Conclusion

It is imperative for system designers and analysts to properly account for all system degradations from ideal performance. This is necessary to insure wise design compromises are made. It will also prevent embarrassment from unexpectedly poor performance when the system is introduced into fleet operations.

TABLE I

Degradation	Estimated Loss (dB)	Remarks	References
1. Beam shading	1.5	D. I. loss	
2. Beam scalloping	1.0	Random bearing target	
3. Quantization	0.1 – 3.0	Equalized spectrum to 20 dB misequalized spectrum at worst frequency	
4. Aliasing	0.0	Assuming good aliasing filter	
5. Filter shading	0.2	Detection threshold loss	1
6. Filter scalloping	0.5	Random frequency target	1
7. Detector	0.2	Linear vs square-law	
8. Integrator sampling	1.6	Could be reduced to 0.25 dB by 50% overlapping of data into FFT.	1
9. Finite Detection Time	0.5	Exponential vs box car integrator	1
10. Normalization	0.1		
11. Bias limit	?	Depends on interference to noise ratio, and side lobe levels.	2
12. "Or"	0.2	Can be much worse if normalization is poor	
13. Display resolution	0.4	Assumes 5 display spots per frequency bin	
14. Operator	3.0 – 6.0	Clutter, overloading, and inattention	3, 4
	TOTAL 9.3 – 15.2		

REFERENCES

1. Pryor, C. N. (1971). "Calculation of the Minimum Detectable Signal for Practical Spectrum Anaiyzers", Naval Ordance Laboratory Technical Report 71-92.
2. Cox, H. (1970). "Performance Prediction for Passive Sonar", unpublished paper.
3. Buckner, D. N., Harabedian, A. and McGrath, J. J. (1960). "A Study of Individual Differences in Vigilance Performance", Human Factors Research, Inc. Technical Report No. 2.
4. Skolnik, M. I. (1962). "Introduction to Radar Systems", McGraw-Hill, New York.

DISCUSSION

D. Nairn: What you have done here is to calculate your degradations on the assumption that if everything is all right you will get the full system gain. We know that this is not the case. Take the shading loss of 1.5 dB for instance. This would assume that if you had no shading you would have full system gain. However, there are practical difficulties in getting this, say due to random phase errors. You may find that once you put in shading the array would not be as sensitive to random phase errors. Also, you cannot separate normalization loss from clipping loss. The point I am making is that there are dangers in calculating these losses because it assumes that if you did not have these losses you would get the full system gain.

Answer: I do not disagree. You must compute all the losses that are involved for a complete system design. If you change one part of the system you may affect other parts and you must work through all the system losses again, e.g. there may be unequal side-lobes in the presence of strong planewave interference. One thing you can do to help is to design your beam patterns for low level uniform side-lobes or at least smoothly varying side-lobes. Another thing you can do is some form of interference cancellation.

H. Cox: To be fair to the audience, who may not have been introduced to this idea, I have some hopes that a paper on this subject will be available in the near future. If there are some specific questions, I will be happy to answer them afterwards.

C. van Schooneveld: When you discuss the losses, you implicitly compare them to the case with zero loss. It is not clear to me whether this case is for noise only or includes reverberation.

Answer: The example I used was for the noise only case. With reverberation, the losses would be different and more difficult to compute.

H. Cox: I would like to make a comment. We make performance predictions based on idealized and simplistic models. To the extent that the real system does not conform to these models, we have to make adjustments. I prefer the word adjustments to losses because losses imply that something evil has happened. In many cases the adjustments have resulted from wise design choices. For instance, increasing your sensitivity to isotropic noise and reducing side-lobe sensitivity by shading is a very intelligent thing to do.

D. Nairn: Would you care to make a statement about how far you can go in specifying a system in terms of dB's? One gets the impression that to begin with one is in the power enhancement business, but towards the end, especially when you come to computer systems, to automatically extract the data, one almost forgets the power and gets into character recognition. It might be better not to match filter the data, but to use a wider filter in order to allow more character in the echo in order to enhance the character recognition at a later stage.

Answer: I agree you cannot specify a system or performance purely in terms of dB's. The point that I was trying to make, as Cdr. Cox said, is that we must use realistic gains and losses when performance predictions are made. In terms of specifying the system and performance to a contractor, of course dB's are not sufficient.

E. J. Risness: The extent to which you can specify the losses in terms of dB depends on the cause of the loss. Take for example the operator loss, which is a major one, if it is due in part to the fact that the operator is only watching the screen 75% of the time, then no matter how many dB's you have, you cannot get the detection probability up to 90%. In other words you can have losses that cannot be expressed in terms of dB's.

Detection of Extended Targets Against Noise and Reverberation

G. A. VAN DER SPEK

Physics Laboratory TNO, The Hague, Netherlands

1. Introduction

The use of pulse-compression in radar and sonar has increased the range resolution of these systems. As a result the range resolution cell may be considerably smaller than the physical dimensions of the target to be detected. Here the effect of increased resolution on target detectability is studied.

A target is assumed which can be divided into n independent resolution cells. The amplitude of the echo of each cell is distributed according to the same Rayleigh distribution. The total average energy reflected from a target does not depend on the resolution. A single pulse is transmitted and the reflected signal is processed by a matched filter and a square-law envelope detector followed by an integrator, which integrates over the total target echo duration.

The background is either white normal noise or reverberation, which is stationary over the duration of the target echo. In both cases the effect of changing the resolution will be studied. Integration loss and fluctuation reduction will be discussed.

2. Detection of a Point Target

First the detection of a non-fluctuating point target (e.g. a corner reflector) in white normal noise is considered. The optimum processor is known to be a matched filter with envelope detector [1]. The energy of the transmitted signal is E, its bandwidth B. A fraction r of the transmitted energy is received.

The detection quality depends only on the signal-to-noise ratio (SNR) rE/N_0, where rE is the echo energy received and N_0 is the noise power density. In Fig. 1 a detection curve (solid line) gives the relation between the probability of detection Pd and the SNR for a false alarm probability $\alpha = 10^{-3}$.

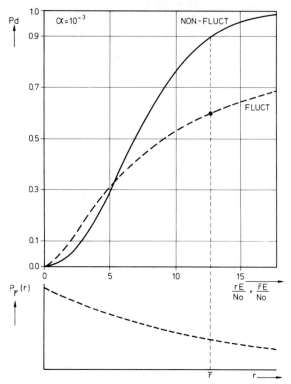

FIG. 1. Probability of detection as a function of (average) SNR for non-fluctuating and fluctuating point target for $\alpha = 10^{-3}$

According to the sampling theorem a signal of bandwidth B can be represented by $2B$ samples per second. The output of the matched filter would require this sampling rate. The envelope detector removes the phase information so that the output of the envelope detector requires B samples/second, which corresponds to a range resolution cell of $1/B$ seconds [2, 3-6]. If P_n is the noise power in the band of the matched filter, B, we can write for the noise power density: $N_0 = P_n/B = P_n \cdot 1/B$ is the average energy of the noise in one resolution cell. So the SNR is the ratio between the energies of the received signal and the noise in the resolution cell that contains the target.

In practice a point target will not reflect the same fraction of the signal energy on each pulse. A target can be thought of as a collection of scatterers. Due to the relative movement of transmitter/receiver platform and target and the effects of changes in the transmission medium, the relative phases of the contributions to the echo of the individual scatterers are different from echo to echo. As a result the fraction r of the signal energy which is received can be considered to be a random variable. The probability density function of r will resemble an exponential distribution (Rayleigh target). The performance of the matched filter detector with envelope detector can be obtained by averaging over the detection curve for a non-fluctuating point target according to a weighting

$$p_{\bar{r}}(r) = \frac{1}{\bar{r}} \exp - \frac{r}{\bar{r}}$$

as shown in Fig. 1 (dashed lines). The parameter along the horizontal axis is the average SNR: the ratio of the average received signal energy to the average noise energy in the target cell. On comparing both detection curves it is seen that the target fluctuation gives rise to a fluctuation loss which increases with Pd for Pd $>$ 0.30, due to the averaging process, which is done mostly over a concave part of the original detection curve. For Pd $<$ 0.30 a small fluctuation gain develops.

When detection in reverberation is considered using a matched filter and an envelope detector the signal-to-reverb ratio determines the detection performance. The average noise energy per cell is now replaced by the average reverberation energy per cell

$$N_r = \rho E(v/2B)$$

where ρE is the average energy received in the reverberation echo per meter when a signal with energy E is transmitted, v is the propagation speed of sound in meters/second, so that $v/2B$ is the length of resolution cell in meters. The average signal-to-reverb ratio is

$$\frac{\bar{r}E}{N_r} = \frac{2\bar{r}B}{\rho v}.$$

The effect of increasing the resolution by increasing the signal bandwidth B is different for noise and reverberation. As long as the target is not re-solved the signal-to-noise ratio is not affected, while the signal-to-reverb ratio increases linearly with B.

3. Detection of Extended Targets

When the resolution cell is decreased further by increasing the signal bandwidth B the cell will become smaller than the target. Suppose the target is equivalent to n cells,

$$n = \frac{L}{v/2B},$$

where L is the length of the target. The problem of detecting a target consisting of n independent identical Rayleigh cells in white normal noise is essentially the same as that of detecting a Swerling II point target by n

incoherent pulses [1]. The optimum detector is known to be a matched filter followed by a square-law envelope detector with linear integrator (Fig. 2).

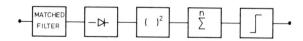

FIG. 2. Optimum detector for n-cell target in white normal noise.

After the envelope detector the average SNR per cell is $\dfrac{(\bar{r}/n)E}{N_0}$. This quantity decreases with increasing resolution. So on increasing n, the average SNR per cell decreases, while the number of cells over which it is integrated increases. When detection in reverberation is considered the average signal-to-reverb ratio per cell is

$$\frac{(\bar{r}/n)E}{\rho E(v/2B)} = \frac{\bar{r}}{\rho L},$$

which, in contrast with the average SNR in the noise case, is independent of the transmitted signal energy and independent of the number of cells on the target n. So on increasing the resolution, the average signal-to-reverb ratio remains the same, while the number of cells over which it is integrated increases.

(a) Extended target in noise

Now let us consider the practical consequences of increasing the resolution, first in the case of a noise background. An instructive way to do so is to compute the relative noise power density, $N_0(n)/N_0(1)$, required to maintain the same detection quality as in the case where the resolution cell matches the target ($n = 1$) as a function of the resolution. Figures 3 and 4 show the results for false alarm probabilities of $\alpha = 10^{-1}$, 10^{-3} and 10^{-6}.

If the target is not resolved the average SNR does not depend on the cell-size, while only one cell is processed. This means that increasing the resolution has no effect on target detectability until $n = 1$. When the resolution is increased further the average SNR per cell decreases linearly with the number of cells n, while integration takes place over the same number of cells. The net effect is seen to depend on the detection quality: for a low probability of detection a constant deterioration rate is present as soon as the target contains more than one cell. For higher detection probabilities the same deterioration rate develops first at a higher resolution. For instance for a detection probability of 0.90 first, an increase in the corresponding noise power density is evident when the target is resolved in more than one cell. The maximum corresponding noise power

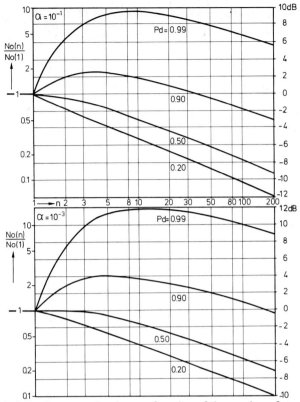

FIG. 3. Relative noise power density as a function of the number of resolution cells for $\alpha = 10^{-1}$, 10^{-3}.

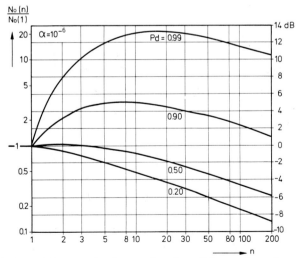

FIG. 4. Relative noise power density as a function of the number of resolution cells for $\alpha = 10^{-6}$.

density is reached for $n = 4$ to $n = 8$, depending on the false alarm probability. The corresponding gain in performance is 2.5 to 5 dB. For increasing resolution the curves all tend to the same deterioration rate of 1.5 dB per doubling of the resolution.

The two trends with increasing resolution: the initial improvement and the eventual degradation, are the result of two effects: fluctuation suppression and integration loss. Both effects will be discussed in more detail.

(b) Fluctuation suppression

In Fig. 1 it was shown what happens when a one-cell target with a fixed cross-section is turned into a one-cell target with an exponentially distributed cross-section. Each point in the second curve can be obtained from the first curve by weighting with the probability density function of the target reflectivity. Now suppose we have an n-cell fluctuating target for which the phase of the return signal, except its initial value, is known. For the processor this would mean that the returns of the individual cells could be integrated coherently, so that the envelope detector and the integrator in Fig. 2 could be exchanged, while the quadrator becomes irrelevant.

Now the output of the envelope detector can be thought of as the output of a processor for a single cell target, where the energy received from the target is wE, in which the random variable w is related to the individual reflectivities of the cells, r_i, according to

$$w = \frac{1}{n} \sum_1^n r_i$$

The random variables r_i are independent and have the same exponential distribution. The random variable w has a probability density function which depends on n, with $\bar{w} = \bar{r}$ and $\sigma_w^2 = \sigma_r^2/n$. The detection curve can be obtained from that of a non-fluctuating one-cell target as before. In Fig. 5 an example of the weighting curve is shown for $n = 8$.

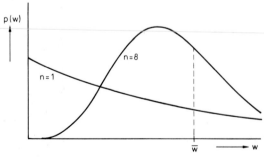

FIG. 5. Probability density function of $w = \dfrac{1}{n} \sum_1^n r_i$.

Since the variance of the weighting curve (= p.d.f. of w) decreases with n, the difference between the performance curves for a non-fluctuating one-cell target and a coherently processed n-cell target decreases with the resolution. This effect is called fluctuation suppression.

(c) Integration loss
The earlier assumption that the phase of the return signal is known is unrealistic, since it is incompatible with the multi-scatterer target model. In fact it should be assumed that the echo from each cell has a random phase. As a result there can be no coherent integration over all target cells and the envelope detector has to precede the integrator. This non-linear part of the processing chain gives rise to the so called integration loss, which in fact is a loss caused by the non-linear character of the envelope detector.

Davenport and Root [3] show that the output signal-to-noise power ratio for any νth-law envelope detector is proportional to the square of the input SNR for small values of the latter and is directly proportional to the input SNR for large values of the input SNR. This is known as the small signal suppression effect. This relation is shown in Fig. 6.

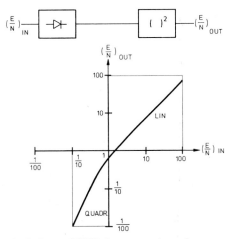

FIG. 6. Loss of SNR due to envelope detector.

The relevance of this relation will be discussed by first considering a non-fluctuating target. On increasing the resolution from $n = 1$ on, initially samples are presented to the envelope detector with a high SNR, at least for a probability of detection which is large enough and a false alarm rate which is small enough. So initially the loss due to the detector non-linearity is small. For increasing resolution the input SNR at the envelope detector decreases so that the output SNR tends to the quadratic part of the curve.

Here a doubling of the resolution gives twice the number of samples, each with an output SNR, which is four times lower, so that the integrated output SNR is reduced by a factor 2. When the input SNR is then increased by a factor $\sqrt{2}$ the same output SNR results. This is equivalent to an increase in the required input SNR of 1.5 dB per doubling of the resolution.

Now if a fluctuating target is assumed both effects take place. When the resolution is increased from $n = 1$ on at first the average input SNR is high so the integrator loss is small. The fluctuation reduction is here important especially for high detection probabilities. On further increasing the resolution the rate of the fluctuation suppression reduces, while the integration loss increases until eventually only an integration loss at a rate of 1.5 dB per doubling of the resolution remains.

(d) Extended target in reverberation

The same processor, consisting of a matched filter, square-law envelope detector and linear integrator, will be used to analyse the influence of increasing the resolution in the case of a reverberation background. Earlier we found that the average signal-to-reverb ratio per cell for a fluctuating target is $\bar{r}/\rho L$, where $\bar{r}$ represents the average received fraction of the transmitted signal energy, L the target length and ρ the received fraction of the transmitted energy due to reflection from the reverberators within a unit of length. In case the target is smaller than the range cell the average

signal-to-reverb ratio for the one-cell target is $\dfrac{\bar{r}}{\rho(v/2B)}$.

As ρ is a measure of the reverberation strength, the influence of resolution on detection of a target in reverberation can be found by computing the value of ρ, as a function of the number of cells on the target, required to maintain the same detection quality as in the case that the resolution cell matches the target ($n = 1$). An example is shown in Fig. 7. As long as the target is not resolved ρ increases linearly with the bandwidth, independent of the required detection quality.

From $n = 1$ on the signal-to-reverb ratio per cell is constant while the number of cells increases. To keep the detection quality constant $\rho(n)$ can be raised with a rate which depends on the detection probability chosen. Both fluctuation suppression and integration loss are encountered as before. In fact all results for the case of normal white noise can be easily translated to the reverberation case by adding a straight line with unity slope. For high resolution the improvement rate will approximate 1.5 dB per doubling of the resolution. Of course this improvement cannot go on indefinitely. The target model will lose its validity when the resolution cell becomes too small and each cell does not contain a number of scatterers any more. Furthermore the point may be reached that the contribution from reverberation in other cells becomes larger than the reverberation contribution from the target cell itself. The main restriction, however, is

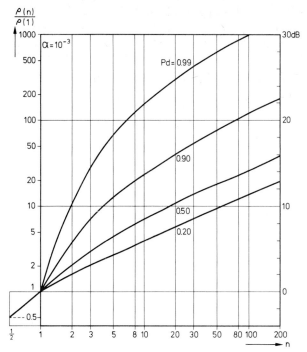

FIG. 7. Relative reverberation strength per meter as a function of the number of resolution cells for $\alpha = 10^{-3}$

that there is always noise present and it is of no use to reduce the reverberation energy per cell beyond that of the noise power density.

It might be argued that a matched filter is not the optimum filter for detection against reverberation. It should be replaced by a properly mismatched filter, consisting of a pre-whitener and a filter matched to the signal processed by the pre-whitener. However when the noise background is white the optimum signal has also a white spectrum and the matched filter processor is optimum [4]. When the signal spectrum is non-white the optimum mismatched filter does in fact create a range resolution cell, which is smaller than that corresponding to a matched filter. Thus integration has to take place over a larger number of cells giving rise to a greater fluctuation suppression and a greater integration loss. In fact the same result would have been obtained by increasing the resolution for a matched filter [5].

4. Conclusions

For single pulse detection of a fluctuating target against white normal noise it is shown that there is an optimum size of the resolution cell, which depends on the detection quality desired. When the resolution is increased further, the gain resulting from the fluctuation suppression of the integrated target echo is exceeded by the increase of the integration loss.

When detection against reverberation is considered the resolution should be increased until the reverberation energy per cell becomes comparable to the noise power density.

REFERENCES

1. DiFranco, J. V. and Rubin, W. L. (1968). "Radar Detection", Prentice Hall, Englewood Cliffs, N.J.
2. Lawson, J. L. and Uhlenbeck, G. E. (1950). "Threshold Signals", McGraw-Hill, New York.
3. Davenport, W. B., Jr. and Root, W. L. (1958). "An Introduction to the Theory of Random Signals and Noise", McGraw-Hill, New York.
4. Kooij, T. (1968). "Optimum Signals in Noise and Reverberation", *In* Proceedings of Nato Advanced Study Institute on Signal Processing with Emphasis on Underwater Acoustics, Twente Institute of Technology, Enschede, Netherlands.
5. Rihaczek, A. W. (1969). "Principles of High-resolution Radar", McGraw-Hill, New York.
6. van der Spek, G. A. (1971). Detection of a Distributed Target, *IEEE Trans AES*. Vol. AES-7, pp. 922–931, September 1971.

DISCUSSION

T. G. Birdsall: If you want to start a campaign for the change of integration loss by loss due to the envelope detector I will join you.

P. M. Schultheiss: Could you specify what signal returns the extended fluctuating target will produce if a sinewave pulse or a swept sinewave pulse is transmitted?

Answer: This depends on the resolution of these signals. If a sufficiently short sinewave is used the return will be a sinewave with an envelope and phase which both vary gradually. Samples of envelope (and phase) will be independent if their separation is more than the length of the transmitted pulse. The return of a swept sinewave of the same duration would be a rather complex waveform. Its envelope will produce independent samples at a rate of B per second.

F. Wiekhorst: Target as well as reverberations are represented by the same model of randomly distributed scatterers. How can they be discriminated in the absence of a Doppler-shift?

Answer: In order to be able to detect a target consisting of scatterers in a background of reverb-scatterers some characteristic should be different. In the case of stationary reverberation this is the difference in average scatter reflectivity between target plus background, and background only.

P. H. Moose: Your target model implies that the reflection coefficient per cell r_i decreases linearly with increasing resolution. Do you have any experimental results on this?

Answer: As far as I know there is no contradictory evidence. However we have not done any special experiment to study this. Both in sonar and radar there is a need for better target models and I guess that experimental data will become available.

G. Vettori: Do you assume that you know the target extent so that you can match the integration time to it? Do you have results which show the influence of a mismatch in this respect?

Answer: It is assumed that the target extent is known. The case of a mismatch has not been studied but could be easily included in the analysis.

H. Mermoz: A target could be considered to be a linear filter, and if we knew this filter we could build the inverse filter. However such a filter would be difficult to obtain while it would depend on the aspect angle. In my opinion the random scatterer target is justified since it confirms our ignorance about the target.

Answer: Perhaps you could use other probabilistic models. A modification of the present model, where some of the cells do not contain any scatterers, has been treated by the author in Reference [6]. Another possibility is to assume some distribution of the reflection strength over the target cells (uniform in the present model) e.g. by assuming strong and weak cells (target with highlights).

P. Raj: How does the spread in Doppler of the reverberation affect the results?

Answer: This paper only considers range resolution. If this resolution is increased by merely expanding the signal bandwidth, without a change in the total signal duration, the results are not influenced by any Doppler spread. This is so because the effective reverberation energy per meter ρE, that is the reverberation energy per meter range in the Doppler cell of the target, is not affected. An increase in Doppler resolution, however, could reduce this quantity.

Some Theoretical Considerations Concerning Time Statistics in Signal Detection

A. R. PRATT

Loughborough University of Technology,
Loughborough, Leicestershire, England

1. Introduction

For many years, the prediction of detection performance of pulse radar or sonar systems has been based on work originated by J. I. Marcum [1]. This has been accepted mainly because good practical agreement with his theory has been found. More recently, C. W. Helstrom [2] has reported another method of performance prediction based on the probability of threshold crossings. The later model has been used to calculate false alarm probabilities but no detailed theory, it appears, has been produced to calculate the probability of detection. Thus, this paper sets down the theory relating to probability of detection at least in the simplified case of base-band signals and low-pass filtered noise. It appears possible that this work may be extended to deal with the more usual envelope-demodulated signals and noise. Experimental work has confirmed the validity of the model even though the performance differences are small. It seems likely that such differences were previously attributed to experimental error.

In the early sections of the paper, we shall be concerned with a review of Marcum's theory and that attributable to Helstrom. Finally, we shall consider the theory relating to the probability of detection and its implications.

2. Marcum's Detection Theory

The theory which Marcum proposed is based on the Neyman–Pearson approach to hypothesis testing which is now generally accepted in the radar and sonar fields. The Neyman–Pearson decision rule is simply that the observer wishes to set an acceptable level of false alarms whilst maximizing probability of detection. The procedure is equivalent to determining a decision threshold by deciding upon an acceptable false alarm probability. This decision rule is particularly useful in some radar and sonar situations because it is difficult to have knowledge of the signal amplitudes, the *a priori* probabilities or the cost of incorrect decisions.

We are only concerned, at this time, with the detection of signals on a one pulse basis, that is there is no integration performed before the decision-making stage. In this simplified case, Marcum argues that we only need to make a decision once for each period of time corresponding to the correlation interval of the signal and noise processes. For intervals shorter than this, we should observe dependence between samples of the noise alone or signal and noise waveforms. If samples are taken at greater intervals, we may miss some important information. The correlation interval is approximately equal to the reciprocal of the I.F. bandwidth of the receiver. The probability that one of these samples lies above the threshold is then given by the area of the relevant probability density function above the threshold. For a Gaussian distribution of noise amplitude, the false alarm probability (FAP) and probability of detection (POD) assuming that the signal is observed as a change in mean are given by:

$$\text{FAP} = \int_k^\infty \frac{1}{\sqrt{2\pi\sigma^2}} \exp\left[-\frac{x^2}{2\sigma^2}\right] dx \tag{2.1}$$

$$= \tfrac{1}{2}[1 - \text{erf}\,(k/\sqrt{2\sigma^2})]$$

$$\text{POD} = \int_k^\infty \frac{1}{\sqrt{2\pi\sigma^2}} \exp\left[-\left(\frac{x-s}{2\sigma^2}\right)^2\right] dx \tag{2.2}$$

$$= \tfrac{1}{2}\left[1 - \text{erf}\left(\frac{k-s}{\sqrt{2\sigma^2}}\right)\right]$$

where s is the change in mean level due to the signal. For noise alone and signal and noise after demodulation, these equations become:

$$\text{FAP} = \int_k^\infty \frac{R}{\sigma^2} \exp\left[-\frac{R^2}{2\sigma^2}\right] dR$$

$$= \exp\left[-\frac{k^2}{2\sigma^2}\right] \tag{2.3}$$

$$\text{POD} = \int_k^\infty \frac{R}{\sigma^2} I_0\left[\frac{sR}{\sigma^2}\right] \exp\left[-\frac{(R^2+s^2)}{2\sigma^2}\right] dR. \tag{2.4}$$

These two cases are graphically illustrated in Fig. 1.

It is true that, in some cases, short samples of the received waveforms are used and here the above equations describe exactly what is happening. However, in many radar and sonar situations, such sampling does not take place and the equations yield incorrect estimates. The equations calculate

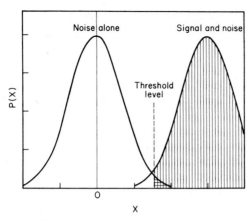

(a) Gaussian variate – low pass filtered noise and signal

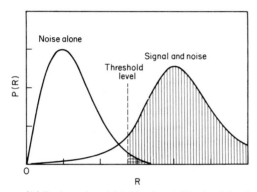

(b) Envelope – demodulated band–pass filtered and signal

FIG. 1. Illustration of detection scheme due to J. J. Marcum.

the total fraction of time spent above the threshold level. If these are to represent the false alarm probability in the case of continuous waveform display then the average time, τ_{av}, spent above the threshold on each crossing must be equal to the correlation interval, τ_c. The number of false alarm samples which are observed in time T is given by:

$$N_s = T \cdot \text{FAP}/\tau_s \tag{2.5}$$

This is the probability that a single sample contains a false alarm multiplied by the total number of samples in T taken τ_s apart. In the continuous case, the number of false alarms is given by the total time spent above the threshold divided by the average time spent above the threshold on each crossing, that is:

$$N_c = T \cdot \text{FAP}/\tau_{av}. \tag{2.6}$$

The sampling interval, τ_s, in equation (2.5) would normally be made equal to the correlation time, when it is clear that use of equations (2.1) and (2.3) assumes that the average time spent above the threshold on each crossing is the correlation interval. In general, this will not be correct.

3. False Alarm Probabilities in the Continuous Case

The false alarm probability in the continuous case is the probability with which a peak of the noise alone waveform appears above the threshold. This is the probability that the noise waveform goes through the threshold with positive slope. We shall proceed initially with the general derivation of the probability for the positive going threshold crossing and then apply it to two cases. One case is that in which the amplitude probability density function is Gaussian and the other is the case where the relevant probability density function is a Rayleigh distribution. The latter case results from envelope demodulation of a narrow band noise process such as occurs in many sonar and radar receivers.

The derivation of the probability of a positive going threshold crossing has been performed by both S. O. Rice [3] (zero crossings only) and M. Kac [4] at least for random waves possessing a Gaussian probability density function. We shall begin the analysis in a rather different way than that of S. O. Rice. Any waveform or function may be represented by the infinite Taylor series expansion:

$$f(t + \tau) = f(t) + f'(t)\tau + f''(t)\frac{\tau^2}{2!} + f'''(t)\frac{\tau^3}{3!} + \cdots \tag{3.1}$$

where $f'(t)$ is the first time derivative of $f(t)$ at t, $f''(t)$ the second derivative, and so on.

Thus, the statistics of a random variable, $f(t)$, may be represented by the probability density function of infinite order of $f(t)$ and its derivatives at one point in time. The requirement that $f(t)$ be infinitely differentiable imposes some limitations on the spectrum of $f(t)$.

If $f(t)$ is to go through the threshold level k with positive slope in the time between t and $t + \Delta t$ then

$$f(t) < k$$
$$f(t + \Delta t) > k. \tag{3.2}$$

For short time intervals, Δt, we may approximate $f(t)$ by:

$$f(t + \Delta \tau) = f(t) + f'(t)\Delta \tau \tag{3.3}$$

Thus:

$$f(t) < k < f(t) + f'(t)\Delta t \tag{3.4}$$

and:

$$k > f(t) > k - f'(t)\Delta t. \tag{3.5}$$

Clearly $f'(t)$ must be positive since $f(t)$ is less than the threshold k. The probability of these values of $f(t)$ and $f'(t)$ occurring is:

$$\int_0^\infty d(x') \int_{k-f'(t)\Delta t}^{k} d(x)P(x, x't). \tag{3.6}$$

$P(x, x', t)$ is the second order joint probability density function for the amplitude and first time derivative of $f(t)$. Providing that $f'(t)\Delta t$ is small, we may assume that $P(x, x', t)$ is constant for small variations of x. The resulting formula after performing the inner integration is:

$$\int_0^\infty dx' \cdot x' \Delta t P(k, x', t). \tag{3.7}$$

Thus the probability with which a random wave crosses the threshold k, with positive slope in time Δt is:

$$\Delta t \int_0^\infty x'P(k, x', t)\, dx'. \tag{3.8}$$

By a similar argument, we may show that the probability with which a random wave crosses the threshold with negative slope in time Δt is:

$$-\Delta t \int_{-\infty}^0 x'p(k, x', t)\, dx'. \tag{3.9}$$

From equation (3.8) we may see that the number of positive-going threshold crossings which occur in time T may be obtained by integration from 0 to T. Thus, the number of false alarms in time T is:

$$Nc = T \int_0^\infty x'P(k, x', t)\, dx'. \tag{3.10}$$

Substitution in equation (2.6) will enable us to estimate the average time τ_{av} spent above the threshold since this is the total fraction of time spent above the threshold divided by the number of threshold crossings in the same time.

$$\tau_{av} = \frac{\displaystyle\int_k^\infty P(x)\, dx}{\displaystyle\int_0^\infty x'P(k, x', t)\, dx}. \tag{3.11}$$

The above formulae are all that are needed to examine the behaviour of the model for false alarms in the continuous case. Two cases are examined. Firstly, the false alarm probabilities for low-pass filter noise which has a Gaussian probability distribution and secondly the false alarm probabilities for envelope demodulated high frequency narrow band noise.

(a) Low-pass filtered Gaussian noise

We may represent the output of the filter when its input is white noise by the Fourier transform relation:

$$f(t) = Re \int_0^\infty c(w) \exp[-jwt]\, dw. \qquad (3.12)$$

The joint probability density function for n normally distributed variables is:

$$P(x_1, x_2, \ldots, x_n) = (2\pi)^{-n/2} |M|^{-1/2} \exp\left[-\tfrac{1}{2}\frac{1}{|M|}\sum_{j=1}^{n}\sum^{n} x_i x_j M_{ij}\right] \qquad (3.13)$$

where M is the determinant of the covariance matrix and M_{ij} is the cofactor of the element μ_{ij} in the matrix. Without loss of generality, it may be assumed that the mean level of $f(t)$ is zero. Then, the probability density function for the amplitude of $f(t)$ is:

$$P(x) = \frac{1}{\sqrt{2\pi}} |M|^{-1/2} \exp\left[-\frac{M_{11} x^2}{2|M|}\right]. \qquad (3.14)$$

The covariance matrix consists of only one element μ_{11}:

$$\mu_{11} = \overline{f(t)^2} = \sigma^2 = \phi(0) \qquad (3.15)$$

where $\phi(0)$ is the value of the autocorrelation function at $t = 0$.

Thus $P(x)$ becomes the well-known form:

$$P(x) = \frac{1}{\sqrt{2\pi\sigma^2}} \exp\left[-\frac{x^2}{2\sigma^2}\right]. \qquad (3.16)$$

The derivation of $P(x, x')$ is only slightly more difficult. The covariance matrix M is:

$$M = \begin{bmatrix} \mu_{11} & \mu_{12} \\ \mu_{21} & \mu_{22} \end{bmatrix}$$

where the order of the elements is (x, x'). Then:

$$\begin{aligned} \mu_{11} &= \overline{x^2} \\ \mu_{12} = \mu_{21} &= \overline{xx'} = \phi'(0) \\ \mu_{22} &= \overline{x'^2} = -\phi''(0). \end{aligned} \qquad (3.17)$$

Since $\mu_{12} = \mu_{21}$ is zero for zero time difference between x and x', the two variables are independent so that we may write:

$$P(x, x', t) = \frac{[-\phi(0)\phi''(0)]^{-1/2}}{2\pi} \exp\left[-\frac{x^2}{2\phi(0)} + \frac{x'^2}{2\phi''(0)}\right] \tag{3.18}$$

which is independent also of t.

Using this result in equations (3.8) and (3.10) to obtain the probability of a positive going threshold crossing $P(+c)$ and the number of such crossing in time T we obtain

$$P(+c) = \Delta t \int_0^\infty x' \exp\left[\frac{x'^2}{2\phi''(0)}\right] dx' \frac{[-\phi(0)\phi''(0)]^{1/2}}{2\pi} \exp\left[-\frac{k^2}{2\phi(0)}\right]$$

$$= \frac{\Delta t}{2\pi}\left[-\frac{\phi''(0)}{\phi(0)}\right]^{1/2} \exp\left[-\frac{k^2}{2\phi(0)}\right]. \tag{3.19}$$

This is essentially the result obtained by S. O. Rice and M. Kac.

$$Nc = \frac{T}{2\pi}\left[-\frac{\phi''(0)}{\phi(0)}\right]^{1/2} \exp\left[-\frac{k^2}{2\phi(0)}\right]. \tag{3.20}$$

Let us take the case of a low pass filter whose frequency response is Gaussian:

$$F(\omega) = \exp[-\omega^2/\omega_0^2].$$

We may assume that the spectral intensity of the input white noise is such that the output power spectral density, $\Phi(\omega)$, is:

$$\Phi(\omega) = \exp[-2\omega^2/\omega_0^2]. \tag{3.21}$$

By suitable application of Fourier transforms, we may find the auto-correlation function:

$$\phi(\tau) = \int_0^\infty 2 \exp[-2\omega^2/\omega_0^2] \cos(\omega\tau) \, d\omega$$

$$= \omega_0 \sqrt{\frac{\pi}{2}} \exp\left[\frac{\omega_0^2\tau^2}{8}\right] \tag{3.22}$$

$$\therefore \quad \phi(0) = \omega_0 \sqrt{\frac{\pi}{2}}.$$

By differentiating $\phi(\tau)$ with respect to τ we may find $\phi'(\tau)$ and $\phi''(\tau)$:

$$\phi'(\tau) = -\omega_0^3 \sqrt{\frac{\pi}{2}}\frac{\tau}{4} \exp\left[-\frac{\omega_0^2\tau^2}{8}\right] \tag{3.23}$$

$$\phi''(\tau) = \omega_0^3 \sqrt{\frac{\pi}{2}}\left[\frac{\omega_0^2\tau^2}{16} - \frac{1}{4}\right] \exp\left[-\frac{\omega_0^2\tau^2}{8}\right]. \tag{3.24}$$

Thus

$$\left[-\frac{\phi''(0)}{\phi(0)} \right]^{1/2} = \frac{\omega_0}{2}$$

and

$$P(+c) = \frac{\omega_0 \Delta t}{4\pi} \exp \left[-\frac{k^2}{2\sigma^2} \right] \tag{3.25}$$

$$Nc = \frac{\omega_0 T}{4\pi} \exp \left[-\frac{k^2}{2\sigma^2} \right]. \tag{3.26}$$

The fraction of time T, spent above the threshold by such a random wave is:

$$\int_k^\infty \frac{T}{\sqrt{2\pi\sigma^2}} \exp \left[-\frac{x^2}{2\sigma^2} \right] dx = \frac{T}{2} \left[1 - \text{erf}\,(k/\sqrt{2\sigma^2}) \right]. \tag{3.27}$$

Thus the average time spent above the threshold on each individual crossing is:

$$\tau_{av} = \frac{1 - \text{erf}\,(k/\sqrt{2\sigma^2})}{\dfrac{\omega_0}{2\pi} \exp \left[-\dfrac{k^2}{2\sigma^2} \right]}. \tag{3.28}$$

Therefore

$$\omega_0 \tau_{av} = 2\pi \left[1 - \text{erf}\,(k/\sqrt{2\sigma^2}) \right] \exp \left[k^2/2\sigma^2 \right].$$

It can be easily shown that both the probability of a positive going threshold crossing and the average time spent above the threshold on each crossing are dependent on the shape of the low pass filter and, in particular, on the way in which the response is terminated. As an example of this, the corresponding formulae for the probability of crossings, the number of crossings and the average time above the threshold in the case of an ideal low-pass filter cutting-off at ω_0 radians per second are:

$$P(+c) = \frac{\omega_0 \Delta t}{2\pi} \exp \left[-\frac{k^2}{2\sigma^2} \right] \tag{3.29}$$

$$Nc = \frac{\omega_0 T}{2\pi} \exp \left[-\frac{k^2}{2\sigma^2} \right] \tag{3.30}$$

$$\omega_0 \tau_{av} = \pi \left[1 - \text{erf}\,(k/\sqrt{2\sigma^2}) \right] \exp \left[k^2/2\sigma^2 \right]. \tag{3.31}$$

When the threshold k is greater than about seven times the r.m.s. noise amplitude, that is when:

$$\frac{k}{\sqrt{2\sigma^2}} > 5$$

the error function, erf(x), may be approximated by:

$$\text{erf}(x) = \text{sign}(x) - \frac{\exp[-x^2]}{x\sqrt{\pi}}, \quad x > 5. \tag{3.32}$$

Thus, at high threshold levels, the average time above the threshold is asymptotic to the following expression:

$$\omega_0 \tau_{av} = \frac{2}{k}\sqrt{2\pi\sigma^2}, \quad k > 7\sigma^2 \tag{3.33}$$

for the Gaussian low pass filter.

Other results may be obtained from the equations derived above. One such result relates to the phase of narrow-band noise with a Gaussian probability density function. It may be shown that in comparison with a sine wave whose frequency is at the centre of the narrow-band noise spectrum, the phase of the noise waveform advances continuously with time. In other words, the effective frequency, as measured by the zero crossing rate of the noise is higher than the centre frequency of noise spectrum. This result is shown in M. I. Skolnik [5], D. Middleton [6] and others.

It is now necessary to interpret the above results in relation to the false alarm probabilities in sonar and radar systems. Equation (3.19) is the probability of a positive going crossing of the threshold k, and as explained earlier is also the probability of a false alarm in a continuous waveform (rather than sampled) detection system. It is clear from this equation that the probability of a false alarm is directly proportional to the factor Δt which is the length of time over which the waveform is observed to make a single decision. We shall call this time the observation interval. Early work in radar and sonar systems suggested the total observation time could be quantized into individual elements each of which was approximately equal to the correlation time of the process. In this way, little information was lost and yet decisions based on the state of adjacent time elements were reasonably independent. Work done during the Second World War and subsequently reported by Lawson and Uhlenbeck has shown that the [7] product of I.F. bandwidth and pulse length should be approximately unity. In particular for a Gaussian shaped pulse and its matched filter, it can be shown

$$\omega_0 T = 0.886\pi = 2.78$$

where

$$F(\omega) = \exp[-\omega^2/\omega_0^2]$$

$$f(t) = \exp\left[\frac{\omega_0^2 t^2}{2}\right]$$

and T relates to the half-power widths of $f(t)$. It will be seen when we discuss the situation which arises when signals are present in the noise that there is some doubt concerning the correctness of this estimate of observation time and that $\omega_0 T = 2\pi = 6.28$ may be more reasonable.

As a comparison of the false alarm probabilities, a graph has been plotted in Fig. 2 of false alarm probability versus the square of the normalized threshold level k/σ for three cases. These all relate to Gaussian distributed

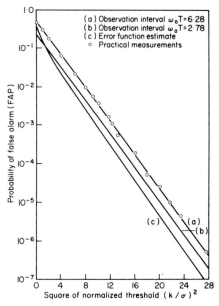

FIG. 2. False alarm probability versus square of the normalized threshold level for Gaussian low-pass filtered noise with a Gaussian power spectral density.

low pass filtered noise. One is the error function estimate of false alarm probability as is given by equation (2.1). The other two are of estimates, based on the probability of a positive going threshold crossing approach for the two observation intervals given above. These may be considered approximately as lower and upper limits. The choice of a satisfactory or optimum observation interval is a difficult problem and relates also to the detection performance of the system when signals are present. It will, therefore, be considered later.

Another factor of interest is the behaviour of τ_{av} with variation of the threshold. This is not immediately obvious because of the presence in equation (3.28) of the error function. A plot of τ_{av} versus threshold level has, therefore, been included in Fig. 3 for the low-pass filtered wave.

From the two graphs it can be seen that at reasonably high thresholds, the false alarm probability calculated by the positive-going threshold crossing method is several times higher than the error function approach at the same threshold level even using the smaller observation interval. As a result, the average time spent above the threshold is several times smaller than the observation interval or the correlation time of the noise process.

Clearly the two theories predict different false alarm behaviour with variation in the threshold. Proof that the probability of a positive going threshold crossing is the correct approach for continuous rather than

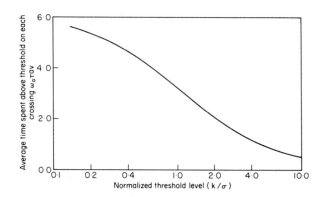

FIG. 3. Variation of average normalized time spent above the threshold on each crossing with normalized threshold level. Gaussian low pass filtered noise with a Gaussian power spectral density.

sampled wave detection has been obtained from practical measurements. The results are plotted in Fig. 2 as circles. They correspond exactly with the behaviour expected using the positive threshold crossing approach. It is interesting to note that at a zero threshold level or mean level, the expected number of crossings per second is:

$$Nc = \frac{\omega_0}{4\pi} = \frac{f_0}{2}.$$

This seems to be a novel way of estimating the cut-off frequency of a Gaussian filter. In general the relationship between Nc and the cut-off frequency will be dependent upon the shape of the filter response but is usually calculable.

(b) Narrow band noise

In most radar and sonar systems signals and noise are received by an aerial array and then passed through R.F. and I.F. amplifiers before detection. The noise at the output of the I.F. amplifier has a high centre frequency and is a relatively narrow band. We shall be concerned here with envelope

demodulation so that the envelope of the noise or signal and noise waveform is used for decision making. It is well known that the probability
density function for the envelope is Rayleigh distributed, that is:

$$P(R) = \frac{R}{\sigma^2} \exp \left[\frac{-R}{2\sigma^2} \right].$$ (3.34)

The probability of a false alarm in a continuous wave decision system is the
probability of a positive-going threshold crossing in a given observation
interval. We shall derive this result together with one for the average time
spent above the threshold on each crossing in a similar manner to that
used in the last section. A comparison between the new estimate of false
alarm probability and the old one, which is based upon the area of the
probability density function $P(R)$ above the threshold can then be made.

First of all, we must find the probability density function for the envelope
and the time derivative of the envelope and evaluate:

$$\int_0^\infty R'P(R, R', t) \, dR'.$$ (3.35)

The required probability density function $P(R, R', t)$, may be obtained
from the joint fourth order Gaussian distribution for two independent
random variables and their time derivatives.

$$P(R, R') = \frac{R}{\sqrt{N2\pi\phi'(0)|M|^{1/2}}} \exp \left[\frac{R^2}{2\phi(0)} \right] \exp \left[-\frac{R'^2 \phi(0)}{2|M|^{1/2}} \right]$$ (3.36)

where

$$|M| = [\phi(0)\phi''(0) - \phi'(0)^2]^2.$$

The probability of a positive going threshold crossing by the envelope R,
in time, Δt, is given by

$$P(+c) = \Delta t \int_0^\infty R'P(R, R') \, dR'$$

$$= \Delta t \int_0^\infty R' \exp \left[-\frac{R'^2 \phi(0)}{2|M|^{1/2}} \right] dR' \frac{R}{\sqrt{2\pi\phi(0)|M|^{1/2}}} \exp \left[-\frac{R^2}{2\phi(0)} \right]$$

$$= \frac{R|M|^{1/4} \Delta t}{\phi(0) \sqrt{2\pi\phi(0)}} \exp \left[-\frac{R^2}{2\phi(0)} \right].$$ (3.37)

At some threshold level, k, this becomes

$$P(+c) = \frac{k}{\sqrt{\phi(0)}} \left[\frac{M}{\phi(0)^4} \right]^{1/4} \exp \left[\frac{-k^2}{2\phi(0)} \right] \frac{\Delta t}{\sqrt{2\pi}}.$$ (3.38)

For filters whose frequency response is symmetrical about the I.F. centre frequency, the first time derivative of the autocorrelation function of the envelope function at zero time is zero, that is

$$\phi'(0) = 0.$$

Therefore:

$$|M| = \phi(0)^2 \phi''(0)^2$$

and

$$P(+c) = \frac{k}{\sqrt{\phi(0)}} \left[\frac{|\phi''(0)|}{\phi(0)} \right]^{1/2} \exp\left[-\frac{k^2}{2\phi(0)} \right] \cdot \frac{\Delta t}{\sqrt{2\pi}}. \qquad (3.39)$$

We will only be considering filters which are symmetrical.

In a sampled-wave decision system, the false alarm probability is just the area of the probability density function of the envelope which lies above the threshold level, k.

$$\text{FAP} = \int_k^\infty \frac{R}{\phi(0)} \exp\left[-\frac{R^2}{2\phi(0)} \right] dR$$

$$= \exp\left[-\frac{k^2}{2\phi(0)} \right]. \qquad (3.40)$$

The average time spent above the threshold is just the quotient of these two:

$$\tau_{\text{av}} = \frac{\int_k^\infty P(R)\, dR \cdot \Delta T}{\int_0^\infty R'P(R, R')\, dR'\Delta T}$$

$$= \frac{\sqrt{2\pi\phi(0)}}{k} \cdot \left[\frac{\phi(0)}{|\phi''(0)|} \right]^{1/2}. \qquad (3.41)$$

For a filter with a Gaussian shaped response around a centre frequency of fc, the above relationships become:

$$F(\omega) = \exp\left[-\left(\frac{\omega - \omega c}{\omega_0} \right)^2 \right], \quad \omega_c \gg \omega_0$$

$$P(+c) = \frac{\Delta T}{\sqrt{2\pi}} \frac{k}{\sqrt{\phi(0)}} \frac{\omega_0}{2} \exp\left[-\frac{k^2}{2\phi(0)} \right], \quad \phi(0) = \sigma^2 \qquad (3.42)$$

$$\omega_0 \tau_{\text{av}} = \frac{2\sqrt{2\pi\phi(0)}}{k}.$$

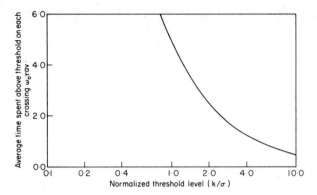

FIG. 4. Variation of average normalized time spent above the threshold on each crossing with normalized threshold level. Envelope demodulated narrow band noise with a Gaussian power spectral density.

The behaviour of the average time spent above the threshold on each crossing, τ_{av}, with variation in the threshold level, k, is shown in Fig. 4. Both estimates of false alarm probability are plotted versus the threshold level in Fig. 5. The probability of a positive going threshold crossing

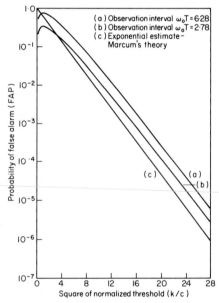

FIG. 5. Variation of false alarm probability with the square of the normalized threshold level for envelope demodulated narrow band noise. Gaussian shaped power spectral density.

approach to false alarm probability has been plotted for two different observation intervals which are:

Upper curve: $\Delta T = 2\pi/\omega_0$

Lower curve: $\Delta T = 2.78\omega_0$.

These are the same observation intervals used in the Gaussian distributed low pass filtered case. It is interesting to note that the theory predicts discrepancies similar to those found in the last section. The same comments and problems apply here as there.

4. Probability of Detection

In the last section it was shown that a method, which has been in use for about twenty years, for estimating the false alarm performance of radar or sonar systems could be in error in some circumstances. These conditions apply to continuous wave decision systems such as P.P.I. displays, "A" scan-displays and some computerized decision systems. The ideas presented in the previous section must now be applied to the situation where signals as well as noise exist in the incoming wave. A comparison between the old and new theories may then be made.

Application of the concept of probabilities of positive going threshold crossings in the signal-present situation allows a new model to be formulated which corresponds closely with the physical situation.

A start will be made to this section by a brief consideration of the models which are used at present for determining the probability of detecting a signal when it is present. In the case of low pass filtered noise, a voltage equal to the peak level of the filtered signal is impressed on the noise probability density function as a simple time-invariant change of mean. The probability of detection is then the area of the noise and signal distribution lying above the threshold level in a similar way to the calculation of false alarm probabilities. There seems to be little justification for using this approach for either the sampled or continuous wave decision cases except that the results it gives are reasonably accurate. Its most serious drawback is that no account of the time-varying nature of the signal is taken. This is possibly important since the signal shape after passage through a matched filter is the same as the autocorrelation function of the output noise from the filter (providing that white noise was fed to the filter input).

A similar technique is used when the signals and noise have been passed through a narrow-band high-frequency filter. The signal is considered as a sine-wave whose amplitude corresponds with the peak signal at the output of the matched filter. The probability distribution of the envelope is modified in shape by the presence of a constant-amplitude sine wave. The new distribution is:

$$P(R) = \frac{R}{\phi(0)} I_0 \left[\frac{RA}{\phi(0)}\right] \exp\left[-\frac{(R^2 + A^2)}{2\phi(0)}\right]$$

where A is the signal amplitude and I_0 is the modified Beszel function zero order and first kind. The area of the new distribution above the threshold level is then used as the probability of detection for the signal amplitude, A. Again the time-varying of the signal has been ignored and there is no justification for the model except that it yields reasonable results.

In deriving the new model based upon the theory of the probability of a positive-going threshold crossing, it is first necessary to consider the effects of the signal on the joint probability density function for the amplitude and first time derivative of the noise process. The noise waveform can be expressed as a Taylor series expansion:

$$f(t + \tau) = f(t) + f'(t)\tau + f''(t)\frac{\tau^2}{2!} + \cdots + f^{(n)}(t)\frac{\tau^n}{n!} + \cdots$$

and the signal waveform can be similarly expressed:

$$s(t + \tau) = s(t) + s'(t)\tau + s''(t)\frac{\tau^2}{2!} + \cdots + s^{(n)}(t)\frac{\tau^n}{n!} + \cdots.$$

The amplifiers and filters which precede the decision making device may be considered linear so the principle of superposition holds. In other words the signal and noise waveform may be expressed as:

$$s(t + \tau) + f(t + \tau) = (s(t) + f(t)) + (s'(t) + f'(t))\tau + (s''(t) + f''(t))\frac{\tau^2}{2!} + \cdots.$$

$$(4.1)$$

In all the following studies, *a priori* knowledge of the presence of the signal is essential because the question to be answered is "what is the probability of detection when a signal is known to be present". In other words the signal $s(t)$ is always present for some value of t but the noise waveform is random. We are really considering an ensemble of random waveforms in each of which a signal is present. By this argument, it can be seen that the signal waveform adds time varying means, to both the distribution of the random wave amplitude and its first time derivative. This applies to both the low-pass and band-pass situations except in the band-pass case four variables are involved instead of two. In this section, we will only be concerned with the low-pass filter case since computation difficulties are associated with the band-pass case.

The probabilities of positive-going threshold crossings when a signal is present may still be calculated using the relationship given in equation (3.8) but it must be realized that the probability is now time dependent. The joint second order probability distribution for the signal and noise amplitude and its first time derivative may be written down immediately from equation (3.18):

$$P(x, x'/s(t), t) = \frac{[-\phi(0)\phi''(0)]^{-1/2}}{2\pi} \exp\left\{-\frac{[x - s(t)]^2}{2\phi(0)}\right.$$

$$\left. + \frac{[x' - s'(t)]^2}{2\phi''(0)}\right\} \tag{4.2}$$

where the values of the autocorrelation function and its derivatives at zero time relate to the noise process only. By substituting $P(x, x'/s(t), t)$ into equation (3.8) the probability of a positive going threshold crossing in a small time increment Δt at time t may be obtained:

$$P(+c, t)\Delta t = \Delta t \int_0^\infty x' \exp\left\{\frac{[x' - s'(t)]^2}{2\phi''(0)}\right\} dx' \cdot \frac{[-\phi(0)\phi''(0)]^{-1/2}}{2\pi}$$

$$x \exp\left\{-\frac{[k - s(t)]^2}{2\phi(0)}\right\} \tag{4.3}$$

Writing $y = x' - s'(t)$, the integral may be shown to be:

$$I = \int_{-s'(t)}^\infty [y + s'(t)] \exp\left[\frac{y^2}{2\phi''(0)}\right] dy.$$

The second time derivative of the autocorrelation function at time $t = 0$ is always negative because the autocorrelation function must always have its maximum value at $t = 0$ and is an even function of time. After integration I becomes:

$$I = -\phi''(0) \exp\left[\frac{s'(t)^2}{2\phi''(0)}\right] + s'(t) \cdot \frac{[-\pi\phi''(0)]^{1/2}}{2}$$

$$x \{1 + \text{erf} [s'(t)/\sqrt{-2\phi''(0)}]\}.$$

Thus, the probability density of a positive threshold crossing at time t is:

$$P(+c, t) = I \frac{[-\phi(0)\phi''(0)]^{-1/2}}{2\pi} \exp\left\{-\frac{[k - s(t)]^2}{2\phi(0)}\right\}. \tag{4.4}$$

This is clearly a complicated function and it is, therefore, difficult to visualize its behaviour with respect to variations in threshold level, k: time; signal shape or peak signal to rms noise ratio. In order that some idea of this behaviour can be obtained, some numerical analysis has been performed for various values of the threshold level, peak signal to rms noise ratio and time. The low-pass filter chosen is the same as that considered in the last section, in other words one with a frequency response:

$$F(\omega) = \exp\left[-\omega^2/\omega_0^2\right].$$

The signal and the low pass filter responses are matched. Thus the signal shape at the filter output is the same as the autocorrelation function of the noise and is

$$s(t) = A \exp\left[-\frac{\omega_0^2 t^2}{8}\right].$$

Some results of the numerical analysis are plotted in Figs 6 and 7 and at the bottom of each graph a plot of the signal shape is included for con-

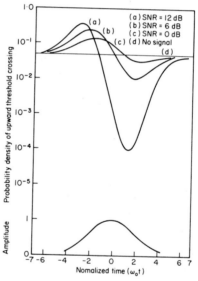

FIG. 6. Curves of probability density of upward threshold crossing for various signal-to-noise ratios against normalized time. Threshold level = 0.990.

venience. As can be seen the general characteristics of the curves are similar and could perhaps have been predicted from a physical considera-tion of the process. The rising edge of the pulse lifts the probability density function for the amplitudes and its first derivative so that positive-going threshold crossings become more probable. After a crossing has most likely occurred, the probability of a crossing drops sharply for high signal to noise ratios and less steeply for lower signal to noise ratios. The dip in the probability density of positive crossing corresponds with the falling

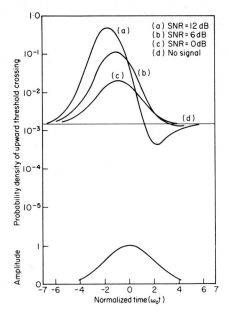

FIG. 7. Curves of probability density of upward threshold crossing for various signal-to-noise ratios against normalized time. Threshold level = 2.890.

edge of the signal pulse when negative going threshold crossings will be more likely.

The probability of detection may now be defined as the probability that a pulse appears above the threshold when the signal is present. This is the probability with which a positive going threshold crossing occurs when the signal is present. In other words, it is the area under a selected portion of the curve of the probability density of a positive-going threshold crossing:

$$\text{POD} = \int_{T_1}^{T_2} \int_0^\infty x' P[x, x', s(t), s'(t)]\, dx'\, dt. \qquad (4.5)$$

One problem relates to the limits T_1, T_2 which are needed for the time integral. As yet no general solution has been found.

Numerical analysis has permitted some solution for the probability of detection for certain signal to noise ratios and threshold levels by an arbitrary choice of an observation interval (T_1, T_2). A sample of the results of calculations for the probability of detection is shown in Fig. 8 as a Receiver Operating Characteristic (ROC) together with a similar curve calculated using Marcum's theory for the same threshold and the results of some experimental trials. The experimental results seem to favour the proposed model. In any event, the performance difference between this

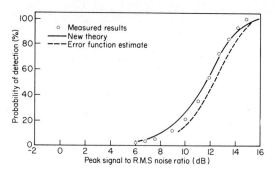

FIG. 8. A plot of probability of detection versus peak signal-to-noise ratio. False alarm probability = 10^{-4}. Threshold level = 4.20.

and Marcum's model is only 0.7 dB and this falls to approximately 0.3 dB if different thresholds are chosen for each model to equate the predicted false alarm probabilities. Apart from the greater probabilities of occurrence of pulses above the threshold when a signal is present, one way in which a further decision concerning whether the pulses are false alarms or signal could be based on the time spent above the threshold by the waveform. First of all it will be useful to find the average time spent above the threshold, τ_{av} when a signal is present. This can be found, as in the last section, by dividing the total time spent above the threshold in T by the number of threshold crossings which have occurred in T. Since signals will only occasionally be present in the received radar or sonar waveform, it is necessary to restrict ourselves to intervals when the signals are present. In other words, we must find the fraction of time of the observation interval which on average is spent above the threshold. This does not take into account the probability of occurrence of a crossing. The probability density function for the amplitude of signal and noise at one point in time is:

$$P(x/s(t), t) = \frac{1}{\sqrt{2\pi\phi(0)}} \exp\left\{-\frac{[x - s(t)]^2}{2\phi(0)}\right\}. \tag{4.6}$$

Thus, the fraction of ensembles for which the signal and noise waveform are above the threshold at one point in time (t) and for one particular signal amplitude, is:

$$P(t) = \frac{1}{\sqrt{2\pi\phi(0)}} \int_k^\infty \exp\left\{-\frac{[x - s(t)]^2}{2\phi(0)}\right\} dx.$$

Over the observation interval from T_1 to T_2 which covers the important variations in signal amplitude, the fraction of time spent above the threshold is:

$$\tau_{av} = \frac{\dfrac{1}{\sqrt{2\pi\phi(0)}} \displaystyle\int_{T_1}^{T_2} \int_k^{\infty} \exp\left\{-\frac{[x-s(t)]^2}{2\phi(0)}\right\} dx}{\left(\begin{array}{l} \displaystyle\int_{T_1}^{T_2}\int_0^{\infty} x' \exp\left\{\frac{[x'-s'(t)]^2}{2\phi''(0)}\right\} dx' \cdot \frac{[-\phi(0)\phi''(0)]^{-1/2}}{2\pi} \\[18pt] \times \exp\left\{-\frac{[k-s(t)]^2}{2\phi(0)}\right\} dt \end{array}\right)} \tag{4.7}$$

These are difficult integrals and numerical analysis has enabled some solutions to be obtained for an arbitrary observation interval, T_1 to T_2. These times were chosen to include the majority of the time variant behaviour of the probability density for a positive-going threshold crossing when signal and noise are present. The graphs in Fig. 9 are plotted as the average time spent above the threshold versus threshold level with signal

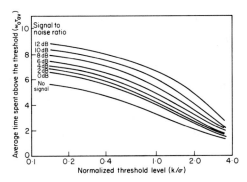

FIG. 9. Normalized average time spent above the threshold plotted against normalized threshold level with peak signal to R.M.S. noise ratio as a parameter.

amplitude as a parameter. The signal shape again is Gaussian and signals and noise are both considered to have passed through the correct matched filter. If we wish to use the average time spent above the threshold on each crossing as a detection parameter, we must clearly choose a threshold or observation interval or both which maximizes the "difference" between signals and noise and noise alone. The word "difference" is used here in a vague sense since there are many parameters which may be used to characterize signals from noise.

One such crude parameter could be the ratio of τ_{av} when a signal is present (τ_{avs}) to τ_{av} when noise alone is present (τ_{avn}) at the input to the threshold device. A curve of this ratio has been plotted in Fig. 10 with signal amplitude as a parameter. It may be important to maximize this ratio if an integration or detection scheme is to be based on the idea of choosing signal from noise on the basis of the time each spends above the threshold.† It is interesting to note that the threshold level which maximizes the ratio of times spent above the threshold with signal and noise

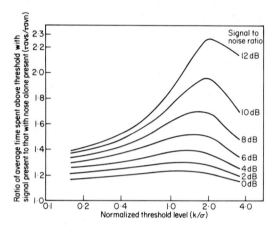

FIG. 10. Ratio of average time spent above the threshold with signal and noise. Present to that with noise alone present as a function of the normalized threshold level and the peak signal to R.M.S. noise ratio.

and then noise alone for small signal to noise ratios is approximately equal to the r.m.s. noise amplitude. This is very similar to Harrington's [8] for the optimum first threshold in a binary integration scheme.

The use of two thresholds, one in amplitude, the other in time, for the calculations of probability of detection and false alarm probability, require knowledge of the distributions for time spent above a threshold by signal and noise. These distributions seem particularly difficult to obtain even using the approximation due to Rice [9] which has been extended to handle the signal present case. These distributions have been obtained (space precludes further details) and the probability of detection obtained for various time and amplitude thresholds for a fixed false alarm rate are insignificantly different to those results obtained for the same false alarm

† A better parameter to optimize would probably be the effective "signal to noise" ratio:

$$\frac{\tau_{avs} - \tau_{avn}}{\sigma_s \sigma_n}$$

where $\sigma_s \sigma_n$ is the geometrical mean of the variances of the relevant crossing time distributions. From other work, it is known that the variances are approximately proportionate to the means and it is this fact that makes τ_{avs}/τ_{avn} a useful measure.

rate using only an amplitude threshold. This is not a surprising result since the amplitude and time responses of such a decision scheme intuitively seem to have high dependence.

5. Observation Interval

One other parameter which appears worth optimising especially in the case of single pulse detection, is that of probability of detection for a fixed alarm rate. It is possible to optimize this parameter because by choosing a threshold level we may select an observation interval to give the required false alarm rate. Thus, for a given false alarm probability, there are an infinite number of satisfactory threshold levels and observation intervals. It is inconceivable that they will all yield the same probability of detection when a signal is present. The results of some computer programs to determine an optimum observation interval are shown in Fig. 11. These show that for smaller false alarm probability a higher threshold level and a

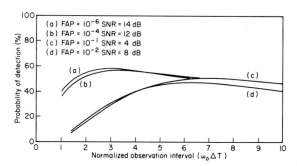

FIG. 11. Variation of probability of detection with observation interval.

smaller observation interval are required. With false alarm probabilities of the order of 10^{-6} the optimum observation interval turns out to approximately satisfy the relationship which suggests that the product of IF bandwidth and pulse length (in this case, the observation interval) is nearly equal to unity:

$$\omega_0(T_2 - T_1) = \omega_0 T_{opt} = 2.5.$$

For much higher false alarm rates, for example, 10^{-1}, the optimum observation interval is much larger—typically double the above figure. A further point of interest is that once the optimum observation time has been achieved, increases in the observation interval do not seriously affect the probability of detection even though the threshold level is being raised to maintain the false alarm probability.

Even though mathematically an optimum observation exists, I do not

believe that this parameter is one whose optimization has any real significance. It seems that a meaningful definition of the observation interval
may be made in the following way. In observing the noise contents of,
say 100 resolution cells, an acceptable probability of a false is 10^{-5}
(corresponding to 10^{-7} per cell), what is the target signal to noise ratio
which gives a probability of a threshold crossing of, say, 0.9 assuming
that the target lies in one of the 100 cells. Thus, the observation interval
corresponds to 100 resolution cells and the extra target strength that is
required over the "optimum" case represents our lack of prior knowledge
of the target position [2]. It seems possible that this concept is also suitable for use with digital processors where the cell structure exists by virtue
of range and bearing gates.

6. Conclusions

The various methods of calculating the outcomes of the detection process
in radar and sonar systems have been reviewed. A proposed model which
involves the time variant behaviour of the signal has been used to find the
receiver operating characteristics under the assumption of base band signal
and noise. The method has involved the concept of decisions based on
positive-going threshold crossing. It is clear that this technique may be
extended to deal with the usual narrow-band case although the calculations
appear to be significantly more difficult. These may be only of academic
interest since, in the case considered the performances of the detection
process are only slightly different by the two methods.

ACKNOWLEDGEMENTS

I would like to acknowledge the facilities placed at my disposal by Prof. D. G. Tucker
and Prof. J. W. R. Griffiths and the many discussions with Dr. D. C. Cooper.

REFERENCES

1. Marcum, J. I. (1960). RAND Research Memo RM-754 Dec. 1947 published in
   IRE Trans. **IT-6**.
2. Helstrom, C. W. (1968). "Statistical Theory of Signal Detection", 2nd Edition,
   Pergamon Press.
3. Rice, S. O. (1945). Mathematical analysis of random noise. *B.S.T.J.* **24**, Sec. 3.
4. Kac, M. (1943). On the distribution of values of trignometic sums with linearly
   independent frequencies. *Amer. Jour. Maths* **65**, 609–615.
5. Skolnik, M. I. (1962). "Introduction to Radar Systems" McGraw-Hill New York.
6. Middleton, D. (1960). "Introduction to Statistical Communication Theory",
   McGraw-Hill, New York.
7. Lawson, J. L. and Uhlenbeck, G. E. (1965). "Threshold Signals", Dover 1965
   (originally M.I.T. Press Radiation Lab. series Vol. 24).
8. Harrington, J. V. (1955). An analysis of the detection of repeated signals in noise
   by binary integration. *IRE Trans.* **IT-1**, 1–9.
9. Rice, S. O. (1958). Distribution of the duration of fades in radio transmission
   Gaussian noise model. *B.S.T.J.* **37**, 581–635.

DISCUSSION

R. R. Rojas: It seems that in the specific formulation of the decision process there is really no problem if it is handled correctly. If one obtains a sampling representation, and goes through the mathematics with the correct statistics, everything falls out.
Answer: In the theoretical formulation using maximum likelihood sampling is an inherent part of the process. As far as I know, there is no decision theory which deals with the strictly continuous waveform situation.

R. R. Rojas: Referring to the diagram showing a linearized version of the noise crossing the threshold is the interval ΔT. Once the number ΔT is fixed, it is also necessary to consider the probability that the process ΔT reaches the vicinity of the threshold. This is a conditional situation.
Answer: It will help to go on a little further because ΔT turns out to be important. In the normal situation, it represents the time to scan the surveillance area in which we wish to determine the presence of one or more targets. In the noise only case, the integral (3.7) has no time dependence and ΔT is just a multiplier. The integral represents a first order average and bandwidth restrictions are not therefore important.

S. S. Wolf: From the forest of hands that shot up when you finished your talk, you must realize that you are in the position of puncturing a very sacred cow. (Laughter.) The threshold crossing theory is due to S. O. Rice.
Answer: I believe that M. Katz started the threshold work. References are given in the paper.

P. M. Schulthiess: I am not clear about the difference between your definition of False Alarms and Marcum's.
Answer: We observe the output of a threshold device at all times, in which no sampling takes place. As soon as the output of the threshold device changes state from "0" to "1", a false alarm (in the case of noise only) is declared present.

P. M. Schultheiss: Your definition of a false alarm relates to the probability of at least one crossing of the threshold.
Answer: Yes.

H. Cox: Because of the similarity between your curves and Marcum's, one suspects that there is an adjustment that one can make to bring these into perfect alignment. I was wondering if this was related to the number of independent samples used in calculating the False Alarm probability using standard techniques.
Answer: That may be the case.

N. C. Prior: In performing similar experiments, I found it necessary, as you did, to distinguish between False Alarm Probability and False Alarm Rate. The latter is, I think, more important operationally. It is curious that the sequential detector allows one some freedom to select a False Alarm Rate for a given False Alarm Probability, so that is one interesting difference between that and the simple single threshold detector.

S. S. Wolf: Marcum's work relates to a signal processing scheme that is utterly unlike the P.P.I. scan, it is a batch processing scheme. The data is observed for a length of time, T, and at the end of that time the output of the matched filter is compared with a threshold. What you are doing is certainly a better and more realistic model of the process which is normally used in radar. Thus, one would expect the results from simulations of actual radar systems to be more nearly like your calculations than Marcum's. I think the essential distinction is between a batch processing scheme and an on-going/continuous time processing scheme.

Robust Many-input Signal Detectors

S. S. WOLFF and F. J. M. SULLIVAN

The Johns Hopkins University, Baltimore, Maryland, U.S.A.

1. Introduction

Robust procedures are appropriate for signal detection problems in which
the probability distribution of the contaminating noise is more or less
unknown in advance and, moreover, may change with time rapidly enough
to render adaptive procedures infeasible or otherwise unattractive. The
design of a robust procedure begins with the choice of a class C of noise
distributions within which the distributions actually to be encountered are
virtually certain to be contained; then a detector is designed to perform as
well as possible uniformly over the chosen class. It is in the choice of the
class C that the signal analyst incorporates all his prior knowledge of the
noise; he may, for example, know that he will encounter the data only after
it has been strongly AGC'd, so that whatever the noise distribution, its
variance is known and constant. Or, he may know from prior measurement
that the distribution function of the noise lies within δ of some known
distribution (e.g., the Gaussian). Thus robust procedures are available to
span the entire spectrum of the analyst's prior information; entering with
almost complete ignorance of the noise, one expects relatively weak
detectors to result; with increasing prior information, the procedures
perform better within C, but at the same time become more sensitive to
deviations from C. At the farthest extreme are procedures designed
explicitly for the Gaussian distribution, with their infamous sensitivity
to "slight" deviations from normality.

In this paper we treat the asymptotic theory of robust detectors—i.e.,
the limiting case of weak signals and large sample sizes. The basic structure
of the problem is as follows: Let $X_1, X_2, \ldots, X_n$ be independent c-
dimensional random variables where $X_t = (X_t^1 X_t^2 \ldots X_t^c)^T$; let X_t have
(c-variate) probability density function $f(\theta s_t, x)$ where f is drawn from a
class C, s is a c-dimensional signal and θ is a scalar signal-to-noise ratio. We
have to test the null hypothesis

$$H_0: \theta = 0$$

against the alternative

$$H_1: \theta > 0$$

in the limit as $n \to \infty$. It is important to note that although in the absence of signal ($\theta = 0$) the data X_t are identically distributed, this assumption is by no means necessary; since we shall design a robust detector, variations of f with time (as long as it remains within C) do not affect the detector structure—only the actual performance attained. Of course, if f were known to vary in some particular fashion, this knowledge could be incorporated into the detector design; the necessary modifications to the development that follows will be apparent.

We shall treat four ways in which the observations X contain the signal s; the four signal models are

M1: $f(\theta s, x) = f(x - \theta s)$

M2: $f(\theta s, x) = \exp\left(\theta \sum_1^c s^j\right) f(x^1 e^{\theta s^1}, \ldots, x^c e^{\theta s^c})$

M3: $f(\theta s, x) = \int_0^{2\pi} f(x - \theta s \sin\phi) \, d\phi/2\pi$

M4: $f(\theta s, x) = \int f(x - \theta s) \, dH(s)$.

M1 and M2 are the usual location and scale problems, M3 is a model for incoherent (envelope) detection, and M4 is the noise-in-noise problem of classical detection theory.

Only detection statistics of the form

$$U_n = (1/n) \sum_1^n W_t(X_t) \tag{1.1}$$

will be considered; since for any given density f the likelihood ratio test has the form (1.1), the assumption is not restrictive.

Section 2 of this paper treats the asymptotic normality of statistics like U_n. In Section 3 we introduce the formal notion of a maximin robust detector and prove a saddlepoint theorem; Section 4 contains the variational arguments that lead to an Euler equation for the robust detector U_n^* and in Section 5 the Euler equation is completed by including the constraints implied by several choices of the class C, and some solutions are obtained.

2. Asymptotic Normality

To test H_0 against H_1, the statistic U_n is computed and compared to a threshold Θ, where Θ is chosen so that the false alarm probability of the detector is some acceptably small number α. But for any $\alpha > 0$ and fixed signal-to-noise ratio $\theta > 0$, the power of any consistent test statistic tends to unity as $n \to \infty$; to be able to make meaningful asymptotic comparisons among detectors, it is customary in analysis, as $n \to \infty$, to let $\theta = \theta_n$ tend to zero in such a way that the increasing sample size is exactly counterbalanced by the weakening signal, so that the power of the test tends to a limit β, $\alpha < \beta < 1$. Thus while the asymptotic normality of a statistic may be easy to establish under H_0: $\theta = 0$, the proof under H_1 is invariably more difficult because the distribution of the statistic changes as θ varies with the sample size n. Our difficulties are compounded because we have not just a single sequence of alternatives $f(\theta_n s, x)$ but a whole family of sequences corresponding to the class C of densities f. The development may be streamlined and all the difficulties collected in one place for disposal once and for all by using the following result due to LeCam [3] (see also Hajek and Sidak [1]).

Lemma (LeCam). Let L_n be the likelihood ratio for $\theta = 0$ against $\theta = \theta_n$, and let $\{T_n\}$ be an arbitrary sequence of statistics. If under H_0 $(\log L_n, T_n)$ is asymptotically bivariate normal with parameters $(\mu_1, \mu_2, \sigma_1^2, \sigma_2^2, \sigma_{12})$ with $\mu_1 = -\frac{1}{2}\sigma_1^2$, then under H_1 T_n is asymptotically normal with mean $\mu_2 + \sigma_{12}$ and variance σ_2^2.

Using Taylor expansion about $\theta = 0$, and assuming enough regularity conditions to allow interchange of θ-differentiation and x-integration, it is straightforward to write down the limiting form of $\log L_n$ and its asymptotic mean and variance under H_0; they are

$$\log L_n \to \theta_n \sum_1^n f_0'(X_t)/f_0(X_t)$$

$$E_0 \log L_n \to -(\theta_n^2/2) \sum_1^n \int f_0'^2(x_t)/f_0(x_t)\, dx_t \qquad (2.1)$$

$$\mathrm{var}_0 \log L_n \to \theta_n^2 \sum_1^n \int f_0'^2(x_t)/f_0(x_t)\, dx_t,$$

where $f_0'(x_t)$ means the partial derivative of $f(\theta s_t, x_t)$ with respect to θ, at $\theta = 0$. While equation (2.1) suffices for models M1 and M2, in M3 and M4

f_0' is identically zero, so that higher order terms in the Taylor expansion are required; equation (2.1) becomes

$$\log L_n \rightarrow (\theta_n^2/2) \sum_1^n f_0''(X_t)/f_0(X_t)$$

$$E_0 \log L_n \rightarrow -(\theta_n^4/8) \sum_1^n \int f_0''^2(x_t)/f_0(x_t)\, dx_t \tag{2.2}$$

$$\text{var}_0 \log L_n \rightarrow (\theta_n^4/4) \sum_1^n \int f_0''^2(x_t)/f_0(x_t)\, dx_t.$$

Immediately we see that in order for $\log L_n$ to be asymptotically non-degenerate we must take $\theta_n = kn^{-1/2}$ for M1 and M2, and $\theta_n = kn^{-1/4}$ for M3 and M4, with $k > 0$ but otherwise arbitrary. When θ_n is chosen as described, with $k = 1$, we define

$$I_f \overset{\Delta}{=} \lim_{n \to \infty} \text{var}_0 \log L_n, \tag{2.3}$$

assume it finite, and, stretching customary usage somewhat, term it "the Fisher information in f about s". We note that I_f depends on s through the chain rule for partial derivatives.

In order for any form of central limit theorem to apply to $\log L_n$, it is clearly necessary that $|s_t^j| < \infty$ for $1 \leqslant j \leqslant c$ and all t, and that s_t not die away too quickly as t increases; it is sufficient to assume that

$$0 < \lim_{n \to \infty} (1/n) \sum_{t=1}^n \sum_{j=1}^c s_t^{j^2} < \infty;$$

i.e., that the signal have finite and non-zero average power in each component. We shall further assume that the terms in the limiting form of $\log L_n$ have finite fourth central moments; this overly restrictive assumption could be considerably weakened, but it seems to be met in many practical cases and it allows the immediate assertion of asymptotic normality for $\log L_n$ by Liapounov's form of the central limit theorem.

Turning our attention now to U_n, we note first that by a law of large numbers U_n is asymptotically degenerate, so we shall prove asymptotic normality of $n^{1/2} U_n = n^{-1/2} \Sigma_1^n W_t(X_t)$. Without further ado we shall assume finite fourth central moments under H_0 of the summands W_t, so that asymptotic normality of $n^{1/2} U_n$ follows immediately. In order that $(\log L_n, n^{1/2} U_n)$ be asymptotically bivariate normal, it is necessary and sufficient that every linear combination

$$\lambda \log L_n + (1 - \lambda)n^{1/2} U_n, \quad 0 \leqslant \lambda \leqslant 1,$$

be asymptotically normal. But this result follows directly from the finite fourth moment assumptions.

Hence LeCam's lemma applies, and the asymptotic power of U_n is a monotone increasing function of

$$G \overset{\Delta}{=} \frac{(\mu_2 + \sigma_{12} - \mu_2)^2}{\sigma_2^2}$$

$$= [\text{cov}_0(\log L_n, n^{1/2} U_n)]^2 / \text{var}_0 n^{1/2} U_n$$

which we term a "processing gain".

3. Maximin Theory

We write $G = G(f, U)$ to emphasize its dual dependence on the statistic U and the noise distribution f in effect; for any fixed U, G is a function of $f \in C$, conversely, for a given f, G depends on the statistic U chosen by the signal analyst. We define the maximin robust detection statistic U^* by the property that it achieves the $\max_{U} \min_{f \in C} G(f, U)$; that is, its worst performance over C is as good as possible.

The theory to follow depends heavily on convexity properties; it is necessary at the outset to assume that C is a convex set of density functions: if $f_1 \in C, f_2 \in C$ then $\lambda f_1 + (1 - \lambda)f_2 \in C$ for all $\lambda \in [0, 1]$. In probabilistic terms, this means that every mixture of densities in C belongs to C. We now establish that G is convex in f.

Lemma 3.1. Let $U_n = (1/n) \Sigma_1^n W_t(X_t)$, and assume that $E_0 W_t = 0$. Then $G(f, U)$ is convex in $f \in C$. The proof follows directly from a lemma of Huber [2].

Lemma 3.2 (Huber). Let $v_1 > 0, v_2 > 0, 0 \leqslant \alpha \leqslant 1$. Then

$$\frac{(\alpha u_1 + (1 - \alpha)u_2)^2}{\alpha v_1 + (1 - \alpha)v_2} \leqslant \alpha \frac{u_1^2}{v_1} + (1 - \alpha) \frac{u_2^2}{v_2}.$$

We can now prove the main theorem of the section.

Theorem 3.1. Let C be a convex set of c-variate density functions such that every f in C is absolutely continuous with finite Fisher information about the signal s (*vide* equation (2.3)).

(i) If there is a density $f^* \in C$ such that $I(f^*) \leqslant I(f)$ for all $f \in C$, then (f^*, U^*) is a saddlepoint for G:

$$G(f^*, U) \leqslant G(f^*, U^*) = I(f^*) \leqslant G(f, U^*) \tag{3.1}$$

where

$$U^* \triangleq - \sum_1^\infty f_0^{*\prime}(X_t)/f_0^*(X_t) \tag{3.2a}$$

for M1 and M2 or

$$U^* \triangleq - \sum_1^\infty f_0^{*\prime\prime}(X_t)/f_0^*(X_t) \tag{3.2b}$$

for M3 and M4.

(ii) If (f^*, U) is a saddlepoint for G, then $I(f^*) \leqslant I(\hat{f})$ for all $\hat{f} \in C$ and U is f^*-equivalent to U^*.

Proof. $G(f^*, U) \leqslant G(f^*, U^*) = I(f^*)$ by the Schwarz inequality. Now for and $\hat{f} \in C$, define

$$J(\delta) \triangleq G((1 - \delta)f^* + \delta\hat{f}, U^*) \tag{3.3}$$

$$K(\delta) \triangleq I((1 - \delta)f^* + \delta\hat{f}). \tag{3.4}$$

Clearly $J(0) = K(0)$, and a little calculation shows that also $J'(0) = K'(0)$. But since f^* minimizes $I(f)$ over C, $K'(0)$ [and hence also $J'(0)$] is non-negative. Then since J is convex in δ by Lemma 3.1, part (i) of the theorem is proved. To prove (ii), we note that if (f^*, U) is a saddlepoint for G then $G(f^*, U^*) \leqslant G(f^*, U)$ by assumption, while $G(f^*, U) \leqslant G(f^*, U^*)$ by the Schwarz inequality as in the proof of part (i); hence $G(f^*, U) = g(f^*, U^*)$ and so $U = kU^*$ for some constant k wherever f^* is non-zero by the condition for equality in the Schwarz inequality. That $I(f^*) \leqslant I(\hat{f})$ follows from the relations

$$I(f^*) = G(f^*, U^*) \leqslant G(\hat{f}, U^*) \leqslant G(\hat{f}, \hat{U}) = I(\hat{f}).$$

The content of the saddlepoint theorem is that the maximin robust detector is the optimum detector for the density in C with minimum Fisher information.

4. Toward the Euler Equation for U^*

The general procedure for finding the density f^* with minimum Fisher information in C is to write down the Fisher information functional (2.3) in the form appropriate to the signal model M_i under consideration, append the constraints imposed by the class C from which f^* may be drawn, together with undetermined multipliers, compute the first variation of the composite functional and (assuming a minimizing density to exist) set it to zero. Here we shall only compute the information functional in its various forms, and defer to Section 5 inclusion of the constraints and some solutions.

From equation (2.3) and the variance equations in (2.1) and (2.2), we see that I_f is a sum of non-negative terms; hence minimization may be done termwise. This simplification is of course a consequence of the assumed time-wise independence of the observed data X_t. Let

$$I_1(f) \triangleq \int f_0'^2(x)/f_0(x) \, dx \tag{4.1a}$$

for M1 and M2, or

$$I_1(f) \triangleq \int f_0''^2(x)/f_0(x) \, dx \tag{4.1b}$$

for M3 and M4, and let

$$K_1(\delta) \triangleq I_1((1 - \delta)f^* + \delta f). \tag{4.2}$$

Then $K_1'(0)$ can be written

$$K_1'(0) = \int (gu^2 - 2g'u) \, dx \tag{4.3}$$

where $u = f_0^*{}'/f_0^*$ (M1, M2) or $u = f_0^*{}''/f_0^*$ (M3, M4) and $g \in C$.

For M1, g is of the form $g(x - \theta s)$, so that after c integrations by parts (and assuming the endpoint contributions vanish)

$$K_1'(0) = \int g(x)(u^2 + 2 \sum_1^c s^j u_j) \, dx \tag{4.4}$$

where u_j is the partial derivative of u with respect to its jth argument. Similarly for M2, after c integrations by parts

$$K_1'(0) = \int g(x)(u^2 + 2 \sum_1^c s^j x^j u_j) \, dx. \tag{4.5}$$

For M3 and M4, $2c$ partial integrations are required; the resulting forms are;

For M3: $\quad K_1'(0) = \int g(x)\left(u^2 - \sum_1^c \sum s^j s^k u_{jk}\right) dx \tag{4.6}$

For M4: $K_1'(0) = \int g(x)\left(u^2 - \sum\limits_1^c\sum u_{jk}R_{jk}\right) dx$ $\qquad$ (4.7)

where R is the covariance matrix of the signal: $R_{jk} = \int s^j s^k\, dH(s)$.

Some insight into the character of the Euler equations that will eventually result can be had by a few easy transformations on the integrands in equations (4.4–4.7). First, since M2 is a classical scale problem, the transformation $z = \log|x|$ renders equation (4.5) formally identical to equation (4.4). Now in equation (4.4) for M1, let $g(x)\,dx = h(y)\,dy$ and $u(x) = v(y)$, where $y = Qx$ and Q is a $(c \times c)$ orthogonal matrix having its first row collinear with the signal s. Then, with $\|s\|^2 = \sum_1^c s^{j^2}$, 4.4 becomes

$$K_1'(0) = \int h(y)(v^2 + 2\|s\|v_1)\, dy. \qquad (4.8)$$

The same rotation works for M3, so that equation (4.6) becomes

$$K_1'(0) = \int h(y)(v^2 - \|s\|^2 v_{11})\, dy. \qquad (4.9)$$

Again, for M4, let the rows of Q include the eigenvectors of the covariance matrix R; then

$$K_1'(0) = \int h(y)(v^2 - 2\sum_1^c \lambda^j v_{jj})\, dy, \qquad (4.10)$$

where $\{\lambda^j\}$ is the set of eigenvectors of R, some of which are zero if R is not of full rank. Equations (4.8) and (4.9) display the essential dependence of the Fisher information functional on only one particular linear combination of the components of the observation X_t in M1 and M3, or their logarithms in M2; that is, the information functional does not uniquely determine the minimum information density f^* on the $(c-1)$-dimensional subspace orthogonal to the signal s, although the maximin robust detector is of course unique.

In a practical realization, the rotation Q would appear to be useful only if all components of the signal have the same time behaviour: $s_t^j = A^j s_t$, for otherwise Q must be time-varying.

The theoretical utility of the rotation Q is real if and only if the constraint of f^* separate in the new coordinates, for then the Euler equation is an ordinary differential equation. If, as is unfortunately often the case, the constraints are naturally phrased in terms of the original coordinates X and do not separate in the Y's, the simplification afforded by Q is only superficial—it merely converts one nonlinear partial differential equation into another.

The distinction between M3 and M4 is that in M3 the "covariance matrix" has rank exactly one: $R_{jk} = s^j s^k$; thus M3 and M4 are equivalent only in univariate problems. In the multivariate case, the rotation Q for M4 is expansion of the observed data X along the principal axes of the signal stochastic process s; this expansion may be useful in practice if s is (at least) wide-sense stationary, for then Q is time independent.

5. Noise Models and Solutions

In this section we complete the Euler equation by including the constraints imposed by some choices of the class C of admissible noise densities, and display a few solutions.

It is interesting that the classical parametric Gaussian detectors are, in some circumstances, maximin robust. In particular, since it is known that the Gaussian density has minimum Fisher information for location among all densities with the same covariance, it follows that for the simple location problem M1 the Gaussian detectors (i.e., those based on correlation or matched filtering) are maximin robust within the class

$$C1 = \{f: \int x^j x^k f(x) \, dx = K_{jk}, \quad \text{given } j, k = 1, \ldots, c\}.$$

Frequently from physical considerations, or on the basis of prior measurement (or simple faith), it is felt that the noise ought to be at least approximately Gaussian; nevertheless it seems prudent to allow for some deviation from normality in greater or less degree. Huber [2] uses two models for classes of such "contaminated" distributions; they include contamination of any underlying distribution, not just the Gaussian. (In what follows, we use upper case letters to refer to distribution functions, lower case for the densities.)

Huber's first model is

$$C2 = \{F: \sup_t |F(t) - B(t)| \le \epsilon\},$$

where ϵ is a given number in the interval $[0, 1)$ and B is a given distribution with finite Fisher information. When the universally necessary constraint $\int f \, dx = 1$ is included, the Euler equation for problem M1 in the univariate case is

$$u^2 + 2su' = \lambda \tag{5.1}$$

(λ is a Lagrange multiplier) as long as the solution to equation (5.1) meets the constraints of C2; on the boundary, no Euler equation is necessary, since $F = B \pm \epsilon$, or $f = b$. The solution to equation (5.1) and the subsequent equation $u = f'/f$ for the minimizing density is

$$f^*(x) = f^*(-x) = \begin{cases} b(x_1)(\cos cx / \cos cx_1)^{2s} & 0 \le |x| \le x_1 \\ b(x) & x_1 \le |x| \le x_2 \\ b(x_2) \exp(-d|x - x_2|) & |x| \ge x_2 \end{cases} \quad (5.2)$$

with the parameters c, d, x_1 and x_2 determined from the requirements that f^* and $f^{*'}$ be continuous, $F^*(x_1) = B(x_1) - \epsilon$ and $F^*(x_2) = B(x_2) - \epsilon$. In the case that B is the unit normal distribution, the maximin robust detector is a peculiar kind of soft limiter:

$$W^*(x) = \begin{cases} 2cs \tan cx & |x| \le x_1 \\ x & x_1 \le |x| \le x_2 \\ d \operatorname{sgn} x & |x| \ge x_2. \end{cases} \quad (5.3)$$

In the multivariate version of this problem, the solution (5.2) continues to apply, providing the constraint C2 applies to the component of F collinear with the signal s (*vide* 4.8); otherwise, the full nonlinear partial differential equation must be attacked numerically.

Huber's second class of contaminated distributions is

$$C3 = \{F: F = (1 - \epsilon)B + \epsilon H\}$$

where ϵ is given in $[0, 1]$, B is a given distribution with finite Fisher information and H is an arbitrary distribution. In the univariate case of M1, the Euler equation is (5.1) as long as $h(x)$ is positive; on the boundary, $h = 0$ and $f^* = (1 - \epsilon)b$. The maximum robust detector is again a soft limiter:

$$W^*(x) = \begin{cases} b'(x)/b(x) & |x| \le x_1 \\ k \operatorname{sgn} x & |x| \ge x_1. \end{cases} \quad (5.4)$$

When the constraint class C3 is applied to M3, additional complications occur; in the univariate case the Euler equation is

$$u^2 - s^2 u'' = \lambda \quad (5.5)$$

as long as h is positive. We might expect a solution like (5.4) to apply, with b' replaced by b''; indeed, Martin and Schwartz [4] conjectured this when b is the Gaussian density. Unfortunately, such a solution cannot satisfy the conditions that f^* and its first two derivatives be continuous; hence we must look further. If the additional constraints $\int xf \, dx = 0$ and $\int x^2 f \, dx = \sigma^2$ are appended, the Euler equation has as a solution a parabolic cylinder function, leading to a robust detector (in the Gaussian case) that is quadratic in x for small x and elsewhere linear in $|x|$.

6. Commentary

The bright hope that the methods of robustness might be able to accommodate all degrees of prior knowledge in the design of a detector is considerably dimmed by the present unavailability of explicit analytic solutions in all but the simplest univariate cases. While further progress along this line may be expected, the prospects of analytic solutions to multivariate problems are dimmer by far (although one has been found [5]), and those of numerical solutions not much brighter if the dimensionality exceeds two or three. What then is the utility of this theory if the practising signal analyst cannot apply it in most cases?

In the first place, we need not dismiss univariate solutions out of hand; as we have said before, they apply if the constraints are themselves one-dimensional. And since in the detection of a particular signal the observation X is scrutinized along a single line in c-space, it may not always be altogether unreasonable to phrase the constraints on the detection procedure in terms of the same single coordinate. Indeed, even M4, which appears irreducibly multidimensional, may in application be of quite small dimensionality; e.g., if it models the detection of a single random signal source by a c-element array.

Second, the theory may be used as a tool of analysis. Any proposed detection statistic W may be substituted in the left-hand side of the Euler equation; the resulting expression must arise from the constraints. Thus the class C for which W is the maximin robust detector may be inferred, and W judged (at least qualitatively) by the appropriateness of C to the signal analysis problem at hand.

REFERENCES

1. Hájek, J. and Šidák, Z. (1967). "Theory of Rank Tests", Academic Press, New York and London.
2. Huber, P. J. (1964). Robust estimation of location. *Ann. Math. Stat.* **35**, 73–101.
3. LeCam, L. (1960). Locally asymptotically normal families of distributions *U. of Calif. Publ. in Stat.* **3**, 37–98.
4. Martin, R. D. and Schwartz, S. C. (1971). Robust detection of a known signal in nearly Gaussian noise. *IEEE Trans.* **IT-17** No. 1, 50–56.
5. Sullivan, F. J. M., VandeLinde, V. D. and Wolff, S. S. (1972). "Densities with Minimum Fisher Information for Location" (in preparation).

Signal Processing in Reverberant Environments

PAUL H. MOOSE

Naval Undersea Research and Development Center, San Diego, California, U.S.A.

1. Introduction

Detection and measurement with active sonar is limited by self-noise, ambient noise, and reverberation noise. Reverberation noise is normally assumed to arise as the result of many signals reflected or scattered from inhomogeneities in the medium and from the boundaries due to the transmitted acoustic pulse. Signals scattered by the boundaries of the sea are caused by roughness or undulations on the boundary. Volume reverberation —signals reflected from within the sea—is a result of the scattering of energy by fish, zooplankton, phytoplankton, particles of inorganic matter, and perhaps thermal or chemical inconsistencies. Both forms of reverberation, however, have properties that set it apart from the other forms of noise. From the point of view of signal detection the two most important are the reverberation noise properties that are functionally related to the transmission waveform and the power of the reverberation that is directly proportional to the energy in the transmission.

These two properties make signal processing in reverberant environments markedly distinct from that in which only nonsignal-related noise is dealt with. In a way, the first property appears very favourable. That is, since the interference characteristics are related to the transmission waveform, it should be possible to adjust the waveform and/or the receiver in a way that will reduce this interference. The second property reinforces the importance of this hypothesis, since it means that it is impossible to simply overpower reverberation noise by increasing the transmitted energy.

The study of optimum signal processing and signal design for this environment depends on accurate mathematical models to characterize the signal and noise processes. The problem of characterizing reverberation is compounded by an incomplete understanding of the random processes at work in the sea that cause the reverberation to vary in space and time. This paper considers reverberation as a random time-varying linear function of the transmitted wave. Although this formulation is appropriate only for weak

413

scattering phenomena, it is sufficiently general to characterize most practical sonar environments and a broad class of transmit waveforms and signal processing techniques.

2. Mathematical Preliminaries

An active sonar transmits a known signal waveform $\chi(t)$ of length T_1 one or more times.

It is desired that target echos are recognized in the return during reception intervals, which commence immediately following each transmit interval and continue until an echo has arrived from a target located at the maximum possible range.

The signal transmission detection problem is considered and the signals are modelled on the interval (O, T_2) as

$$\chi(t) = Re[x(t) \, e^{2\pi j f_0 t}]; \quad t \in [O, T_1] \tag{1}$$

$$\zeta(t) = Re[z(t) \, e^{2\pi j f_0 t}]; \quad t \in [T_1, T_2], \tag{2}$$

where $x(t)$ and $z(t)$ are the complex modulation envelopes of the transmit and receive waveforms, respectively, f_0 is the carrier frequency, and additive noise

$$z(t) = s(t) + n_a(t) + n_r(t); \quad t \in [T_1, T_2] \tag{3}$$

due to ambient $n_a(t)$ and reverberation $n_r(t)$ is assumed to interfere with the ability to detect the target echo $s(t)$. It will be convenient to consider signals with normalized energy

$$x(t) = E_x^{1/2} \bar{x}(t); \quad E_x = \int_0^{T_1} |x(t)|^2 \, dt \tag{4}$$

$$s(t) = E_s^{1/2} \bar{s}(t); \quad E_s = \int_{T_1}^{T_2} |s(t)|^2 \, dt \tag{5}$$

If $\bar{s}(t)$ is a known signal waveform, except for an assumed uniformly distributed random phase ϕ, $\bar{s}(t) = |\bar{s}(t)| \, e^{j\phi}$, $\phi: U[-\pi, \pi]$, then a test for detection based on the likelihood ratio amounts to the computation of

$$L_0 = \int_{T_1}^{T_2} z(t) q^*(t) \, dt \tag{6}$$

and comparison of $|L_0|$ to a threshold. If it exceeds the threshold, an echo is declared present, if it does not, noise alone is assumed to have caused the observed process [1]. In equation (6) the optimum reference function $q(t)$ is a solution of the integral equation

$$\bar{s}(t) = \int_{T_1}^{T_2} K(t, u)q(u)\, du \tag{7}$$

and the optimality of the procedure depends on the fact that the noise processes are jointly Gaussian random processes with covariance function $K(t, u)$. In the event $K(t, u)$ is not sufficiently well known or sufficiently stable, the expected echo signal can be chosen for a reference and the magnitude of the

$$L_m = \int_{T_1}^{T_2} z(t)\bar{s}^*(t)\, dt, \tag{8}$$

test statistics compared to a threshold.

Equation (8) is commonly referred to as the matched filter and the processing as matched filter detection. For either type of processor the detection performance monotonically increases with the detection index d^2 defined by

$$d^2 = \frac{E^2\,[L\colon \text{signal present}]}{\text{Var}\,[L\colon \text{signal absent}]}, \tag{9}$$

which can easily be found for the test statistics defined above as

$$d_0^2 = E_s \int_{T_1}^{T_2} \bar{s}(t)q^*(t)\, dt \tag{10}$$

and

$$d_m^2 = \frac{E_s}{\displaystyle\int_{T_1}^{T_2}\int_{T_1}^{T_2} \bar{s}^*(t)K(t, u)\bar{s}(u)\, dt\, du}, \tag{11}$$

respectively.

3. Noise Models

The generally well known results presented above clearly indicate the importance of the noise covariance function in the development of processor designs and performance predictions. For the purposes of this discussion assume that the ambient and reverberant noise processes are independent and that the ambient is white with spectral density N_0. If the reverberation is due to weak scattering it may be represented as a linear time-varying functional of the transmit wave. Thus,

$$K(t, u) = N_0 \delta(t - u) + K_r(t, u) \tag{12}$$

$$K_r(t, u) = E_x \int_{t-T_1}^{t} \int_{s-T_1}^{s} K_h(t, u; t', t'') \bar{x}(t - t') \bar{x}^*(u - t'') \, dt' \, dt'', \tag{13}$$

where $K_h(t, u; t', t'')$ is the four-dimensional correlation functional of the stochastic time-varying linear filter $h(t, t')$. Consider the Fourier transform of this filter on the time-varying parameter t

$$r(f, t') = \int_{-\infty}^{\infty} h(t, t') \, e^{-2\pi j f t} \, dt \tag{14}$$

since

$$n_r(t) = \int_{t-T_1}^{t} h(t, t') x(t - t') \, dt' \tag{15}$$

then it is also representable in terms of $r(f, t')$ as

$$n_r(t) = \int_{-\infty}^{\infty} \int_{t-T_1}^{t} r(f, t') x(t - t') \, e^{2\pi j f t} \, df \, dt' \tag{16}$$

From equation (16) $r(f, t')$ can be interpreted as a stochastic reflectance function which describes the density of scattering at a range delay time t' and doppler shift frequency f. It is clear that the time-varying nature of the impulse response $h(t, t')$ is created by scattering at non-zero doppler velocities, since when only zero doppler scattering is present

$$r(f, t') = r(t') \delta(f) \langle \overset{F}{=} \rangle h(t, t') = r(t') \tag{17}$$

The four-dimensional correlation function $\sigma(f, v; t', t'')$ of the stochastic reflectance function is related by two-dimensional Fourier transform to $K_h(t, u; t', t'')$, i.e.,

$$\sigma(f, v; t', t'') = \int_{-\infty}^{\infty} \int_{-\infty}^{\infty} K_h(t, u; t', t'') \, e^{-2\pi j (ft - vu)} \, dt \, du \tag{18}$$

Therefore, specification of the second order statistics of either $r(f, t')$ or $h(t, t')$ is sufficient, along with equation (13) to determine $K_r(t, u)$.

Assumptions which lead to a reduction in dimension and physical interpretation follow.

(a) Stationary reverberation process

The reverberation process $n_r(t)$ can be covariance stationary, i.e., $K_r(t, u) = K_r(t - u)$, only if the covariance function of the random time-varying impulse response $h(t, t')$ is doubly stationary, i.e., $K_h(t, u; t', t'') = K_h(t - u; t' - t'')$. If this is true, then

$$K_r(t - u) = E_x \int_{(t-u)-T_1}^{(t-u)+T_1} K_h(t - u; t' - t'')R_x\left[(t - u) - (t' - t'')\right]$$

$$\times \, d(t' - t''), \tag{19}$$

where

$$R_x(\tau) = \int_0^{T_1} \bar{x}(t)\bar{x}^*(t + \tau) \, dt,$$

which is the auto-correlation function of $\bar{x}(t)$.

The reverberation spectrum, $S_r(f) = F[K_r(\tau)]$, is similarly related according to

$$S_r(f) = E_x \int_{-\infty}^{\infty} S_h(f - f'; f')S_x(f') \, df', \tag{20}$$

where

$$S_h(v, f) = \int_{-\infty}^{\infty} \int_{-\infty}^{\infty} K_h(t - u; t' - t'') \, e^{-2\pi j[v(t-u)+f(t'-t'')]}$$

$$\times \, d(t - u) \, d(t' - t'')$$

which is the double transform of the doubly stationary covariance kernal K_h, and $S_x(f)$ is the power spectrum for $\bar{x}(t)$. If the random impulse response is not time-varying (all scatterers at zero relative velocity), the covariance kernal K_h depends only on $(t' - t'')$ and equation (20) simplifies to

$$S_r(f) = E_x S_h(f)S_x(f), \tag{21}$$

i.e., the product of the statistical transfer function and the power spectrum of the transmit wave.

A physical interpretation of the conditions implied by stationary scattering can be obtained when the form of the stochastic reflectance function's covariance kernal, $\sigma(f, v; t', t'')$ is considered. From equation (18)

$$\sigma(f, v; t', t'') = \left[\int K_h(t - u; t' - t'') e^{-2\pi j f(t - u)} \, d(t - u) \right]$$

$$\times \, \delta(f - v) = \sigma(f, t' - t'')\delta(f - v) \tag{22}$$

This implies scattering with different doppler shifts must be uncorrelated but its magnitude may vary with doppler frequency whereas scattering from different ranges may be correlated but the correlation may only vary with range separation and its magnitude may not vary with range.† Strict adherence to these requirements seems clearly inconsistent with present knowledge of sea reverberation. Partial relief, at least for short duration signals, can be obtained when quasi-stationary reverberation is defined. Quasi-stationary reverberation allows slow changes in $K_h(t - u; t', t'')$ with t' (range) but for any ranges within ± a pulse length of t', $K_h(t - u; t', t'')$ must depend, to good approximation, only on the difference $t' - t''$ (see equations (13) and (19)). The shorter the signal the more likely this condition will be fulfilled. It is also possible that it may be met over some portions of the reception interval but not others. With simple inverse square spreading losses, for example, the reverberation is very nonstationary in the early part of the interval but quasi-stationary at longer delay times.

(b) *Poisson reverberation processes*
Recent authors [2, 3, 4], following earlier treatments of radar [5], have modelled sea reverberation as a superposition of stochastic perturbations

$$n_r(t) = E_x^{1/2} \sum_{i=1}^{N(t)} a_i \bar{x}(t - t_i') e^{2\pi j f_i t}, \tag{23}$$

where $N(t)$ is a Poisson random variable which governs the number of reflections that contribute to the sum at time t and the a_i are ubiquitous random coefficients which include all such factors as transducer patterns, propagation loss, and scatterer target strengths that affect the individual

---

† The normal range/velocity mapping is assumed into time delay $t' \sim 2R/c$ and frequency shift $f \sim 2f_0 V/C$. Similar arguments can be developed for more complicated models, including nonisovelocity propagation (long range), high velocity and/or accelerating bodies and wide-band signalling waveforms. The complexity of these more complete models can be important for simulations and system calculations but seems unnecessarily cumbersome for the expository nature of this paper.

signal levels. If the a_i are zero-mean statistically independent random variables, then it has been shown [6]† that the reverberation covariance function takes the form

$$K_r(t, u) = E_x \int\limits_{-\infty}^{\infty} \int\limits_{t-T_1}^{t} \sigma(f, t')\bar{x}(t - t')\bar{x}^*(u - t') \, e^{2\pi j f(t-u)} \, dt' \, df, \quad (24)$$

where $\sigma(f, t') = E[\,|a(f, t')|^2]\rho(f, t')$ is called the scattering function, and $E\,|a(f, t')|^2 \sim E\,|a_i|^2$ for scatterers with ranges and doppler shifts near (f, t') and $\rho(t', f)$ is the Poisson parameter which describes the density of scatterers near (t', f). That is,

$$P(N; f, t') = \frac{[\rho(f, t') \, \Delta f \, \Delta t]^N}{N!} \, e^{-[\rho(f, t') \, \Delta f \, \Delta t]} \quad (25)$$

gives the probability of exactly N scatterers that have doppler shifts between f and $f + \Delta f$ and range time delays between t' and $t' + \Delta t$. $\rho(f, t') \, \Delta f \, \Delta t$ is the expected number of scatterers in the area $\Delta f \, \Delta t$.

Equation (24) can also be obtained from equations (13) and (18) if

$$\sigma(f, v, t', t'') = \sigma(f, t')\delta(f - v)\delta(t' - t'') \quad (26)$$

The interpretation here is consistent with the physical model, i.e., the scattering at any range and doppler is independent of that at any other range and doppler but its magnitude may vary with both range and doppler. For example, the noise power at any time t,

$$K_r(t, t) = E_x \int\limits_{-\infty}^{\infty} \int\limits_{t-T_1}^{t} \sigma(f, t')\,|\bar{x}(t - t')|^2 \, dt' \, df \quad (27)$$

is clearly dependent on range delay time t. In the special case of $\sigma(f, t') = \sigma(f)$, i.e., scattering function independent of range delay time, the reverberation is stationary since now there is a special case of equation (22) so that

$$\sigma(f, v, t', t'') = \sigma(f, t' - t'')\delta(f - v) = \sigma(f)\delta(t' - t'')\delta(f - v), \quad (28)$$

and the reverberation spectrum may be found directly from equation (20)

$$S_r(f) = E_x \int\limits_{-\infty}^{\infty} \sigma(f - f')S_x(f') \, df' \quad (29)$$

† The deviations of References [3] and [4], respectively, lead to the same result but are somewhat different on certain technical points as is that of Reference [6].

as the convolution of the doppler distribution of the scatterers with the power spectrum of the transmitted wave. Note the basic difference in the underlying assumptions and resultant forms of equations (21) and (29), both of which describe stationary reverberation noise processes.

4. Implications of the Noise Models

When the noise is stationary or quasi-stationary, the optimum reference signal $q(t)$ has a spectrum $Q(f)$ found from equation (7) of

$$Q(f) = \frac{\bar{S}(f)}{N_0 + S_r(f)} = \frac{\bar{S}(f)}{N_0 + E_x \int_{-\infty}^{\infty} S_h(f - f'; f') S_x(f') \, df'} \tag{30}$$

where $\bar{S}(f) = F \, | \, \bar{s}(t) \, |$ is the spectrum of the anticipated echo.

Also, from equation (10) and Parseval's theorem the detection index for the optimum receiver is

$$d_0^2 = \left(\frac{E_s}{N_o} \right) \int_{-\infty}^{\infty} \frac{S_s(f) \, df}{1 + \dfrac{S_r(f)}{N_0}} \tag{31}$$

It is interesting to contrast the matched filter detection index with equation (31):

$$d_m^2 = \frac{E_s}{N_0} \frac{1}{1 + \dfrac{1}{N_0} \int_{-\infty}^{\infty} S_r(f) S_s(f) \, df} \tag{32}$$

Bounds for performance follow easily if it is noted that the denominators of equations (31) and (32) are $\geqslant 1$. Thus,

$$\frac{\dfrac{E_s}{N_0}}{1 + \dfrac{M}{N_0}} \leqslant d_m^2 \leqslant d_0^2 \leqslant \frac{E_s}{N_0}, \tag{33}$$

where $M = \text{Max} \{ E_x S_h(v, f) \text{ for all } v \text{ and } f \} = \text{Max} \{ S_r(f) \}$.

The problem of signal design is to choose $S_x(f)$ to maximize either equations (31) or (32), depending on the receiver employed. Under certain rather restrictive conditions this has proven possible. For example, when the assumptions that lead to equation (21) hold and the echo is a time-invariant filtered version of the transmit wave (the time-invariant condition is the restrictive part since it excludes doppler shifted echos) such that

$$S_s(f) = | \, G(f) \, |^2 S_x(f), \tag{34}$$

then, as was shown at the previous institute [7], equation (31) is maximized with the selection of

$$S_x(f) = \frac{\text{const} \, | \, G(f) \, | N_0^{1/2} - N_0}{S_h(f)} \tag{35}$$

for the power spectrum of the transmit wave. It is also easy to show that

$$S_x(f) = \frac{N_0 \cdot \text{const}}{S_h(f) | \, G(f) \, |^2} \tag{36}$$

maximizes d_m^2 for the same conditions. In equations (35) and (36) the constants are chosen to meet an energy constraint such that

$$\int_{-\infty}^{\infty} S_x(f) \, df = 1$$

Note that if $| \, G(f) \, | = 1$, i.e., no filtering of the transmit wave is caused by the sonar equipment, medium or target except for the pure range delay, then equations (35) and (36) yield identical results. The problem with these solutions is that they are only pertinent to the case when there is no relative motion between source receiver, target and scatterers, and experience has demonstrated that doppler separation of echos and reverberation greatly assists detection.

The model of equation (29), derivable from both the Poisson and stationary scattering models, along with a quasi-stationary restriction on range variations, is more suitable to practical sonar needs. Unfortunately closed form solutions for $S_x(f)$ for arbitrary doppler distributions have not yet been found. An approximate solution for a restricted class of $x(t)$'s can be found as follows.

Construct a periodic $\bar{x}(t)$ from $N = T_1 W$ harmonics of $1/T_1$ so that

$$S_x(f) = \sum_{-N/2}^{N/2} | \, x_n \, |^2 \delta \left(f - \frac{n}{T_1} \right) \tag{37}$$

It can be assumed that $S_s(f) = | \, G(f) \, |^2 S_x(f)$ and still allow target doppler if the frequency origin for $\sigma(f)$ is defined so that it corresponds to the desired target doppler. Note that by the energy constraint

$$\sum_{-N/2}^{N/2} | \, x_n \, |^2 = 1 \tag{38}$$

With the reverberation spectrum from equation (29) and with equation

(37) for $S_x(f)$, the optimum and matched filter detection indexes take, respectively, the forms

$$d_0^2 = \left(\frac{E_s}{N_0}\right) \sum_{-N/2}^{N/2} \left[\frac{|x_n|^2 |G_n|^2}{\left(1 + \dfrac{E_x}{N_0} \displaystyle\sum_{-N/2}^{N/2} |x_m|^2 \sigma_{n-m}\right)} \right] \tag{39}$$

and

$$d_m^2 = \frac{\left(\dfrac{E_s}{N_0}\right)}{\left[1 + \left(\dfrac{E_x}{N_0}\right) \displaystyle\sum_{-N/2}^{N/2} \displaystyle\sum_{-N/2}^{N/2} |x_n|^2 |x_m|^2 |G_n|^2 \sigma_{n-m}\right]} \tag{40}$$

where $|G_n| = |G(n/T_1)|$ and $\sigma_n = \sigma(n/T_1)$ are sample values of, respectively, the echo-filter transfer and scattering functions.

Equation (39) or equation (40), when combined with the linear equality constraint of equation (38) and the $N + 1$ inequality constraint equations

$$|x_n|^2 \geqslant 0 \text{ and } n = \frac{-N}{2}, \ldots, \frac{N}{2}, \tag{41}$$

specify a nonlinear programming problem for which a number of numerical solution techniques are well developed [8]. The optimum solution, that is, the $\{|x_n|^2\}$ that maximize equation (39), has been investigated with Rosen's gradient projection method [8]. Results of this investigation for a particular scattering function, $\sigma(f) = \sigma_0 e^{-fT_1}$, $|f| \leqslant W$, and $|G(f)| = 1$, show that the optimum signal waveform is independent of reverberation/noise ratio, and that it varies from a uniform spectrum at zero doppler shift to a signal much like a tone ping for doppler shifts large compared to the width of $\sigma(f)$. Figure 1 shows the relative distribution of energy in a signal synthesize from 13 tone pulses spaced equally over a bandwidth $2W$ for target doppler shifts of 0, $1/3W$, $2/3W$, and W. The scattering function, which would also be the reverberation spectrum of a single tone ping is superimposed on these plots with a doppler shift to indicate the separation between target and reverberation in frequency for each case.

In order to evaluate the gains to be achieved with the optimum signal waveform, plots of relative detection index d_0^2/d_{max}^2, where $d_{max}^2 = E_s/N_0$, versus doppler shift are derived for the same case of 13 signal components and the exponential scattering function, as shown in Figs 2 and 3. A 10 dB and a 0 dB reverberation-to-noise ratio is shown in, respectively, Figs 2

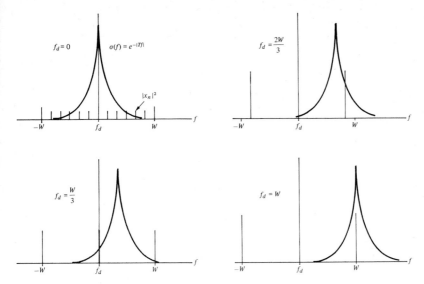

FIG. 1. Optimum signal spectrum for $N + 1 = 13$.

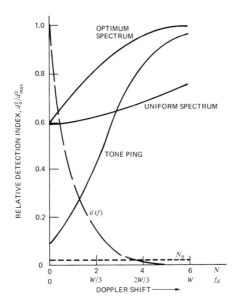

FIG. 2. Relative detection index versus target doppler shift; reverberation/noise ratio = 10 dB.

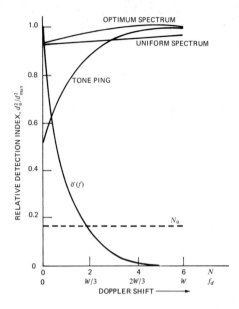

FIG. 3. Relative detection index versus target doppler shift; reverberation/noise ratio = 0 dB.

and 3. The reverberation-to-noise ratio is defined as $10 \log_{10}$ of the quantity

$$\frac{E_x}{N_0} \sum_{n=-N/2}^{N/2} \sigma_n$$

and would be the actual received reverberation-to-noise ratio for a single tone ping. In this formulation, the reverberation-to-noise ratio establishes the scale factor σ_0 on the scattering function.

Also shown in Figs 2 and 3 are the relative detection indexes for a single tone ping and a uniform spectrum signal along with the curve for the optimum signal. Several features are evident. The optimum signal spectrum has little advantage over a uniform spectrum at very low doppler shifts. The optimum signal spectrum has very little advantage over a tone ping at high doppler shifts. All three curves move toward unity for all doppler shifts at low reverberation-to-noise ratios, although the tone ping is still 3 dB below a uniform spectrum signal at zero dB reverberation-to-noise. At 10 dB reverberation-to-noise the best tone ping or uniform spectrum performance is within about 2 dB of the optimum for all doppler shifts investigated.

5. Conclusions

If reverberation noise is modelled by random linear time-varying filtering of the transmit wave, then it will be covariance stationary only if the four-dimensional covariance function of the random time-varying impulse is doubly stationary. The random impulse response can be considered time-invariant only if no relative motion exists between the transmitter/receiver and the sources of reverberation. Furthermore, strict covariance stationarity requires that the reverberation be constant with range. This restriction can partially be removed by defining quasi-stationary reverberation, which requires that the medium impulse response be stationary over only two pulse lengths.

A review of the Poisson weak scattering model shows that it is contained in the time-varying linear filter model as a special case. If the scattering function $\sigma(f, t')$ which characterizes this Poisson model is independent or quasi-independent of the range delay variable t', then it produces a stationary reverberation noise process, although the equivalent linear filter model is still time-varying.

When these models are investigated for the effect they have on the performance of likelihood ratio and matched filter receivers, it is found that their performance, bounded above by E_s/N_0 and below by $(E_s/N_0)/(1 + M/N_0)$, where M is the peak of the reverberation spectrum, may be maximized by proper choice of the transmit signal spectrum. The optimum transmit spectrum for two types of stationary reverberation noise is found from the solution of constrained extremization problems by variational methods.

When the reverberation and target spectra are non-time varying filtered versions of the transmit wave, the solution is available in closed form (equation (35) for the likelihood ratio receiver and equation (36) for the matched receiver). However, when doppler shift is present as the result of relative scatterer and/or target motion, the filter models become time-varying and solutions must be found at the present time with numerical methods such as nonlinear programming.

For example, it is shown that for the likelihood ratio receiver there is little gain over a tone ping at high doppler shifts or over a uniform spectrum signal (e.g., an FM slide) at zero or very small doppler shifts from the use of an "optimally" designed waveform. Further investigation into the optimum signal for a matched filter and optimizing the signal for both filter types for a variety of scattering functions is still underway. Additional results may indicate cases in which substantial gains are to be had through signal design; however, to date no cases of more than 2 dB gain have been observed.

ACKNOWLEDGMENTS

The author thanks Mr. Riley Cruse, NUC, San Diego, for his contributions of nonlinear programming and numerical results presented in this paper.

REFERENCES

1. Helstrom, C. W. (1968). *Statistical Theory of Signal Detection*, Pergamon Press.
2. Faure, P. (1964). Theoretical models of reverberation noise. *JASA* **36**, 259–268.
3. Middleton, D. (1967). A Statistical theory of reverberation and similar first-order scattered fields, Part I. *IEEE Trans.* **IT-13**, 372–392.
4. O'lshevskii, V. (1967). "Characteristics of Sea Reverberation", *Plenum Pub. Corp., Consultant's Bureau, New York.* (Translated into English).
5. Kelly, E. J. and Lerner, E. C. (1956). A mathematical model for a radar echo from a collection of random scatterers. M.I.T. Lincoln Lab., Lexington, Mass. T.R. 123.
6. Moose, P. H. (1970). "On the Detection of Signals in Reverberation", Ph.D. Thesis, Univ. of Wash.
7. Kooij, T. (1968). "Optimum Signals in Noise and Reverberation", NATO Advanced Study Institute, Enschede, Netherlands, Paper 17.
8. Himmelblau, David M. (1972). "Applied Nonlinear Programming", McGraw Hill, New York.

DISCUSSION

N. Owsley: Is this a technique you could do *in situ*, i.e., you have a platform and you want to design your signal on board, so to speak, by adaptively measuring your reverberation correlation function, or it is really worth it? Should we just take CW followed by an FM slide and live with it?

Answer: Several factors must be considered. Note that the examples of the curves presented show results for the optimum receiver and not the matched filter. Van Trees conjectured in a 1968 paper that it was not really worth doing the optimum filter because the gain in performance over the matched filter was very small. However, he also conjectured that a good deal of performance might be gained with optimum signal design. It is planned to evaluate the matched filter performance for the same cases that have been looked at for the optimum receiver. However, there is a theorem that says if one can find a signal so that the optimum receiver entirely eliminates the reverberation noise, then the optimum filter is just the matched filter. Therefore, in the area of the results in which the performance of the optimum receiver is near E_s/N_0, it is known that the matched filter will do about the same. This means that it is probably of little value to consider adaptively adjusting the receiver according to *in situ* measurements. (This should not be construed to eliminate the idea of adaptive normalization or adaptive TVG in order to reduce dynamic range and/or stabilize the thresholds.) On the other hand, it is worthwhile to measure the doppler spread *in situ* or on-line, so to speak, from tone ping data to determine the doppler range for which a broadband signal should be superior to a narrowband signal in performance. A problem arises when different signals are needed for different doppler regions. If a tone ping is transmitted and followed by FM, the energy is halved for each one plus reverberation is smeared from one into the other. And, if transmitted alternately, the search time is doubled.

T. Kooij: Although you have assumed some randomness in your target model, do you think if you had made the same calculation with an assumed random model for target response, say a sum of random points instead of a known filter response—which is crucial to the whole thing for a matched filter—that you would have arrived at similar conclusions?

Answer: In this work the echo was assumed to be an arbitrarily filtered version of the transmit wave, which was known except for a constant phase assumed to be uniformly distributed over $-\pi$ to π and a scale factor $E_s^{1/2}$ of arbitrary distribution. This, of course, makes it reasonable to do the mathematics. If a more realistic target is considered, then the analysis will be more difficult. Some results were seen in Dr. Van der Spek's paper yesterday for a fixed length line target of random point reflectors. However, his work does not consider doppler shift for either the target or for the scatterers. An analysis of this target model which includes the doppler effects would be of much interest because it would give additional insight into the problem, but the target model of randomly distributed point sources on a line is not the final answer to the target model problem.

F. Wiekhorst: In your model you have a target filter in parallel with a reverberation filter. Now this model imposes certain restrictions and nature may not always behave the way you have assumed. It seems to me that there should be another filter in front of the target filter that depends on the reverberation filter. The reason is that if you have no reverberation at all you get just the target filter, which is clear; however, if you do have reverberation, you will not excite that target filter by a clear signal. It will be distorted by multipath, by forward scattering, and what not. The cause for the distortion is also the cause of the reverberation; that is, they are not independent. My question is have you made any calculations that include this effect, or do you have any thoughts about how significant an effect it may be?

Answer: Thus far only weak scattering has been considered. However, as Dr. Wiekhorst has pointed out, there is more to it than that. It is known that the signal which reaches the target is distorted in time and frequency by the effects of the medium, including scattering. Although this problem has not been personally explored, there is an interesting problem which is related. There are situations in which the reverberation cannot be legitimately considered Gaussian. All the types of calculations done in this paper really depend on this Gaussian assumption. There is presently considerable effort in the area of optimum processing for point processes. If a model had strong interactive scatterers, but scatterers sparsely distributed in space, then the target would be included as one of these strong scatterers and it would provide an excellent method to include multipath. Here, the false target problem would actually be the focal point, which is a serious problem, indeed. However, these lines have not as yet been pursued.

D. Winfield: You mentioned that the crossover point between tone ping and uniform spectrum performance versus doppler shift did not change appreciably with reverberation-to-noise ratio in your example. Did you assume that the tone pulse spectrum was much less than your scattering function spectrum?

Answer: Reverberation-to-noise ratio is defined as the reverberation power in the band, due to a T second long tone ping of energy E_x to the white noise power in the band, which defines the scale of the scattering function. In fact, the reverberation spectrum of a unit energy tone ping would just reproduce the scattering functions. In the numerical calculations the reverberation-to-noise ratio was varied from about -3 dB to $+20$ dB. The crossover changed a little bit over that range, but not much.

D. Winfield: If you transmitted a very short tone pulse, you should get a crossover point that differs from that of an infinitely long tone pulse when just the scattering function controls the reverberation spectrum.

Answer: The pulse length was fixed for all signals used.

D. Winfield: Then its reciprocal was much less than the scattering function of the reverberation?

Answer: The reciprocal was about the same as the scattering function of the reverberatio example presented. The discreet equations were only approximations, as you will recall, in an attempt to construct a strictly bandlimited signal from a T second long signal, which is not possible. The approximation becomes better for larger (TW) numbers of components; however, only 13 components are used in the example. Your point is well taken, however, and probably more degrees of freedom than 13 should be used in signal design in order to substantiate the conclusion that the crossover point is independent of reverberation-to-noise ratio.

A Comparison of Several Data–Rate–Reduction Techniques for Sonar†

J. J. DOW, B. M. BROWN and H. A. REEDER

TRACOR, Inc., Austin, Texas, U.S.A.

1. Statement of the Problem

As hardware technology advances in the direction of performing more functions per unit volume of equipment, sonar designers find it possible to provide many of the things that signal processors have been advocating, not the least of which is increased resolution in time (range), frequency, and space. In search sonars, a problem which results from greatly increased resolution is that the quantity of data that occurs per unit time is far in excess of both our capability for displaying it and the sonar operator's ability to assimilate it. In short, there is a data-rate mismatch between the sonar receiver and display/operator sub-system. A simple example will serve to illustrate the magnitude of the problem. Consider a preformed beam search receiver which has 50 azimuthal beams and on each, a spectrum analyser (or equivalently, a comb filter bank) with 1000 contiguous 1.0 Hz filters. This receiver would generate approximately 5×10^4 independent samples per second or per display update, a rate which is far in excess of the rate at which sonar data are presently displayed to operators.

Our approach to this problem is to investigate the performance of several data–rate–reduction (DRR) techniques. Each of these devices reduces N channels of data in one of the dimensions of the sonar to 1 channel so that in the example the $N = 1000$ spectral channels are reduced to 1 sample per second per beam. In principle, these samples can be used for alerting the operator to those beams which are most likely to contain target signals or alternately, the DRR output samples themselves can be used for display (for example in a bearing-time recording (BTR) format). The analysis presented here will pertain to the alerting performance of the DRR devices and will be presented in the context of the earlier example—i.e., the case where the DRR device acts as an automatic detector of signals of unknown

† This work was supported by the Sonar Technology Office (Code 302-4) of the Naval Ship Systems Command under Contract N00024-69-C-1317.

centre frequency, f_0, and bandwidth $b \ll B$ where B is the bandwidth of uncertainty concerning f_0.

Although this work is couched in these terms, the results are obviously valid for receiving systems where the data–rate–reduction is performed in any other dimension. All that is required is that the amplitude distributions of the data being operated on by the DRR device be the same as those treated here.

Briefly, the DRR techniques investigated here are the following.

(1) A post-OR integrator, i.e., a processor which OR-gates† the parallel outputs of the filter bank, integrates M temporal samples of the OR-gate output, and thresholds the integrated data for alerting purposes;

(2) A maximum likelihood ratio device, i.e. a processor which forms likelihood ratios on the sequence of samples from each of the parallel outputs of the filter bank, OR-gates the likelihood ratios, and then thresholds the OR-gated output; and

(3) A Kolmogorov–Smirnov (nonparametric) device, i.e., a processor which estimates the difference in the distribution function of the amplitude of noise alone and the distribution function of the amplitudes of the samples in each channel of the filter bank, then OR-gates these estimated differences and finally thresholds the OR-gated output to alert the operator.

We are led to consider these three processing schemes for the following reasons. The post-OR integrator is equivalent to a system which OR-gates the filter bank outputs on each beam and displays the OR-gated beam outputs to an operator, probably on a bearing-time recording. This system is simple, inexpensive and readily implemented so that its performance warrants our attention. The maximum likelihood ratio processor, on the other hand, has been shown to be an optimum‡ processing technique for detecting signals of unknown carrier frequency in the presence of broadband noise [1]. Thus, because of its optimum property this processor also deserves consideration.

The third data–rate–reduction technique, the Kolmogorov–Smirnov device, has received our attention because of its nonparametric behaviour. This simply means that the probability of an incorrect alert is constant regardless of changes in the statistical behaviour of the background noise. It is thus worthwhile to examine the behaviour of this system relative to that of the optimum system.

The performance analyses presented here follow the general pattern of first obtaining the distribution functions and moments of the samples at the output of OR-gates for both signal-plus-noise and noise alone, then from these, obtaining performance curves in the form of graphs of correct-alert probability as a function of signal-to-noise ratio for a constant false-

† An OR-gate is a device which selects the maximum of a set of samples.
‡ By optimum, we mean that the probability of error is minimum.

alert probability. Also, we investigate the sensitivity of performance with respect to various system parameters.

2. Description of the Data–Rate–Reduction Techniques

The block diagrams of the three receiver configurations to be compared are shown in Fig. 1. There are L preformed beams, each driving a bank of contiguous filters or equivalently, a spectrum analyser. The analyser for the lth beam is shown. There are N outputs from the lth spectrum analyser. The N outputs are the detected waveforms of N filters of a comb filter bank with spectral resolution and tooth separation approximating the reciprocal of the duration between output sample times. For 1 Hz resolution (as suggested for illustration in Fig. 1), the input waveform is analysed in 1.0 second and each of the N analyser outputs provides an independent output sample at $\Delta = 1.0$ second intervals.

The N outputs of the spectrum analyser are fed to each of the three processors listed at the beginning of this section. The three processors modify the data and produce a significantly reduced amount of data for alerting or presentation to the operator. For each processor a single output sample occurs per beam each $M\Delta$ seconds. A commutator circulating around the L beams after each $M\Delta$ seconds could provide to an alerting display the thresholded outputs of the data reduction processors for use

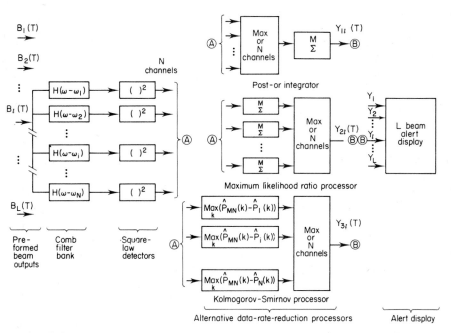

FIG. 1. Simplified block diagram of three search receiver configurations.

in alerting the operator to a beam containing a target. This alerting display could be a bank of lights (one for each beam), or a panel of meters or for the sake of preserving long time histories, a BTR.

For the first processor an independent sample emerges at the OR-gate output from each of the L beams each second. The operational display for this processor could very likely be a bearing-time recording (BTR). For the purpose of carrying out a processor comparison, it is necessary to describe analytically a processor which would have the same performance as the operator viewing the BTR display. As is shown elsewhere [2], this equivalent processor is a square-law detector followed by an averager. Hence, the descriptive title "post-OR integrator" for this equivalent data reduction processor. Thus, the post-OR integrator OR-gates the N spectrum analyser (squared envelope) outputs and then integrates M temporal samples. The sequence of samples from the integrator is thresholded and used to alert the operator.

The second data–rate–reduction technique, the maximum likelihood ratio processor, operates in the following manner. With the signal and noise forms that are assumed here—namely Gaussian—the likelihood ratio is the sum of the squares of the envelope samples from each analyser channel. Thus, this complete processor consists of taking squared envelope samples from each channel, integrating these, and OR-gating the N parallel likelihood ratios. Since OR-gating selects the maximum of all these likelihood ratios, this processor is appropriately called the maximum likelihood ratio processor. This receiver has been treated by Kelly *et al* [1], and has been shown to be an optimum receiver for detecting single component signals of unknown frequency.

The nonparametric processor is a novel application of a statistical method that has been in use in other fields for some time. This device selects and outputs a single analyser channel per beam by first computing the Kolmogorov–Smirnov statistic, δ_i, for each analyser channel and then selecting the maximum of the set of δ_i's, i.e. OR-gating the δ_i's on each beam. The δ_i for each channel (i indexes the frequency channel) is produced by first taking M temporal samples from each of the N spectral channels and obtaining an estimate of the noise-alone distribution function, $\hat{P}_{MN}(Y_k; 0)$.† Next, M samples from each channel are used to obtain an estimate of the distribution function, $P_{iM}(Y_k; \rho)$, of the data in each channel. Finally, the Kolmogorov–Smirnov statistic, δ_i, is formed for each channel by the following operation,

$$\delta_i = \max_{Y_k} | \hat{P}_{MN}(Y_k; 0) - \hat{P}_{iM}(Y_k; \rho) |,$$

---

† The presence of a signal in one of the N channels creates only a slight deviation in this distribution function from the actual noise-alone distribution function estimate.

where ρ denotes the signal-to-noise ratio. These N values of δ_i are then OR-gated and the maximum δ_i is passed to the alerting display.

A point about these processors concerns the number of temporal samples that each is allowed to process. We have kept this number of samples, M, equal for the three processors when making performance comparisons so that each processor can be evaluated on equal terms.

3. Results

As mentioned earlier, the alerting displays of Fig. 1 can take on numerous forms from binary to multilevel. For purposes of this analysis, we will assume a binary form—i.e. alert, no alert. If, using the statistics[†] of the waveforms which drive the display, performance curves in the form of probability that signal plus noise will exceed a threshold (probability of correct alert) as a function of the signal-to-noise ratio for a fixed false alert probability are obtained for each processor, a valid performance comparison of the effectiveness of the three processors can be made. The remainder of this section consists of a summary of the theoretical comparisons which have been made.

(a) Theoretical results
The post-OR integrator The OR-gate processor shown in Fig. 1 operates by taking the largest sample from each set of N spectral samples (from each beam) and using the L samples to drive L integrators, one for each beam, which maintain a running sum of the M most recent samples from each beam.

The complete statistical behaviour of the waveform at the output of the N channel OR-gate processor has been developed [2]. The results are presented here. The probability, P_n, that the OR-gate output will exceed a threshold is

$$P_n = 1 - (1 - e^{-z})^N. \tag{1}$$

In this expression z is the threshold in units of the average noise power in a single resolution band of the spectrum analyser. Similarly for signal plus noise the probability of the OR-gate output exceeding threshold is

$$P_{s+n} = 1 - (1 - e^{-z})^{N-1}[1 - e^{-z/(1+\rho)}] \tag{2}$$

where

ρ = signal-to-noise ratio measured in resolution band.

The processor consisting of an integrator which sums M OR-gate output samples yields a new random variable with distribution function, $G(y; \rho,$

[†] By "statistics" we mean families of probability distribution functions or their complements for noise alone and signal plus noise for various signal-to-noise ratios.

N, M). $G(y; \rho, N, M)$ is obtained by M convolutions of $P(z; \rho)$ with itself. The direct analytical calculation of $G(y; \rho, N, M)$ by means of the convolution process where $M = 30$ or 300 is not attractive although numerical convolution by use of the computer may be possible. A simple alternative is possible if it is assumed that by virtue of the central limit theorem $G(y; \rho, N, M)$ is Gaussian. When this is done, the statistics of the waveform at the post-OR integrator output will be described by

$$G(y; \rho, N, M) = \frac{1}{\sqrt{2\pi}\sigma} \int_{-\infty}^{y} e^{-(x-\mu)^2/2\sigma^2} \, dx, \tag{3}$$

where μ and σ^2 are given by

$$\mu = M\mu_1 \quad \text{and} \quad \sigma^2 = M\sigma_1^2, \tag{4}$$

in which μ_1 and σ_1 are the mean and standard deviation of the distribution associated with equation (2). It has been shown [2] that these moments are given by

$$\mu_1(\rho) = \sum_{k=1}^{N} (-1)^{k+1} \binom{N}{k} \frac{1}{k} + \sum_{k=0}^{N-1} (-1)^{k+1} \binom{N-1}{k} \left[\frac{1}{k+1} \right.$$

$$\left. - \frac{1}{(k+1)/(1+\rho)} \right] \tag{5}$$

and

$$\sigma_1^2(\rho) = 2! \left\{ \sum_{k=0}^{N} (-1)^{k+1} \binom{N}{k} \frac{1}{k^2} + \sum_{k=0}^{N-1} (-1)^{k+1} \binom{N-1}{k} \times \right.$$

$$\left. \times \left[\frac{1}{(k+1)^2} - \frac{1}{\left(k + \frac{1}{1+\rho}\right)^2} \right] - \mu_1^2(\rho) \right\} \tag{6}$$

The mean and variance, defined by equations (5) and (6) can be evaluated by numerical techniques and used to determine the distribution $G(y; \rho, N, M)$ defined by equation (3). These distributions have been used to construct the performance curves shown in Fig. 2. This plot gives the probability that the waveform in a beam containing single plus noise will exceed threshold as a function of the signal-to-noise ratio in a single frequency resolution channel when, for example, the probability of false alert is 0.001.

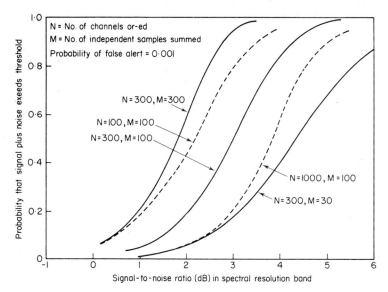

FIG. 2. Performance characteristics of the post-OR integrator.

Figure 2 shows that as N increases the signal-to-noise ratio for equal performance† must also be increased. At 0.5 correct alert probability approximately 1.7 dB of input signal-to-noise ratio is required per factor 10 increase in N. The dependence upon M in the vicinity of $M = 100$ is somewhat stronger and in the opposite direction. A factor 10 increase in M allows the processor to operate with signals which are 2.5 dB lower in signal-to-noise ratio.

Such curves are useful for predicting the performance of the post-OR integrator, but the primary use that is made in this paper is to obtain the desired performance comparison. Thus, we present next the results of similar analyses performed on the maximum likelihood ratio and non-parametric processors.

The maximum likelihood ratio processor. The block diagram of this processor is shown in Fig. 1. The likelihood ratio generation process and the OR-gate provide an output sample each M seconds as in the previous processor. The theoretical aspects of this processor are discussed fully in

---

† By equal performance we mean that the probability of signal plus noise exceeding threshold, i.e., probability of correct alert, is equal to 0.5. This occurs at the expense of an increased signal-to-noise ratio.

[2] where it is shown that the distribution function, $G(T'; \rho, N, M)$, of this processor's output is,

$$G(T'; \rho, N, M) = \left[1 - \int_{T'}^{\infty} \frac{w^{M-1}}{(M-1)!} e^{-w} dw \right]^{N-1}$$

$$\left[1 - \int_{T'/(1+\rho)}^{\infty} \frac{w^{M-1}}{(M-1)!} e^{-w} dw \right]. \tag{7}$$

In this expression T' is the threshold measured in units of the noise power in a resolution filter band. When signal is absent, $\rho = 0$ and equation (7) becomes

$$G(T'; 0, N, M) = \left[1 - \int_{T'}^{\infty} \frac{w^{M-1}}{(M-1)!} e^{-w} dw \right]^{N}. \tag{8}$$

These distributions have been evaluated using digital techniques and performance curves similar to those previously given were obtained. An example derived from the distributions for $N = 100$, 300, and 1000 and $M = 30$, 100, and 300 is shown in Fig. 3. It is clear from this figure that the performance is rather insensitive to the number of spectral channels, N. A factor 10 increase in N causes about 0.5 dB increase in the signal-to-noise ratio that is required for the signal plus noise, threshold-exceeding

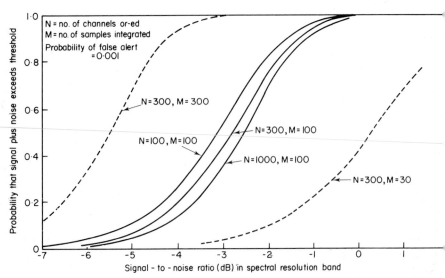

FIG. 3. Performance characteristics of the maximum likelihood ratio processor.

probability to be maintained at 0.5. However, an order of magnitude change in M, the number of samples integrated causes a change in required signal-to-noise ratio of approximately 5 dB.

The nonparametric processor. The nonparametric processor is quite different from the usual processing methods employed in locating the signal channel. In this processor the probability distribution $P_{iM}(Y_k; \rho)$ is estimated for the ith spectral channel, using M independent waveform samples from the channel.

If there are N identical, spectral channels present and all contain noise, it is possible to describe a distribution $\hat{P}_{MN}(Y_k; 0)$ which is the estimated probability that a noise-alone sample will be less than or equal to Y_k. M times N samples are used in obtaining this sample distribution. Simply because there are more samples, this distribution will be a better estimate of the underlying distribution than the single-channel distribution estimate, $\hat{P}_{iM}(Y_k; \rho)$, which is obtained with M waveform samples. The nonparametric processor compares the distribution estimates $\hat{P}_{iM}(Y_k; \rho)$ from the $i = 1$, $2, \ldots, N$ channels with $\hat{P}_{MN}(Y_k; \rho)$ and determines the maximum difference δ_i between $\hat{P}_{MN}(Y_k; 0)$ and $\hat{P}_{iM}(Y_k; \rho)$:

$$\delta_i = \max_{Y_k} [\hat{P}_{MN}(Y_k; 0) - \hat{P}_{iM}(Y_k; \rho)] \tag{9}$$

The distribution functions of δ_i for noise alone and signal-plus-noise have been determined in [2]. Although the nonparametric processor is based on the well-known Kolmogorov–Smirnov test, it deviates from the original form by determining the estimates of the probability distribution function by sorting the data into discrete bins or intervals. The analysis [2] treats the deviations from known results caused by the sorting process. This theoretical framework also allows the treatment of specified signal-plus-noise alternatives. Knowing these functions it is possible to proceed as we have done for the first two processors to get expressions for the probability of exceeding threshold after the OR-gate. The resulting performance curves are plotted for fixed N, M and NB, the number of bins† in Fig. 4. From this figure we find that to maintain the same performance—i.e., keep P_{CA} constant—when N changes over 1 decade, the signal-to-noise ratio must be increased by 0.6 dB. By the same token as M is increased by a decade the signal-to-noise ratio may be decreased by 7 dB and maintain the same false alert and correct alert probabilities.

---

† The number of bins, NB, is actually the number of values of the threshold Y_k which are employed by the system in computing δ_i. It has been shown [2] that for $NB \geqslant 5$ there is no significant performance variation in the nonparametric processor as a function of NB. The results given in this section are prepared for $NB = 10$.

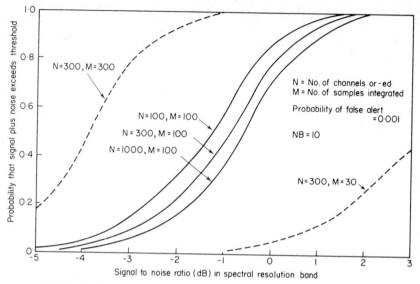

FIG. 4. Performance characteristics of the nonparametric processor.

(b) Performance comparison

Constant centre frequency signals. The primary emphasis of this study is on the relative performance of the three data–rate–reduction processors in the presence of a signal with unknown but constant centre frequency. However, since there are conditions under which the centre frequency of the signal may exhibit variation, we also have looked briefly at the relative performance of the three processors under this condition. In the following performance comparison the results for constant centre frequency signals are presented and discussed first, then we examine performance for variable centre frequency signals.

Figure 5 shows the comparative performance of the three processors for $M = 300$, $N = 400$ at 0.001 false alert probability. The comparison takes the form of a plot of the probability that a correct alert will occur as a function of signal-to-noise ratio in a single spectral resolution channel. For this particular set of parameters it is found that the nonparametric processor requires approximately 2 dB greater input signal-to-noise ratio than the maximum likelihood ratio processor for probablities of correct alert and false alert equal to 0.5 and 0.001, respectively. Similarly the post-OR integrator requires almost 7.5 dB greater input signal-to-noise ratio to have performance equivalent to the maximum likelihood ratio processor. Finally, the nonparametric processor exhibits a 5.6 dB advantage over the post-OR integrator.

Another form of comparison is shown in Fig. 6. In this plot the signal-to-noise ratio required for 0.5 probability of correct alert is plotted as a

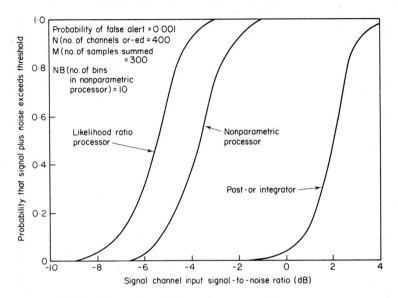

FIG. 5. Comparison of the performances of three processors.

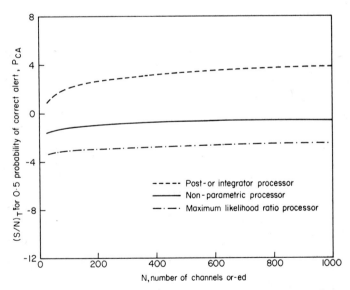

FIG. 6. Variation in required signal-to-noise ratio as a function of $N(M = 100$, $P_{FA} = 0.001)$.

function of the number of samples OR-gated for each of the processors for false alert probability equal to 0.001 and $M = 100$. This figure shows that for $N > 100$ there is little variation in performance with changes in

N, for each data–rate–reduction technique. This was also found to be true for a wide range of values of M.

The signal-to-noise ratio required for 0.5 probability of correct alert may be thought of as a threshold signal-to-noise ratio, $(S/N)_T$. The difference, $\Delta(S/N)_T$, in required signal-to-noise ratio is presented in Fig. 7. $\Delta(S/N)_{T1}$ is the difference in $(S/N)_T$ between the post-OR integrator and the maximum likelihood ratio processor. $\Delta(S/N)_{T2}$ is the difference in

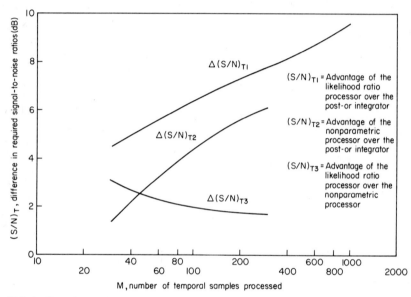

FIG. 7. Relative performance of the three data–rate–reduction techniques ($N = 1000$, $P_{FA} = 0.001$).

$(S/N)_T$ for the post-OR integrator and the nonparametric processor. Finally, $\Delta(S/N)_{T3}$ is the difference in performance between the maximum likelihood ratio processor and the nonparametric processor. These curves are valid for $N = 1000$ and $P_{FA} = 0.001$. It follows from Figs 6 and 7 that the significant factor in performance difference variations is M, the number of temporal samples processed by each processor. As shown in Fig. 7 the superiority of the likelihood ratio technique over the post-OR integrator ranges from approximately $+4.5$ dB to $+9$ dB as M changes from 30 to 1000. The nonparametric processor outperforms the post-OR integrator by anywhere from approximately 1.5 dB to 6.0 dB as M changes from 30 to 300. Finally, the maximum likelihood ratio processor is found to be superior to the nonparametric processor by only 3 dB to 1.75 dB as M changes from 30 to 300. Thus for signals with constant centre frequencies the maximum likelihood ratio processor is clearly the superior performer.

Frequency variable signals. There are several effects that can arise from variations in the centre frequency of a signal. First, if the variation is sufficiently rapid, the signal power will pass into and out of the resolution filter before that filter has reached a steady-state output. This of course, causes a loss in signal-to-noise ratio. Second, assuming that the variation in frequency is slow enough to preclude the first effect, changes in the frequency location of the signal can cause performance degradation in any post detection processing scheme which is designed to process M temporal samples in each analysis channel but which receives $M^* \neq M$ consecutive samples of signal in a given channel. That is, if the processor integrates M samples per channel but receives only M^* signal-plus-noise samples, then there is a processing mismatch. Specifically, if $M^* \neq M$, there will be a loss in performance because of a loss in signal-to-noise ratio simply because the integration time is not matched to the signal duration in a single channel.

When the number of available samples, M^*, is less than the number of samples processed, M, degradation in performance will occur in the maximum likelihood ratio and nonparametric processors. This is true because the best design of these processors dictates that they operate on only M^* samples instead of M samples.† Admittedly there will be M/M^* opportunities to obtain a correct alarm in $M\Delta$ seconds, and multiple opportunities will partially compensate for reduced integration time.

It is important to note, however, that the performance of the post-OR integrator is completely unaffected by variations in the centre frequency of the signal. This is true because the OR-gate will select and pass on to the integrator the maximum spectral sample regardless of the channel in which it may occur.

To obtain a quantitative comparison of the performance of these three processors under the conditions described above we proceed as follows. If $n = M/M^*$ is the number of opportunities to obtain an alarm, then the probability of correct alert is given by

$$D_n = 1 - (1 - d_1)^n, \tag{10}$$

where d_1 is the single opportunity probability of correct alert. Taking $D_n = 0.5$, we have

$$d_1 = 1 - (0.5)^{1/n} \tag{11}$$

as the value of d_1 required for $D_n = 0.5$.

---

† M may be thought of as the number of independent samples available in the tactically allowed detection time while M^* is the number of samples available while the spectral line remains in a single spectral channel.

It is also true that the probability of false alert P_n is altered accordingly. Thus, for small single opportunity probability of false alert, p_1,

$$P_n = 1 - (1 - p_1)^n \approx np_1 \tag{12}$$

Now by setting P_n to the desired false alert probability we can find p_1. This allows us to find the proper threshold from the distribution functions which in turn provides the required signal-to-noise ratio, $(S/N)_T$, for $D_n = 0.5 = 1 - (1 - d_1)^n$ to hold. By plotting $(S/N)_T$ versus P_n for various values of n, we get a quantitative assessment of the performance of the three processors for various degrees of mismatch in M and M^*. These results are shown in Fig. 8.

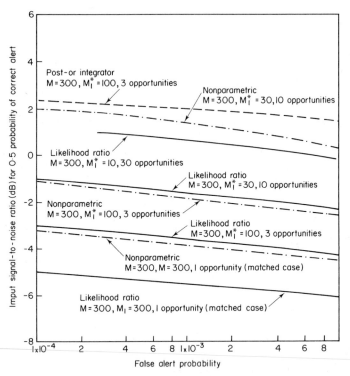

FIG. 8. Performance comparison with limited channel occupancy, $N = 400$.

It can be seen that the nonparametric processor, matched to M^* samples requires about 6 dB less signal-to-noise ratio than the corresponding post-OR integrator processor to provide equivalent performance at $N = 400$, $M^* = 300$. A decrease in $M^* = 300$ to $M^* = 30$ essentially destroys any advantage the nonparametric processor has under ideal conditions over the post-OR integrator processor.

The same reduction from $M^* = 300$ to $M^* = 30$ for the maximum likelihood ratio processor reduces its advantage over the post-OR integrator processor from approximately 7.5 dB to 3.5 dB.

As the mismatch becomes greater, the advantage of the likelihood ratio and nonparametric processors over the post-OR integrator becomes less. This is shown in Fig. 9. By extrapolation (shown by dashed lines) it can be seen that when M^* is equal to approximately 5, or $M/M^* = 60$, the performances of the likelihood ratio processor and the post-OR integrator become equivalent. Similarly, when M^* equals approximately 25, or $M/M^* = 12$, the performances of the nonparametric and post-OR integrator

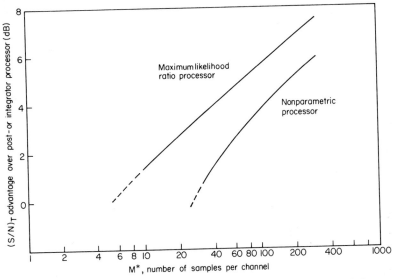

FIG. 9. Comparison of performance of the three data–rate–reduction techniques as a function of M, M^* Mismatch ($P_{FA} = 0.001, M = 300, N = 400$).

processors become equivalent. These curves point out that when the mismatch in M^* and M is great enough, all of the advantage of the maximum likelihood ratio and nonparametric processors over the post-OR integrator is lost.

To summarize these results we have prepared Table I. The figures given in this Table are valid for values of $M, N,$ and P_{FA} which are considered most interesting. The value chosen for M is 300. If 1 Hz resolution is employed, $M = 300$ provides 5 minutes processing time. In addition, a typical value for the number of spectrum analyser channels is 400. The choice of false-alert probability of 0.001 would give an overall false-alert probablity of 0.05 if the passive search receiver were composed of 50 preformed beams. Under these conditions, we find that the threshold signal-to-noise ratios are $+2$ dB, -5.4 dB, and -3.6 dB for the post-OR

TABLE I. Summary of results

Data-rate-reduction technique	Threshold signal-to-noise ratio. $(S/N)_T$ for $M = 300$ (5 min) $N = 400$ $P_{FA} = 0.001$	Nominal sensitivity to N $\Delta(S/N)_T/\Delta N$ $P_{FA} = 0.001$	Nominal sensitivity to M $\Delta(S/N)/\Delta M$ $P_{FA} = 0.001$	Superiority over post-OR integrator (fixed frequency signals) $P_{FA} = 0.001$ $M = 300, N = 400$	Superiority over post-OR integrator (mismatched M^* and M) $P_{FA} = 0.001$ $M = 300, N = 400$
Post-OR Integrator	$+2.0$ dB	≈ 1.7 dB/ decade for $M = 30, 100, 300$	≈ 2.5 dB/ decade for $N = 100, 400, 1000$	—	—
Maximum Likelihood Ratio Processor	-5.4 dB	≈ 0.5 dB/ decade for $M = 30, 100, 300$	≈ 5.0 dB/ decade for $N = 100, 400, 1000$	7.4 dB	7.4 dB, $\dfrac{M^*}{M} = 1$ 3.0 dB, $\dfrac{M^*}{M} \approx 1/12$ 0.0 dB, $\dfrac{M^*}{M} = 1/60$
Nonparametric Processor	-3.6 dB	≈ 0.5 dB/ decade for $M = 30, 100, 300$	≈ 7.5 dB/ decade for $N = 100, 400, 1000$	5.6 dB	5.6 dB, $\dfrac{M^*}{M} = 1$ 3.0 dB, $\dfrac{M^*}{M} \approx 1/4$ 0.0 dB, $\dfrac{M^*}{M} = 1/12$

M = number of temporal channels processed; N = number of channels OR-ed.

integrator, the maximum likelihood ratio processor, and the nonparametric processor. In this same order, the sensitivity of the processors to changes in N is 1.7 dB/decade, 0.5 dB/decade, and 0.5 dB/decade. Thus, the post-OR processor is approximately 3 times as sensitive to increases in the number of analyser channels as are the likelihood ratio and nonparametric processors.

With regard to the sensitivity to changes in the number of temporal channels processed, M, we find that the likelihood ratio and nonparametric processors are more than twice as sensitive as is the post-OR integrator.

Insofar as relative performance is concerned, the likelihood ratio and nonparametric processors require 7.4 dB and 5.6 dB less signal-to-noise ratio than does the post-OR integrator to yield equal correct alert and false-alert probabilities. This is true when the signal exhibits constant carrier frequency or when line tracking is employed successfully.

When the signal exhibits frequency variations and when no tracking is employed, then the deterioration in performance of the likelihood ratio and nonparametric processors which occurs is given in column 5 of Table I.

4. Summary

For signals with steady frequency the maximum likelihood ratio processor is clearly the superior performer. Its superiority over the post-OR integrator ranges from 4 dB to 9 dB depending on the choice of M, N, and P_{FA}. Under equal conditions, the likelihood ratio processor is a uniformly better performer than the nonparametric processor by 1 dB to 3 dB. With steady frequency signals, the nonparametric processor outperforms the post-OR integrator by 1 to 6 dB, again depending on the choice of M, N, and P_{FA}.

For $N > 100$, the performance of all three processors is relatively insensitive to variations in N, with the post-OR integrator being the most sensitive. Because the sensitivity of all three processors is greater for variation in M than in N, it is possible to regain losses that result from increasing N by less than a one-for-one increase in M.

In the presence of signals whose centre frequencies vary so that single channel occupancy is less than the processing time, the performances of the likelihood ratio and nonparametric processors deteriorate from their best values while the performance of the post-OR integrator is unaffected. When the mismatch, M^*/M equals 1/12, the performance of the non-parametric processor becomes equivalent to that of the post-OR integrator. Similarly, when $M^*/M = 1/60$, the likelihood ratio processor performance degrades to that of the post-OR integrator.

REFERENCES

1. Kelly, E. J., Reed, I. S. and Root, W. L. (1960). The detection of radar echoes in noise, I. *J. Soc. Indust. Appl. Math*, **8**, No. 2, 309–341.

2. Dow, J. J., Brown, B. M. and Reeder, H. A. (1972). "Theoretical Analysis of Three Statistical Decision Methods Employing Maximum OR-gating", TRACOR Document T72-AU-9556-U, Unclassified.

DISCUSSION

T. G. Birdsall: Is the "ORing" done within a beam and across frequency?

Answer: Yes, that is correct. The OR-gating is performed across frequency channels on each of the spectrum analysed preformed beam outputs.

T. G. Birdsall: Is not the primary application of your system found where signals are sparse?

Answer: In a sense, yes. If for some reason, *finding* targets is no problem, e.g. because there are many targets close by or because of large gains from perhaps very large arrays we *find* targets on nearly all beams, then the primary use of these types of systems would be to allow one to rank the data on the various beams. That is, each of the data–rate–reduction devices yields an output which is proportional to the *a posteriori* probability of a target signal being present; thus, the system outputs provide a means of obtaining guidance concerning which of the beams to investigate first, then second, etc.

However, the utilization of this system in a binary decision sense does make sense only when there is a detection problem. This tends to be the case when the incidence of targets is low. This is certainly the case with mobile sonar platforms over vast ocean areas. Therefore, this is when these data rate reduction systems would find application in the binary decision sense.

D. Narin: Have you processed real data with these systems?

Answer: Yes. We have processed recorded sea data through digital computer implementations of all three of the data–rate–reduction processors. Let me clarify one point. The data was indeed recorded sea data insofar as the noise background was concerned; however, the signal was synthetic and was inserted in the data. We compared the performance of the systems predicted by the theory with that obtained in the experiments. The result of the comparison was that the experimental performance was nearly always from 1.0 dB to 2.0 dB poorer than that by theory. This consistent difference could probably be accounted for by such careful considerations as those described by Dr. Prior in his paper.

D. Narin: Does your theory account for the large signal fluctuations that sometimes occur because of variations in such things as target aspect angle?

Answer: No, this theory is based on the assumption that the signal is a random process whose amplitude statistics are Gaussian. No multiplicative processes have been included. However, had such multiplicative effects been included the performance curves would likely have been modified in much the same way as indicated by Swerling's well known analysis which treats fluctuating targets.

E. J. Risness: I would like to raise the question of the whole philosophy underlying your processor. In your delightful slides the message seemed to be that sonar displays, shortly after World War II, were about matched to the operators' capacity, and now that we have made the systems and displays more complicated we have got to get them back to the previous state. I would question whether this is right on two grounds. First, in practice we are trying to make unalerted detections, and there is no evidence whatever that even with old-type sonars with long intervals between detections, operators almost always tend to miss targets when they really come on. The second thing is when the operator is alerted to something he can tap in great quantities of information. So it seems to me that the aim is not to reduce sonars down to their previous state, but to find some way of fully automatically alerting the man, and in parallel with this, another system (display) that would allow him to see all of the information associated with the alert.

Answer: If I gave the impression that I advocated a return to the state of affairs that existed 10 to 15 years ago, let me correct that now. I do not, of course. I would advocate exactly the use of automatic detection to which you referred at the conclusion of your comment. There are situations where the *search* system generates so much data that it is impossible for the operator to do the actual search. This situation would best be treated by utilizing the data–rate-reduction system as an automatic binary decision maker for each beam, and then upon occurrence of several alarms, the associated values of the data–rate-reduction system outputs should be presented to the operator since these quantities are proportional to the *a posteriori* probability that a target signal exists on the beam in question. In this way the operator can perform his examination of the detailed information on a priority basis instead of an aimless choice process.

H. Cox: The notion of a nonparametric processor implies robustness. By its construction, this processor is sensitive to the assumption that we have effected normalization or prewhiting. In your particular case you are very sensitive to the assumption that you have prewhitened or effected normalization, and if you did not have that, it would not work at all.

As regards the maximum likelihood ratio processor, once you have the structure built even though you have got the maximum likelihood processor from a parametric approach, what you have built is no more parametric than anything else. Once you have it in that form, you have an opportunity at the output to do some more normalization. For example, if there are some trends in frequency within a given beam, there is an opportunity to take them out. So I really feel in practice that your conclusion, that if things are not Gaussian, that one should go to the nonparametric processor is not fully justified.

Answer: Concerning the sensitivity of the nonparametric processor to time stationarity, whiteness, and independence across frequency, let me make two points. First, the device is sensitive to deviations in stationarity—however, we have developed an algorithm which operates in the two dimensions of time and frequency and performs a normalization of the data at the spectrum analyser output. This device has been used on sea data and has produced excellent background noise normalization in both the dimensions of time and frequency. That is, in these two dimensions the output power was constant to at least within ±0.5 dB. This is accomplished by the algorithm automatically and without prior knowledge of the detailed characteristics of the non-stationarity.

Second, with regard to independence the experiments that we conducted with the nonparametric device included data at the spectrum analyser output which was highly correlated between adjacent frequency channels. In spite of this, the theoretical and experimental performance results for this processor were within about 1.5 dB of each other. This is due to the insensitivity of the nonparametric device to N, the number of *independent* frequency channels processed.

Concerning your comment about the parametric nature of the maximum likelihood ratio processor, your point is quite valid. It may very well turn out that this processor is just as robust as the nonparametric device with respect to changes in input distribution functions. We have seen that this is indeed the case with the performance of maximum likelihood processor with respect to the number of components in the target signal. That is, this "parametric" device is quite robust with respect to mismatches in the number of components in the signal (the device is optimum only when the number of components is equal to one).

Graphical Comparison of Broadband and Tone-pulse Sonar Transmissions

D. NAIRN

Admiralty Underwater Weapons Estab., Portland, Dorset, England

1. Introduction

Although an ideal performance comparison between broadband and tone-pulse sonar transmissions only involves simple calculations, the number of possible permutations is very large and the difficulty lies in arranging the data so as to make apparent the overall picture. Almost of equal importance is to be able to readily assess the effect of the departures from the theoretical performance which are found to occur in practice. This note will describe a method of plotting the data to enable a sonar engineer to tell at a glance the relative performance of this or that transmission type, and the influence in practical terms of any variations or degradations.

The approach here is to choose a simple datum against which the ideal performances of the various transmissions are compared. Once the overall picture is grasped, experimental data can then be used piece-meal to apply "corrections" to these idealized plots. In many cases it will be seen that these discrepancies are swamped by the extent to which the environmental parameters vary, so that a fair understanding of the problem can often be had from a relatively simple analysis.

This paper will consider only a simple type of propagation, i.e. one which consists of a reverberation-limited phase followed by one where detection is noise limited.

2. Initial Assumptions Which Simplify Idealized Plot

The following assumptions are made initially in order to simplify the problem:

(a) *Target strength falls off with range at same rate as reverbs.* The S/N ratio from any one target-echo is therefore constant for all ranges where detection is reverberation limited.

449

(b) *Perfect matched processing for all signals.*

(c) *Broadband signals are assumed to be doppler insensitive.* If separate doppler matching is not included, some performance degradation is bound to occur.

(d) *Same beamwidths and source intensities are assumed for all transmissions.* Allowances must obviously be made if this is not the case in some practical situations.

(e) *Comparison is made only of output S/N.* Differences in recognition differential and o/p statistics are not used.

3. Construction of Idealized Performance Plot

Plot *ABC* in Fig. 1 represents the performance of a 150 millisecond zero doppler tone-pulse transmission, plot *DEBC* is the same pulse with an 8 knot echo doppler shift, and plot *FGH* is that of a 300 Hz, $\frac{1}{2}$ second transmission:

(a) Zero doppler tone pulse
The datum chosen for Fig. 1 is seen to be the reverberation-limited performance of a zero doppler 150 millisecond tone-pulse. This performance is independent of range in the reverberation-limited region *AB*, by virtue of assumption that the echo and reverberations fall off with range at the same rate. The range at which the background changes from reverberation to ambient noise, point *B*, has arbitrarily been taken as 10 kiloyards in this example. Obviously other cases involve only a simple change of scale.

Region *BC* shows the signal sinking into the ambient noise background. The slope of this line is taken to be the same as the reverberation decay rate with range, in accordance with the above-mentioned assumption. A typical figure for this rate, 9 dB per range doubling, is taken for Fig. 1; other values are considered presently.

(b) Tone-pulse with doppler shift
From the sin x/x reverberation spectrum of a square 150 millisecond tone-pulse at 7.5 kHz centre frequency, the *background* reverberation level for, say, an 8 knot target is some 20 dB *lower* than that at the centre frequency. This fact is shown in plot *DE* on Fig. 1, where the reverberation-limited performance for the 8 knot doppler tone-pulse is shown to be +20 dB better than the zero doppler channel.

The doppler-shifted channel is seen to become noise limited (point *E*) considerably before the centre channel; not surprising since reverberations in all channels decay at the same rate and those in outer channels are always lower so that they are first to sink into the noise. It is seen that when both channels are noise limited, region *BC*, they have identical performances. This is as expected, since they have the same energy (pulse-length).

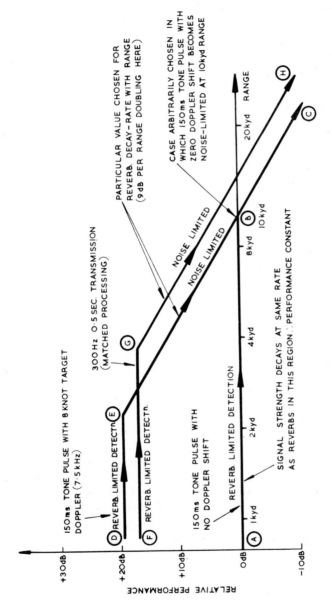

FIG. 1. Construction of relative performance plot.

(c) *Long broadband pulse*

Plot *FGH* on Fig. 1 shows the relative performance of the 300 Hz, $\frac{1}{2}$ second broadband pulse. Since matched processing is assumed, the actual form of this pulse is of no importance, e.g. it could be a frequency-modulated sweep, pseudo-random noise,† or even a very short tone pulse (although cavitation would severely limit the energy-output of the latter in practice). Under matched conditions, the reverberation-limited performance of a pulse is dependent only on its bandwidth, as this determines the average number of background scatterers in a resolution cell, given the same beam-widths etc. Thus the 300 Hz pulse is seen to have a reverberation-limited performance some $16\frac{1}{2}$ dB better than the zero-doppler tone-pulse, this figure being 10 log ratio of their bandwidths.

Point *G* shows the range where the broadband pulse becomes noise limited, and the performance falls off along *GH* as the echo-signals sink more deeply into the noise, at a rate equal to that chosen for the reverberation fall-off with range. In fact line *GH* produced shows the maximum reverberation limited range for *any* 0.5 second transmission, i.e. any 0.5 second transmission will fall-off along this line regardless of where it meets it. This being the case, it is seen that another 0.5 second pulse with a broader bandwidth, say 600 Hz, would meet this line above point *G*, and so would become noise limited at a closer range than the 300 Hz pulse. With the broader bandwidth, the *same energy* is spread over a *wider spectrum*, so that each individual frequency component now has a lower initial energy level.

4. Use of Plot to Compare Performances

Figure 2 shows the complete performance plot relative to a 7.5 kHz, 150 milli-second tone pulse transmission. A maximum reverberation-limited range of 10 kiloyards has arbitrarily been chosen for the zero-doppler channel, and a typical value of 9 dB per range doubling has been taken for the reverberation decay rate as above. Experimental data have been used to plot the spread in the doppler-shifted echo performance.

Figure 2 could have been plotted in more general parametric terms but it is hoped that it will be more instructive to follow through a typical example. To this end, Figs 2 and 3 will now be used to compare the performance of the *150 millisecond tone pulse* with that of the *300 Hz broadband pulse* under semi-ideal conditions:

(a) The "break-even" echo doppler is about 6 knots, below which the broadband pulse *always* has a better performance.

† PRN sidelobes can lose up to 3 dB.

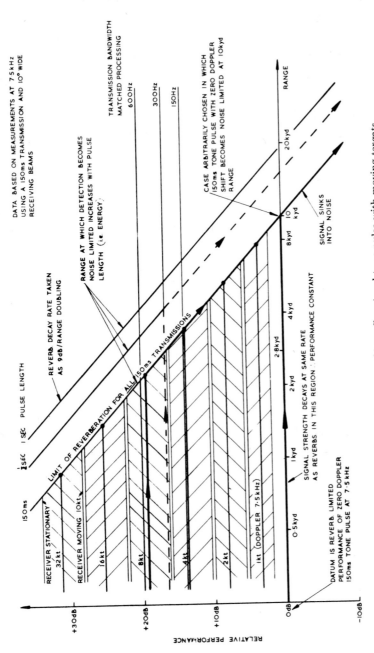

FIG. 2. Relative performance of broadband and tone-pulse with moving targets.

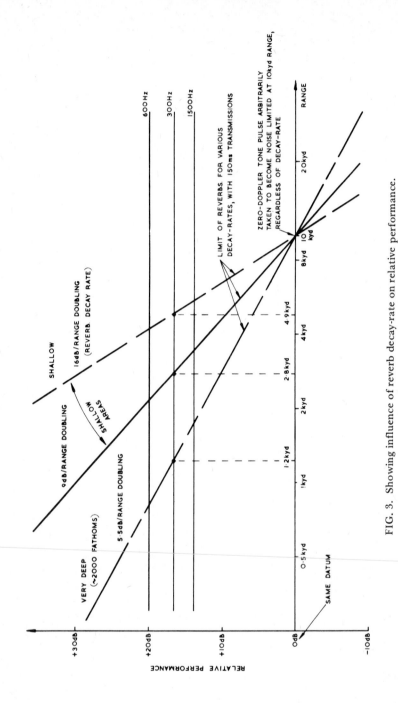

FIG. 3.   Showing influence of reverb decay-rate on relative performance.

(b) For doppler-shifts in excess of 6 knots, the tone-pulse has a better performance only for the first 28% of the centre-channel-reverberation-limited-range (2.8 kyd in Fig. 2).

(c) Figure 3 shows that this figure can vary between 12% and 49% for deep water and very shallow water respectively (i.e., for reverberation decay rates of 5.5 dB and 16 dB per range doubling).

(d) Even at relatively short ranges, detections of the highest doppler tone-pulse echoes are likely to be noise limited.

(e) Similarly, maximum reverberation-limited ranges are seen to come down quite markedly as the transmission bandwidth is increased. This can significantly curtail the benefits to be had from increasing a transmission bandwidth while maintaining the same pulse-length (although changes in recognition differential can complicate matters).

5. Conclusion

In Fig. 2 the comparison was made for the most part between the *ideal* performances of the various pulse types. More experimental data can obviously be introduced, e.g. to allow for echo-fragmentation of broadband pulses, in much the same way as the measured doppler-spread information has been used. Also, if the echo S/N ratio was found to vary significantly with range in the reverberation-limited region, the relative performance datum can be defined as that of the zero-doppler tone pulse at some specific range. By continuing this modification process, the final plot yields a performance comparison to the full accuracy of the available information. Furthermore the information is in a form which readily shows the price of any one departure from the ideal.

As stated initially, performance comparisons, even on the basis of idealized behaviour, are likely in many cases to be a fair approximation to the actual situation. This is because the variation possible in such parameters as the reverberation decay rate with range often have effects which are large in comparison to those due to differences between ideal and practical behaviour.

DISCUSSION

J. J. Dow: Is there some technique for including the effect of the background noise statistics in these comparisons?

Answer: Allowance can be made by expressing such differences in terms of dB change in the relative performance scale. However the main use of the diagram is to grasp the overall situation and such variations are unlikely to affect this to any great extent.

P. H. Moose: Is it realistic to assume that the signals decay at the same rate as the reverberation, as used here for the datum?

Answer: The main point about the plot is that no assumptions have to be made if there is experimental data available. In this case, the datum could be taken as the reverb/noise transition-point, with the relative performance of even the zero-doppler tone pulse changing with range.

Signal Processing Device Technology

G. W. BYRAM, J. M. ALSUP, J. M. SPEISER and H. J. WHITEHOUSE

*Naval Undersea Research and Development Center, San Diego,
California, U.S.A.*

1. Introduction

Many real time signal processing requirements impose a computational
load which exceeds the capabilities of even a large general purpose computer.
A major portion of this load arises from the large number of linear and
bilinear operations required. The use of special purpose devices to perform
these operations can reduce system size, cost, and processing time.

The special purpose devices examined here are based on the use of delay
lines for the storage and shifting of data, which requires parallel access to
the entire contents of the line in most cases. The parallel access requirement
is the single most important factor that will determine the technology
required to implement these devices.

Transversal filters, including the special case of serial access memory,
may be formed from any type of signal propagation which has low loss
and low dispersion, provided there is a means of constructing lightly loading,
nonreacting taps. Also, a crossconvolution can be formed when the pointwise product of the signals present in two delay lines is integrated along
the length of the lines.

Two natural figures of merit for the comparison of different types of
delay lines are the bandwidth (W) and delay-bandwidth (TW) product. In
the serial access memory application the bandwidth determines the
allowable data rate and the delay-bandwidth product determines the
storage capacity of the line. For transversal filters or crossconvolvers, the
bandwidth determines the processing speed and the delay-bandwidth
product determines, respectively, the number of degrees of freedom in
the filter's impulse response or the number of independent signal samples
that can be contained in the device.

The speed advantage indicated by the high TW product arises from two
factors, the most important of which is the highly parallel structure,
although the most conspicuous is the high data rate that corresponds to
high bandwidth.

2. Structures

(a) *General purpose linear processor*
In many applications it is desirable to combine the central control
afforded by a general purpose computer with the efficiency provided
by special purpose devices. Such applications might require operations
which include (a) matched filtering for echo-ranging or for coherent
communications systems, (b) spectrum analysis for passive detection and
classification, (c) crosscorrelation for interferometric analysis, and (d)
beamforming for large sensor arrays.

The elements required for such a General Purpose Linear Processor
(GPLP) can be collected in four groups: (a) transversal filters, (b) time
compressors, (c) frequency shifters, and (d) auxiliary memory [1]. The
structure of such a processor is shown in Fig. 1. Maximum flexibility is
obtained if the transversal filters are programmable, so that any desired

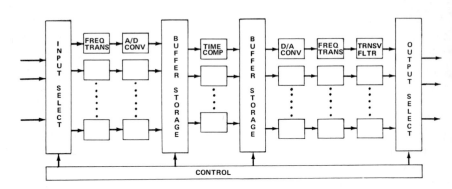

FIG. 1. General purpose linear processor.

reference function can be introduced upon command (including the case
in which the reference function is another real time signal); however, a
large GPLP might also have at its command some fixed reference transversal filters. A time compression capability is often required so that
signals can be processed linearly against several different reference functions
in real time, with a small number of transversal filters used repeatedly in
compressed time. This requires that the transversal filters be capable of
handling bandwidths larger than the signal bandwidth by the same factor
used to describe the amount of time compression. Frequency shifters,
preferably single sideband modulators, are required to shift signal and
compressed-signal spectra to match the different passbands of the transversal filter and the signals. Ideally, A GPLP would be subject to computer
control in the modular arrangement of its elements so that larger or smaller
TW-product capability could be synthesized, a larger or smaller amount of

time-compression could be used, or different amounts of frequency shifting could be implemented, depending upon which signal processing task was given priority by the computer control. The computer would then be given the tasks of thresholding, sorting, decision-making, and display formatting the outputs of the GPLP, in addition to executive control of the GPLP—tasks for which the computer is ideally suited.

(b) Serial access Fourier transform device
The flexibility of the GPLP structure permits not only the efficient utilization of time-invariant components, but it also permits the incorporation of these compounds into time-varying filters when needed. One frequently needed time-varying filter is a serial access Fourier transform device. A GPLP could be programmed to implement such a device, thus providing a complete (amplitude and phase) Fourier transform. Similar components have been previously used to implement an amplitude-only Fourier transform [2]. The GPLP lets the same group of components be restructured to form the more complicated amplitude-and-phase Fourier transform when needed.

If the Fourier transform $G(f)$ of the signal $g(t)$ is defined by

$$G(f) = \int g(t) \, e^{-j2\pi ft} \, dt$$

then

$$G(f) = e^{j\pi f^2} \int e^{-j\pi(f+t)^2} \left[g(t) \, e^{j\pi t^2} \right] \, dt$$

Mertz [3] observed that the above identity corresponds to the realization of a Fourier transform by the combination of multiplication by a complex chirp, filtering by a complex chirp, and a second multiplication by a complex chirp. Such a complex time-varying filter is shown in Fig. 2.

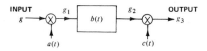

FIG. 2. Complex time-varying filter.

To implement this idea, the complex operations can be replaced by the corresponding pairs of real operations, and a modified form of the identity can be used which will permit the use of bandpass convolvers or chirp filters.

The complex operations which will implement the time-varying filter are $g_1(t) = a(t)g(t)$, $g_2(t) = b(t) * g_1(t)$, and $g_3(t) = c(t)g_2(t)$, as shown in Fig. 2. With the subscripts R and I used to denote the real and imaginary parts, the above complex operations can be decomposed into real operations:

$$g_{1R}(t) = a_R(t)g_R(t) - a_I(t)g_I(t)$$
$$g_{1I}(t) = a_R(t)g_I(t) + a_I(t)g_R(t)$$

Complex multiplication

$$g_{2R}(t) = b_R(t) * g_{1R}(t) - b_I(t) * g_{1I}(t)$$
$$g_{2I}(t) = b_R(t) * g_{1I}(t) + b_I(t) * g_{1R}(t)$$

Complex filter

$$g_{3R}(t) = c_R(t)g_{2R}(t) - c_I(t)g_{2I}(t)$$
$$g_{3I}(t) = c_R(t)g_{2I}(t) + c_I(t)g_{2R}(t)$$

Complex multiplication.

For a Fourier transform device, the functions $a(t)$, $b(t)$, and $c(t)$ are specialized to

$$a(t) = e^{-j2\pi[ut^2 + v_1 t]}$$
$$b(t) = e^{j2\pi[ut^2 + v_2 t]}$$

and

$$c(t) = e^{-j2\pi[ut^2 + v_2 t]}$$

With this choice of a, b, and c, the output of the structure is

$$g_3(t) = G(ut + v_1 + v_2)$$

The corresponding implementation of a serial access Fourier transform device with bandpass components is shown in Fig. 3. The performance of the complex device will be limited generally by implementation of the chirp filters rather than the other components.

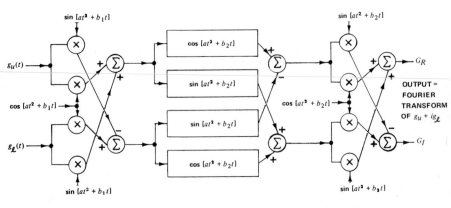

FIG. 3. Bandpass implementation of dual channel Fourier transform device with interlaced outputs.

(c) Cascade filter synthesis

A transversal filter or cascade of transversal filters with uniform tap spacing will have an impulse response of

$$h(t) = \sum_k h_k \delta(t - kd)$$

and a transfer function of

$$H(f) = \sum_k h_k e^{-j2\pi fkd}$$

where the only constraint for realizability is that the tap weights are all real, or equivalently $H(-f) = H^*(f)$, and $H(f)$ has period d^{-1}, where d is the delay between taps.

For simplicity let $d = 1$; then

$$H(f) = e^{Q(f)+jP(f)}$$

where $Q(f) = \log |H(f)|$, and $P(f) =$ phase of $H(f)$. The only realizability constraints on $P(f)$ and $Q(f)$ are that both are real functions with period 1 and that $P(f)$ is odd (modulo 2π) and $Q(f)$ is even.

When $P(f)$ and $Q(f)$ are represented by, respectively, their Fourier sine and cosine series,

$$H(f) = e^{\Sigma Q_s \cos 2\pi sf + j\Sigma P_m \sin 2\pi mf}$$

a cascade realization of the filters is obtained such that

$$H(f) = c \prod_{s=1}^{\infty} e^{Q_s \cos 2\pi sf} \prod_{m=1}^{\infty} e^{jP_m \sin 2\pi mf}$$

where c is a constant gain.

The individual filters correspond to individual terms of the amplitude or phase ripple in the desired transfer function. The number of taps required in a given filter in the cascade is small if the corresponding order of amplitude ripple or phase ripple is small. If several of the filters require only a few taps each, they may be replaced by their cascade equivalent: a single filter whose tap weight function is the convolution of their tap weight functions.

The tap weights required for each of the filters to be cascaded are the coefficients in the Fourier series expansions

$$e^{jP \sin 2\pi f} = \sum_{n=-\infty}^{\infty} J_{-n}(P) e^{-j2\pi nf}$$

$$e^{Q \cos 2\pi f} = \sum_{n=-\infty}^{\infty} I_n(Q) e^{-j2\pi nf}$$

where J_n and I_n denotes Bessel functions of the first kind.

The tap weights for the first order filters are $J_{-n}(P_1)$ in the phase compensation cascade and $I_n(Q_1)$ in the amplitude compensation cascade. For the kth order terms in the respective cascades the tap weights are $J_{-n}(P_k)$ spaced by $k-1$ zeros and $I_n(Q_k)$ spaced by $k-1$ zeros.

The cascade realization is particularly desirable for a network used to correct small errors in an existing filter, since a relatively small total number of taps is needed to compensate a highly erratic amplitude and phase function, provided that the maximum deviations in phase and logarithmic amplitude are small.

3. Technology

(a) Digital correlators

The first device considered will be the Large Scale Integrated Circuit (LSI) version of a conventional discrete component digital correlator. In its simplest form a binary correlator consists of two shift registers, many exclusive ORs and a resistive summing network. Such a correlator can be assembled from conventional medium scale integrated circuit logic modules or can be obtained in LSI form.† The LSI unit shown in Fig. 4 is composed

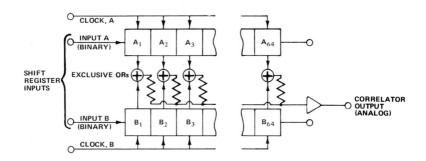

FIG. 4. Binary correlator.

of two 64-bit shift registers, 64 exclusive ORs, and an analogue network which sums the outputs of the exclusive ORs. Each shift register can be operated independently at rates up to 20 MHz; the outputs from corresponding stages in the two shift registers are two inputs of an exclusive OR, so that the correlator output represents the number of agreements minus the number of disagreements for the contents of the two shift registers, and changes whenever new information is shifted into either register [4].

† TRW Systems Group, Redondo Beach, California, U.S.A.

A binary correlator can be operated as either a fixed reference correlator or a matched filter. In this case, the desired reference is shifted into one shift register and left undisturbed for the duration of the required signal processing. This corresponds to the implementation described by R. H. Barker [5].

The binary correlator is particularly suited for situations in which the reference function must frequently be changed and constitutes a programmable matched filter for binary signals. The change is made by shifting a new sequence into one shift register on command and then introducing the signal to be correlated into the other shift register.

To calculate the crosscorrelation function of two signals for a single value of relative delay, the two shift registers are shifted simultaneously in the same direction. If the crosscorrelation must be evaluated for many values of delay, it is advantageous to hold one shift register fixed while the other register is shifted for a number of shifts equal to the number of delays desired. Then the roles are interchanged, so that the two signals are alternately shifted past each other. With this technique the correlation value for each value of delay is updated a number of times inversely proportional to the number of delays for which correlation values are calculated.

Multilevel operation can be achieved in either the signal channel, the reference channel, or both of these channels with several binary correlators used in parallel. The case of a 3-bit-quantized signal correlated against a binary reference is illustrated in Fig. 5 in which a total of 3 binary correlators are operated in parallel. The outputs are summed in a weighting network such that the correlator outputs are binary weighted with the most significant digit (MDS) given the greatest weight.

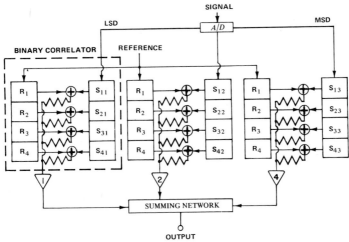

FIG. 5. Correlator for 3-bit signal *versus* binary reference.

In order to correlate an M-bit quantized signal against an N-bit quantized reference, the samples may be represented in the form

$$S(k) = \sum_{m=1}^{M} (-1)^{S_m(k)} 2^m$$

and

$$R(k) = \sum_{n=1}^{N} (-1)^{R_n(k)} 2^n$$

where $S_m(k)$, $R_n(k) = 0$ or 1.

Apart from a constant bias, the desired inner product $\Sigma_k S(k) R(k)$ is proportional to

$$\sum_{m=1}^{M} \sum_{n=1}^{N} 2^{m+n} \left[\sum_k (S_m(k) \oplus R_n(k)) \right],$$

where $\oplus$ denotes a modulo 2 sum, and each of the terms $\Sigma_k (S_m(k) \oplus R_n(k))$ may be computed by a binary correlator. In general MN binary correlators will be required to correlate an M-bit-quantized signal against an N-bit quantized reference. This involves the use of 2 MN shift registers and MN groups of exclusive ORs. The number of shift registers required could be reduced to M + N if the individual shift register stages were capable of driving a sufficient number of inputs in parallel. Such a reduction is illustrated in Fig. 6 for the case M = N = 3 but is apparently slightly beyond the present state-of-the-art for high-speed LSI implementation.

A modified reduction in redundancy is accomplished when only one

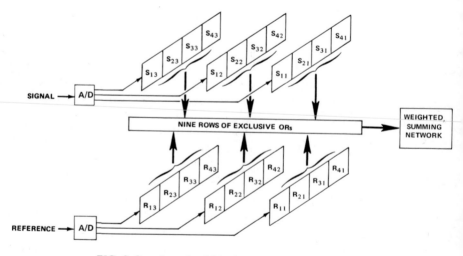

FIG. 6. Correlator for 3-bit signal *versus* 3-bit reference.

input is multilevel quantized. This requires only a single digital correlator composed of M + 1 shift registers and M groups of exclusive ORs instead of M binary correlators required to correlate an M-bit quantized signal against a binary reference. An attempt to manufacture such a unit to operate at 5 MHz, 4 bits by 1 bit (64 bits in length), was undertaken in 1969 but was unsuccessful because of difficulties in the fabrication of discretionary interconnections [6].

Most systems presently under consideration in the NUC laboratory are intended to process signals with TW products from 200 to 2000. If correlators with the desired speed are available in 64-bit unit lengths, the units can be combined to achieve the desired TW product within a multiple of 64. For instance, a binary correlator of length 1024 could be implemented if 16 of the LSI binary correlators were attached in sequence and the 16 correlator outputs summed. This approach could be combined with the multilevel approach to obtain both the desired TW product and the desired quantization.

An alternate procedure for implementing either approach can be used if there is sufficient time to repeatedly use a single binary correlator. For multilevel operation subsequent binary correlator outputs are weighted according to their equivalent bit position and added to previous correlation functions stored in memory via an A/D converter and a programmable digital adder. For operation with larger TW products the correlation functions generated for each 64-bit portion of the reference are similarly combined, with proper overlap, so that the total correlation function is synthesized. Both procedures follow as a direct consequence of the bilinear nature of correlation. The trade-off between time-shared operation and operation with parallel units is generally determined by the time available for processing.

(b) Charge transfer devices

Charge transfer devices [7], both bucket brigade [8a] and charge coupled [8b], are the semiconductor equivalent of a volatile, nondispersive, variable velocity delay line. These devices utilize the analogue storage of charge in potential wells as the means of storing information. They can propagate analogue signals both in the audio and the video frequency range and can be implemented as integrated circuits.

The delay line is the simplest signal processing function realized. The fact that the memory is both serial and volatile is seldom important for signal processing applications since real time operation is usually desired. If the signal processing is under computer control, the variable shift rate is especially useful since the serial memory can be momentarily stopped if a higher priority function interrupts the processing.

The critical parameters are the dynamic range of the memory or shift register and the attenuation and smearing introduced by incomplete charge transfer. The dynamic range of 40 to 60 dB is significant, since the corresponding digital shift registers would require 7 to 10 bits of quantization. The incomplete transfer of charge limits present charge transfer devices to at most one hundred or two hundred stages before smeared charge becomes significant.

Although the bucket brigade devices currently available commercially operate at the relatively low clock rate of 200 kHz, laboratory charge transfer devices operate at clock rates above 1 MHz and it is anticipated that devices operating above 10 MHz will soon become available.

The ability to contiguously tap the charge transfer devices is the most important consideration for signal processing, since transversal filters can easily be constructed if the propagating charge can be sensed without being altered. A 13-bit Barker code matched filter with bucket brigade technology [9] and a tapped delay line with charge coupled technology [10] have been made.

(c) Magnetostrictive devices

Two interactions can be used to tap a magnetostrictive delay line. Both involve interaction between local magnetization of the line and a propagating torsional strain. Since the interaction is multiplicative and linear in both variables, superposition holds and transversal filters can be constructed. In the Wertheim effect [11] a solenoidal field interacts with the strain to produce an axial voltage on the wire. In the inverse Wiedmann effect [12] an axial magnetic field in the wire interacts with the strain to produce a voltage in a solenoid which encloses the wire.

A pulse generator [13] that used the Wertheim effect was one of the first successful applications of a multiply tapped acoustic delay line. This device has been modified to form a delay line matched filter [14]. Further developments that have used the inverse Wiedmann effect have led to a commercially available solenoidally tapped line.†

Torsional wave propagation provides nondispersive delays up to 10 msec at bandwidths up to 1 MHz to yield TW products up to 10 000. The propagation velocity variation with temperature can be chosen in certain alloys to cancel the change in length of the wire with temperature and the delay lines are temperature stable [15].

Either tapping interaction, by reciprocity, can also be used for input transduction. Groups of such taps can provide spatially distributed input transducers. In addition, desired magnetic and electrical properties can be achieved simultaneously when a conductor is provided down the centre of the magnetostrictive cylinder. The magnetic fields used to tap the line along its length may be supplied by electromagnets or solenoids so that the individual taps may be electrically altered.

† Andersen Laboratories, Inc., Bloomfield, Connecticut, U.S.A.

Extensional propagation in a wire is also nondispersive and can be used as the basis for a programmable memory [16]. Recent development of this memory using plated wire suggests its consideration for a programmable matched filter.

(d) Surface wave devices

Another form of nondispersive propagation is provided by Rayleigh waves on the free surface of a solid. In suitable materials these waves may be launched and received by direct electrical transduction [17, 18]. In single crystal piezoelectric materials the propagation is nondispersive over bandwidths of the order of 1 GHz, and tapping by interdigitated metal electrode structures may be accomplished over bandwidths of a few hundred MHz.

A finger pair forms a lightly coupled broadband tap with approximately unity bandwidth at a centre frequency for which the centre-to-centre finger spacing is a half wavelength [19]. Such a tap will typically be many acoustic wavelengths wide and will act as a broadside radiator, forming a well collimated acoustic beam. An extensive description of materials, physical phenomena, and methods of tapping and guiding surface waves is available [20].

The taps may be amplitude weighted when the finger length is varied or phase inverted if the sign is changed. In the latter case precautions should be taken to control fringing fields. One method of control is the use of the field delineated transducer element, shown as the dashed portion of the coded transducer of Fig. 7. Another method of control is the use of a tap spacing large relative to the finger separation.

The reciprocity of transduction permits designing of the launch and receive transducers according to transversal filter principles. If the design impulse response of the cascade of launch and receive transducers is an impulse, it is possible to obtain the high fractional bandwidth of a single tap, together with the low insertion loss of an array of transducers. Such a configuration has been described [21] and methods for generating the required codes are available [22].

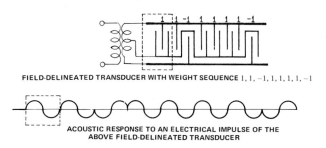

FIELD-DELINEATED TRANSDUCER WITH WEIGHT SEQUENCE 1, 1, −1, 1, 1, 1, 1, −1

ACOUSTIC RESPONSE TO AN ELECTRICAL IMPULSE OF THE
ABOVE FIELD-DELINEATED TRANSDUCER

FIG. 7. Field-delineated transducer and its acoustic response.

A coded delay line on lithium niobate with length 64 Golay complementary sequences exhibited a fractional bandwidth of about 70% with impulse response sidelobes about 21 dB down from the main lobe. Since launch and receive transducers were each of length 64, and the signal had to propagate under both transducer structures, this may be regarded as equivalent to a transversal filter with a time-bandwidth product of 128. Double these lengths have been achieved on quartz.

Noninteracting Golay sequences have been used to fabricate a surface wave recirculating delay line memory with a bit rate of 220 MHz and a storage capacity of 1280 bits per recirculating loop. Two pairs of noninteracting sequences are used for a total of four recirculating loops, or 5000 bits of memory [23].

Most devices reported to date have been chirp filters and dispersive delay lines. These may be fabricated by the use of uniformly spaced taps, by dispersive propagation, or by multiple propagation paths. A chirp matched filter with a design time-bandwidth product of 1000 and a realized time-bandwidth product of about 900 has been reported [24]. Lardat, Maerfeld, and Tournois have described methods for fabricating dispersive delay lines [25]. Delay lines with very small departures from linear delay versus frequency, bandwidth of 30 MHz, and delay variation of 8 microseconds have been reported.

Programmable transversal filters have been fabricated with diode [26] or transistor switching [27] of the individual taps or groups of taps. Currently reported tap switching rates have been quite low compared to the data rate, and the power required for a switchable tap is comparable to that required for one stage of a corresponding shift register device.

The parametric interaction between a pair of propagating waves in a nonlinear piezoelectric material may be used to provide a distributed multiplier. A convolver is formed with the addition of a distributed electrode to spatially integrate the electric field produced by the nonlinear interaction. Using this technique, Quate and Thompson built a bulk wave convolver [28] in lithium niobate, and Luukala and Kino similarly built a surface wave convolver [29] on lithium niobate. The use of ferroelectric ceramics of the lead zirconate titanate family has been proposed [30] to provide a stronger nonlinear interaction. However, attenuation and dispersion would limit the maximum operating frequency in PZT-8 to about 50 MHz, as compared to an upper limit of about 1.5 GHz for lithium niobate. A convolver on lithium niobate with a bandwidth of 9 MHz and an integration time of 8 microseconds has been reported [31]. The above surface wave convolvers have low fractional bandwidth. Their bandwidths were restricted by their periodic launch transducers. Convolver structures which use broadband (Golay coded) launch transducers have been proposed [32], and a chirp coded convolver with a fractional bandwidth of 70% has been reported [33].

ST-cut quartz is the only known material which has a zero temperature coefficient of delay at room temperature [34]. For lithium niobate electrical compensation via heating of a thin conducting film has been demonstrated [35]. This technique should significantly extend the usefulness of this and other high coupling materials.

(e) Acoustic-optic devices

An optical processing system provides a large number of parallel channels since the input and output data can be two-dimensional. Systems based on photographic film for input and output have attained an advanced state of development and can simultaneously correlate and beamform on a large number of channels [36]. Application of similar techniques to real time processing has been severely handicapped by a lack of good methods for converting a time function into the required spatial input.

Use of an acoustic wave in a delay line to modulate a light beam will provide a spatial replica of a portion of the time history of an input signal and also provide shifting operations required for correlation. A wide variety of techniques have been used to achieve acousto-optic interaction [37]. Although significant progress has been made in acoustic light modulators and deflectors, the achievable interactions are generally very weak or impose geometric restrictions which limit the potential time-bandwidth product achieved in a correlator. Examination of some representative interactions will illustrate the limitations which arise.

In an ideal acousto-optic correlator, two zero-thickness acoustic regions in contact would successively amplitude modulate light. Collection of the twice modulated light would then provide the required integration. These ideal conditions are rarely approximated in practice.

Bulk waves in a transparent solid cause local variations in refractive index which phase modulate light. Two successive phase modulations would produce a sum rather than the required product. One method to achieve amplitude modulation is to use spatial filtering to separate out amplitude modulated diffraction components. The separation of the two modulating regions required to provide room for spatial filtering generally necessitates the use of imaging optics to preserve a one-to-one relationship between corresponding portions of the two delay lines. It is generally difficult to match the imaging system and the acoustic cell dimensions efficiently. As a result the attainable TW product is limited either by field of view or by resolution. Other methods of converting from phase to amplitude modulation produce similar problems. Interferometry, for example, imposes severe stability requirements. The use of polarization techniques to detect strain induced birefringence which results from the acoustic wave [37] imposes material constraints.

In all of the above methods the acoustic beam must be many acoustic wavelengths across to prevent it from spreading. Diffraction spreading of light within a thick acoustic beam may cause light sidebands to enter regions in which the refractive index changes cancel. This will occur [38] if the thickness of the acoustic beam is greater than

$$\frac{\lambda^2 \text{ acoustic}}{4\lambda \text{ optical}} \text{ (Upper boundary of Debye–Sears region)}$$

For thicker beams the sidebands disappear through destructive interference. This is a serious limitation since it corresponds to an upper frequency limit of 30 MHz for longitudinal waves and 60 MHz for shear waves in an acoustic beam 25 wavelengths thick in fused quartz. Alternatively, it implies that at a frequency of 100 MHz the beam could be only 15 wavelengths thick. Since a thin beam will spread rapidly, the total length of sound path is limited. This again restricts the TW product.

If the Bragg condition [39]

$$\text{angle of incidence} = \frac{\lambda \text{ optical}}{2\lambda \text{ acoustic}}$$

is met, one sideband is constructively reinforced and the rest vanish. The required angle is a function of signal frequency and imposes bandwidth limitations for a fixed geometry. A wider bandwidth is possible if a phased transducer is used to tilt the acoustic beam [38, 40] and preserve the Bragg condition. This method becomes awkward if two successive interactions are required. A Bragg device has been proposed [1, 41] which uses two successive Bragg diffractions to cancel the geometric dependence.

Acoustic surface waves [20] exist only within a few acoustic wavelengths of the surface on which they travel. This avoids the limitations due to the inherent thickness of bulk waves. Although diffraction efficiencies near 1% can be achieved in reflection from a surface wave [42], the path length changes and refractive index changes tend to cancel for a transmitted beam. This complicates successive interactions. A very large interaction can be achieved between an acoustic surface wave and a guided optical wave coplanar with it [43], but the geometry does not utilize the limited depth of the surface waves.

Amplitude modulation of light by an acoustic surface wave can be achieved if the surface wave perturbs boundary conditions on an interface at which light is being totally internally reflected [44]. The geometry of this method is shown in Fig. 8. This method provides a sensitive interaction since the division of light between the reflected and transmitted components is extremely sensitive to changes in spacing of prism faces which form the reflecting boundary. The potential sensitivity of the method is illustrated in Fig. 9. The extremely small spacing required and the difficulty of generating large surface waves complicate implementation of this method. Use in a correlator that requires two successive interactions is complicated by the 45° tilt of the interaction regions with respect to the light path.

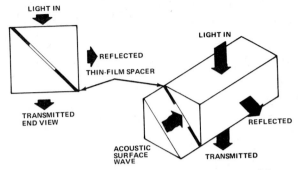

FIG. 8. Geometry of an interval reflection light modulator.

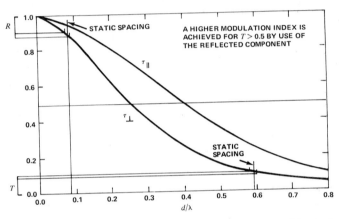

FIG. 9. Sensitivity curves for an internal reflection light modulator. (Transmission curves after Court, *et al.* [45]).

Mechanical problems preclude use of two modulating regions in close contact.

Real time optical processing can be summarized as having great promise and formidable problems. The considerable progress that has been made has been summarized by Preston [46].

4. Summary

Several major families of signal processing devices have been examined. Their states of development range from proposed interactions to devices which are now commercially available. The entries in Table I are representative of reported results. It should be noted, however, that initial results are often limited by experimental conditions rather than intrinsic limitations of the device. Many of these devices and others still under investigation

TABLE 1. Representative reported signal processing device parameters.

Device	W	$2TW^a$	Ref.	Input Quantization	Output Quantization	Distinctive Features	State of Development
LSI Chip Correlator	10 MHz	64/chip	[4]	binary	binary	volatile phase programmable variable rate	commercially available
Bucket Brigade	4 MHz	13	[9]	analogue	binary	volatile phase programmable variable rate	preliminary
Charge Coupled	4 MHz	64	[10]	analogue	binary	volatile phase programmable variable rate	proposed
Magnetostrictive (Wiedmann)	2 MHz	1024/output	[14]	analogue	analogue	multiple output	commercially available
Ferroacoustic	8 MHz	175/output	[16]	analogue	binary	multiple output phase programmable	proposed
Surface Wave Transversal Filter	220 MHz	16	[23]	analogue	analogue		preliminary
	12 MHz	128	[41]	analogue	analogue		preliminary
	10 MHz	16	[47]	analogue	analogue		preliminary
Surface Wave Chirp Filter	18 MHz	1800	[24]	analogue	analogue	phase programmable dispersive fixed reference	preliminary
Surface Wave Convolver	9 MHz	144	[31]	analogue	analogue	transducers easily fabricated below 1 GHz	preliminary
Bulk Wave Convolver	100 MHz^b	c	[48]	analogue	analogue	transducers easily fabricated above 1 GHz	preliminary
Acousto-optic Convolver	c	c	[49]	analogue	analogue	highly folded acoustic path holds promise of TW up to 10^4	preliminary

[a] Number of taps or number of independent samples. [b] Bandwidth of the input acoustic port. Bandwidth of the output port limited by test equipment. [c] Results limited by available test equipment.

hold the promise of substantial improvement. As the achievable bandwidths increase, the problem of interfacing with conventional techniques becomes more severe. There is an inherent mismatch between sonar data in the kHz range, digital logic in the MHz range, and fast special purpose devices which may extend into the GHz range. This indicates that the requirements of the special purpose device should be taken into account in the earliest system design stages. The charge transfer devices should provide a convenient interface between sonar data and digital logic since analogue samples can be shifted at rates controlled by digital logic. The tap switchable surface wave devices permit digital logic to control the structure and operation of high bandwidth processing devices. The flexibility of input/output afforded by operation under control of a general purpose computer will be essential in most cases.

REFERENCES

1. Squire, W. D., Whitehouse, H. J. and Alsup, J. M. (1971). "Acousto-Optic Transversal Filters as Linear Signal Processors", *Proc. Conf. Electro-Optics '71 International*, pp. 153–163.
2. Tournois, P. and Bertheas, J. (1968). "The Use of Dispersive Delay-Lines for Signal Processing in Underwater Acoustics", Proc. Conf. Nato Advanced Study Institute on Signal Processing, Enschede, Netherlands.
3. Mertz, L. *Transformations in Optics*, pp. 94–96. Wiley, New York.
4. Buie, J. L. and Breuer, D. R. (1972). "An LSI Correlator", TRW Systems Group, Redondo Beach, California, U.S.A.
5. Barker, R. H. (1953). Group synchronizing of binary digital systems. *in* "Communications Theory" (W. Jackson, Ed.), Butterworth; London.
6. Hochman, H. T. and Hogan, D. L. (1970). Technological advances in large-scale integration. *IEEE Spectrum* 5, 50–58.
7. Berglund, C. N. and Strain, R. J. (1972). Fabrication and performance considerations of charge transfer dynamics shift register. *Bell System Tech. J.* 51, 655–703.
8a. Altman, L. (1972). Bucket brigade devices pass from principle to prototype. *Electronics* 45, 62–71.
8b. Altman, L. (1971). The new concept for memory and imaging: charge coupling. *Electronics* 44, 50–59.
9. Buss, D. D., Bailey, W. H. and Collins, D. R. (1972). Bucket-brigade analog matched filters. Proc. Conf. IEEE International Solid-State Circuits, Univ. of Pennsylvania, Phil., PA, pp. 250–251.
10. Tiemann, J. J. (1971), "Surface Charge Devices for Analog Signals" (General Elec. Corp. Res. & Dev., Schenectady, New York), 1971 IEEE Ultrasonics Symposium, Miami Beach, Florida. (To be published; will be available as a Special Publication from IEEE, New York, N.Y. 10017, U.S.A.)
11. Wertheim, W. (1857). "Note sur des courants d'induction produits par la torsion du fer", Compt. Rend. 35, 702–704 (1852): "Memoire sur la torsion: 2e Partie, Sur les effets magnétiques de la torsion", *Ann. Chim. Phys.* 50, 385–432.
12. Smith, I. R. and Overshott, K. J. (1965). The Wiedmann Effect: a theoretical and experimental comparison. *Brit. J. Appl. Phys.* 16, 1247–1250.
13. Perzley, W. and Fishbein, M. (1962). Digital function generation with torsional delay lines. *Electronics* 35, 62–65.

14. Whitehouse, H. J. (1963). Parallel digital delay-line correlator. *Proc. IEEE* **51**, 237–238.
15. Clark, C. A. (1962). Alloys for electromechanical filters and precison springs. *Proc. IEEE* **109B**, 389–394 (suppl. 22).
16. Gratian, J. W. and Freytag, R. W. (1964). Ultrasonic approach to data storage. *Electronics* **37**, 67–72.
17. White, R. M. and Voltmer, F. W. (1965). Direct piezoelectric coupling to surface elastic waves. *Appl. Phys. Letters* **7**, 314–316.
18. Hickernell, F. S. (1972). "Piezoelectric Film Surface Wave Transducers" (Motorola Government Elec. Div., Scottsdale, Arizona), 1971 IEEE Ultrasonics Symposium, Miami Beach, Florida. (To be published; will be available as a Special Publication from IEEE, 345 E. 47th St., New York, N.Y. 10017.)
19. Coquin, G. A. and Tiersten, H. F. (1967). Analysis of the excitation and detection of piezoelectric surface waves in quartz by means of surface electrodes. *J. Acoust. Soc. Am.* **41**, 921–939 (pt. 2).
20. White, R. M. (1970). Surface elastic waves. *Proc IEEE* **58**, 1238–1276.
21. Tseng, C. C. (1971). Signal Multiplexing in surface-wave delay lines using orthogonal pairs of Golay's complementary sequences. *IEEE Trans. Sonics and Ultrasonics* **SU-18**, 103–107.
22. Schweitzer, B. P. (1971). "Generalized Complementary Code Sets", Ph.D. thesis, U.C.L.A., Westwood, CA.
23. van de Vaart, H. and Schissler, L. R. Acoustic surface wave recirculating memory. (To be published in complete form in the April 1973 combined issue of the *IEEE Trans. Sonics Ultrasonics* and the *IEEE Trans. Microwave Theory and Techniques*; to be published in brief form in *Appl. Phys. Letters*.)
24. Burnsweig, J. and Arneson, S. H. (1972). Surface-wave dispersion with a time-bandwidth of 1000. *IEEE J. Solid-State Circuits* **SC-7**, 38–42.
25. Lardat, C., Maerfeld, C. and Tournois, P. (1971). Theory and performance of acoustical dispersive surface wave delay lines. *Proc. IEEE* **59**, 355–367.
26. Hunsinger, B. J. and Franck, A. R. (1971). Programmable surface-wave tapped delay line. *IEEE Trans. Sonics Ultrasonics* **SU-18**, 152–154.
27. Claiborne, L. T., Staples, E. J. and Harris, J. L. (1971). MOSFET ultrasonic surface-wave detectors for programmable matched filters. *Appl. Phys. Letters* **19**, 55–60.
28. Quate, C. F. and Thompson, R. B. (1970). Convolution and correlation in real time with nonlinear acoustics. *Appl. Phys. Letters* **16**, 494–496.
29. Luukala, M. and Kino, G. S. (1970). Convolution and time inversion using parametric interactions of acoustic surface waves. *Appl. Phys. Letters* **18**, 393–394.
30. Lim, T. C., Kraut, E. A. and Thompson, R. B. (1972). Nonlinear materials for acoustic-surface-wave convolver. *Appl. Phys. Letters* **20**, 127–129.
31. Boniganni, W. L. (1971). Pulse compression using nonlinear interactions in a surface acoustic wave correlator. *Proc. IEEE* **59**, 713–714.
32. Speiser, J. M. and Whitehouse, H. J. (1971). Surface wave transducer array design using transversal filter concepts. *In* "Acoustic Surface Wave and Acoustic-Optic Devices" (T. Kallard, Ed.), Optosonic Press, New York, pp. 81–90.
33. Waldner, M., Pedinoff, M. E. and Gerard, H. M. (1971). Broadband surface wave nonlinear convolution filters. 1971 IEEE Ultrasonics Symposium, Miami Beach, Florida.
34. Schulz, M. B., Matsinger, B. J. and Holland, M. G. (1970). Temperature dependence of surface acoustic wave velocity on a quartz. *J. Appl. Phys.* **41**, 2755–2765.
35. White, R. M. and Goyal, R. C. (1972). "Stabilizing Surface-Wave Devices Against Temperature Variations" (to be published).
36. Williams, R. E. and von Bieren, K. (1971). Combined beam forming and cross

correlation of broadband signals from a multidimensional array using coherent optics. *Appl. Opt.* **10**, 1386–1392.

37. Damon, R. W., Maloney, W. T. and McMahon, D. H. (1970). "Interaction of Light with Ultrasound: Phenomena and Applications in Physical Acoustics" (W. P. Mason and R. N. Thurston, Eds.). Academic Press, New York and London.

38. Adler, R. (1967). Interactions between light and sound. *IEEE Spectrum* **4**, 42–54.

39. Quate, C. F., Wilkerson, C. D. W. and Winslow, D. K. (1965). Interaction of light and microwave sound. *Proc. IEEE* **53**, 1604–1623.

40. Pinnow, D. A. (1971). Acousto-optics light deflection: design considerations for first order beam steering transducers. *IEEE Trans. Sonics Ultrasonics* **SU-18**, 209–214.

41. Squire, W. D., Whitehouse, H. J. and Alsup, J. M. (1969). Linear signal processing and ultrasonic transversal filters. *IEEE Trans. Microwave Theory Techniques* **MIT-17**, 1020–1040.

42. Zory, P. and Powell, C. (1971). Light diffraction efficiency of acoustic surface waves. *Appl. Opt.* **10**, 2104–2106.

43. Kuhn, L. (1971). "Interaction of Acoustic Surface Waves With Optical Guided Waves in Thin Films". 1971 IEEE Ultrasonics Symposium, Miami Beach, Florida. (To be published; will be available as a Special Publication from IEEE, New York, N.Y. 10017, U.S.A.).

44. Byram, G. W. (1971). Opto-acoustic correlation devices. *NUC Tech. Pub.* **161**.

45. Court, J. N. and Von Willisen, F. K. (1964). Frustrated total internal reflection and application of its principle to laser cavity design. *Appl. Opt.* **3**, 719–726.

46. Preston, K. Jr. (1972). "Coherent Optical Computers" McGraw-Hill, New York.

47. "Signal Processor Speeds Code Change," (1971). *Electronics* **44**, 30.

48. Quate, C. F. private communication.

49. Gottlieb, M., Conroy, J. J. and Foster, T. (1972). Optoacoustic processing of large time-bandwidth signals. *Appl. Opt.* **5**, 1068–1077.

DISCUSSION

W. S. Liggett: Seems to me I did not hear anything about reliability or physical ruggedness in your talk. This might be difficult to cover in a short talk of course.

Answer: The crossconvolver I showed is quite fragile and is usually kept on a foam rubber pad. It is however a laboratory prototype. The electrode attachment and metalization techniques would be quite different in a production model which should be as sturdy as the basic crystal from which they are made.

The magnetostrictive devices are extremely rugged. There is not too much you can do to a piece of wire.

The LSI chips will be roughly as reliable as other semiconductor devices. Your problem there will be in getting a reasonable yield in the initial manufacture. As a result, if a company is selling 64 bit correlators, you may find they have quite a few "bargain specials" of 63 bits—62 bits, etc.

S. W. Autrey: I would like to make one minor comment. Even though in sonar processing we are concerned mainly with serial operations and can use shift registers effectively, it turns out in practice, at least in the near term, we are going to see a lot of random access memories used simply because the major users of semiconductor devices now are using them. As a result, when we go to our implementors and say here is the process we want to implement, I think shift registers would be ideal they say "forget it" the random access memory is cheaper.

Answer: I think you are absolutely correct. However, there you must think of your overall cost which will include the overhead of additional control structure required to organize data within your random access memory. Your random access memory for example is particularly well adapted to doing a FFT in place, but where you do have an inherently serial organization of your input information as in correlation it may well be worth avoiding the additional overhead cost.

S. W. Autrey: I agree with that but it turns out the circuit designers are quite different types of people.

Answer: *Quite.*

H. Mermoz: Could you express the cost of these devices as a function of TW product?

Answer: The surface wave devices—or any of them?

H. Mermoz: Any of them.

Answer: Those 64 bit chips come from TRW. At the moment, they are sort of prototypes and cost about US $1500 each. But if you were to buy one or two hundred, the cost would probably be less than US $200 each. That is an informal estimate. The magnetostrictive device cost is a small cost for materials and a few man days to assemble it unless an elaborate mounting is required. The surface wave devices cost us the price of the quartz or lithium niobate crystal plus roughly a man month for fabrication. You do, however, have a rather heavy overhead in the cost of the facility [showed 3 slides of surface wave device facility]. This facility represents an investment of roughly US $200 000.

J. W. R. Griffiths: To continue on that, these non-programmable surface wave devices can be quite cheap. They are being used in television sets.

Answer: Yes, they have the potential of being very cheap. But in research when you need one of anything, it is exceedingly expensive. I hesitate to emphasize the low cost of these devices so soon after our heavy expenditure for the facility.

Van Schooneveld: When you are not designing for mass production but when you have a one time experimental application and when you are restricted to commercially available devices, which do you think we will be using over the near term and forseeable future?

Answer: I would say that the charge transfer devices hold more promise than just about anything that has come along yet.

T. Curtiss: I noticed you did not mention magnetic bubble devices.

Answer: I think that those do hold some promise. When they first came out they were oversold a bit and this caused interest to subside when some of the problems became apparent. I think that they may become a very useful alternative to the other devices. It will ultimately boil down to the type of access you need to the information you have stored. If you can achieve parallel optical access for example that would be extremely useful.

Sampled-analogue Signal Processing

R. K. P. GALPIN

Plessey Telecommunications Research, Maidenhead, Berkshire, U.K.

1. Introduction

A major contributory factor to recent rapid advances in the art of underwater communication has been the availability of complex microelectronic
circuits for performing sophisticated processing of signals obtained from
large and complex arrays of transducers. Digital processing techniques are
particularly attractive when the latest MSI and LSI techniques are compared
with the massive LC delay lines of a few years ago. Even modern active RC
and active RLC delay lines, although much smaller than their passive LC
counterparts, appear clumsy against the versatility of digital shift registers.

However, one feature of the analogue circuit, namely its capability for
handling wide variations in signal level, is particularly valuable in underwater
acoustics where very large variations in signal level are encountered. The
range of variation or dynamic range is determined at the high end by the
onset of non-linear distortion and at the low end by the system noise.
Linear active circuits, restricted to working between the supply voltages
and suffering from amplifier noise, offer a range of from about 100 dB to
80 dB, depending upon the internal gain or "Q" factors in the network.
Passive circuits can offer much more than this.

With digital circuits, the dynamic range is directly related to the number
of levels or bits into which the input signal is quantized. A rough figure of
6 dB per bit means that 10 bits are needed for 60 dB and 16 bits for 96 dB.
The cost of providing wide-range analogue to digital conversion for the
many elements of modern transducer arrays is therefore very high, and for
this reason analogue beam-forming is sometimes used to provide a smaller
number of input signals for subsequent digital processing.

This paper presents an alternative approach in which some of the versatility
of digital methods is combined with the merits of analogue signal circuits.
The salient advantages and limiting factors of this approach are illustrated
by a description of their use in beam-forming arrangements, in sampled-
data filtering, and in the adaptive equalization of pulse distortion. At
present the last is applied to line and radio communication; but its possible

477

use in conjunction with other techniques, for improving underwater communication in conditions of multipath propagation is being studied.

2. Sampled Analogue Delay Element

The basic element of this technique is the simple sample-and-hold delay network [1, 2] shown in Fig. 1. It comprises an input commutating switch, a number of capacitors (N), an output commutating switch, and a voltage-follower amplifier. The switches, which are conveniently realized by m.o.s. transistors, are operated synchronously but out of phase, the

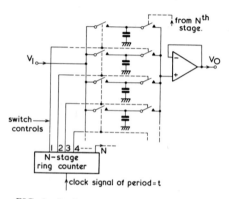

FIG. 1. Basic sample-and-hold network.

output switch being one step ahead of the input switch. The operation of the delay network is as follows:

The input signal is applied to the first capacitor by the associated input switch for the duration of the dwell time, t. The switch is then stepped on to the next capacitor for a further time, t, while the first capacitor holds the input voltage as sampled when the switch stepped on. This is repeated until, $(N - 1)t$ later, the signal is applied to the last capacitor. At the same time the output switch for the first capacitor is switched on and the first voltage sample is applied to the voltage-follower. Thus the network provides, for an analogue signal, a delay of $(N - 1)t$ which can be varied as a function both of N and of t. The choice of N is determined by the required maximum delay-bandwidth product, as discussed below, but t can be varied continuously by varying the period of the clock to the ring counter driving the switches. The network may therefore be considered as an analogue shift register element.

There is a superficial resemblance between this network and the so-called "bucket-brigade" delay lines of Sangster [3] and others [4]. The fundamental difference is that, whereas the latter provide a clocked delay by transferring a charge deficit representing the input signal sample from stage to stage, the network described here is a voltage-transfer circuit in which the signal sample is regenerated by each voltage-follower with very low loss.

The salient features of the network are briefly listed below. Several points are discussed in more detail in the content of specific arrangements later.

(a) Since the signal cannot be sampled at less than twice the highest signal frequency, $\hat{f}_s$, $t \leqslant 1/2\hat{f}_s$ and the maximum delay bandwidth product is given by $(N-1)/2$. However, in practice, to allow for realizable filters to prevent aliasing problems, $(N-1)/3$ is a better design criterion.

(b) The basic network may be adapted to provide for adding-in signals by replacing the voltage-follower by a differential summing amplifier as shown in Fig. 2, the gain with respect to the non-inverting input being compensated-for by an attenuator in the output.

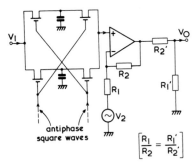

FIG. 2. $N = 2$ delay network with add-in facility.

(c) The value of the storage capacitance is not critical.

(d) The network is particularly suitable for hybrid circuit realization, since the only critical components are resistors, which can be produced cheaply with high accuracy.

3. Beamformer for 12-stave Array

The first application of this network was for providing a continuously-steerable beam from a 12-stave array. An $N = 3$ network was used to provide an interstave delay of up to 67 μS for a maximum signal frequency of 6 kHz. First demonstrated at the 1968 Physics Exhibition in London [5] the system, outlined in Fig. 3, comprised a sector switch, which reversed the sequence of connections from the tranducers to the delay line to provide steerage into either quadrant, a line of twelve $N = 3$ networks realized as tantalum thin film networks with μA 709c amplifiers, and an 8th-order Chebyscheff low-pass RC active filter. The switching waveforms were generated in a 3-stage ring counter driven from a clock generator with a sinusoidal frequency characteristic to give linear angular control of steer.

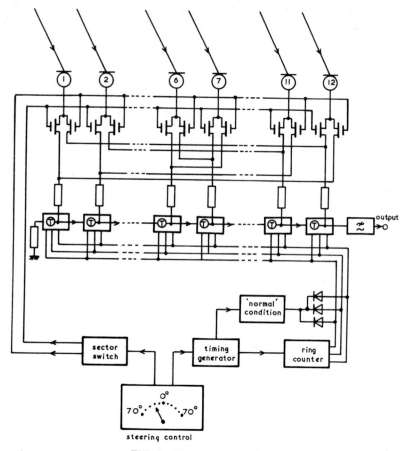

FIG. 3. 12-stave beamformer.

The normal (perpendicular) direction was obtained by turning all switches on, to give a direct connection down the line. The in-band noise at the output of the delay line was less than 10 μV with the maximum delay setting. The maximum signal level was about 2V, so the potential dynamic range was greater than 100 dB. (In fact this range was reduced by the RC active low-pass filter.) A source of noise which has to be considered was that arising from subharmonics of the switching frequency, f_c, the principal component being at f_c/N. For the delay setting of 67 μS, this component was less than 2 mV at 10 kHz, so the output low-pass filter was designed to provide at least 50 dB suppression at this frequency and above.

4. Beamformer for 28-stave Array

Following the evaluation of the 12-stave system a more ambitious specification was drawn up to provide from 4 μS up to about 200 μS of interstave delay in increments of $\frac{1}{2}$ μS, for a signal frequency range of up to 15 kHz, a maximum delay-bandwidth product of 3. In principle this could have been met with an $N = 10$ network of the type already discussed, but the subharmonics of the sampling frequency for the larger delay values would have fallen well within the signal band and could not have been suppressed by filtering. The interstave delay network was therefore designed as nine cascaded $N = 2$ sections. Other problems imposed by the specification were a loss variation of about ± 0.02 dB on each 9-section interstave network, the need to equalize the $\sin x/x$ distortion inherent in the sampling process, a fast sampling rate for the 4 μS delay, and a usable dynamic range approaching 80 dB.

The circuit adopted is shown in Fig. 4. To meet the stringent loss

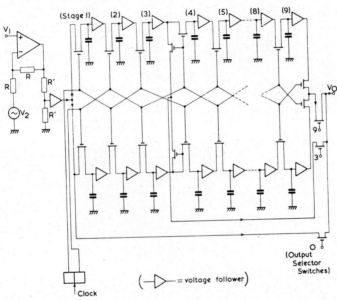

FIG. 4. Schematic circuit of 9-stage delay network.

requirements, each capacitor was buffered with a voltage-follower with a fast slew-rate (initially LM 302, subsequently LM 310) and the two paths were not recombined until the output stage. The delay range was split into two ranges by taking an output after the third stage for the lower delay values. Zero delay was provided by taking the output from the summing amplifier. Sector-switching was provided as before with a 28-way changeover switch comprising 28 dual m.o.s. transistors (ML 102B). Delay was controlled digitally in $\frac{1}{2}$ μS steps by counting down from an 18 MHz crystal oscillator using Signetics M.S.I. to provide increments of $\frac{1}{6}$ μS for the three-section network and $\frac{1}{18}$ μS for the nine-section network, the countdown being controlled by the range switch so that the delay setting switch was correct for either range. Equalization of the sin x/x distortion was provided by a 1st-order sampled-data filter [6] immediately preceding the output filter, an 8th-order low-pass active network using low-noise LM 307 amplifiers with frequency compensation [7]. The input summing amplifier on each section was a μA 740c, which offered high input impedance, high slew rate, and low noise (compared with other devices with high slew rate).

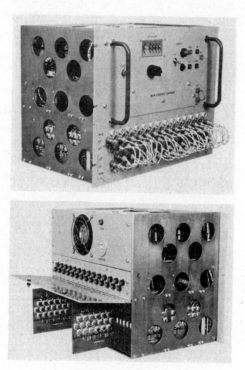

FIG. 5. 28-stave beamformer.

The equipment, which in practice provides the required delay variations with an overall loss variation which increases to a total of about 1.5 dB at the extreme ranges, provides a usable signal dynamic range of more than 70 dB at maximum delay with all delay networks in use. The lower limit is set by the system noise, most of which arises from the demodulation into the signal band of the wide-band noise of the voltage-followers by all the harmonic components of the square-wave sampling pulses. For small delays the sampling frequency is higher and the demodulated noise decreases until it falls below the noise of the active filter, which imposes an ultimate dynamic range of about 80 dB. The equipment is shown in Fig. 5, with two of the interstave networks partially withdrawn.

5. New 28-stave Beamformer

For a new development in which a maximum of only 45 μS of interstave delay is required for a 10 kHz signal band, an $N = 2$ network of the original type is used, but with more complex switching waveforms. The smaller number of amplifiers permits (on economic grounds) the use of μA 740c amplifiers throughout, which leads to a much lower system noise. A dynamic range approaching 90 dB is expected, which will be preserved by the use of part active and part-passive filtering in the output stage.

6. Conclusions—Beamforming

The equipments described above demonstrate that the use of sampled-analogue techniques offers the flexibility normally associated with digital techniques while providing a usable dynamic range that would be very difficult to achieve with digital processing. The use of devices like the μA 740c is costly at present; but considering that the μA 709c amplifiers, which cost £10 each when the initial experiments were conducted, are now available for 20p, a more liberal use of such devices is hopefully anticipated.

7. Sampled Data Filtering

Sampled data filtering is now becoming commonly used, as is shown by the frequent references to digital filtering in the papers presented at this Institute; but it appears that most filtering is done by processing in digital computers in conjunction with other operations. References to real-time digital filtering are not so common. The reason for this is the present high cost of the digital hardware to realize even simple filter functions, LSI being economical only when applied to large volume requirements. Until the cost of LSI does come down, sampled-analogue filters offer an inexpensive interim solution to the small-quantity high quality sampled-data filter requirement.

A typical structure is shown in Fig. 6. where the delay functions are provided by $N = 2$ networks of the Fig. 1 type, and the differential operational amplifiers and resistors provide the multiplication and addition functions. This is a general 6th-order structure for a transfer function factorized into three biquadratic factors. This form of realization offers:

    (a) quantization in time: the filters can be "tuned" by means of a clock frequency just as a digital filter. Using the devices mentioned previously, accurate filters can be made with clock rates up to 100 kHz.

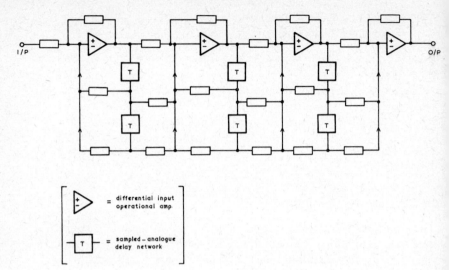

FIG. 6. 6th order sampled-data filter (three 2nd order factors).

(b) high accuracy: 0.1% accuracy (10 bits) is easy with current thin or thick film resistors.

(c) wide dynamic range: the delay network has a dynamic range of about 110 dB; a complete filter offers about 80 dB, depending upon "Q" factors and complexity.

(d) cheap multiplication and addition.

The use of multiplexing is possible, but economically it is not worth the trouble.

A recent application of this technique was for a tuneable 6th-order bandpass filter. The filter was designed to have an arithmetically-symmetrical pass band 2 kHz wide at 10 kHz, using a 50 kHz sampling rate, and a minimum stop band attenuation of about 22 dB. The amplitude frequency response is shown in Fig. 7. An earlier application was the sin x/x equalizer referred to on page 482. Here the particular merit was that the equalization was "timed" to the distortion, since both were functions of the clock frequency.

8. Adaptive Equalization

In the field of line (and to a lesser extent radio) communication, adaptive equalization is being used in conjunction with sophisticated modulation techniques to provide more efficient use of the available channel capacity [8]. For example, it is now possible to transmit data at 4800 bits/s where previously only 600 bits/s was possible.

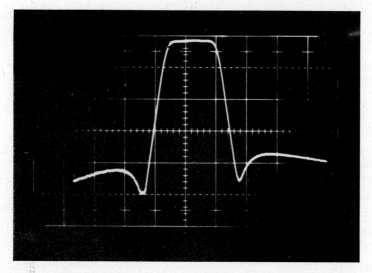

FIG. 7. Amplitude response of 6th order bandpass filter.

The basic unit of the adaptive equalizer is the transversal filter of Fig. 8. It comprises a tapped delay line, the taps of which are provided with adjustable gain. Early versions of this network were built with LC delay networks which, for voice-band data transmission, were large and expensive. Alternative versions have comprized an input a-d converter followed by mutilevel digital processing, but such techniques await a dramatic drop in the cost of LSI before they become economically

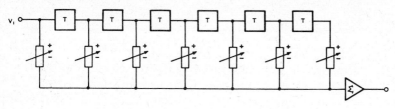

FIG. 8. Schematic circuit of transversal filter.

attractive. Other approaches have used non-linear quantization methods to avoid the problems of the analogue line.

The sampled-analogue delay network has proved to be very attractive for this application, since it provides an inexpensive but accurate and low-loss linear delay. Moreover, the inter-tap delay can be accurately synchronized with the data symbol rate for optimum performance. Being an analogue network, it operates equally well with binary or multilevel signals, whereas the quantization and hence the complexity required of digital delay lines is necessarily a function of the signal characteristics.

The operation of the equalizer is illustrated in Fig. 9. The upper trace shows (on a slow time-base of 100 mS/division) a 4-level pseudo-random data signal at about 6000 symbols/sec (12 000 bits/sec) entering the adaptive equalizer. The lower trace shows the equalizer output. In the middle of the picture, very severe distortion is introduced into the line, sufficient to close a 2-level pattern. The adaptive equalizer, operating on an estimated reference, is seen in the lower trace to converge within about 75 mS.

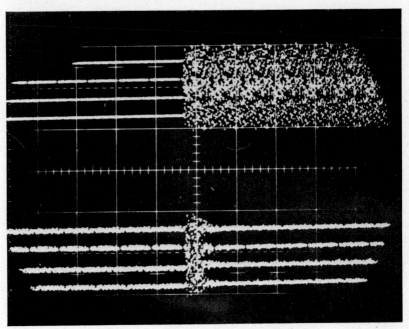

FIG. 9.   Convergence of adaptive equalizer.

The comparatively static and well-defined characteristics of line communications are, of course, not to be compared with an underwater channel under conditions of severe reverberation; but used in conjunction with other techniques, adaptive equalization will provide additional flexibility in dealing with time-varying multipath distortion of digital transmission. Although the wide signal dynamic range of the sampled-analogue network is not so relevant here, the ability to process non-quantized signals on a flexible time axis will be a significant asset to the system designer of the future.

REFERENCES

1. Galpin, R. K. P. (1968). Variable electronic all-pass delay networks. *Elect. Lett.* 4, p. 137.

2. Saraga, W. (1968). Low frequency analogue micro-circuits for acoustic processing. Proc. NATO Advanced Study Institute, Enschede, The Netherlands.
3. Sangster, F. L. J. and Steer, K. (1969). Bucket-brigade electronics . . ., *IEEE. J. Solid State Circuits*, **SC**-4, p. 131.
4. Feature articles, *Electronics*, Feb. 28th, 1972.
5. Exhibition Report, *Electronics*, April 1st, 1968, p. 145.
6. Galpin, R. K. P. (1969). First-order correction for zero-order hold distortion. *Elect. Lett.* **5**, p. 574.
7. Lim, J. T. Private communication.
8. Conference Record. Digital Processing of Signals in Communications, I.E.R.E., Loughborough, England, April 1972.

DISCUSSION

C. N. Pryor: You say that in your adaptive system you have no reference waveform; do you generate your error function from the nearest of the four levels that you expect?
Answer: Yes, if the probability of error is less than $\frac{1}{2}$, the excess of correct decisions will start the convergence. With very severe distortion the equalizer hesitates until a favourable sequence in the randomized data stream occurs which starts the convergence. Once started, the rate of convergence rapidly increases as the error probability decreases.
C. N. Pryor: Your error function is just the distance from the nearest of the four levels?
Answer: Yes.
T. Curtiss: With regard to Sangster's "bucket-brigade" line, which is fully integrateable in monolithic form, have you considered the use of bipolar switches in your circuit?
Answer: Yes, we have. Our problem is that the high accuracy of this circuit exploits the particular merits of the m.o.s.t. switch and of the bipolar operational amplifier. On one hand the bipolar switch has offset and a limited signal handling capacity the offset being a variable quantity, compared with the zero offset and wide dynamic range of the m.o.s.t. switch. On the other hand, an m.o.s.t. amplifier, even if one were possible with adequate performance, would not be compatible with the m.o.s. process required to produce a good m.o.s.t. switch.

A further problem is that we need access to every tap and so would have a large number of external connections for a comparatively small integrated circuit.

Acoustic Signal Processing in Fast Unmanned Underwater Vehicles

E. J. RISNESS

Admiralty Underwater Weapons Establishment,
Portland, Dorset, England.

1. Introduction

Most, if not all, of the papers presented at this symposium are aimed at discussing signal processing techniques which could be of use in underwater acoustic applications. The applications themselves are not discussed at length because a discussion of those of interest in the defence field can lead on to classified areas rather quickly. However, participants at the Institute have a general picture of one of the major uses of underwater acoustics, namely the detection, location and classification of objects at the maximum possible range from a mobile manned platform—a definition which covers the use of sonar for fish-finding from ships, or for detecting and tracking submarines from ships or other submarines. This leads, for example, to a focusing of interest on such problems as the optimum detection and extraction of signals buried in noise, problems of display and man-machine interaction, and the propagation of acoustic signals to long distances.

The purpose of this paper is to point out another major area of application for underwater acoustics which is less often discussed, but is nevertheless important in defence and may have wider interest too. This is the use of acoustic signal processing in fast unmanned underwater vehicles. The acoustic homing torpedo is the most obvious example of such a vehicle, but the class also includes such devices as autonomous acoustic "targets" which can be used to test sonar detection and weapon homing systems. Such vehicles, though currently developed primarily for military purposes, could have civil applications for sea-bed exploration and other oceanological activities.

In the following discussion, some of the characteristics and constraints of underwater vehicles of this type will be discussed. It is hoped that this discussion will remind those concerned with developing new acoustic signal processing techniques of the possibility of their application in such vehicles.

2. Need for Acoustic Communication

The first point to be made is the obvious one that in-water vehicles can normally communicate with their environment only by acoustic means. There are two significant exceptions to this generalization. One is that electromagnetic communication can occur over very short ranges (a few tens of feet at most) and with very restricted bandwidth. The second is that a vehicle can sometimes be connected to its launching platform by a wire which can carry a useful bandwidth of some kHz. But apart from these exceptions, all information passing to the vehicle (e.g. the position and characteristics of other vehicles, distance from the sea bed, etc.) or from the vehicle (e.g. the position and movement of the vehicle itself) must use acoustic communication.

The limitations on acoustic communication imposed by the medium are already well known to you. However there are further limitations imposed by the vehicle itself, which will now be briefly discussed.

3. Size and Shape of the Vehicle

Underwater vehicles of the type considered here are very restricted in size and shape. Vehicles carried by, and dropped from, aircraft and air-flight missiles are constrained to be as light as possible and usually as small as possible also. Such vehicles are typically around 0.3 metres in diameter and 2 to 3 metres long. Vehicles launched from submarines are constrained by the size of the launching tube through the submarine hull. The standard launching tube is designed for vehicles of about 0.5 metre (21 inches) diameter and up to about 6.5 metres long, and although there are moves to be able to handle vehicles of larger diameter, the diameter is still rigorously constrained by the need to launch vehicles through the pressure hull.

These size and shape constraints place strict limitations on the size of acoustic arrays for transmission or reception. In ship-mounted sonar design, trade-offs can be made between sonar size (and hence cost) and performance, but with unmanned vehicle sonars the size cannot be extended beyond strict limits.

4. Effect of Self Noise

A further constraint on acoustic arrays in a fast vehicle is imposed by flow noise generated by the vehicle itself (self noise). As the vehicle moves through the water, the water flow past the body results in a turbulent boundary layer all over the surface except perhaps for a limited area at the nose. The turbulence generates so much noise that the reception of weak signals is possible only at the front of the torpedo—and even then great care is needed in designing the nose shape and the transducer array to minimize this noise. The propulsor which drives the vehicle through the water, and the machinery which drives the propulsor, also produce noise which couples into any sensitive acoustic receiving system and is a further reason for mounting receivers as far away as possible in the noise of the vehicle.

This means that a sensitive acoustic receiving array on such a vehicle is normally confined to an area at the nose of the vehicle of diameter less than the vehicle diameter.

These constraints on acoustic array sizes, particularly for reception, imply that if a system is required of a given beamwidth (or spatial resolution) then it must be designed at or above a given minimum frequency. In practice this means frequencies in the tens of kHz region are of interest, whereas lower frequencies can often be used on larger sonars on ships. Techniques for narrowing beamwidths, such as the use of non-linear acoustics of super directive arrays, can be more attractive than with ship sonars, because the alternative of building a bigger array is completely impossible.

It should be pointed out that with a listening array in the nose of the vehicle the sources of self noise are behind the array whereas the direction from which the signals of interest are most likely to come are ahead or to either side. Thus array directivity can give substantial gains, perhaps more than predicted by the simple theory based on isotropic background noise.

5. Detection and Location of Targets

A prime function of the vehicle is likely to be to detect, identify (if necessary), locate, and converge on to a "target" using either active or passive sonar. If the vehicle runs wholly in the water, as when launched from a ship or submarine, the direction of the target is usually approximately known, and the vehicle is set to run out along the bearing line. The main criterion for the vehicle's own sonar is then that the width of the path, stretching on either side of the vehicle's track, within which the target can be acquired by the vehicle should be as large as possible. This can be achieved by spreading acoustic beams over a fan of bearings ahead and to either side of the vehicle. The swept path can be further enlarged by programming the vehicle to sinuate about its mean line of advance at an appropriate amplitude and periodicity. However this results in some loss in mean speed of advance.

If the vehicle is first carried through, and then dropped from, the air it is normally released over the most likely position of the target. Consequently the vehicle on entering the water is programmed to carry out one or more circular search patterns aimed at locating the target if within a certain cylindrical volume whose axis is vertically below the dropping point and whose height covers the full water depth which the target is able to reach. The sonar is then designed, in conjunction with the vehicle's trajectory, to maximize the radius of the cylinder.

6. Converging on to a Target ("Homing")

Having located a target the vehicle is normally expected to converge or "home" on to it. The change from initial detection at long range, when the received signals (active or passive) are only just above threshold, to a point of close approach to a large distributed target implies a very wide change in sonar operating conditions. The mode of operation of the sonar itself, and the way in which its information is used by the vehicle's control system, may have to change very significantly during this sequence. The point is well illustrated by the way bats vary their active sonar transmissions as they converge on to their prey. It should be noted, however, that in this case the prey is much smaller than the bat itself, whereas the reverse is generally the case with underwater vehicles and their targets. Thus with underwater vehicles accurate final navigation may be less important, but the target may show greater variation in its sonar characteristics as its apparent angular size increases as seen from the vehicle's sonar.

It is obvious that in the whole process from initial search, detection, location and convergence, through to the final stages of homing, there is a very close interaction between the vehicle's dynamics and the way it interacts with its acoustic environment. Background noise varies very markedly with vehicle speed and manoeuvre. Reverberation from sea surface and sea bed depend critically on where the vehicle is and how it is orientated in relation to these surfaces. The way signals from the target behave as a function of time also depend on vehicle manoeuvre in a partly predictable way. The fact that the whole process is not even quasi-stationary, but is undergoing dynamic change, may make techniques such as Kalman filtering of particular interest to underwater vehicle engineers.

7. Other Acoustic Functions

Underwater acoustics can be used for other functions besides detection and location of targets. A major use is to pass accurate position information on the vehicle itself and other significant in-water objects during trials and exercises. This is normally done by transmitting from each vehicle a pulse or series of pulses at accurately known intervals, receiving the signals on spaced hydrophones at known positions, and calculating the vehicle's position from the times of arrival of the pulses.

One important feature of this and other supporting uses of underwater acoustics is that the total acoustic bandwidth is extremely limited, and all functions have to be carefully interleaved to maximize the information transmitted and minimize mutual interference. If as science and technology progresses it appears desirable, for example, to design a homing system at a different frequency, this may have repercussions on other functions and may necessitate the re-development of, for example, the vehicle location system.

8. Effect of Modern Technology

A final point worth making is that, in spite of the strictly limited space inside an underwater vehicle, modern electronic and associated technology has made such progress in the miniaturization of equipment that complex electronic processes can now be carried out in vehicles which a decade ago could not be fitted in. Other constraints, such as cost or reliability, may mitigate against incorporating complex processes; but in general there is now a *prima facie* case for considering quite complex signal processing and related techniques which could be of value to the development of improved underwater vehicles.

9. Conclusions

In summary, the aim of this paper has been to draw attention to the application of acoustic signal processing techniques to fast unmanned, underwater vehicles. Some of the more significant properties of these vehicles are:

(a) Strict size and weight limitations.
(b) Self-noise, with some directionality, due to fluid flow and to the propulsor and its machinery.
(c) The complex sequence of activity from initial search and detection through to terminal homing, and the close interaction between acoustic information and vehicle manoeuvre throughout this sequence.
(d) The need for other acoustic functions (e.g. vehicle location) and the problems of interference between different acoustic activities.
(e) The possibility of complex signal processing and other functions, given modern micro-electronic techniques.

DISCUSSION

H. Mermoz: Have you made any measurements of the correlation of the background noise between hydrophone receivers on a torpedo, and if so do they show that optimum beamforming techniques might be of value?
Answer: I cannot give details of any measurements, but it is clear from the geometry of the situation that there is a real possibility of such techniques being applicable.
J. J. Dow: If the pulse repetition frequency of a homing system is varied, this must surely raise some very difficult problems in normalization of the received signals.

Conventional TVG (time-varying gain) may no longer be possible. Do you agree?

Answer: A torpedo homing system has to cope with a very large dynamic range of signals and background since it needs to perform from maximum range (distances) right in to nearly zero range. Thus the normalization is very difficult in any case, irrespective of whether changes occur in pulse repetition frequency.

A. Khandekar: Some torpedos are liable to "bottom capture" in shallow water. Can you comment?

Answer: Not in detail, obviously but the point you raise illustrates the complex environmental situation facing the acoustic sensors in a torpedo, which have to distinguish between wanted targets, other targets, sea bed and sea surface over the full homing sequence of operations.

P. H. Moose: You mentioned the possible application of nonlinear acoustics. Are you interested in both parametric transmission and parametric reception?

Answer: As I said, interest stems from the fact that array sizes are strictly limited and hence parametric transmission is worth considering. Parametric reception clearly poses problems in how to accommodate the equipment required for the acoustic interaction within the overall size and shape constraints of the torpedo.

N. L. Owsley: Since self-noise is highly directional there should be a real possibility of a relatively cheap adaptive beamforming (null steering) system to give worthwhile gains.

Results on Active Sonar Optimum Array Processing

C. GIRAUDON

CIT-Alcatel, Arcueil, France

1. Introduction

During the last decade, a new concept has evolved in antenna processing: the optimum adaptive system (OAS). This concept applies to the detection by an N element array of a signal issued from a point source imbedded in a non-stationary noise field.

According to the theory developed by Mermoz (Reference [1, 2]), the basic scheme which is obtained, using a maximum signal over noise ratio criterion, is the following:

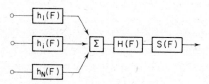

Assumption: the wanted signal components are equal on all input branches.

Each elementary filter $h_i(F)$ depends on a submatrix of the noise cross-spectral matrix M. Filter $H(F)$ also depends on M and is undefined when M is singular i.e. $D(M) = 0$. Filter $S(F)$ only depends on the form of the signal and is in fact the usual matched filter. The adaptivity is obtained by "refreshing" the $h_i(F)$ and $H(F)$ filters according to the supposedly slowly varying noise field. Thus defined, this scheme is applicable to an un-restricted bandwidth, which however leads to a peculiar case [$H(F)$ undefined] when M gets singular. This peculiar case corresponds to an n discrete noise point source distribution ($n < N$) for which it can be shown that $H(F)$ need not be defined and that it is always possible to find $Nh_i(F)$ filters that completely cancel a number of noise makers up to $N - 1$.

Under a narrow bandwidth assumption, the above scheme can again be simplified by omitting the $H(F)$ filter which then reduces to a complex scaling factor. One obtains a system working correctly irrespectively of the noise source distribution, i.e. discrete or diffuse. The OAS when confronted with discrete noise jammers eliminates $(N - 1)$ jammers; in all other cases it gives the highest attainable signal over noise ratio at the output.

This simplified scheme allows us to imagine numerous practical combinations, one of which is sketched below.

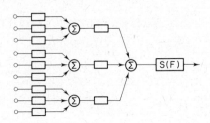

These last systems, although generally suboptimum (SOAS) allow antenna processing devices to be built which are amenable to a simple and repetitive technology and which however give interesting results.

Within these guidelines, we are developing SOAS using pairwise and triplewise combinations of hydrophones. In the following paragraphs we give experimental results so far obtained in a laboratory environment on an analogue version of these doublets and triplets.

2. Doublet: Pairwise Combination of Hydrophones

(a) Analytical description
Working under the narrow bandwidth assumption we use the complex amplitude representation where U stands for a real steady signal $u(t) =$ Real$[U \exp(j2\pi F_0 t)]$. For a noise term, or a pulse, complex U would be a slowly varying function of time.

The general solution of optimum filtering is of the form:

$$h_j(F) = \sum_j M_{jk}(F).$$

The cross-spectral matrix being

$$M(F) = \begin{bmatrix} C_{11}(F) & C_{12}(F) \\ C_{21}(F) & C_{22}(F) \end{bmatrix}$$

we get

$$\left\| \begin{array}{l} h_1(F) = C_{22}(F) - C_{12}(F) \\ h_2(F) = C_{11}(F) - C_{21}(F). \end{array} \right.$$

The system structure is then:

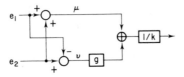

One can show that an equivalent structure is the following

where

$$u = e_1 + e_2, \quad v = e_1 - e_2, \quad k = \frac{2}{h_1 + h_2}, \quad g = \frac{h_1 - h_2}{h_1 + h_2}.$$

One can rewrite $h_1 + h_2 = C_{11} + C_{22} - 2 \, \text{Real} \, (C_{12}) = \text{real}$; k being merely a real scaling factor can be neglected.

The system thus reduces to a single transfer function

$$g(F) = -\frac{C_{11}(F) - C_{22}(F) + 2j \, \text{Im} \, [C_{12}(F)]}{C_{11}(F) + C_{22}(F) - 2 \, \text{Real} \, [C_{12}(F)]}.$$

Let us call F_0 the centre frequency, ϕ the bandwidth and let $\phi \to 0$. We get:

$$E[b_j(t)b_k(t + \tau)] = 2\phi \mid C_{jk}(F_0) \mid \cos \, [2\pi F_0 \tau + \arg \, C_{jk}(F_0)]$$

and after some mathematical manipulation

$$E \, \frac{[(v(t))^2]}{2\phi} = C_{11}(F_0) + C_{22}(F_0) - 2 \, \text{Real} \, [C_{12}(F_0)]$$

$$E \, \frac{[u(t) \cdot v(t)]}{2\phi} = C_{11}(F_0) - C_{22}(F_0)$$

$$E \, \frac{[u(t) \cdot v(t) + \pi/2]}{2\phi} = 2 \, \text{Im} \, [C_{12}(F_0)].$$

The transfer function g thus reduces to a complex scaling factor and can be written as

$$g = -\frac{\overline{UV} + j\overline{UV}\pi/2}{\overline{V^2}}.$$

The output of filter g is then

$$\rho = Vg = \frac{V}{\sigma_V} \left[U \cdot \frac{\overline{V}}{\sigma_V} + jU \cdot \frac{\overline{V\pi/2}}{\sigma_V} \right].$$

The structure immediately follows

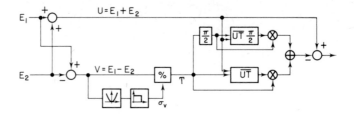

(b) Measurements on a doublet

The experiment was set as sketched in Fig. 3. We dispose of the parameters

$$K = \frac{\sigma B_1^2}{\sigma B_2^2} = \text{noise components power ratio}$$

C: noise components correlation coefficient (normalized)

φ_B: phase angle between the noise components

We synthesize

$$\left| \begin{array}{l} E_1 = S_1 \sin wt + B_1(t) \sin [w_1 t + \theta(t)] \\ E_2 = S_2 \sin wt + B_2(t) \sin [w_1 t + \theta(t) + \varphi_B] \end{array} \right.$$

The noise generators give wide band noise in the 4900 ± 400 Hz. We measure a gain defined as:

$$G = \frac{[S + N/N] \text{ doublet}}{[S + N/N] \text{ direct } \Sigma}$$

in the 4900 ± 400 Hz. This is the gain of the doublet as compared to the simple sum of E_1 and E_2 as would be done in a usual beamformer. The wanted signal is a steady sinusoid at 4900 Hz.

Figure 1 gives the results.

On Fig. 1(a) it is noticeable that $C = 1$ gives a drastic improvement althou the theoretical value of G would be infinity (cancellation of noise makers). It is also noticeable that $G = 0$ for $\varphi_B = 0$ degree and 180 degrees, which seems normal since:

(i) at $\varphi_B = 0$ and $K = 0$ dB nothing can distinguish the wanted signal from the undesirable components,

(ii) at $\varphi_B = 180°$ and $K = 0$ dB, the noise components are readily eliminated by the direct sum.

Most remarkable are the improvements at small values of C which appear from Fig. 1(b)–(d). It is also interesting to notice on these figures that when $C = 1$ even at $\varphi_B = 0$, a difference in noise components power ratio

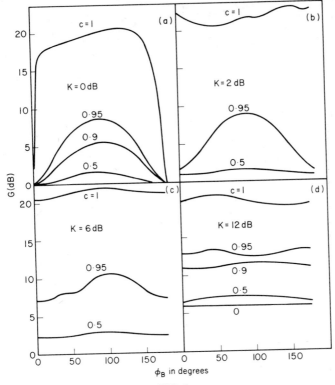

FIG. 1.

can drive the system towards a cancellation of these noise components.
Figure 2 is a resumé of the former cases.

Other measurements using CW pulses instead of steady CW were made
which confirmed the preceding results.

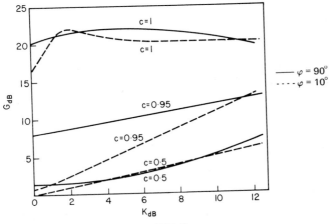

FIG. 2.

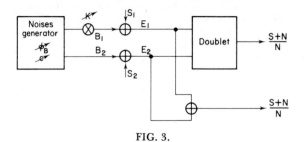

FIG. 3.

3. Triplet: Triplewise Combination of Hydrophones

(a) Description of a triplet

We do not give the analytical description of the triplet due to the lengthy derivation. The reader will realize from the sketch on Fig. 4 that the triplet somehow combines two doublets to reach its goal.

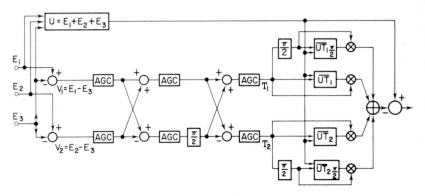

FIG. 4. Triplet.

One can distinguish 3 kinds of operations namely:

  (i) combinations: sums or differences,

 (ii) normalization: on this sketch "AGC" stands for the following operation:

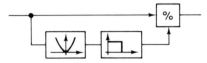

(iii) correlation: one finds here twice the preceding doublet correlation block.

This triplet must eliminate 2 discrete noise makers, and otherwise gives the best output signal over noise ratio.

(b) Measurements on a triplet

These measurements were undertaken with a simulator shown in Fig. 5.

The desired signal is a 100 ms pulse of 4900 Hz CW. Two wide band (4900 ± 400 Hz) independent noise makers are simulated which can sweep the horizon with variable phase angles on each hydrophone (simulating two

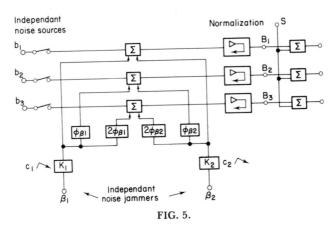

FIG. 5.

rotating, remote, independent noise sources). In addition to the noise makers, one can add to each hydrophone calibrated independant noises which make it possible to vary the global noise correlation coefficient between adjacent hydrophones. The integration times were the following: triplet's AGC: 20 ms; triplet's correlator: 200 ms; signal detection at the triplet's output: 100 ms.

We evaluated a gain defined by:

$$G = \frac{S + N/N \text{ triplet}}{S + N/N \text{ direct } \underset{3}{\Sigma}}$$

Figure 6 gives the values of G obtained when both equal-power noise jammers were the only disturbance superimposed on the desired signal (i.e. correlation coefficient of 1 between hydrophones) with variable phase angles φ_{B_1}, φ_{B_2}.

One observes that the maximum gain is obtained when both jammers issue from a same direction: $\varphi_{B_1} = \varphi_{B_2}$. A singular point: $G = 0$ is noticeable at $\varphi_{B_1} = \varphi_{B_2} = 120°$.

Figure 7 gives the same picture when the total noise power on all hydrophones were set unequal (a 2 dB difference between adjacent hydrophones).

Figure 8 is a synthetic view of the doublet and triplet performances.

For each device we pictured the situation (i.e. $\varphi_B = 90°$ for the doublet; $\varphi_{B_1} = \varphi_{B_2} = 60°$ for the triplet) of maximum gain as above defined against

the adjacent hydrophones correlation coefficient. In both cases we took
equal values for the global noise powers on each hydrophone.

These measurements are a few samples of the innumerable situations one
can imagine on a triplet. They show however that although the triplet's
ultimate maximum gain is slightly less than the one for the doublet
certainly due to system noise, in every other less critical situation
$(C_i, i +1 < 1)$ the triplet performance is better. Other measurements were
undertaken with narrow band and sinusoidal jammers which gave equally
satisfying results. One has also noticed that the triplet performance is
tolerant of a slight desired-signal unbalance in amplitude and phase
between the hydrophones. This last point is most important with
regard to the desirable quality of the preceding beam-former (quantization
of amplitudes and time delays).

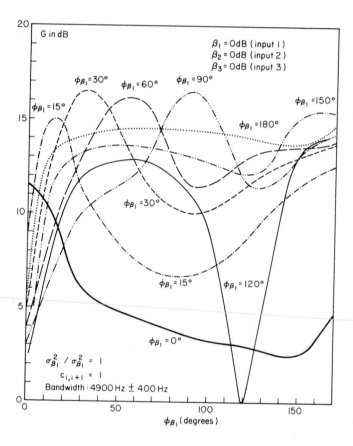

FIG. 6.

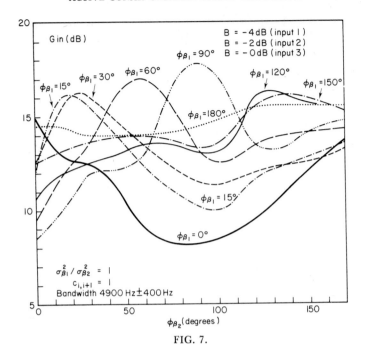

FIG. 7.

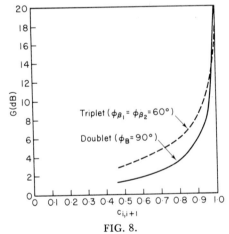

FIG. 8.

4. Application to Active Sonar Design

The interest in SOAS stems directly from the experimental fact that on sea trials one often measures correlation coefficients, between adjacent hydrophones, lying somewhere between 0.65 and 0.9, on reverberation returns, with frequent power unbalance. Therefore, in view of the surprising results so far obtained with an SOAS sonar using simplified doublets (performing less than the one herein described) we expect good results of an SOAS sonar using triplets.

If the breadboard model herein described operates in analogue form, a practical realization would obviously be in digital form. The digital realization of an SOAS system using triplets as a basic building block is to be undertaken on a global system point of view, so as to take advantage of the maximum time multiplexing possibilities. The principle of the digital triplet would remain unchanged except for the wide use in analytic signal low frequency components. There is a variety of digital complex demodulation algorithms which can be used.

This means that the beamforming, to be performed before the SOAS (excepting the final sum), should be realized on the low frequency components of the complex amplitudes. Therefore a possible building block diagram for the SOAS sonar is the following (Fig. 9):

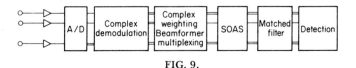

FIG. 9.

REFERENCES

1. Mermoz, H. (1964). Proceedings of the NATO Advanced Study Institute, Grenoble, France, 1964.
2. Mermoz, H. (1968). Proceedings of the NATO Advanced Study Institute, Enschede, Netherlands, 1968.

DISCUSSION

N. L. Owsley: Did you notice a time constant effect i.e. a signal suppression when using short time constants?
Answer: In the doublet and triplet just described the adaptation takes place in a noise only branch free of any signal if present.
H. Mermoz: It means that even in the case of a very long signal the system adapts to the noise only.
N. L. Owsley: When you cascade triplets with adaptive weights have you not some signal suppression?
Answer: There are no weights between cascaded triplets. These devices have no effect on the signal, they only work on the noise.
P. Heimdal: Did you think of extending these devices to 4 elements?
Answer: Yes we have, however the practical gain will not increase with the number of elements as the theoretical one.
H. Mermoz: In optimal processing, the higher is the cross correlation of noise between sensors the higher is the gain you may expect from the processing. Therefore you would rather process small groups of clustered sensors than more widely separated sensors.

P. M. Schultheiss: In the case of interference suppression would it not be better to use more widely spaced sensors?

M. Mermoz: As was said before the performance only depends on noise cross correlations, as for the sensitivity to signal deviation it is comparable to that of a conventional beam former.

P. Heimdal: What are the effects of discrepancies between channel gains and phases?

Answer: These devices are tolerant enough to channel gains and phases unbalance as in a conventional beamformer.

R. B. Gilchrist: Did you try any experiment on cascaded triplets?

Answer: Not yet.

G. Vettori: How would you choose the locations of the doublets on a transparent circular array?

Answer: We would choose to treat by doublet couples of adjacent sensors.

Design of Arrays to Achieve Specified Spatial Characteristics over Broad Bands

S. W. AUTREY

Hughes Aircraft Company, Fullerton, California, U.S.A.

1. Introduction

This paper deals with methods of achieving specified array spatial characteristics over a broad frequency band. Two general techniques will be postulated, and the one more suitable to subsequent broad band processing of the array output will be examined in moderate detail. This technique is based on the use of frequency dependent array weighting (shading) networks. Much of what follows is concerned with approximation of the desired spatial characteristic, approximation of the frequency dependent weighting functions that yield that approximate spatial pattern, and the realization of these frequency functions in practical network structures, namely transversal filters.

A uniformly spaced line array with additive processing will be taken as the model, and an example will be presented for achieving an approximately constant beamwidth over a broad band. Extension to other models will then be addressed briefly. These include general considerations for other arrays, e.g. non-uniformly spaced line arrays, planar arrays, and other forms of processing for uniformly spaced line arrays, e.g., supergain and multiplicative processing.

2. Prior Effort

Many authors have dealt with the problem of achieving a specified beam pattern with a line array [1], basing their syntheses on the Fourier series nature of the patterns obtained with equally spaced array elements [2, 3], or using the Fourier transform pair relationship of the beam pattern and the illumination in continuous arrays [4, 5, 6].

These and later works have generally been of a theoretical nature with the design frequency fixed, i.e., they are implicit single frequency studies.

When a broad band problem is studied the theoretical aspects of the Fourier transform pair are generally overlooked in favour of apparent and attractive approximating techniques. In one such technique a broad beam is generated by summing a large number of beams steered to different directions [7, 8, 9]. In its simplest form the scheme yields beamwidths that are not constant with frequency, but with modifications equivalent to varying the steering angles of the constituent beams as functions of frequency, the summed beamwidth can be maintained approximately constant over several octaves [8]. The formed beams must be properly delayed before summation to relate them all to the same reference, i.e. they must appear to have been formed from arrays with coincident phase centres. In another study the same basic idea is implemented theoretically with a number of line arrays all lying in the same plane and having the same midpoint [10]. This technique is then extended to the formation of beams essentially constant in width in two dimensions with a twisted planar array.

Other techniques are referred to, such as shaping the surface of the transducer to produce the desired pattern directly (which is actually what is done with the twisted planar array [10] and varying the effective aperture as with low pass filters or a multiplicity of resonant arrays [9, 11]. The latter technique can be extremely effective over an octave band using very simple crossover networks having all their poles and zeros outside the pass band. [11]. Problems are quickly encountered when an attempt is made to apply this technique to wider bands or more than two constituent arrays, however.

3. Approach Overview

The approach taken here is a return to application of the Fourier transform pair, synthesizing the beam pattern directly through design of frequency dependent illumination of the array.

Specification of the desired beam pattern at any single frequency automatically specifies the required illumination or, with discrete array elements, the element weighting at that frequency. Hence, specification of the desired beam pattern as a function of frequency automatically specifies the element weightings as functions of frequency. The problem is thus the determination of suitable approximations for the desired beam pattern, computation of the associated weighting functions, and approximate approximation and realization of those weighting functions.

The desired beam pattern may be approximated by any of a number of techniques, the choice being determined by the design goals and the array structure. When the on-axis beam shape is to be controlled as well as the sidelobe structure and the array consists of equally spaced colinear elements the Fourier series approximation to the beam pattern is probably the most convenient and is also reasonably efficient. This particular approximating technique is discussed at some length in a subsequent section. It will be seen at that time that the process of approximating the beam pattern leads directly to the determination of the frequency dependent weighting (shading) functions of all of the array elements. Realization of these weighting functions, and hence the desired beam patterns, can then be accomplished in two general ways, either by a sequence of narrow band processes or by a single broad band process.

4. Pattern Synthesis Through Narrow Band Processing

To synthesize array patterns by narrow band processing, the output of each of the array elements is first filtered into a large number of narrow bands spanning the total system bandwidth, with estimates of the real and imaginary components of the energy in each band being determined, as in a discrete transform such as the FFT. If it is a uniformly spaced line array that is being processed, beam steering can be achieved by use of a discrete spatial Fourier transform in each of the narrow bands, the progressive phase of the transform providing the beam steering and the amplitude shading determining the array pattern. By designing each of the narrow band patterns to be the pattern desired at that frequency, the composite coverage will provide the desired spatial characteristic over the system bandwidth. If an estimate of the waveform of the beam output is desired, it can be derived from still a third transform. This general technique can be applied to virtually any array, except that the beam pattern design in the second phase of processing will generally be much more involved than the foregoing, and the beam steering will generally not be a progressive phase shift.

This narrow band processing scheme is efficient if all that is required is an estimate of the spectral density of energy incident on the array having the spatial weighting described by the array beam pattern. If estimates of the broad band incident energy are required, however, this process may not be suitable. An example is in the measurement of transient signals in an acoustic calibration range.

5. Pattern Synthesis Using Transversal Filters

Estimates of the incident waveform may be obtained directly in a single broad band process by steering to the desired direction and using frequency dependent array weighting functions to achieve the beam patterns desired across the entire system bandwidth. In typical problems these weighting functions will be found to be real but have amplitude characteristics that are highly frequency dependent. Thus, they cannot be realized with minimum phase networks. The conventional approach to their realization would be to design networks approximating their amplitude characteristics and then to provide the required phase compensation with appropriate all-pass networks. In addition to requiring a monumental design effort, it is thought likely that this design approach would be rendered unworkable by the effects of dispersion and dissipation except possibly for the simplest cases.

A somewhat more unconventional approach would be to realize approximations to the required weighting functions directly with non-minimum phase networks. One type satisfying the requirements directly is the transversal filter, which is just a uniformly tapped delay line without feedback. [12] Its output can be purely real, purely imaginary, or complex, and each component is a separately controllable trigonometric series in phase (or frequency) and hence, can be used to obtain a Fourier series approximation of the desired element weighting function. Of course, a separate filter must be used for each array element. The transversal filter for a single frequency and having $2s + 1$ taps is shown in Fig. 1.

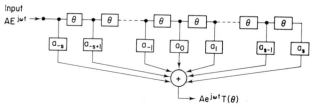

FIG. 1. Transversal filter.

$$T(\theta) = \left[a_0 + \sum_{k=1}^{s} a_{-k} e^{jk\theta} + a_k e^{-jk\theta} \right] e^{-js\theta} \tag{1}$$

$$T(\theta) = \left[a_0 + \sum_{k=1}^{s} (a_k + a_{-k}) \cos k\theta - j(a_k - a_{-k}) \sin k\theta \right] e^{-js\theta} \tag{2}$$

Here the input is progressively shifted in phase and the output of each tap is multipled by the tap weighting, a_k, before summing. When the filter is partitioned into symmetric and antimetric (equal but opposite) components, as in equation (2), it is seen that the symmetric component of the tap weighting produces the real part of the transmission, and the antimetric component of the tap weighting produces the imaginary part of the transmission, neglecting the phase shift $s\theta$ which is independent of the weighting.

Now, if the blocks of progressive phase shift, θ in Fig. 1, are replaced by blocks of delay, τ, as in a delay line or shift register, then θ of equations (1) and (2) is replaced by $\omega\tau$, and the filter is seen to be a broad band device with a constant delay and with a transmission characteristic that is.

expressed as two independent trigonometric series in the frequency domain.

$$T(\omega) = \left[a_0 + \sum_{k=1}^{s} (a_k + a_{-k}) \cos (k\omega\tau) - j(a_k - a_{-k}) \sin (k\omega\tau) \right] e^{-js\omega\tau} \tag{3}$$

The transversal filter with symmetric taps yields the characteristic desired of the array weighting networks in the problem at hand, i.e. that of approximating a real but highly frequency dependent function. In the examples to be pursued later in this study the approximations will be made in the Fourier sense, i.e. that of providing the least mean-square deviation over the band of approximation. It should be noted, however, that other approximations are possible using the same general structure but different trigonometric series.

At this point proper homage must be paid to engineering reality; the effects of dissipation can be accounted for by modifying the tap values, even though dispersion will tend to scuttle the analogue transversal filter as a design approach just as it would in the previously considered approach using minimum phase networks with all pass phase compensation. A digital approach appears desirable, if not mandatory, and is also more in tune with current trends in implementation.

The problem of realizing specified beam characteristics over a broad band of frequencies is thus solved in three steps.

(a) The specified beam characteristic is suitably approximated. With uniformly spaced line arrays the approximation is a trigonometric series whose coefficients are the required element weighting functions.

(b) The required weighting functions are suitably approximated by a trigonometric series whose coefficients are the weightings of the taps of a transversal filter.

(c) The signal as cast in a form which permits accurate realization of the transversal filters.

6. Considerations in Line Array Beam Pattern Approximation

Beam formation occurs at the point where the signals from the array transducers are summed. If, at a single frequency, the ith input signal has amplitude c_i and phase ψ_i, then the summed signal, A, is as given by equation (4).

$$A = \sum_{i=1}^{N} c_i \, e^{j\psi_i} \tag{4}$$

If the transducers themselves are omnidirectional, then the c_i are constants and A can then be a function of only the ψ_i, which in turn are functions of the direction from which the energy is arriving.

As an aside, it is noted that if the elements and if the array are colinear and there are no phase shifts except those in the medium, then the ψ_i will be odd functions of ϕ, the energy arrival angle measured from broadside.

$$\psi_i(-\phi) = -\psi_i(\phi) \tag{5}$$

For this condition, equation (6) follows directly, showing that line arrays with arbitrary (real) element weighting and element spacing still yield amplitude patterns symmetric about broadside.

$$A(-\phi) = \sum_{i=1}^{N} c_i \, e^{j\psi_i(-\phi)} = \sum_{i=1}^{N} c_i \, e^{-j\psi_i(\phi)} = \overline{A(\phi)} \tag{6}$$

Here, the bar denotes the complex conjugate. Note also that if the spacing and element weighting are also symmetrical about the midpoint, then for every input with a positive phase shift (measured from the midpoint) there is an equally weighted input with the same, but negative, phase shift, and the array factor is real as well as symmetric, the array factor being defined as the sum normalized to unity in the direction to which the array is steered.

Of more interest is the case where the ψ_i are all multiples of some minimum value, ψ. This can be achieved if the array element spacings are made commensurate, and then the array factor, $A(\psi)$ can be expressed as a trigonometric series in ψ. Since the assumption that the element spacings are commensurate is equivalent to the assumption that the elements are equally spaced, though perhaps some may be inoperative, the latter assumption is used here. This allows ψ to be re-defined conveniently as the *progressive* phase shift between elements at the point of summation; it is comprised of a phase shift in the medium and a (steering) phase shift in the beamformer.

$$\psi = \pi \frac{f}{f_d} \, [\sin \phi - \sin \phi_i] \tag{7}$$

Here f is the frequency, ϕ_i is the ith steering angle, numbered and measured from broadside, ϕ is the energy arrival angle also measured from broadside, and f_d is the design frequency, i.e., the lowest frequency for which all of the element spacings are an integral number of half wavelengths; for the assumption of equal spacing, the spacing is half a wavelength at f_d. Note that the beamformer phase shift is $\pi f/f_d \sin \phi_i$ and that when energy arrives from the direction ϕ_i, then ψ is zero. Equation (7) may also be expanded to show that ψ is approximately proportional to $\sin(\phi - \phi_i) \cos \phi_i$ for small values of $\phi - \phi_i$; a given value of ψ corresponds to successively larger values of $\Delta\phi$ as the beam is steered progressively further from broadside. Since the beam patterns are a function of the magnitude and phase of the signals at the point of summation only, they can be written as functions of ψ only and then be related to frequency or geometry through equation (7). Since the patterns are trigonometric functions of ψ, $F(\psi + 2N\pi)$ must be identical to $F(\psi)$.† This means that if spurious beams are to be avoided, the range of ψ must be restricted to something less than 2π radians. Inspection of equation (7), however, shows that at the design frequency ψ will equal 2π for energy coming 180° from the direction an endfire beam is steered. At higher frequencies ψ can equal 2π for beams steered off of endfire and for arrival angles other than the opposite endfire, giving rise to spurious major lobes.

Every finite set of equally spaced line array elements has an array factor, $A(\psi)$, which can be written as a finite trigonometric series in ψ. Similarly, every desired array factor, $B(\psi)$, can be approximated by many different trigonometric series in ψ. The particular rth order trigonometric series for which the mean-square-error between the approximating function and the desired function is a minimum is the Fourier series, $F(\psi)$. (The approximation is made for ψ ranging from $-\pi$ to π. Thus, $F(\psi)$ will have a period of 2π, in accordance with the preceding paragraph.)

$$F(\psi) = \frac{a_0}{2} + \sum_{n=1}^{r} [a_n \cos n\psi + b_n \sin n\psi] \tag{8}$$

$$a_n = \frac{1}{\pi} \int_{-\pi}^{\pi} B(\psi) \cos n\psi \, d\psi \tag{9}$$

$$b_n = \frac{1}{\pi} \int_{-\pi}^{\pi} B(\psi) \sin n\psi \, d\psi \tag{10}$$

The function $F(\psi)$ approximating $B(\psi)$ can be realized with an equally spaced line array of $N = 2r + 1$ elements. To approximate $B(\psi)$ with an array having an even number of elements, $N = 2r$; the corresponding relations are as follows:

$$F(\psi) = \sum_{n=1}^{r} \left[a_n \cos \left(\frac{2n-1}{2} \psi \right) + b_n \sin \left(\frac{2n-1}{2} \psi \right) \right] \tag{11}$$

$$a_n = \frac{1}{\pi} \int_{-\pi}^{\pi} B(\psi) \cos \left(\frac{2n-1}{2} \psi \right) \tag{12}$$

$$b_n = \frac{1}{\pi} \int_{-\pi}^{\pi} B(\psi) \sin \left(\frac{2n-1}{2} \psi \right) \tag{13}$$

† Note, however, that if the midpoint of the array is taken as the reference and if there are an even number of elements, then the pattern is comprised of the odd harmonics of $\psi/2$, and $F(\psi + 2N\pi)$ is equal to $-F(\psi)$. This is not at odds with the statement above, which assumes the reference to be at an allowable array element position.

Numbering the array elements from $-r$ to $+r$ for either an odd or even number of elements, and with the centre element being element number zero if it exists, the element shading coefficients, c_n, are given by equation (14).

$$c_0 = a_0, \quad c_{-n} = a_n + jb_n, \quad c_n = a_n - jb_n \tag{14}$$

The a_n and b_n are the desired shading functions and are derived from equations (9) and (10) or (12) and (13). They will generally be found to be functions of both n and f, i.e. position and frequency. It is noted from equations (8) and (11), however, that if the desired beam pattern is to be symmetric with respect to ψ (and to ϕ for the broadside beam), then all of the b_n are zero, and the array weighting is real and symmetric. This is in agreement with the more general discussion linking real weighting to beam pattern symmetry and element weighting and spacing symmetry to real beam patterns. In view of the fact that real beam patterns are the result of symmetric weighting and element spacing, which generally pose no implementation problems, it appears that no benefits would accrue from making a different choice for the form of the approximating function. Apparently there is no penalty for choosing it to be real, as there would be in designing network functions.

As an illustration of this approximating procedure, consider the problem of obtaining a constant beamwidth over a wide range of frequencies. Ideally the array factor has the form shown in Fig. 2; it is called a zero order pattern here because of the discontinuities in the function.

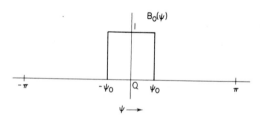

FIG. 2. Zero order beam pattern.

The element weighting functions for this desired array factor are determined from equation (9), and the array factor approximating $B_0(\psi)$ follows immediately, assuming an array with an odd number of elements.

$$a_n = \frac{\sin n\psi_0}{n\psi_0} \tag{15}$$

$$F_0(\psi) = \frac{\psi_0}{\pi} \left[1 + 2 \sum_{n=1}^{r} \frac{\sin n\psi_0}{n\psi_0} \cos n\psi \right] \tag{16}$$

This apparently easily realized design is deceptive; the a_n are actually functions of frequency because ψ_0 is proportional to the frequency, as shown in equation (17) for the broadside case.

$$\psi_0 = \pi \frac{f}{f_d} \sin \phi_0 \qquad (17)$$

From equations (15) and (17) it is seen that each element weighting function is a $(\sin x)/x$ function of frequency as shown in Fig. 3; for a given frequency these functions are also seen to vary as $(\sin y)/y$ with position.

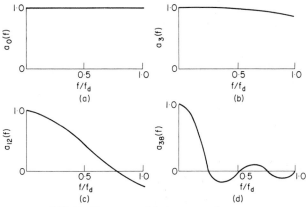

FIG. 3. Element weighting versus frequency.

Because the approximating function, $F(\psi)$, is frequency dependent, it is clear that it will not approximate the desired function, $B(\psi)$, equally well at all frequencies, assuming that only a finite number of terms are used. At very low frequencies, for example, it is clear that the array will be omnidirectional regardless of how the array elements are weighted. This is shown in Fig. 4, which is a plot of the envelope of the amplitude of each input versus the input number, measured from the centre of the array. This is a plot of amplitude versus input number for a fixed value of ψ_0, i.e., frequency. Identical plots could be made for amplitude versus frequency for each fixed input number; only the scales would change. For the case illustrated, ψ_0 has been chosen equal to $\pi/4$, which leads to every fourth input being zero (except for the centre input). If the array has $2r + 1$ elements and $\psi_0 = \pi/r$, then all element weightings will fall within the main peak of this weighting function. This occurs at a frequency determined from equation (18).

$$f = \frac{f_d}{r \sin \phi_0} \qquad (18)$$

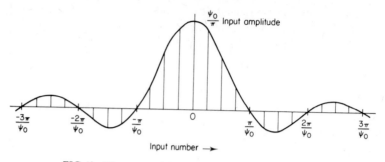

FIG. 4. Element weighting versus position at $\psi_0 = \pi/4$.

If the array length were to be doubled, then this frequency would be halved, and one would expect that the desired function, $B(\psi)$, would be approximated equally well at the original length and frequency and at the revised length and frequency. This illustrates the fact that the lowest frequency at which the desired pattern may be approximated to some specified accuracy is inversely proportional to the aperture. Note that at frequencies somewhat lower than that given by equation (18) all of the array inputs are bunched up toward the middle of the main loop of Fig. 4 and converge to the ordinate axis at *dc*.

At low frequencies the approximations become poor. Further, they become poor in a discouraging manner if the intent is to maintain coverage in the desired direction. As the frequency decreases the approximation degrades as in Fig. 5(a), rather than as shown in Fig. 5(b). That is, the low frequency approximation degrades the on-axis part of the beam rather than the sidelobe structure. The reason is that the Fourier series is a least-squares approximation; the integral of the square of the difference between $B_0(\psi)$ and $F_0(\psi)$ is minimum when the coefficients of the trigonometric series for $F_0(\psi)$ are the Fourier coefficients, as they have been chosen. Since the integrated square of the error is clearly less for Fig. 5(a) than 5(b), this is the form the approximation takes. At a somewhat lower frequency the characteristic will look more like 5(b).

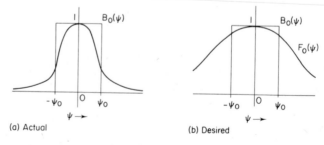

FIG. 5. Low frequency approximation.

There are several ways to reduce this effect of poor on-axis approximation at low to intermediate frequencies and also to eliminate the Gibbs phenomena due to the discontinuities in the function approximated. Techniques derived for the design of transversal filters, e.g. windowing the truncated Fourier series with an appropriate function [13] or using a frequency sampling method [14], are applicable to array pattern approximation and should yield superior single frequency results. Since they do not yield closed form solutions, however, they are computationally more difficult to apply on a broadband basis than the method to be pursued here. This method, which is simply to modify the pattern being approximated to more closely match the characteristics of the

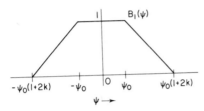

FIG. 6. First order beam pattern.

truncated Fourier series, is illustrated in Fig. 6, where the array factor being approximated is chosen to have a linear transition zone between the uniform gain zone and the zero sidelobe zone. It is approximated by $F_1(\psi)$ as in equation (19).

$$F_1(\psi) = \frac{\psi_0(1+k)}{\pi} \left[1 + 2 \sum_{n=1}^{r} \frac{\sin n\psi_0(1+k)}{n\psi_0(1+k)} \frac{\sin n\psi_0 k}{n\psi_0 k} \cos n\psi \right] \quad (19)$$

The envelope of the element weighting for $F_0(\psi)$ possesses zeros for input numbers that are multiples of π/ψ_0, while the corresponding function for $F_1(\psi)$ has zeros at multiples of $\pi/\psi_0(1+k)$ and $\pi/\psi_0 k$. If the array has $2r+1$ elements and $\psi_0 = \pi/r(1+k)$, then all of the elements will be included in the main peak. This will occur at the frequency given by equation (20).

$$f = \frac{f_d}{r(1+k)\sin\phi_0} \quad (20)$$

Hence, for k equal to unity, approximately equal results are obtained for $B_1(\psi)$ to about half the frequency as for $B_0(\psi)$. The results may be somewhat better than this because of the reduction in the magnitude of the envelope in the loops further removed from the centre of the array. This effect has its greatest impact at the higher frequencies, i.e. when ψ_0 is relatively large.

The next logical step in design is to remove the sharp corner in the

$B_1(\psi)$ and eliminate the discontinuities in the derivative. This is done in $B_2(\psi)$, shown in Fig. 7. Here the discontinuities have been banished to the second derivative; the transition region is spanned by two parabolic sections with a continuous slope.

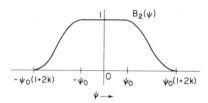

FIG. 7. Second order beam pattern.

This function has the approximating function, $F_2(\psi)$, given in equation (21).

$$F_2(\psi) = \frac{\psi_0(1+k)}{\pi}\left[1 + 2\sum_{n=1}^{r}\frac{\sin n\psi_0(1+k)}{n\psi_0(1+k)}\left[\frac{\sin\dfrac{n\psi_0 k}{2}}{\dfrac{n\psi_0 k}{2}}\right]^2\cos n\psi\right]$$

(21)

The envelope of element weights differs from that of $F_1(\psi)$ mainly in the rapidity of the decay, which is a little slower out to about the second zero and quite a bit faster beyond about the third zero. Thus, $F_2(\psi)$ will give excellent approximations for large values of ψ_0, i.e. at high frequencies.

It may be desirable to use an even number of elements in the array, as in equation (11). At first glance it would appear that the lack of a *dc* term would preclude approximations of characteristics with non-zero average values. This is indeed the case, and it is guaranteed by the fact that $F(\psi + 2\pi) = -F(\psi)$, as can be seen by inspection of each term of equation (19). Within the principal interval of approximation, however, i.e. for $-\pi \leqslant \psi \leqslant \pi$, the restrictions on $F(\psi)$ are the same as for the case with an odd number of elements.

7. Design Example

A set of weighting networks was designed for a uniformly spaced 30 element vertical line array to be used in a broadside or near broadside mode in an acoustic calibration range. Directive elements were employed to suppress endfire energy, permitting the array to be used at up to about 2.4 times the design frequency, i.e., the frequency for which the element spacing is one half wavelength. The design goal was to achieve a uniform response with a total variation of 1 dB over a beamwidth of 6.0° and over a frequency band extending from very low frequencies to 2.4 times the design frequency. Additional design goals were the achievement of an abrupt transition from the main lobe to the sidelobe region and that the sidelobes be generally low.

A set of transversal filters was designed in accordance with the foregoing discussions, but employing some design refinements, notably a frequency dependent design transition interval to take advantage of the additional directivity available at the higher frequencies. The centre filter has a single tap, and those toward the array edges have successively more, with a maximum of 31 taps being employed for the end elements. The symmetry of the array and of the filters permits a twofold collapsing of the filters, however, so that a total of only 156 taps are employed in the entire shading network.

The response achieved at ±3.0°, the nominal 1 dB down points, is −1.0 dB ± 0.3 dB, over the band from 0.32 to 2.4 times the design frequency, as shown in Fig. 8. At lower frequencies the aperture limitations

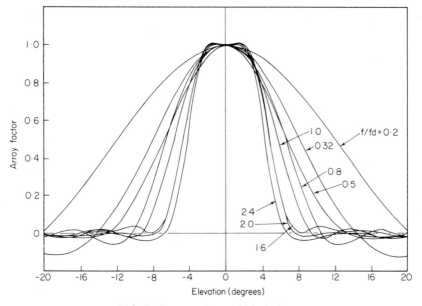

FIG. 8. Constant beamwidth design.

result in beam broadening, and since the array shading is essentially uniform at the lower frequencies, the largest sidelobes at low frequencies approaches −13 dB. Above about 0.4 times the design frequency the sidelobes are all smaller than −24 dB.

8. Non-uniformly Spaced Line Arrays

A crude extension of the technique described herein may be made to the non-uniformly spaced array by performing a design for a densely packed uniformly spaced reference array and then interpolating to determine the element weighting for the actual array. At any single frequency this interpolation can be performed with a fair degree of accuracy by first interpolating between the nearest two uniformly spaced elements and then weighting the results in proportion to the number of uniformly spaced elements represented by the actual element, thus maintaining the overall array illumination approximately correct. Closed form solutions will generally not be obtained for the frequency functions to be approximated, but this is of little consequence in determining the transversal filter tap values since numerical integrations would normally be performed anyway.

9. Planar Arrays

The simplest approach to the design of planar arrays is to perform a sequence of designs as if the final structure were a line array of line arrays, with the overall response being obtained by pattern multiplication. This, however, will result in compounding the approximations, so that a 1 dB roll-off in each dimension will yield a total roll-off of 2 dB, and the overall quality of the approximations will be degraded. This may be acceptable in some situations and may even be a desirable design technique for roughly rectangular beam patterns. This technique is not applicable, however, for the design of general two-dimensional patterns, i.e. those which cannot be obtained from pattern multiplication; these must be synthesized directly [15]. In either case the array shading function must be described as a function of frequency, and then the frequency dependent filter functions can be approximated and realized by the methods described herein.

10. Supergain

It is theoretically possible to achieve substantially greater directivity than that normally attributed to a given aperture by forming the difference of two arrays, one nested within the other [16, 17]. This phenomena is called supergain and is effective against noise that is not locally generated, the limitations in receiving systems being established by the tolerances on the array element sensitivities and element weightings. Two dB or so improvement may be practical in rigidly controlled implementations of small arrays, corresponding to operating with a conventional array about 60% longer.

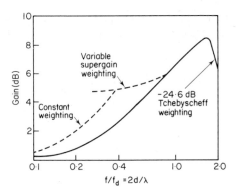

FIG. 9. Gain of a 5 element array.

An example of the array gain that might be achieved is shown in Fig. 9. Here, following Prichard, a 5 element array with −24.6 dB sidelobes has been designed. Below the design frequency the weighting is permitted to vary, as shown in Fig. 10, to maintain an essentially constant sidelobe structure and main lobe response, and hence, an essentially constant gain. This could be carried to substantially lower frequencies in theory, but in practice a number of effects will combine to limit the extent to which this technique can be carried. One such limitation comes from the gain and phase errors associated with the array element locations, the gain and phase characteristics of the data links, and the tolerances of the weighting networks. Another limitation is inherent in the nonisotropicity of the noise field; a locally generated noise field cannot be processed against by this technique.

In the example shown, the array shading is permitted to be frequency dependent down to about 0.4 times the design frequency, at which point the on-axis amplitude response is only about one-tenth what it is at the design frequency, and it degenerates even more rapidly at lower frequencies.

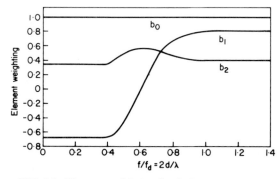

FIG. 10. Element weighting for 5 element array.

This frequency is estimated to be about the minimum at which the frequency dependent weighting should be applied, so the weighting factors of Fig. 10 are shown as constant below this frequency and above the design frequency. This leads to about a 2 dB increase in gain over the octave ranging from one-quarter to one-half the design frequency. This is clearly not a solution to major system design problems, but it may be of some value in applications of small arrays. The sharpness of the frequency variations required also suggest that this approach is probably most readily implemented in narrow bands, i.e. using the FFT approach described earlier.

11. Multiplicative Array Patterns

For some purposes, as in determining unbiased estimates of target energy, it may be desirable to design the array spatial pattern in terms of the product of two separately shaded array halves, rather than as the weighted sum of all of the elements. For uniformly spaced line arrays this turns out to be a relatively simple process, since the product pattern can be expressed as a trigonometric series in the progressive phase shift, ψ, in terms of combinations of the element weighting, much as in the additive array. Hence, all of the procedures outlined in the body of this note are applicable. The desired pattern is approximated by a Fourier series, the coefficients of which are equated to the coefficients of the trigonometric series of multiplicative array, permitting an approximate solution for the element weightings. If the desired patterns are frequency dependent, then the element weightings are realized in transversal filters, as before.

Space limitations prevent this topic from being explored in depth in this note. However, there are only two major differences between this technique and that followed for additive arrays. First, an array with a total of $2n$ elements has a pattern without a constant term, and second, the pattern has a total of $2n - 1$ harmonic terms in ψ. If the array is assumed to be symmetrically weighted, then the first n of these terms may be matched to those of the desired pattern, and the remaining terms contribute an error that is in addition to the truncation error of the pattern being approximated. The effect of the constant term is to add a bias to the overall pattern, assuming the desired pattern to be a major lobe and essentially zero response elsewhere. Lacking a constant term the (negative) area of the nominally zero response region must equal the area of the main lobe.

12. Summary

The use of frequency dependent shading can improve the spatial characteristics of many types of arrays. This shading can be applied either in narrow bands with subsequent combining of the bands, as with multiple discrete Fourier transforms, or in broad bands by the use of transversal filters. These filters can be designed to approximate arbitrary real, imaginary, or complex frequency functions arbitrarily closely by Fourier series approximation techniques.

REFERENCES

1. Wolff, Irving (1937). Determination of the radiating system which produces a specified directional characteristic. *Proc. I.R.E.*, 25, 630–643.
2. Shelkunoff, S. A. (1943). A mathematical theory of linear arrays. *B.S.T.J.*, 22, 80–107.
3. Kraus, J. D. (1950). "Antennaes", pp. 97–106. McGraw-Hill Book Company, Inc., New York.
4. Spencer, R. C. (1946). Fourier integral methods of pattern analysis. *MIT Rad. Lab. Rpt* 762–1, 21.
5. Ramsay, J. F. (1946). Fourier transforms in aerial theory. *Marconi Review*, 9, 139–145.
6. Bracewell, R. (1965). "The Fourier Transform and Its Applications", McGraw-Hill, New York.
7. Tucker, D. G. (1956). Some aspects in the design of strip arrays. *Acustica*, 6, (1956). 403–411.
8. Tucker, D. G. (1957). Arrays with constant beam width over a wide frequency range. *Nature*, 180, 496–497.
9. Morris, J. C. and Hands, E. (1961). Constant-beamwidth arrays for wide frequency bands. *Acustica*, 11, 341–347.
10. Morris, J. C. (1964). Broadband constant beam-width transducers. *J. Sound Vib.* 1, 28–40.
11. Smith, R. R. (1970). Constant beamwidth receiving arrays for broadband sonar systems. *Acustica*, 23, 21–26.
12. Zadeh, L. A. and Desoer, C. A. (1963). "Linear System Theory", pp. 447–448. McGraw-Hill Book Company, Inc., New York.
13. Helms, H. D. (1968). Nonrecursive digital filters: design methods for achieving specifications on frequency response. *IEEE Trans. Audio Electroacoustics*, AU-16, 336–342.
14. Rabiner, L. R., Gold, B. and McGonegal, C. A. (1970). An approach to the approximation problem for nonrecursive digital filters. *IEEE Trans. Audio Electroacoustics*, AU-18, 83–106.
15. Collins, R. E. (1964). Pattern synthesis with non-separable aperture fields. *IEEE Trans. Antennaes Propagations*, AP-12, 502–503.
16. Riblet, H. J. (1947). Discussion on "A current distribution for broadside arrays which optimizes the relationship between beamwidth and sidelobe level". *Proc. IRE* 35, 489–492.
17. Prichard, R. L. (1953). Optimum directivity patterns for linear point arrays. *J. Acoust. Soc. Am.* 25, No. 5, 879–89.

DISCUSSION

P. L. Stocklin: In the slide of the five element supergain array, if you had continued the curves of the weighting functions you probably would have shown that, relative to that of the centre element weight unity, the amplitudes would approach a number like the fraction of a half wavelength separation between elements, raised to the power of the number of array elements. So at a frequency for which the spacing is a twentieth of a wavelength and there are five elements then the weighting is ten to the fifth, which yields horrible efficiency.

Answer: Yes, it really is foolish to talk about achieving more than a dB or two of supergain, I think, even for very small arrays. When you talk about using supergain at much below about four-tenths of the half wave length spacing frequency as in the example, that by then the response is diminished to the point where very small deviations in the element positions or in the weighting functions or in the transmission from the hydrophones to the actual weighting point destroy the whole response pattern.

E. R. Thomas: Does steering of broadband arrays upset the sidelobe response, or do you have to re-optimize if you want to steer the beam?

Answer: First of all, there is no optimization involved, we take the sidelobes we get, and they decrease with increasing frequency. The pattern is invariant with psi, the progressive phase shift at the point of summation. Regardless of what pattern we have chosen to design, we get that same pattern in the psi plane regardless of whether we have steered to broadside, endfire or in between. The beamformer merely translates the designed pattern to the steered direction. If we go back to the phi plane, the azimuthal plane, the endfire beam will be considerably fatter, the transform being through the sine of phi. So the sidelobes amplitudes will remain the same, but they will be at different locations relative to the main beam.

E. R. Thomas: Is the frequency dependence slightly different for it then?

Answer: No, it only changes with phi.

J. W. R. Griffiths: The diffraction secondaries will then come up because you cannot get rid of them.

Answer: That is correct, the grating lobes still remain.

G. Pearce: Were the results that you showed measured or calculated?

Answer: The results I showed were calculations. The verification of the design was established in a system that was actually built about five years ago in which the responses achieved were comparable to those presented here. In actual practice it turned out that, to reduce the complexity of the hardware, we arbitrarily threw away all transversal filter tap values that were negative, and we threw away all tap values that were less than 5% of the maximum tap value of any one filter. We also evaluated the performance with a uniform random variation up to ±5% on all tap values that remained. It turns out that, because of the fact that we use first of all a large number of elements and secondly a large number of taps, that the average errors tend to zero, and the overall response that we achieved was quite comparable to what I have been citing here. We achieved sidelobe levels that were, in general, into the noise of our instrumentation. They were lower than −24 or −25 dB by an unknown amount.

G. Pearce: There was no effect due to element interaction at the lower frequencies?

Answer: This is a receiving system. It could have been a transmitting system, but we are talking about an acoustic calibration system.

P. Heimdal: Of the three array responses you looked at for the additive array, which is optimum—the square pattern, the sloped pattern or the one with parabolic roll-offs? For which is the error most independent?

Answer: If you have an infinite number of elements they will all work equally well, except for the Gibbs phenomena in the square pattern. Other than that they are about equal except that the square patterns have more on-axis deviation for low to intermediate frequencies. The one with the parabolic roll-off is somewhat more efficient in terms of having the least deviation in on-axis response for a given number of filter taps.

Broadband Hydrophone Arrays for Use With Explosive Sound Sources

ARISTIDES A. G. REQUICHA

SACLANT ASW Research Centre,
La Spezia, Italy

1. Introduction

In underwater acoustics research work involving explosive sound sources, it is sometimes desirable to use directional receiving elements, for the purposes of rejecting noise and/or obtaining bearing information. This paper deals with the design and implementation of broadband hydrophone arrays for such applications. Much of the discussion, however, is general and can be applied to any wideband array of sensors, be they acoustical, seismic or electromagnetic.

The performance goals to be achieved can be broadly described as follows:

(a) The arrays should have "good" directional characteristics throughout the frequency band of operation. Beamwidth and maximum side-lobe level will be used as measures of goodness.
(b) Signal distortion in the band must be kept low. This is necessary when the arrays are used in the measurement of impulse responses.

Although beam steering will not be treated in detail in this paper, it is helpful to keep in mind that the arrays will have to be steered for certain applications.

It will become apparent in the sequel that the choice of processing techniques influences the design at an early stage. Digital implementation of the array processor in a general purpose minicomputer will be considered. Because of the reliability, accuracy and flexibility of digital techniques a large proportion of underwater acoustics research data is now being acquired and analysed digitally.† Digital processing facilitates the designer's task, in that he is no longer constrained to specify processor

† At SACLANTCEN analogue data acquisition and processing are presently almost nil.

elements, such as filters, that can be realized by analogue means at a reasonable cost. It is felt that flexibility is a very important characteristic of a research tool. This explains the choice of a software processor that can be changed with relative ease when a different performance is required. Implementation of the processor in a general purpose minicomputer is rather attractive since the computer can be time-shared with other activities, and effectively perform array processing and various other acquisition and analysis functions during sea trials. It should be noted, however, that a trade-off exists between flexibility and efficiency. A hardware processor designed for a specific job will generally have better performance than a general purpose machine. An interesting compromise can be achieved by using microprogrammed wired units, such as a fast Fourier transform (FFT) processor, as peripherals to a general purpose computer.

The frequency band of interest in work involving underwater explosions extends from close to DC to the tens of kHz. For typical applications a bandwidth of a few octaves is required, with a low frequency cut-off that can be in the order of the hundreds of Hz or considerably lower (nearly 0 Hz for seismic work). Because of the large array dimensions needed to ensure directivity at such low frequencies it is unpractical to operate from a ship anything but linear arrays. Two types of linear arrays can be handled from ships with relative ease: (a) long (1 mile, say) flexible arrays, which can be towed, and (b) rigid arrays of length not over about thirty metres, which can be lowered into position. Furthermore, strip, or continuous, arrays are not feasible for these frequencies and therefore discrete hydrophones must be used. For these reasons only linear arrays of discrete hydrophones will be considered in this paper. The design principles discussed are, however, applicable to other configurations.

The width of the frequency band of interest presents an immediate difficulty for array design, because the directional properties of most commonly-used arrays are strongly frequency dependent.

Figure 1 illustrates a typical example of this frequency dependence for a broadside linear array of eight equally-spaced omnidirectional elements. The directional pattern for a frequency f_0 that corresponds to half-wavelength spacing of the sensors is depicted, together with the patterns for $f_0/2$ and $2f_0$. The array's output in relative amplitude is plotted as a function of the angle θ, in degrees, measured from broadside. Because the patterns are symmetrical about $\theta = 0$, only the region $(0°, 90°)$ is shown. It can be seen that the beamwidth roughly doubles when the frequency is halved and that additional "main" lobes—grating lobes—may move into the region of interest when the frequency increases.

Since the signal distortion impaired by the array must be kept very small, the linear distortion caused by pointing errors has also to be taken into account. This effect is illustrated in Fig. 2, which depicts the frequency

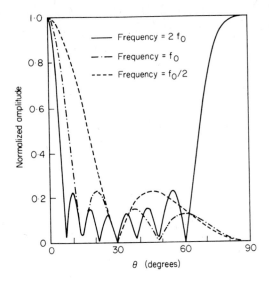

FIG. 1. Beam patterns for an 8-element uniform array.

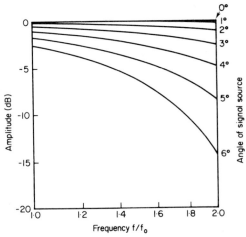

FIG. 2. Distortion due to pointing error for an 8-element uniform array.

response of the eight-element array previously considered for signal sources located at various angles within the array's main beam. For given source angles the amplitude of the array's response is shown as a function of frequency on a logarithmic scale over the octave $(f_0, 2f_0)$. It is clear that reception is undistorted only when the source is exactly broadside. If the exact angular location of the source were known, one could compensate for this type of distortion. However, in practice it is only known that the source is somewhere in the main beam. Therefore, for undistorted signal reception, it is necessary to keep the shape of the main beam constant throughout the frequency band of operation.

A less stringent requirement, which often leads to acceptable designs, is to maintain the beamwidth (measured, for example, at the 3 dB points) constant throughout the band. Arrays with constant beamwidth are also advantageous when the signal sources are distributed over an angular region. Reception with a constant-beamwidth array ensures that the contributions from the various parts of the target contained in the main beam do not undergo a frequency-dependent distortion. (There will always be, of course, an angle-dependent distortion due to the beam shape.)

The large signal bandwidth leads still to another technical problem. Because the output of all the sensors must be accessible for beamsteering,[†] and the sampling frequency must be three or four times the signal bandwidth, very high data acquisition rates are involved. For example, 20 hydrophones sampled at 12 kHz correspond to a data acquisition rate of 240 kHz, which is the maximum value that can be accepted by a state-of-the-art system currently being used at SACLANTCEN [19]. High data rates can only be achieved at considerable equipment cost and complexity. It is therefore important to design an array with the smallest possible number of hydrophones compatible with a desired performance.

This paper deals specifically with design techniques for achieving constant beamwidth over a frequency band of a few octaves. Section 2 is a general discussion of the problems involved in constant-beamwidth design and a review of the relevant literature. A particular technique, based on combining the outputs of two arrays, is treated in detail in Section 3. It is shown, by means of examples, that this technique leads to very acceptable designs. Implementation and test of a broadband 20-hydrophone array built at SACLANTCEN is discussed in Section 4. Finally, digital beam forming is very briefly considered in Section 5.

2. Methods for Achieving Constant Beamwidth

This section surveys the work on constant-beamwidth arrays published in the open literature,[‡] and discusses the general problems encountered in the design of such arrays.

Early work on this subject was performed at the University of Birmingham, England. Tucker [13] and Morris and Hands [5] studied constant beamwidth arrays based on what they call a "synthesis" procedure, which can be viewed as a technique for defocusing the array at high frequencies, thereby obtaining a wider beam. These authors constructed a strip array

---

† If only a small angular sector is to be scanned the hydrophones can be added in groups of 2 or 4, say, prior to digitizing.

‡ Autrey's work, to be presented at this Institute, will not be discussed here since it was not known in sufficient detail to the present author at the time of writing.

that operated over a 9:1 frequency range. Although the beamwidth was nearly constant, the shape of the beam varied considerably; sidelobe levels were not reported. Later on, Morris [14] exploited the fact that the beamwidth of a curved array depends on its actual curvature. He built several curved arrays that did exhibit fairly constant beamwidth, although the beam shape was again rather variable.

More recently Smith [1] in New Zealand, and Hixson and Au [3] at the University of Texas independently proposed a technique (hereafter referred to as the "SHA technique") for achieving constant beamwidth over an octave by suitably combining the outputs of two arrays. Smith reports good experimental results obtained with a strip array operating over an octave. Arrays with discrete elements were considered in [3]. Hixson and Au's attempt to verify the theory by constructing a two-octave array encountered several technical difficulties. Their array did not perform very well but showed that the technique warrants further study.

A geometrically linear array of N omnidirectional hydrophones with an associated multichannel processor that is linear in the circuit theory sense (i.e. such that the principle of superposition holds) can always be modelled as shown in Fig. 3. The most general design problem for such an array consists in finding the sensor locations z_i and the shading coefficients w_i, which may be frequency dependent, that will ensure the desired performance over the frequency band of interest. The joint optimization of

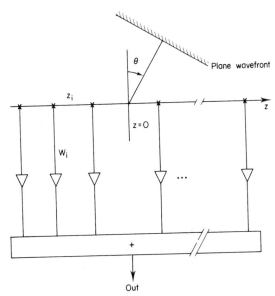

FIG. 3. Array model.

element locations and frequency-dependent weights is a rather formidable problem. Numerical minimization techniques are, in principle, applicable [6] although the design of arrays to cover a few octaves by such techniques has never been attempted, to the author's knowledge. At present, it is unlikely that acceptable solutions will be reached with reasonable computational effort because of the very large number of variables involved. (Note that the weights w_i must be computed for a large number of frequency points for the filter characteristics to be known with sufficient accuracy.)

Because the beam pattern of an array is a highly non-linear function of the element locations, it is a difficult matter to position the sensors so as to achieve a given performance. A large body of literature exists on the subject (see [9] for a survey of work published before 1966), but it deals mainly with the design of arrays with no shading and for a single frequency. Since the optimal sensor locations are frequency dependent, this work is of little help in the design of broadband arrays.

To avoid the inherent difficulties of non-linear optimization methods one can start by specifying a "reasonable" array geometry, and then try to determine the filter characteristics $w_i(f)$ as a function of the frequency f. This is a more manageable problem because the beam pattern depends linearly on the shading coefficients. To facilitate further the design it would be desirable to have an array with the simplest possible configuration, i.e. uniform sensor spacing. However, when the number of hydrophones is to be minimized such a geometry cannot be used for the following reasons. The beamwidth of a linear array is roughly inversely proportional to its acoustic length (length measured in wavelengths). To keep the beamwidth constant the effective length of the array must decrease when the frequency increases. Placing all the hydrophones at equal distances is wasteful, since some of them will only be operating at low frequencies. Intuitively, a logarithmic spacing appears to be more appropriate. This hypothesis is corroborated by the successful log-periodic electromagnetic antenna designs in existence [7, 8]. For unequally-spaced sensors, however, finding the frequency-dependent weights necessary to achieve constant beamwidth and low sidelobe level throughout the frequency band of operation is by no means trivial. A possible approach consists in using linear programming techniques for Chebyshev-type designs. The shading problem can be viewed as a problem in the approximation of functions by trigonometric polynomials. If a minimax criterion is chosen, it is well-known [10, 11, 12, 21], that the approximation can be formulated as a linear program, which can be solved by standard techniques (see [22] for a recent application to array design). For broadband arrays a linear program must be solved for each of a sufficiently large number of frequency points and the computational effort is therefore rather large. (Dimensional problems are not as bad as in the nonlinear optimization methods.)

It is apparent from the above discussion that it is rather difficult to optimize a broadband array. Simple suboptimal techniques are therefore of interest.

The promising results obtained with the SHA method have prompted the present author to investigate the possibility of using it to design arrays with larger frequency bands and lower sidelobes [2]. This work is summarized in the following sections which show that the technique can provide a systematic way of controlling beamwidth and sidelobe level and yield acceptable designs.

It should be noted that it is easy to obtain an exact theoretical solution to the problem of achieving a constant beam pattern for continuous line arrays. It suffices to require that the shading be the same at all frequencies. From this condition filter characteristics can be computed for each point along the aperture. Another suboptimal technique for designing constant beamwidth arrays of discrete hydrophones consists in approximating a continuous array designed by the above procedure. This approximation involves sampling the aperture, and, if equally-spaced hydrophones are used, the effects are quite easy to predict via classical sampling theory. For a minimal number of sensors, however, non-uniform sampling must be used. This immediately raises the problem of choosing sensor locations and filter characteristics. How to make these choices without appreciably degrading the array performance is not clear-cut. Examples presented in [2] show that a straightforward approach to this problem leads to performances inferior to those obtained by the SHA technique.

3. Constant-beamwidth Design by the SHA Technique

(a) Theory
Consider the design of a constant beamwidth array for the frequency band (f_L, f_H), where $f_H = K \cdot f_L$. The SHA procedure is as follows:

(i) Design an array for the given beamwidth and sidelobe level at frequency f_L. This array will have a beamwidth that decreases as the frequency increases.

(ii) Construct an homothetic replica of the previous array K times "smaller". This array will have the required beamwidth at f_H, but larger beamwidth at lower frequencies.

(iii) Form a linear combination of the output of the two arrays so as to keep the (3 dB, say) beamwidth constant over the band.

Let the complex amplitude of the responses of the "high frequency" and "low frequency" arrays to a plane wave of unit amplitude, frequency f, and angle of incidence θ be denoted, respectively, by $A_H(\theta, f)$ and

$A_L(\theta, f)$. According to step (iii) above one must find the weights $H(f)$ and $L(f)$ for the resulting output

$$A_R(\theta, f) = H(f) \cdot A_H(\theta, f) + L(f) \cdot A_L(\theta, f) \tag{1}$$

to exhibit constant beamwidth. For a broadside (non-steered) array this amounts to solving

$$A_R(\theta_B, f)/A_R(0, f) = 1/\sqrt{2}, \tag{2}$$

where θ_B is the design half-beamwidth. For unique H and L another condition is needed. It is natural to require that broadside signals be received undistorted, by imposing

$$H(f) + L(f) = 1, \quad f_L \leqslant f \leqslant f_H. \tag{3}$$

H and L are uniquely determined by equations (2) and (3).

Note that the SHA technique only ensures that the beamwidth is constant. There is no guarantee that the beam pattern will not vary considerably and that the sidelobes will not increase.

The main interest of the method lies in its simplicity. Indeed, it essentially reduces the design of a wideband array to the design of an array for a single frequency, which is a well understood problem. It is convenient to start with a uniform array as basic building block because design procedures are simplest, and because the number of hydrophones can be reduced when $K = 2$ (one-octave band), as shown in the following sections.

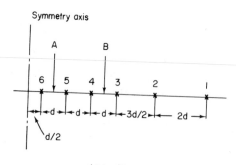

$$d = \lambda_H/2$$

$$\lambda_H = \text{wavelength for } f = f_H$$

FIG. 4. One-octave arrays geometry.

(b) Examples of one-octave designs

Two one-octave arrays with the same geometry (Fig. 4) were designed by the SHA technique, using as basic building block a uniform array of eight elements half-a-wavelength apart at the upper edge of the frequency band f_H. In the first example no shading was applied, while in the second a -26 dB Chebyshev shading was considered.

Note that two additional hydrophones (denoted by A and B in Fig. 4) should belong to the low frequency array. These were not used in the design, on the grounds that adding hydrophones 5 and 6 should be approximately

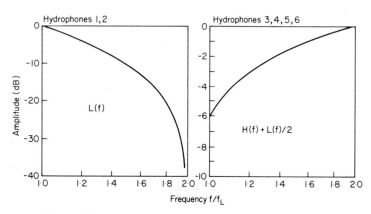

FIG. 5. Filter characteristics for one-octave unshaded array.

equivalent to having a hydrophone at A, at least in the lower part of the band.† Filter characteristics for the unshaded array are shown in Fig. 5. (The phase response is identically zero.) For the Chebyshev array different linear combinations of the corresponding functions $H(f)$ and $L(f)$ must be used for each pair of elements [2].

Beam patterns computed for 21 different frequencies are shown in Fig. 6 for the unshaded array, and in Fig. 7 for the Chebyshev array. (Note the different scales for the array factor.) The first and last patterns in each figure correspond respectively to frequencies f_L and f_H. The frequency increases at equal logarithmic increments throughout the octave.

The 3 dB beamwidths were found to be constant to within $0.1°$ at $12.8°$ and $15.8°$ for the unshaded and the Chebyshev designs, respectively. In both cases the sidelobe level throughout the octave never exceeded the design values of -13 and -26 dB.

† For arrays with an odd number of elements some of the sensors in the component arrays coincide geometrically, and one can save hydrophones without resorting to the approximation above [2].

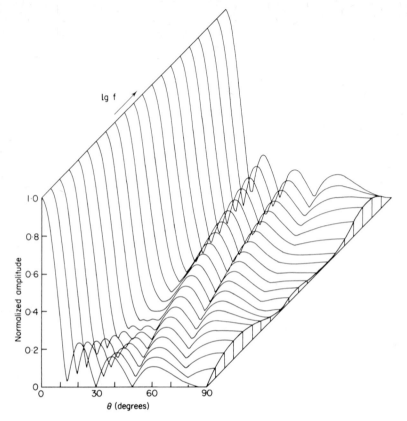

FIG. 6. Beam patterns for a one-octave unshaded array.

Computed frequency responses for sources in the main beam were flat to within 0.05 dB for both arrays [2]. Therefore, distortion with pointing errors is practically non-existent for such arrays.

(c) A $3\frac{1}{2}$ octave design

An array operating over the band 500 Hz to 6 kHz was designed by a simple extension of the SHA technique.

An eight-element unshaded array with sensors located half-a-wavelength apart at 500 Hz was used as starting array. By constructing a scaled down replica of this array and by suitable filtering, as in Section 3(b) above, the first octave was covered. The second array was then considered as the low frequency array for the next octave, and the procedure continued until three octaves were covered. The final configuration, using twenty hydrophones, is shown in Fig. 8; the filter characteristics are sketched in Fig. 9.

In principle, the array should only operate over the three octaves from

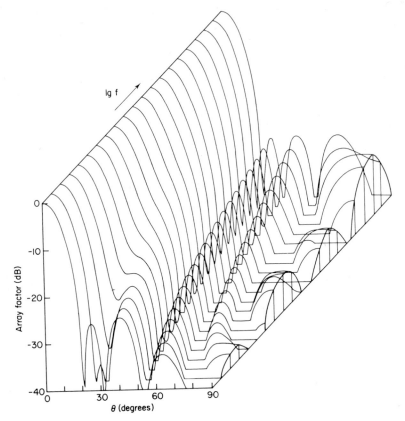

FIG. 7. Beam patterns for a one-octave Chebyshev array.

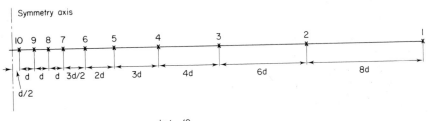

$d = \lambda_H/2$

λ_H = wavelength for $f = 8f_0$

FIG. 8. $3\frac{1}{2}$ octave array geometry.

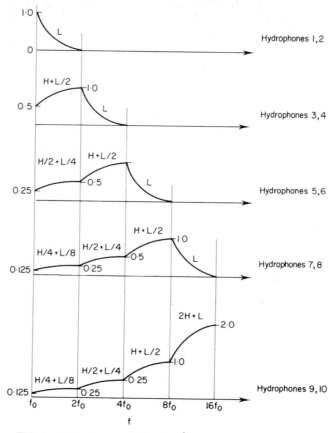

FIG. 9. Filter characteristics for $3\frac{1}{2}$ octave array (f_0 = 500 Hz).

500 Hz to 4 kHz. However, by using the filters shown for the central hydrophones acceptable performance can be obtained up to 6 kHz. For higher frequencies a grating lobe comes into the "visible" region [2]. The band of operation could be extended up to at least 8 kHz (4 octaves) if the grating lobe were eliminated by using directional hydrophones.

4. Implementation and Test of a $3\frac{1}{2}$ Octave Array

The $3\frac{1}{2}$ octave array described in the previous section was built and tested at SACLANTCEN. The hydrophones had a flat response over the frequency band of interest and were omnidirectional to within ±0.5 dB.

Implementation of the array processor involves essentially the design of filters to approximate the characteristics shown in Fig. 9. The approximation problem is difficult to solve for analogue filters, especially because

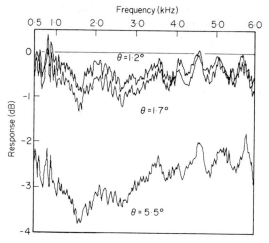

FIG. 10. Experimental curves for the distortion due to pointing errors.

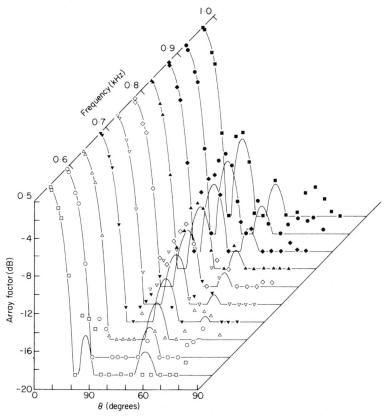

FIG. 11. Experimental and theoretical beam patterns for the 1st octave.

their phase characteristics must also be accurately controlled. Finite-impulse-response digital filters, however, permit separate control of the magnitude and phase characteristics, and can be easily designed with a purely real frequency response so that undesirable phaseshifts are not introduced.

For the $3\frac{1}{2}$ octave array processor digital filters with an impulse response length (filter order) of 256 were used. The peak error in the magnitude response of these filters is in the order of 4% for the low frequency filters,

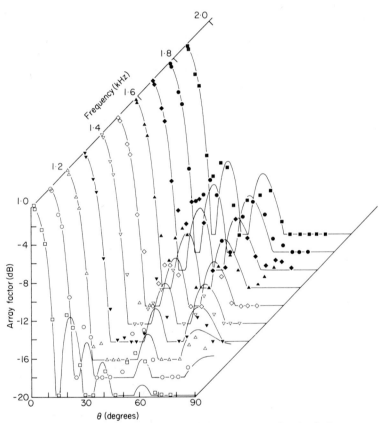

FIG. 12. Experimental and theoretical beam patterns for the 2nd octave.

and considerably less for the others. Although rather sophisticated techniques are available for the design of finite-impulse-response filters [12, 20, 23], a very simple frequency sampling procedure [23] was used. Because the computing time in fast Fourier transform (FFT) implementation of such filters is weakly dependent on the filter order, it was not judged necessary to try to reduce the filter order.

The digital array processor was programmed in a Hewlett-Packard 2116B mini-computer using standard FFT techniques and the HP 5450A Fourier Analyser software in the SACLANTCEN modified version. The data are acquired, edited, and transferred to the computer with the SPADA system, and control is subsequently given to an ITSA program that performs the actual processing. (SPADA and ITSA are described in papers presented at this Institute.)

The array was tested at sea by using explosive sound sources fired at a range of about a mile and various bearings. The direct arrival for each shot was processed, and the array output was deconvolved against the signal received at a single hydrophone. The impulse response of the array alone was thus obtained.

Experimental curves for the array frequency response when the sources are in the main beam are shown in Fig. 10. The spectral distortion is in the order of ±0.5 dB for the shots close to broadside, and ±1 dB for the shot at the beam edge.

For each octave, theoretical beam patterns were computed at ten equally spaced frequencies and are shown as solid lines in Figs. 11 to 14, together with the experimental points.

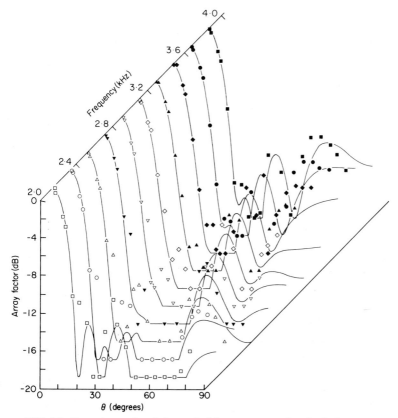

FIG. 13. Experimental and theoretical beam patterns for the 3rd octave.

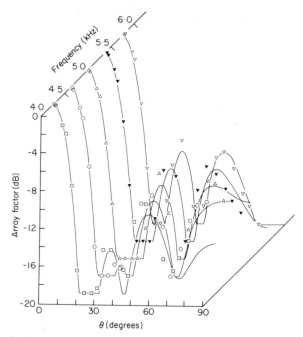

FIG. 14. Experimental and theoretical beam patterns for the 4th octave.

In the test one hydrophone failed. The theoretical patterns shown were computed assuming a 19 hydrophone array, and exhibit a sidelobe level of -12 dB instead of -13 dB. The experimental results are in good agreement with the theoretical predictions, especially near broadside. The experimental sidelobe level is about 1 dB higher than the theoretical value.

5. Broadband Digital Beam Steering

Steering a broadband array consists essentially in compensating for the travel time differences between arrivals at the different hydrophones by suitably time-shifting their outputs.

For the data acquisition rate not to become unduly high, in broadband digital array processing it is necessary to use a sampling frequency close to its minimum theoretical value. This implies that accurate beam forming will usually involve interpolating the signals' samples. Consider for example an array with sensors spaced half a wavelength apart at a frequency f. If a sampling frequency $F_s = 3f$ is chosen, a delay of one sample corresponds to a phase shift of $120°$ at the frequency f, while the phase shift between sensors needed to steer the array at $15°$ from broadside is $45°$.

It is helpful to consider digital beam steering as a problem in multi-channel filtering [4]. This is particularly true for arrays, such as the one described in Section 4, that require complicated filtering to achieve constant beamwidth, and interpolation of the central sensors' samples for beam forming.

Finite-impulse-response filters provide convenient means for digital time-shifting, and can be designed by standard techniques for amplitude and phase tolerances specified in the frequency domain. The filters can be implemented in the time domain by direct convolution (for delays of an integral number of samples time-domain implementation is trivial), or in the frequency domain by FFT techniques.

Estimates for the computing effort in general purpose computer implementations of time-domain and frequency-domain processors are derived in [4]. The computing time depends on the number of sensors, the number of beams, the percentage of sensors that require fractional delays, and the order of the filters. Assuming that one wants to form five beams for the array of Section 4, it was found [4] that frequency-domain processing leads to savings in computing time in the order of 10:1.

It was also shown in [4] that FFT techniques can be competitive even when no constant-beamwidth filtering is involved.

6. Concluding Remarks

The SHA design technique, based on combining the output of homothetic arrays, leads to good performance for arrays operating over frequency bands in the order of four octaves, and with beamwidths in the order of 13°. It is likely that the results for larger arrays will also be acceptable.

The technique itself can be viewed as an artifice to help a designer in finding the non-uniform sensor spacing and the filter characteristics that will ensure the required performance throughout the band of operation. It is very easy to apply since it amounts to designing a narrowband array for the given performance and then extending its operation to the whole band by a straightforward procedure. It seems doubtful that numerical optimization techniques will yield designs with much better performances than those of the examples presented.

Array processors designed by the SHA technique involve complicated filtering operations. This makes the implementation by analogue means difficult, but is not a serious problem in digital implementations that use FFT methods. An attractive feature of FFT processing is that the array performance can be modified, e.g. to obtain lower sidelobes, by simply reading in different filter characteristics.

Software array processors programmed in general purpose digital computers seem to be a good solution for arrays used in research work. For operational or commercial applications (if any) some flexibility and/

or accuracy could probably be traded-off for cost and/or speed, by using hardware.

The major problem encountered with the "first-generation" processor used to obtain the experimental results reported is its speed. Presently, it takes in the order of 5 minutes to process a shot of length $4k$ samples (about 150 msec). Much of this time is spent in unclever file-handling (e.g., reading signals from magnetic tape). Better file-handling and the use of a commercially available hardware FFT processor are expected to considerably reduce the computing time.

REFERENCES

1. Smith, R. P. (1970). Constant beamwidth receiving arrays for broadband sonar systems. *Acustica* **23**, 21–26.
2. Requicha, A. A. G. (1972). "Design of Wideband Constant-Beamwidth Acoustic Arrays", SACLANTCEN Technical Report No. 205, NATO Unclassified.
3. Hixson, E. L. and Au, K. T. (1970). "Broadband Constant Beamwidth Acoustical Arrays", Tech. Memo. No. 19, Acoustics Research Lab., University of Texas at Austin.
4. Requicha, A. A. G. (1972). "Digital Filtering Techniques for Broadband Beamforming", SACLANTCEN Technical Memorandum No. 179, NATO Unclassified.
5. Morris, J. C. and Hands, E. (1961). Constant beamwidth arrays for wide frequency bands. *Acustica* **11**, 341–347.
6. Perini, J. and Idselis, M. (1971). Note on antenna pattern synthesis using numerical iterative methods. *IEEE Trans. Antennas Propagation* **AP-19**, 284–286.
7. Isbell, D. E. (1960). Log periodic dipole arrays. *IRE Trans. Antennas Propagation* **AP-8**, 260–267.
8. Rumsey, V. H. (1966). "Frequency Independent Antennas", Academic Press, New York and London.
9. Lo, Y. T. and Lee, S. W. (1966). A study of space-tapered arrays. *IEEE Trans. Antennas Propagation* **AP-14**, 22–30.
10. Stiefel, E. L. (1963). "An Introduction to Numerical Mathematics", Academic Press, New York and London.
11. Rabinowitz, P. (1968). Applications of linear programming to numerical analysis. *SIAM Rev.* **10**, 121–159.
12. Helms, H. D. (1971). Digital filters with equiripple or minimax responses. *IEEE Trans. Audio Electroacoustics* **AU-19**, 87–92.
13. Tucker, D. G. (1957). Arrays with constant beamwidth over a wide frequency range. *Nature* **180**, 496.
14. Morris, J. C. (1964). Broadband constant beamwidth transducers. *J. Sound Vib.* 28–40.
15. Stockham, T. G. Jr., High speed convolution and correlation. AFIPS Conf. Proc., Vol. 28, Spring Joint Computer Conf., 1966.
16. Stockham, T. G. Jr. (1969). High speed convolution and correlation with applications to digital filtering. "Digital Processing of Signals" (B. Gold and C. R. Rader, Eds.), McGraw-Hill, New York.
17. Helms, H. D. (1968). Non-recursive digital filters: design methods for achieving specifications on frequency response. *IEEE Trans. Audio Electroacoustics* **AU-16**, 336–342.

18. Helms, H. D. (1967). Fast Fourier Transform method of computing difference equations and simulating filters. *IEEE Trans. Audio Electroacoustics* **AU-15**, 85–90.
19. Barbagelata, A., Castanet, A., Laval, R. and Pazzini, M. (1970). "A High-Density Digital Recording System for Underwater Sound Studies", SACLANTCEN Technical Report No. 170, NATO Unclassified.
20. Rabiner, L. R. (1971). Techniques for designing finite-duration impulse-response digital filters. *IEEE Trans. Comm. Tech.* **COM-19**, 188–195.
21. Hersey, H. S., Tufts, D. W. and Lewis, J. T. (1972). Interactive minimax design of linear-phase nonrecursive digital filters subject to upper and lower function constraints. *IEEE Trans. Audio Electroacoustics* **AU-20**, 171–173.
22. McMahon, G. W., Hubley, B. and Mohammed, A. (1972). Design of optimum directional arrays using linear programming techniques, *J.A.S.A.* **51**, No. 1 (Part 2), 304–309.
23. Rabiner, L. R. and Schafer, R. W. (1971). Recursive and nonrecursive realizations of digital filters designed by frequency sampling techniques. *IEEE Trans. Audio Electroacoustics* **AU-19**, 200–207. (Corrections appeared in Vol. **AU-20**, March 1972, pp. 104–105.)

DISCUSSION

P. H. Moose: How do you modify this technique to steer the beam off broadside?
Answer: Steering is done as usual by introducing time delays. The filtering operations described in the paper are independent of the steering angle and always ensure that the beamwidth is constant *with frequency* for each steering angle. As in single frequency arrays the beamwidth will, however, change with steering angle, the beam broadening as one steers away from broadside.

C. N. Pryor: You seem to arbitrarily have chosen octaves as the steps at which you increase array size. If you chose a different number, a factor of 3, say, would this just increase the distortion?
Answer: The technique apparently works because one is combining two patterns that are not too different. For larger values of the scale factor one will always have a fixed 3 dB point but I suspect that the beam shape will not be constant and the sidelobes will increase. One could try other numbers and perhaps save hydrophones, but I do not think that it will work for values much larger than 2. A factor of 2 is also advantageous because some of the hydrophones in the different arrays coincide geometrically.

P. L. Stocklin: Commented on the problem of optimal placing of sensors and conjectured that they should be placed at the points where the spatial likelihood ratio is larger.

C. J. M. Wolff: Radio people use log-periodic antennas which have wide bandwidths. Is there anything comparable to this in sonar problems?
Answer: I do not know of any log-periodic sonar arrays. The array I described has a log-periodic geometry and exploits the same basic principle as log-periodic arrays, namely, the invariance of array response when the array dimensions are scaled down and the frequency increased by the same factor.

A. Stansfield: Log-periodic arrays are endfire arrays. Have you used yours as an endfire array?
Answer: No, but I do not see any reason why one could not.

Applications of Holographic Interferometry in Underwater Acoustics Research

C. D. JOHNSON and G. M. MAYER

*Naval Underwater Systems Center, New London
Connecticut, U.S.A.*

1. Introduction

Holographic interferometry as a measurement technique has been in existence since 1965. Since that time, it has gained wide acceptance in non-destructive testing and vibration analysis and is presently practised by a great many organizations. The purpose of this paper is to highlight some of the applications of holographic interferometry that are particularly relevant to underwater acoustics work.

2. Background

Photography by wavefront reconstruction was first developed by Gabor [1] in 1947. He proposed that this process could be used to achieve better resolution in electron microscopes by using a hologram to decrease astigmatism. By making the hologram with radiation of very short wavelength and reconstructing with visible light, he expected to achieve resolutions in the order of 1 Å. Gabor experimented with an optical analogue of his system, but practical application of his proposal was hampered by the lack of a good coherent light source.

With the introduction of the laser in 1960, interest in holography was revived, and in 1963 Leith and Upatnieks [2] reported obtaining successful holograms of three-dimensional diffusely illuminated objects. Whereas Gabor's experimental setup required only limited coherence length because of the short path length difference associated with a transparency, the work of Leith and Upatnieks required the longer coherence length available only from a laser. Shortly thereafter, Haines and Hildebrand [3] used holography to perform interferometric measurements on diffusely reflecting objects, and Powell and Stetson [4] discovered that holographic interferometry could be used to observe vibrating surfaces. Refinements of these basic techniques have greatly increased the versatility of holographic interferometry. The basic holography setups used during the development of this technology are shown in Fig. 1.

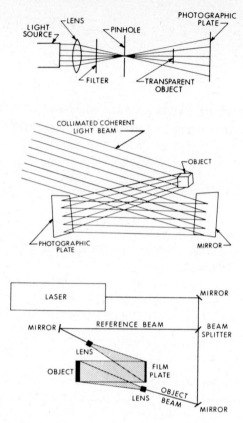

FIG. 1. Holography apparatus. Top: Gabor's Optical Analogue; Centre: Leith and Upatnieks' setup. Bottom: Typical currently used setup.

3. The Holography Process

Holography is a photographic process carried out entirely with coherent illumination. It differs from conventional photography in that it records light fields rather than images. A hologram is made by splitting a laser beam into two parts by means of a partially reflecting mirror. One part, known as the reference beam, is expanded and directed onto a high-resolution photographic plate. The other part, known as the object beam, is expanded and used to illuminate the object. Since both fields are made up of coherent light, optical interference occurs in the photographic emulsion. The recording will diffract light into a replica of the object beam when it is illuminated by a replica of the reference beam, thereby producing an accurate three-dimensional representation of the object in its original position.

Holographic interferometry is accomplished in the same manner as ordinary holography, except that double or multiple object positions are recorded on the same photographic plate. If the change in object position is in the form of a surface deformation and if the magnitude of the change is approximately several wavelengths of light, optical interference will occur between the two images reconstructed upon viewing the hologram. The object will appear in its original position, but will have a fringe pattern on its surface. The fringe pattern is actually a contour map of the surface deformation that occurred between exposures. The contour interval represents changes in object position that caused a net change of one wavelength in the path of the light originating in the object beam and falling on the photographic plate.

The fact that it is possible to record interference fringes simply by holographically recording a scene in which variable path length changes have occurred in the object beam is of considerable significance in making engineering measurements. The case discussed above was a simple double exposure of an object that incurred surface deformations between exposures, but any disturbance that results in a variable path-length change in the object beam will produce similar results. Some of the more important physical disturbances that can be holographically recorded are localized changes in the index of refraction of a transparent medium, surface deformation, angular displacement, and periodic vibration.

Periodic vibration is of fundamental importance to underwater acoustics, and most of the relevant applications involve the measurement of such vibration by various techniques. The most widely used technique is that of time-average holography devised by Powell and Stetson. This technique involves the continuous recording over a time interval of many cycles of vibration of the continually changing positions of a vibrating surface. The resultant fringe pattern comes about because the amplitude distribution dictates that the vibrating surface spends more time near the extreme positions than in transit between the extremes. The fringe contrast in the time-average case is related to the square of the J_0 Bessel function, a relationship that proves to be a mixed blessing. On the positive side, the J_0^2 dependence results in easily identified nodal regions, inasmuch as stationary regions are intensely bright, and the brightness of successive fringes diminishes rapidly with increasing amplitude. Unfortunately, the diminishing fringe contrast also imposes a limitation on the amplitude that can be readily observed. A second limitation is the loss of all information on the spatial phase distribution of the vibration, inasmuch as the time averaging process provides only amplitude envelope information.

In many cases, a time-average hologram provides sufficient information to answer the questions at hand. In other cases, such as array interaction studies, it is sometimes necessary to obtain information on spatial phase distribution, or information at high vibration amplitudes. In such cases, it is possible to supplement the time-average holograms with holograms made with stroboscopic illumination. Stroboscopic illumination accomplishes two significant things—it "freezes" the vibration amplitude at some given point in the vibration cycle, thus making possible relative phase measurement, and it eliminates the J_0^2 dependence thereby providing constant fringe contrast, which permits investigation of high amplitude vibration. Stroboscopic techniques were developed concurrently in the United Kingdom by Archibold and Ennos [5] and in the United States by Shajenko and Johnson [6]. A comparison of the relative fringe visibility between time-average and stroboscopic holograms is shown in Fig. 2.

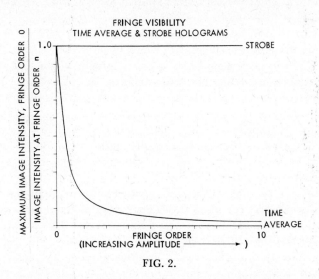

FIG. 2.

   The preceding sections briefly describe the fundamental tools available in holographic interferometry. Subsequent sections will describe some of the underwater acoustics related applications that have been achieved using these tools. The applications described are intended to be illustrative of what can be done. They do not presume to cover all possible applications and are offered by the authors to stimulate thought regarding how this new technology can be more fully exploited.

4. Measurements in Air

There are many applications of holographic interferometry that can be quickly and conveniently carried out using a table top setup in an optical laboratory. Such measurements are primarily the double-exposure and time-average techniques, with variations in the method of loading or excitation. They are accomplished using holography because it offers technical advantages over other measurement techniques both in the quality of data and the ease in obtaining it. The applications described below cover non-destructive testing, validation of transducer element design programs, experimental design analysis of prototype transducers, experimental inputs to computational programs, and determination of material parameters.

(a) Non-destructive testing

Non-destructive tests have been carried out successfully on a number of specimens using both double-exposure and time-average holographic interferometry. One example is the inspection of large moulded rubber fairings for air voids. The fairings were intended for use in the immediate vicinity of hydrophones, and air voids in the rubber would have a detrimental effect on hydrophone performance. This case is interesting in that several non-destructive test methods were used on the same specimens, after which the specimens were cut apart to determine which method provided the most reliable results. In addition to holography, the specimens were inspected by means of ultrasonic testing, X-ray radiography, and surface temperature distribution. Holography proved to be the most reliable indication of size and location of voids.

The ultrasonic method indicated the existence of several additional voids when none were in fact present. X-ray radiography did not work well because the specimens had a tapering cross section, giving an unacceptably large density gradient across the specimen. A second shortcoming of the X-ray technique was the failure to detect flaws of a laminar nature. The measurement of surface temperature distribution with a transient heat input was not sufficiently sensitive to detect flaws an appreciable distance beneath the surface. Only holography gave a reliable indication of the number and actual locations of the voids.

The holography technique used was double-exposure interferometry. The specimen was first subjected to hydrostatic pressure. The pressure was then removed, and two holograms exposed with a time lapse of one or two minutes between exposures. The areas in the vicinity of flaws exhibited creep at a different rate than the rest of the specimen and were clearly indicated by anomalies in an otherwise regular pattern of interference fringes.

A second example is the detection of cracks in the fibreglass prestress layer of a segmented ceramic ring transducer. Double-exposure holographic interferometry was again used, with the differential loading accomplished by imposing different levels of dc voltage across the ceramic for each exposure. The crack was readily detected by a discontinuity in an otherwise regular pattern of fringes.

A final example is the detection of unbonded areas in the rubber covering the radiating face of a transducer. Several techniques have been tried on this problem with varying degrees of success. Although the optimum technique is known, it is not yet in regular use. Double-exposure

holograms have been made using both vacuum loading and transient
thermal inputs to generate differential displacements on the transducer
face. Both techniques have been used successfully, but are not reliable
indicators. Time-average holograms show debonded areas very nicely,
if one is fortunate enough to be near a resonance frequency where the
unbonded region vibrates differently from the rest of the transducer face.
The obvious answer is to scan a wide range of frequencies to ensure that
the resonance of the unbonded area is excited. That unfortunately would
require a prohibitively large number of time-average holograms and is not
an economically sound technical approach. The answer lies in real-time
stroboscopic holography, which, as the name implies, permits the continuous observation in real time of the vibration of a transducer face
during a frequency sweep. The real-time capability comes about from
making a reference hologram of a stationary object and observing the
object through the reference hologram while the load on the object is varied.
At zero load, the holographic image and the actual object coincide exactly,
and no interference fringes are formed. As the actual object is deformed,
however, interference fringes related to the deformation are formed
between the actual object and the holographic image. The same phenomenon occurs when the motion of a vibrating surface is frozen by stroboscopic illumination and compared with a hologram of the surface at rest.
It is a simple extension of the concept, then, to synchronize the stroboscopic illumination with the frequency sweep and observe the vibration
behaviour in real time. Limitations of available light-modulating equipment
have kept this technique from being widely employed, but the situation
is rapidly improving.

(b) Validation of transducer element design programs
A great deal of effort has been spent in recent years in developing sophisticated computational models for the design of transducers. The new
technology of finite element structural analysis has been heavily used, and
although it provided a level of detail in design analysis that was not
possible with continuum solutions, the models became quite complex and
could not be used with confidence without experimental validation. A
validation program was undertaken during which theoretical and experimental results were compared first for individual discrete transducer
components vibrating under free-free conditions, and finally for the
individual components assembled together into a complete transducer and
excited electrically. The results of the predicted and measured displacement
contours for the complete transducer are shown in Fig. 3. The predicted
contours shown at the bottom of the figure are derived entirely from the
theoretical transducer model, whereas the experimental results shown
are taken from time-average holograms of the same transducer. Mirrors
were used in the hologram to obtain simultaneous data on the radiating

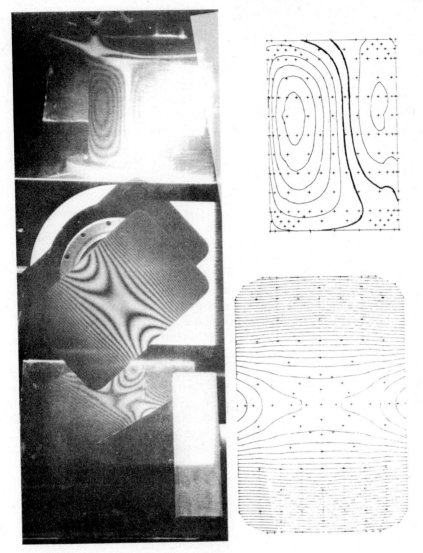

FIG. 3.

face and the ceramic drive assembly. Seven independent vibration modes were measured by using the same experimental arrangement. In all cases, the agreement between predicted and measured resonance frequencies was better than 5%, and the predicted contour plots are virtual overlays of the holographic data. This work was carried out jointly by the Naval Underwater Systems Center and the Naval Undersea Research and Development Center. The authors are indebted to Dr. John Hunt, of NURDC, for providing the theoretical contour plots.

(c) Experimental design analysis
Time-average holographic interferometry has been used for routine
measurements during the transducer development process. Comparative
evaluations have been performed on various radiating face configurations
to determine differences in vibration characteristics. The junctions between
the ceramic drive assembly and radiating face, and the ceramic drive
assembly and reaction mass have been studied qualitatively. The behaviour
of the hinge region of flexural disk transducers has been studied in some
detail. Cylinder vibration modes of the ceramic drive assembly have been
examined. Such measurements are routine in every sense. They are
accomplished quickly, at low cost, and require no special preparation of the
test subject.

(d) Experimental inputs to computational programs
Experimental inputs can sometimes be used to advantage in improving
the accuracy of mathematical analyses. A case in point is the shock
analysis of a flexural disk transducer. A normal mode shock analysis of a
continuous structure may be made by defining an equivalent discrete
mass mathematical model of the structure, determining the mass and
stiffness matrices, solving the free-vibration problem for frequencies and
mode shapes, and applying the desired transient input by means of the
convolution integral. The primary source of error in such an analysis is
the mathematical modelling; that is, a valid definition of equivalent mass
and stiffness matrices. The modelling process can be carried out with good
accuracy for certain types of structures characterized by isotropic materials
and known boundary conditions. Other structures, particularly those with
anisotropic or composite materials and unknown boundary conditions,
can be very difficult to model with accuracy. It is in analysing this type
of structure that the holography technique for solution of the free-
vibration problem is most useful.

Time-average holography was used to experimentally determine the
resonance frequencies and displacements associated with the mode shapes
of the particular discrete mass model chosen. This in effect completely
defines the eigenvalues and eigenvectors of interest and obviates the need
for determining the equivalent mass and stiffness matrices. By use of this
method, excellent agreement was achieved between computed shock
accelerations and corresponding values measured during a shock test. The
recent introduction of sophisticated finite element structural analysis
techniques has lessened the importance of this holographic application.

(e) Determination of material parameters
A very useful application is the indirect determination of piezoelectric
material parameters in practical transducer configurations. Piezoelectric

properties are determined by making a series of prescribed measurements
on small standard test specimens. The measured piezoelectric properties,
unfortunately, do not remain constant when the same basic ceramic
materials are fabricated into shapes required for use in transducers. These
properties, however, must be known accurately if mathematical models are
to be used successfully in transducer design. The problem, then, is one of
accurately measuring the required material parameters.

This problem can be approached indirectly through the use of holographic measurements and finite element structural modelling. Finite
element models are remarkably accurate when the subject of the analysis
is constructed of materials with well defined properties. This has been
repeatedly demonstrated through the comparison of predicted results
with experimental measurements, and holds true even for very complex
structures. The holographic validation of a transducer element design
program described earlier is one example. There is a high level of confidence, then, in the finite element modelling procedure.

Similarly, there is a high level of confidence in the accuracy of holographic interferometry measurements. The physical nature of optical
interference is well understood, and the wavelength of the laser illumination is known to a high degree of accuracy. It follows then that the
predicted behaviour and the holographic measurements should agree very
closely. If it is assumed that the cause for any discrepancies lies in the
incorrect values of material parameters, the material properties can be
adjusted to force the predictions from the mathematical model to match
the experimental observations. This procedure is not as arbitrary as it first
appears, inasmuch as the adjustments amount to small changes from the
values measured on the standard test specimens. The procedure has been
used on large ceramic cylinders with excellent results.

The holographic technique used for the large-cylinder measurements is
somewhat unique. The fringe structure on a cylindrical surface is very
difficult to interpret quantitatively, because the actual displacement
associated with a fringe is dependent on the angle of the illuminating
beam and the angle of observation (with respect to the normal) to the
surface. In the case of a cylinder, these angles are continually changing
as one moves around the cylinder. A second disadvantage is that only
a sector of the cylinder of somewhat less than 180 degrees can be directly
observed. Both of these limitations were overcome by placing the cylinder
inside a large conical mirror such that the axis of the cylinder coincides
with the axis of the mirror. With this arrangement, the image of the
cylinder appears as a circle, thereby giving 360-degree coverage, and
constant angles of illumination and observation. An example comparing a
direct hologram of the cylinder with one taken inside the conical mirror
is shown in Fig. 4. In the conical mirror picture, axial nodes appear as
radial lines, and circumferential nodes appear as concentric circles.

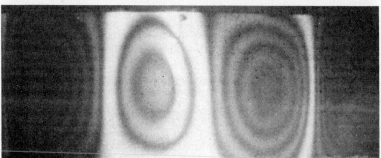

FIG. 4.

5. Measurements in Water

Holographic measurements underwater are not particularly difficult. The making of an underwater hologram was reported in 1966 by Grant, Lillie and Barnett [7]. The most straightforward way to accomplish such measurements is to place the test subject in a tank of water and keep the optical setup in air outside the tank, viewing the subject through a transparent window. The only correction necessary is for the index of refraction difference between air and water. This technique has been applied with tanks ranging in size from a small aquarium to a 15-metre-diameter container, the latter being large enough to permit free-field measurements through the use of synchronized optical and acoustical gating.

(a) Underwater plate measurements

The simplest type of underwater measurement involves placing the test specimen in a small container of water in a conventional laboratory holography setup. This method yields valid results when the characteristics being measured are not adversely affected by the acoustic environment of the container. An example is the measurement of the vibration modes of a thin plate. Such an experiment was performed to compare the in-air and underwater vibration modes of a thin plate to determine the effect of hydrodynamic mass loading on the resonance frequency associated with particular vibration modes. The use of holography made it possible to positively identify comparable modes despite frequency shifts of greater than $2:1$. By examining several modes, it was possible to determine the frequency dependence of hydrodynamic mass loading.

(b) Transducer array measurements

The ultimate measurement in water involves test specimens that are sensitive not only to the hydrodynamic mass loading, but also to the acoustic characteristics of the medium. Sonar transducers are in this category. An example is a measurement program undertaken on a planar array of 25 transducer elements under various drive and element termination conditions. These measurements are unique and potentially of great value in supporting underwater acoustics research.

The objective of the measurements was to provide experimental verification of the predicted acoustically induced mutual coupling effects between sonar transducer elements operating in multi-element arrays. This required obtaining holograms of the array operating in a body of water large enough to minimize the effects of acoustic reflections from the boundaries. The water tank chosen to satisfy this condition was 15 metres in diameter and permitted operation of the transducer array at a depth of 30 metres. Measurements were made with the array driven in a steady-state condition, and with the array and laser synchronously gated so that the exposure was made during the interval before the first reflection from the boundary had arrived back at the array.

The laser and optical apparatus were set up outside the tank. An argon-ion laser operating at 5145 Å was used and was fitted with an etalon for coherence extension. Output power was approximately 1 watt. Holograms were recorded on film plates with exposure times of $\frac{1}{8}$ second. The reference beam was passed through a porthole and returned from a corner reflector mounted on the array. The illuminating beam was passed through a second porthole, which was also used for viewing and photographic recording. The array surface was covered with a retro-reflecting tape to increase the amount of light returned from the subject. The setup used subjected the illuminating beam and the reference beam to the same underwater path to minimize effects of pressure and temperature gradients between the porthole and the array. The distance from porthole to array was 7.5 metres. No attempt was made to isolate either the array structure or the optical table. A schematic representation of the setup used is shown in Fig. 5.

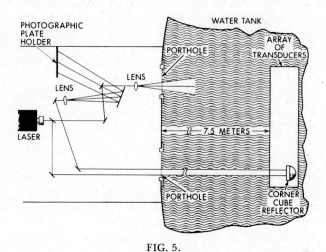

FIG. 5.

Little difficulty was experienced in obtaining good quality holograms under these conditions. By taking reasonable care in photographing the reconstructions, it was possible to resolve fringes spaced 0.6 mm apart on the surface of the array. This resolution is more than adequate for the engineering analysis to be performed using these holograms.

An example of a representative hologram obtained in this experiment and an artist's concept of the amplitude envelope described by the holographic data are shown in Fig. 6. Only the centre element was electrically driven, with the other elements terminated in an open circuit impedance. The displacement noted on the adjacent elements is attributed to coupling through the water.

This experiment has demonstrated the feasibility of obtaining holographic interferometry measurements of large subjects vibrating underwater with considerable distances between subject and optical apparatus. Moreover, it is possible to achieve fringe resolution fine enough to permit

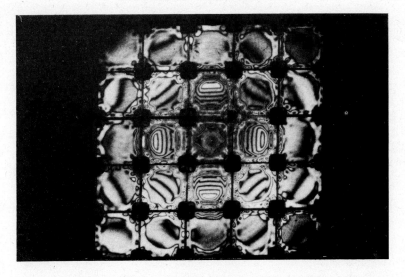

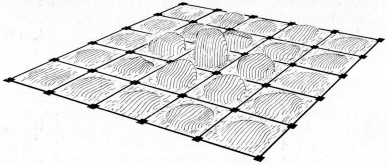

FIG. 6. Top: A simulation of the displacement magnitude derived from a holographic interferogram of a 5 x 5 sonar transducer array. Bottom: The centre element is being driven. The others are terminated.

detailed engineering analyses to be performed. Underwater holographic interferometry provides more complete data on the nature of underwater vibrations than could be obtained by any other means and should prove to be of significant value to those involved in underwater acoustics research.

6. Measurements Planned in the Future

Plans for the immediate future centre on exploiting the technique of stroboscopic holography and automating the processing of holographic data from large transducer arrays. The array measurements described above are incomplete inasmuch as they contain only amplitude envelope information; the spatial phase distribution must also be determined to adequately define the array characteristics. This can be accomplished by making stroboscopic exposures at known phase increments from an arbitrary reference, such as the displacement of the centre element. Phase information can then be extracted from relative amplitudes, and rate of change of the relative amplitudes.

Extraction and processing of holographic data from large transducer arrays are formidable tasks because of the intricate fringe structure. Fortunately, the data appears to lend itself well to automated processing. This can be accomplished by digitizing grey scale information obtained with a scanning optical sensor and processing that information on a digital computer.

7. Potential Future Applications

There are many potential applications of holographic interferometry in underwater acoustics research that have not been investigated. Some are possible with existing technology; others require further development of holographic techniques. Several of these potential applications are briefly described below to stimulate thought about which are worthy of further pursuit.

(a) Investigation of sound transmission through domes
Sonar domes are typically very complex structures consisting of extensive internal trusswork and sheet metal skins. Their design is largely determined by structural considerations, and the acoustic characteristics are less than optimum. Attempts to include the dome in mathematical models of large arrays involve gross assumptions, and unresolved discrepancies between theory and validation experiments are known to exist. Holographic interferometry offers the potential of providing a qualitative assessment of the sound transmission mechanism as a first step in refining existing mathematical models.

(b) Investigation of cavitation phenomena
Cavitation in an array of transducers is relatively easy to detect by using a probe hydrophone, but localization of the point of inception of cavitation is considerably more difficult. With the development of good high-amplitude, high-resolution techniques, probably using stroboscopic holography, cavitation can be observed visually and the points of inception easily identified.

(c) Investigation of particle velocity distribution in interstitial areas
Array performance calculations indicate that the nature of the interstitial areas has a pronounced effect on computed array performance. By assuming that particle velocity continuity is maintained across a thin

compliant membrane, it is possible to holographically observe the velocity distribution in the interstices by spanning the interstitial area with a thin layer of reflective material.

(d) Investigation of noise sources on underwater structures
Assuming there is a correlation between vibrational displacement and radiated noise, complex underwater structures can be qualitatively analysed for the presence of vibration "hot spots" that could indicate local noise sources.

(e) Shock wave propagation through acoustic materials
Shock wave propagation can be observed holographically by using a pulsed laser that emits two pulses separated by a variable time delay and triggered by an impulsive force that initiates a shock wave in the material of interest.

(f) In situ investigations of large underwater structures
It is technically feasible to make holographic measurements of large structures in their actual operating environment. However, such measurements would entail a significant expenditure for development of the optical equipment and considerable inconvenience in obtaining the measurements. At the present time, undertaking such a large-scale holography effort is considered unreasonable from an economic standpoint.

8. Conclusion

The material presented in this paper has been arranged to give the reader some background in holography and to suggest applications of potential interest in underwater acoustics research. The literature available in the field of holography has become voluminous, and detailed technical information on many of the techniques discussed is readily available.

REFERENCES

1. Gabor, D. (1948). A new microscopic principle. *Nature* **161**, 777–778.
2. Leith, E. and Upatnieks, J. (1963). Wavefront reconstruction with continuous tone objects. *J. Opt. Soc. Am.* **53**, No. 12, 1377–1381.
3. Haines, K. and Hildebrand, B. (1966). Interoferometric measurements on diffuse surfaces by holographic techniques. *IEEE Trans. Instrumentation Measurements*, **IM-15**, 149–161.
4. Powell, R. and Stetson, K. (1965). Interferometric vibration analysis by wavefront reconstruction. *J. Opt. Soc. Am.* **55**, 1953–1958.
5. Archbold, E. and Ennos, A. (1968). Observation of surface vibration nodes by stroboscopic hologram interferometry. *Nature* **217**, 942–943.

6. Shajenko, P. and Johnson, C. (1968). Stroboscopic holographic interferometry. *Appl. Phys. Lett.* **13**, 44–46.
7. Grant, R., Lillie, R. and Barnett, H. (1966). Underwater holography. *J. Opt. Soc. Am.* **56**, No. 8, 1142.

DISCUSSION

P. L. Stocklin: Could holography be used to look at receiving hydrophones and the complex structures that support them, to see what has happened to incoming acoustic waves near the hydrophones?

Answer: The vibration response of the supporting structure to an incoming acoustic wave could easily be observed, but observing the acoustic field is more difficult. We have observed fringes suspended in space which can be attributed to acoustically induced index of refraction changes in the water, but this occurs at pressure levels very much higher than would be expected at a receiving array.

A. Stanfield: Have holographic measurements been made on an array that has been steered, or have all measurements been made broadside?

Answer: All measurements to date have been made broadside; that is, with no phase difference between the electrical signals driving the transducers.

A. Stansfield: Are any measurements on steered arrays planned in the future?

Answer: It will be some time before they are attempted. There are more immediate problems to be solved first for the simple case. We intend to achieve a capability for spatial phase measurement before attempting measurements on a steered array.

K. Bakke: What is the highest frequency that can be used in the stroboscopic technique?

Answer: I believe we can easily achieve 50 kHz, perhaps higher. The limiting factor is usually achieving the minimum displacement required to observe interference fringes.

Comment: That is quite the whole range of acoustical frequencies we are accustomed to.

J. W. R. Griffiths: How would the sensitivity of holography compare with other means of visualizing what is going on acoustically, such as the Schlieren effect, which has been used for looking at the actual beams in the water?

Answer: The Schlieren techniques would probably be more desirable for observing effects in the water itself. The holographic techniques are best for observing effects on a solid surface.

E. J. Risness: When you mentioned that considerable expense could be involved in developing further techniques, I envisioned a self-contained holograph apparatus that one might take in the field, perhaps even underwater. Is this in fact a feasible concept?

Answer: Yes. That is in fact what I was referring to. Such a device does not exist now, packaged in a form that could be taken underwater. There is no reason why it could not be done if the need were there, and the money available to pursue the equipment development.

Holographic Processing of Acoustic Data Obtained by a Linear Array

R. DIEHL,

Krupp Atlas-Elektronik, Bremen, West Germany

1. Introduction

Acoustic holography is a special case of the well-known holographic process with the hologram being produced by means of coherent sound waves. As soon as the holographic information is exposed in an optical hologram, the object wave can be reconstructed in terms of coherent light waves [1]. The special problem of this method in underwater sound technique is the conversion of the sound information into a hologram which has to be suitable for reconstruction. In order to obtain the desired high resolution in bearing, large apertures must be used. Consequently, the receiving hydrophone system has to consist of a large number of discrete elements [2]. Since, on the other hand, the receiving signals are extremely weak, each hydrophone must be followed by its own individual amplifier in order to bring its output voltage to a reasonable level. For these reasons a holographic receiver for underwater sound signals becomes very expanded. As an example a holographic display consisting of 100 lines and 100 columns would require $2 \cdot 10^4$ receiving channels. In the literature various methods are described to reduce the number of hydrophones [3]. They are generally based on the fact that the hologram information is built up sequentially, i.e. the receiving system being shifted slowly from ping to ping over the full aperture. In underwater sound technology, however, relative motion between object and hologram aperture must always be taken into account. This requires a short time of reception. In the following, a method is described which makes use of a linear aperture. Since the one-dimensional holographic information is not sufficient for producing a two-dimensional display, wideband sound signals are transmitted. The echo range is taken then as a second dimension for producing the display. At first sight, this method differs only slightly from usual active sonar techniques.

Holographic signal processing, however, produces—unlike the simple addition of the individual signals—a focused display both in the far field and in the Fresnel region.

561

Therefore, the holographic method yields higher resolution in the Fresnel region and should be mainly applied where high angular resolution is wanted at comparatively short ranges.

2. Principle of One-dimensional Pulse Holography

Figure 1 shows a set-up for the reception of one-dimensional holograms as a function of the echo delay. A transmitter sends out a sinusoidal pulse of ultrasonic frequency. In complex notation this signal is

$$S_e(t) \cdot e^{j2\pi f_0 t}, \quad j = \sqrt{-1}.$$

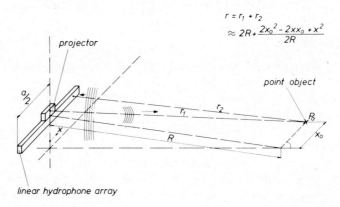

FIG. 1. Arrangement for the construction of one-dimensional holograms.

Let the envelope $S_e(t)$ be a real time function with the bandwidth w. On the assumption that P_0 is an ideal point reflector, the following receiving voltage is obtained

$$u(x, t) = \frac{\text{const.}}{r} S_e(t - t_0) \cdot e^{j2\pi(f_0 t - r/\lambda)} \tag{1}$$

where

$$r = [R^2 + (x - x_0)^2]^{1/2} + [R^2 + x_0^2]^{1/2}, \quad t = r/c,$$

c being the sound velocity.

In practice, $u(x, t)$ can only be determined at discrete points. If the sampling distance is so small that the spatial sampling theorem is not violated, $u(x, t)$ can be obtained from the discretely sampled function by spatial filtering. All terms not depending on x are combined in a complex constant $\gamma(r)$. Moreover, if the Fresnel approximation is applied, the square root for r can be replaced by a series expansion discontinued after the first term. The resulting error of the exponent in equation (1) should be less than

$\pi/4$. The limits of the approximation are illustrated by the following example: At a wavelength of 0.02 metres, an aperture length of 2 metres and a maximum beamwidth of 30 degrees, the Fresnel region extends from about 10 to 2000 metres. For larger distances, far field conditions apply.

The expression thus simplified for the receiving signal is the following:

$$u(x) = \gamma(r) \cdot S_e(t - t_0) \cdot \exp\left(-j\frac{\pi x^2}{\lambda R}\right) \cdot \exp\left(j\frac{2\pi x x_0}{\lambda R}\right) \cdot e^{j2\pi f_0 t}. \qquad (2)$$

In the receiver, the complex modulation function of this signal is obtained by multiplication with a reference signal and subsequent low-pass filtering (Fig. 2). The simplest way to generate the reference signal is to take the

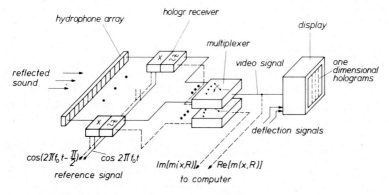

FIG. 2. Principle of one-dimensional holographic signal processing.

oscillator signal of the transmitter itself. If, however, a reference signal is used which is phase-shifted by $\pi/2$ then two output signals are obtained representing the real and the imaginary components of the complex spatial modulation function $m(x)$:

$$m(x, t) = \gamma(r) \cdot S_e(t - t_0) \cdot \exp\left(j\frac{2\pi x x_0}{\lambda R}\right) \cdot \exp\left(-j\frac{\pi x^2}{\lambda R}\right) \cdot \text{rect}(x/a). \quad (3)$$

At a fixed time this function is a one-dimensional section of a Fresnel zone plate, known as the hologram of a point-shaped object. This means that from the information given by $m(x)$, a one-dimensional hologram can be obtained. The multiplication of $u(t)$ with the reference signal simulates the normal incidence of a plane acoustic reference wave relative to the receiving aperture. The term $\text{rect}(x/a)$ (Woodward's notation) takes into account that the sound field is known only along the linear aperture of length a.

Supposing the envelope $S_e(t)$ is a short rectangular pulse of duration T, R can be replaced approximately by $t_0 \cdot c/2$. Now the function m can be expressed with the echo delay t_0 and the coordinate x as follows:

$$m(x, t_0) = \gamma(r) \cdot \exp\left[-j\frac{2\pi}{\lambda t_0 c}(x^2 - 2xx_0)\right] \cdot \text{rect}\,(x/a). \qquad (4)$$

An arbitrary shaped area can be described by a generally complex image function $\psi_0(x_0, R)$. This function describes the space-dependent reflection coefficient projected onto the horizontal plane. Because of the linearity of the sound propagation, the general shape of the complex envelope is then:

$$m(x, t_0) = \gamma(t_0) \cdot \int_{-\infty}^{\infty} \psi_0(x_0, t_0) \cdot \exp\left[-j\frac{2\pi}{t_0 \cdot \lambda \cdot c}(x - x_0)^2\right]$$

$$\cdot \text{rect}\,(x/a)\,dx_0. \qquad (5)$$

Apart from the factor rect (x/a), this equation describes a linear, space-invariant system with the echo delay as a parameter. The information contained in the image function $\psi_B(x, R)$ can be retained from the function $m(x, t_0)$. This is possible under the assumption that t_0 can be determined exactly and the aperture is infinitely large. In practice, of course, these conditions cannot be fulfilled. The image function $\psi_B(t_0, x_0)$, computed from $m(x, t)$ has, therefore, an uncertainty in t_0 and x_0. It may also be mentioned here that in the close near field region where the Fresnel approximation is no longer valid, a description can be given by means of a linear system as well.

In the special case that only the amplitude of ψ_B is of interest, the signal processing becomes very simple. The modulation function is then:

$$m(x, t_0) = \gamma(t_0) \cdot \exp\left(-j\frac{2\pi x^2}{\lambda t_0 c}\right) \cdot \int_{-\infty}^{\infty} \psi_0'(x_0, t_0)$$

$$\cdot \exp\left(j\frac{4\pi x x_0}{\lambda t_0 c}\right) dx_0 \qquad (6)$$

with

$$\psi_0' = \psi_0 \cdot \exp\left(-j\frac{2\pi x_0^2}{\lambda t_0 c}\right).$$

The integral is a Fourier transform of the modified image function ψ_0'. The reconstruction can, therefore, be performed by multiplication with a factor $\exp[j2\pi x^2/(\lambda \cdot t_0 \cdot c)]$ and by a subsequent Fourier transform. The result will be an image function ψ_B, which in the case of an infinitely large aperture is proportional in amplitude to ψ_0. Taking into account the limited aperture of a rectangular dimension, the result is a convolution with the Fourier transform $\mathrm{sinc}\,[(x \cdot a)/\lambda R]$. One can see now that the resolution limit in x-direction is $\lambda \cdot R/a$.

In radial direction, the resolution does not depend on the range. It is $T \cdot c/2$ (Fig. 3) and depends only on the pulse length T of the sinusoidal transmitting pulse.

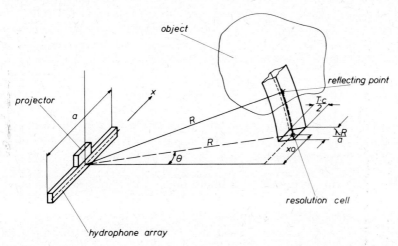

FIG. 3. Resolution properties of the one-dimensional holographic method.

Pulses with a great time-bandwidth product can also be used provided the phase information is not distorted [4].

In practice, the reconstruction can be performed either by a digital computer or by means of a coherent optical system. In the first instance the real and imaginary parts of the signals are sampled and digitized. The necessary sampling frequency depends on the number N of the receiving channels and on the bandwidth w of the transmitted pulse. It is $2 \cdot N \cdot w$. If possible, the digital information is fed directly into a computer. Since usually the sampling frequency is very high, the computers commonly in use, are not capable of taking in the data. Consequently, the analogue or digital values must first be buffered into an intermediate store, and from there passed to the computer at a reduced rate. Inside the computer the values of each sampling cycle are then multiplied by the exponential factor and Fourier transformed. The absolute value or intensity of the result is shown on a display unit (Fig. 4). If real-time operation is wanted and the computer channel is fast enough, the corresponding FFT-processor has to be capable of computing an N-point transformation

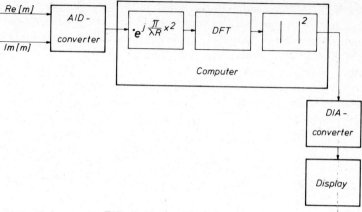

FIG. 4. Digital reconstruction.

in less than half the transmission pulse duration. In the event of near-field holograms, a convolution with the conjugate complex transfer function of the before-mentioned linear system has to be performed instead of the Fourier transform.

The processing of the sound field information by means of coherent optical methods is very similar to methods used with side-looking radars [5]. As practical methods for phase and amplitude modulation of a coherent light wave are not yet available, only the real part can be processed instead of the complex function $m(x, t_0)$. As known from the theory of holography, this will result in an ambiguous display.

Both the object wave and the corresponding conjugate wave occur at the same time. For separation, the optical holography uses a reference wave of oblique incidence. In the array shown in Fig. 5 this effect is reached by phase-shifting the reference signal for each individual receiver proportionally with x.

In this way, an acoustic reference wave of oblique incidence is electronically simulated. The conversion of the modulation function thus obtained into

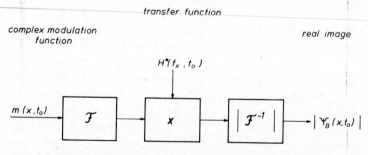

FIG. 5. Digital reconstruction of short range holograms.

an optical form which is suitable for reconstruction can best be performed by means of a cathode-ray tube. Since only positive brightness values are displayed, a sufficiently high DC voltage must be added to the electrical signal. Subsequently, the hologram is recorded on a film. After development it can be used for the modulation of a laser beam. This time-consuming procedure can be avoided by using a real-time conversion tube as described in reference [6]. The video signal is used to modulate the charge distribution on a DKDP crystal by means of an electron beam modulated with the video signal. Laser light transmitted through the crystal is thus modulated in accordance with the charge distribution. The principle of further optical processing is illustrated in Fig. 6. It shows that at first the term $\exp[-j2\pi x^2/(\lambda t_0 c)]$ is compensated by a conical lens. Via the combination

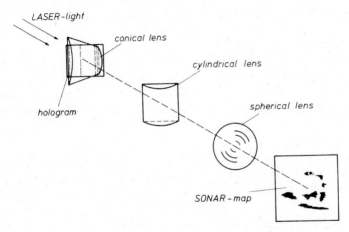

FIG. 6. Principle of optical reconstruction.

of a cylindrical lens and a spherical lens, a sonar map is finally displayed. A completely distortion-free display requires much more complicated techniques, which shall not be dealt with in this paper.

The optical method offers the advantage that large quantities of data can easily be processed. Since suitable laser modulators will soon be available, a real-time reconstruction will certainly be possible in the future. Another advantage is that pulse compression and reconstruction can be performed in one operation [4]. On the other hand, the reconstruction by a digital computer offers other possibilities. The main advantage is that the complex sound field information can be processed. In addition, system errors can easily be corrected by the computer program. These errors, for instance, are different phase responses of the receivers or mounting of the hydrophones on a curved surface (ship's side). If the quantity of incoming data is not too large, even a reconstruction in real-time could be possible.

A third advantage of the computer use is the ability of performing the proper reconstruction also in the close near field, i.e. if the Fresnel approximation is no longer valid. In this case, however, the computer work increases by a factor of two. Altogether two Fourier transformations and a complex multiplication are required.

The basic difference between the signal processing as described here and the conventional active Sonar technique is that by the holographic signal-processing an angular resolution of λ/a is achieved both in the far field region as well as in the Fresnel field. Since the Sonar technique does not regard the quadratic phase factor in Fig. 6, a defocused display at close range is obtained. Roughly speaking the minimum resolution that can be achieved is never less than the physical width of the aperture.

3. Experiments

To acquire further information on the characteristics and the practicability of the described method, experiments were carried out with a test equipment. Its principle of operation is shown in a simplified block diagram (Fig. 2). The ultrasonic frequency used is approx. 100 kHz. The received electrical signals which are proportional to the sound pressure, are separately preamplified. The output of each amplifier is connected to a holographic detector, which as already described, electronically simulates a reference wave of oblique incidence. The outputs of the individual receiving channels are multiplexed. At the output of the multiplexer a wideband video signal is obtained. This signal is connected to the z-axis of an oscilloscope and causes an intensity modulation. The time base of the oscilloscope is triggered by the transmitted pulse. Finally, the vertical amplifier is supplied with a saw-tooth voltage synchronized by the multiplexer. In this way a set of one-dimensional holograms with the distance R as a parameter is displayed.

Since an on-line device for the optical real-time reconstructions was not yet available, the displayed hologram information was first photographed and then via an optical sample fed into a computer. The digital reconstruction was performed in accordance with the principle described above.

The aperture width of approx. 0.8 metres gives a resolution angle of 1.2°. As the sound transmitter emits pulses of 0.3 ms duration, the radial resolution is approx. 0.15 metres.

A typical one-dimensional pulse hologram, which was obtained by this set-up is shown in Fig. 7. As objects some air-filled glass-balls with a diameter of 5 cm were arranged in a small lake of about 14 metres depth. They were fixed in position by means of weights and marked by small buoys. On the hologram several one-dimensional zone plates resulting from the objects and buoys can be seen. The angle of incidence of the electronically simulated reference wave was −18°. The structure on the right is due to bottom and surface reverberation. Figure 8 shows the one-dimensional pulse hologram of a line target. One can see that the spatial

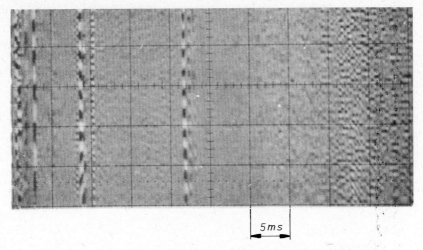

5 ms

FIG. 7. One-dimensional pulse hologram of point-shaped objects.

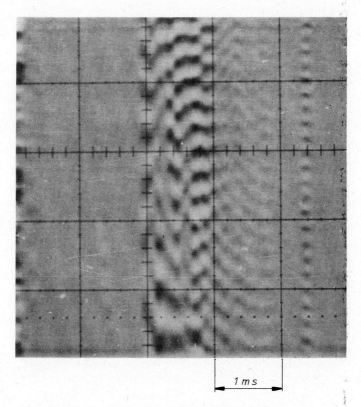

1 ms

FIG. 8. One-dimensional pulse hologram of line target.

frequency is a function of the aximuth angle of the scattering elements.

The reconstruction was done via the photographed hologram, which was scanned by an opto-electronic device. The image of the point objects is shown in Fig. 9. The theoretically predicted azimuthal resolution is

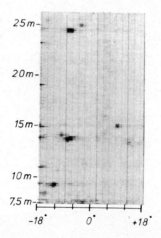

FIG. 9. Digital reconstruction of the hologram shown in Fig. 7.

achieved both in the near field and in the far field region. The focused image is compared with the result of unfocused reconstruction, i.e. the quandratic phase factor being replaced by unity (Fig. 10). The difference in azimuthal resolution in the near field is clearly to be seen.

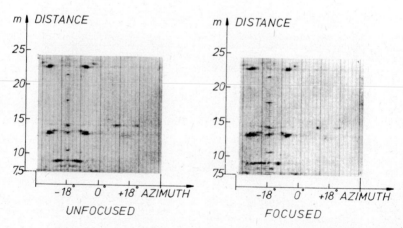

FIG. 10. Comparison of focused and unfocused processing.

4. Method with Synthetic Aperture

As the resolution, normal to the direction of propagation, is mainly influenced by the aperture width, it is suggested that an enlarged aperture could be established synthetically by means of a linear system. This method is employed in the side-looking radar technique. Since the generation of the holographic information requires more time for this instance, only stationary targets or bottom structures can be displayed. The principle may be illustrated by means of Fig. 7.

The phase of the sound pulses reflected by an arbitrary point P in space is as follows:

$$\varphi_E = \frac{2\pi}{\lambda} r \approx \frac{2\pi}{\lambda} \cdot \frac{(x_E - x_0)^2}{R} + \varphi_0. \qquad (7)$$

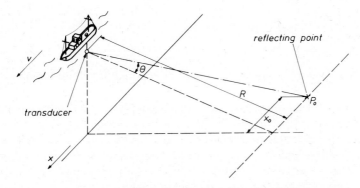

FIG. 11. Synthetic aperture holographic method.

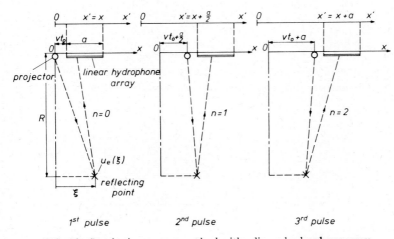

FIG. 12. Synthetic aperture method with a linear hydrophone array.

Transmitter and receiver are assumed to be point shaped and move in x-direction.

Phase φ_0 only depends on the coordinates of the reflector and is of no importance for the further signal-processing. The Fresnel approximation may apply again. It can now be seen that the expression (7) agrees with equation (4) for a stationary linear antenna except of a factor of two. It must be taken into account, however, that between succeeding transmission pulses a delay of Δt occurs which depends on the maximum range R max to be covered

Δt is:

$$\Delta t = \frac{2R \text{ max}}{c}. \tag{8}$$

As a consequence the sound field can be measured only with a minimum sampling distance of

$$\Delta x \cdot v \cdot \Delta t = \frac{2R \text{ max} \cdot v}{c}. \tag{9}$$

Already at low ship's speed, the sampling theorem is violated. Because of the much higher propagation velocity of electromagnetic waves, this problem is of no importance with side-looking radars. To overcome this problem Pekau [7] proposes that a linear hydrophone line of length a is used. The length a has to be twice as long as the distance covered between two transmitted pulses. As in general the transmitter is also moving, the phase relations are more complicated than with the fixed array. According to Fig. 7, the phase of the nth received pulse is given by

$$\varphi_E = \frac{\pi}{\lambda R} \left[(\xi - v \cdot n \cdot t_1)^2 + (\xi - x)^2 \right] + \varphi_0,$$

$$n \cdot v \cdot t_0 \leqslant x \leqslant n \cdot v \cdot t_0 + a. \tag{10}$$

The constant part φ_0 is of no importance in the following and, therefore, deleted. After performing a transformation of coordinate system

$$x' = x + \frac{a \cdot n}{2},$$

$$\varphi_E = \frac{\pi}{\lambda R} \left[2\xi^2 - 2x'\xi + x'^2 + \frac{n^2 a^2}{2} - a \cdot n \cdot x' \right], \tag{11}$$

$$v \cdot t_0 + n \cdot a \leqslant x' \leqslant v \cdot t_0 + (n + 1)a$$

is obtained.

Apart from the two last terms of the sum, the distribution of the receiving phase on the entire synthetic aperture corresponds to a one-dimensional zone plate. To perform a correct reconstruction, therefore, the two disturbing terms will have to be compensated by multiplication with a factor:

$$\exp\left[-j\,\frac{\pi}{\lambda R}\left(\frac{n^2 a^2}{2} - a \cdot n \cdot x'\right)\right],\tag{12}$$

which is easily possible with a computer reconstruction. For an optical reconstruction, however, a phase hologram with a complex transparency function given by equation (12) will be necessary.

The realization of the synthetic aperture method, however, can become rather difficult. The greatest problem will be to maintain a linear, uniform motion of the array with respect to phase sensitivity. Deviations from a straight line must not exceed a quarter of the acoustic wavelength. The only possibility we see now is that the whole equipment is fitted in a towed submerged body. Even then such large apertures as with side-looking radars. can never be realized. Another problem arises with the spatial coherence which is limited by inhomogeneities of the medium itself. We can divide the complex sound pressure amplitude at the receiver into a coherent and an incoherent part p_c and p_i, the mean value of the incoherent part being zero. Then the spatial coherence function provides a measure for the average influence on the holographic information. If with increasing distance this function drops only slightly below the maximum value, the coherent part will be prevailing. The angular resolution is only slightly influenced. With greater distances the incoherent part increases relatively to the total sound pressure. The spatial coherence function now shows a strong maximum, its correlation width is called coherence length. The aperture should not be enlarged beyond this coherence length because the resolution is limited by the incoherence of the medium.

It can be presumed, for the reasons listed above, that only relatively small apertures can be synthetically built up. The gain that can be achieved in angular resolution, however, justifies further investigations in this field.

5. Summary

The holographic signal-processing of the acoustic information received by a linear antenna makes it possible to display a sonar map of high resolution. The radial resolution is proportional to the reciprocal value of the band-width of the transmitted pulses. Unlike ordinary sonar techniques, also in the Fresnel region an angular resolution is achieved, which is an agreement with the Rayleigh criterion. The received signals are regarded as one-dimensional holograms and are a function of the echo delay. The holographic information can be processed to a sonar map either optically or by means of a digital computer. The advantage of the optical reconstruction is that real-time processing is possible. Appropriate systems are in the state of development. With the computer reconstruction, on the other hand, system errors can easily be corrected. However, in most cases the computing speed will not be high enough to perform real-time processing. Experiments with digital reconstruction carried out at an ultrasonic frequency of approximately 100 kHz have shown that the method is feasible. Another version, in which a synthetic aperture is generated by a linear motion of the transmitter-receiver equipment yields much better angular resolution. Because of the other geometry a compensation hologram will be necessary. The usefulness of the method with synthetic aperture widely depends on how well a linear motion can be maintained and on the other hand it depends on the coherence of the sound propagation in the medium.

REFERENCES

1. Greguss, P. (1965). "Ultraschallhologramme," Research Film 5 (4), pp. 330–337.
2. Metherell, A. F. (1971). "Acoustical Holography", Vol. 3, pp. 211–223, Plenum Press, New York.
3. [2], pp. 287–315.
4. Leith, E. N. (1971). Quasi-holographic techniques in the microwave region. *Proc. IEEE*, **59** (9) 1305–1318.
5. Brown, W. M. and Porcello, L. J. (1969). An introduction to synthetic-aperture radar. *IEEE Spectrum*, **6**, 52–62.
6. [2], pp. 207–208.
7. Pekau, D. and Diehl, R. (1970). Recording of One-Dimensional Holograms as a Function of the Object Range, Proceedings of the International Symposium of Holography Besançon, 6–11, Juli 1970, Service de Reproduction de la Faculté des Sciences de Besançon.

DISCUSSION

S. S. Wolff: You made some suggestions on synthetic aperture sonar, is it only a matter of thinking or have you also done experiments?
Answer: Experiments have been carried out only with a fixed linear array. The synthetic aperture sonar has not been realized experimentally.
S. S. Wolff: I expect some problems in towing a submerged vehicle in shallow water because one can use only short tows. That means that the irregularities of the ship's motion are transferred to the submerged vehicle.
Answer: I think there are many problems concerning the linearity of the motion of the array. In the best case one cannot hope that the synthetic aperture would be much longer than about ten times the physical aperture length.
S. S. Wolff: You have worked with a wavelength of 1.5 cm. In comparison a ship motion of a few metres is not rare.
Answer: The experiments were intended to show the practicability of the method. We did not match the frequency to real applications.
P. L. Stocklin: We have seen pictures of side-looking sonar (Gloria) which works at a

frequency of 6.4 kHz. This seems to me a good system for doing that holographic work. Which problems do arise, if you use considerably lower acoustic frequencies?

Answer: At low frequencies, the limitations concerning the linear motion of the vehicle are not as severe as in the high frequency case. In many applications, however, one is interested in very high azimuthal resolution at close distances. This requires a high frequency.

M. L. Somers: It is possible to imagine that the array can be compensated for. This would be very difficult because real measurements require high accuracy. A hopeful direction to go is to use low frequencies and to confine yourself to very short ranges and actually use the spatial sampling frequency which means a very slow motion.

Answer: The maximum velocity even for relatively low frequencies is very small if the spatial sampling theorem shall not be violated.

M. L. Somers: Another problem is that the range and the azimuth resolution are related by the pulse length because one must have a sufficient pulse length to make an estimate of the phase. At low frequencies I think the pay-off is more favourable.

Answer: At low frequencies comparatively long pulses have to be used for that reason. That means, at the same number of periods in a tone pulse one will have better range resolution at high frequencies.

H. Cox: Normally in holography we think of this process as a single frequency process. But it seems to me there will exist a possibility by using bandwidth in the transmitted signal to overcome some of the ambiguities associated with spatial sampling of the array.

Answer: In principle you are right, but I think it will become difficult to realize such a device. One reason is that in the one-dimensional pulse holography the bandwidth is used to achieve the range resolution.

Space–Time Processing for Optimal Parameter Estimation

WILLIAM J. BANGS

William and Mary College, Williamsburg, Virginia, U.S.A.

and

PETER M. SCHULTHEISS

Yale University, New Haven, Connecticut, U.S.A.

1. Introduction

A great deal of attention has been devoted in recent years to the problem of detecting random space–time signals such as those presented to a passive sonar array. The structure and performance capabilities of the optimum detector have been determined and compared with those of suboptimal configurations attractive from the point of view of implementation. This paper deals with the second stage of the signal processing problem, the extraction of information concerning parameters such as bearing and range.

From the most general point of view detection and parameter estimation are not qualitatively distinct. We may view both as parameter estimation problems. If we are interested in target bearing, we are estimating a parameter continuously variable from 0 to 2π. If we are interested in detection, we are estimating a binary parameter (target present or absent). Alternatively, we may view both problems as decision problems. Detection is the classical binary decision problem. Bearing determination is a decision between a multiplicity of possible alternatives. This point of view immediately brings out a practically important distinction: basic statistical theory [1] says that there is always a sufficient statistic (the likelihood ratio) for the binary decision problem. In general there is no sufficient statistic (or uniformly most powerful test) for the multiple alternative problem. As a result, parameter estimation procedures always have an element of arbitrariness not present in detection. The "optimum detector" is a well defined entity. It may be too complicated for practical use, but it is realizable in principle and serves as a realistic standard against which practical instrumentations can be compared. The notion of "optimum bearing estimator" is much less well defined, even if one interprets "optimum" in a very specific sense such "minimum mean square error". What is often possible in the parameter estimation problem is to set an absolute upper bound (not necessary realizable) on possible performance and to compare that bound with the performance of realizable instrumentations that are appealing for a variety of practical or analytical reasons. Our discussions in later sections will reflect this dual approach.

2. Review of Detection Theory

In the light of the comments made above it is not at all surprising that the data processing procedures for detection and parameter estimation have much in common. For that reason we find it useful to begin with a brief review of well-known results from optimal detection theory.

Our signal is the noise spontaneously generated by the source, in our case a ship. In general, the signal will contain both coherent components (e.g. from machinery noise) and incoherent components (e.g. flow noise). We assume that both signal and noise are Gaussian random processes with known spectral properties. This assumption is open to serious question only for the coherent signal components and even here it may be tolerable unless the frequency stability of the signal components is extreme and propagation conditions are near perfect. We further assume that signal and noise are statistically independent.

Among the basic results of detection theory of interest to us are the following.

(1) The best detector forms the likelihood ratio

$$\text{L.R.} = \frac{p(\mathbf{x}/\text{signal present})}{p(\mathbf{x}/\text{signal absent})} \tag{1}$$

where $\mathbf{x}$ is the received data vector and $p(\mathbf{x}/\ )$ is the conditional probability density of $\mathbf{x}$ under the indicated hypothesis.

(2) L.R. or any monotone function thereof is a sufficient statistic for detection. The monotone function favoured for Gaussian statistics is the quadratic form

$$Y_0 = -\mathbf{x}^*[(P+Q)^{-1} - Q^{-1}]\mathbf{x} \tag{2}$$

P and Q are the signal and noise covariance matrices respectively. $\mathbf{x}^*$ is the conjugate transpose of the vector $\mathbf{x}$.

(3) If the observation time is large compared with the correlation times of signal and noise as well as with the travel time of sound across the array, a representation of $\mathbf{x}$ in terms of Fourier coefficients leads to block diagonal P and Q matrices (dependence only in the space dimension). If,

in addition, the signal wavefront is coherent (point target), P_k has rank one and Y_0 can be written in the form

$$Y_0 = \sum_k h_k | V_k^* Q_k^{-1} \mathbf{x}_k |^2 \qquad (3)$$

The subscript k refers to the kth Fourier coefficient (frequency ω_k). Q_k is normalized so that its diagonal elements are unity.

$\mathbf{v}_k$ is the "steering vector"

$$\mathbf{v}_k^* = e^{-i\omega_k \tau_1} \dots e^{-i\omega_k \tau_M} \qquad (4)$$

$\tau_1 \dots \tau_M$ being the delays required to align the signal components at the output of the M sensors. h_k is a frequency filter (Eckart filter) given by

$$h_k = \frac{S(\omega_k)/N^2(\omega_k)}{1 + G(\omega_k)\dfrac{S(\omega_k)}{N(\omega_k)}} \qquad (5)$$

In equation (5), $S(\omega)$ and $N(\omega)$ represent the power spectra of signal and noise at any point near the receiver. $G(\omega)$ is the "array gain", formally given by

$$G(\omega_k) = \mathbf{v}_k^* Q_k^{-1} \mathbf{v}_k = \mathrm{Tr}\,(P_k Q_k^{-1}) \qquad (6)$$

$\mathrm{Tr}(\ )$ stands for the trace of the bracketed quantity.

(4) Equation (3) specifies the structure of the optimum detector. It consists in essence of a spatial filtering operation given by $V_k^* Q_k^{-1}(\ )$ and a frequency filtering operation h_k. A convenient equivalent form can be obtained by writing

$$\mathbf{v}_k \equiv V_k \mathbf{1} = \begin{bmatrix} e^{i\omega_k \tau_1} & & 0 \\ & \ddots & \\ 0 & & e^{i\omega_k \tau_M} \end{bmatrix} \begin{bmatrix} 1 \\ 1 \\ \vdots \\ 1 \end{bmatrix} \qquad (7)$$

Thus the steering *matrix* V_k and the column vector of ones $\mathbf{1}$ together replace the steering *vector* $\mathbf{v}_k$. The optimum detector structure then assumes the form shown in Fig. 1.

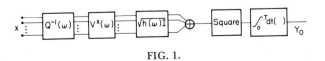

FIG. 1.

Time integration has replaced the frequency sum (Parseval's theorem) and the frequency filtering operation h has been shifted to the matrix portion of the processor by introducing the matrix filter $\sqrt{h}I$, I being the identity matrix.

(5) We note that Fig. 1 reduces to the conventional sum and square detector with Eckart filter when $Q(\omega) = I$, i.e. the noise is spatially incoherent.

3. Maximum Likelihood Estimation. General Theory

An estimator widely accepted as "good" selects the parameter θ which maximizes the "likelihood function" $p(\mathbf{x}/\theta)$. For computational reasons it is generally preferable to work with $\ln p(\ )$. Hence one seeks solutions to

$$\frac{\partial}{\partial \theta} \ln p(\mathbf{x}/\theta) = 0 \tag{8}$$

There are two obvious problems with the use of equation (8),

(a) It may have many solutions, not all of which need to be maxima.
(b) In all but the simplest cases no explicit solution can be found.

Actually neither of these objections is as serious as it may appear on first glance. Concerning (a): when the signal-to-noise ratio is high enough so that meaningful parameter estimates can be made, the effective uncertainty is usually confined to a range sufficiently narrow so that there is only one solution of equation (8) within that range. Concerning (b), it is a simple matter to form the likelihood functional

$$y(\theta) = \frac{\partial}{\partial \theta} \ln p(\mathbf{x}/\theta) \tag{9}$$

A simple null seeking (feedback) system can then determine the maximum likelihood estimate θ_{ML}.

For observation times large compared with the correlation time of $\mathbf{x}$, $y(\theta)$ (considered as a function of θ) will be an approximately Gaussian random process. It is then possible to relate the statistics of θ_{ML} to those of y. Specifically

$$\bar{\theta}_{ML} = \theta_0 - \frac{\bar{y}_0}{\partial \bar{y}_0 / \partial \theta_0} \tag{10}$$

and

$$D^2(\theta_{ML}) = \frac{D^2(y)}{(\partial \bar{y}_0 / \partial \theta_0)^2} \tag{11}$$

θ_0 is the true parameter value and the subscript 0 elsewhere indicates evaluation of the indicated quantity at $\theta = \theta_0$. The overbar designates a statistical average and $D^2(\ )$ stands for variance.

When the signal and noise components of x are complex Gaussians, $p(x/\theta)$ has the form

$$p(x/\theta) = \frac{1}{\text{Det}\,(\pi k)}e^{-x*k^{-1}x} \tag{12}$$

where $k = P + Q$ in terms of the nomenclature of Section 2.

A straightforward computation yields the likelihood functional

$$y(\theta) = x*k^{-1}\frac{\partial k}{\partial\theta}k^{-1}x - \text{Tr}\left(k^{-1}\frac{\partial k}{\partial\theta}\right) \tag{13}$$

From equation (13) and the known statistics of x one can compute the statistical properties of y.

$$\bar{y}(\theta) = \text{Tr}\left\{k^{-1}\frac{\partial k}{\partial\theta}\,(k^{-1}k_0 - I)\right\} \tag{14}$$

and

$$D^2(y) = \text{Tr}\left\{\left(k^{-1}\frac{\partial k}{\partial\theta}k^{-1}k_0\right)^2\right\} \tag{15}$$

In particular, when $\theta = \theta_0$ (on target estimate)

$$\bar{y}_0 = 0$$

$$D_0^2(y) = \text{Tr}\left\{\left(k_0^{-1}\frac{\partial k_0}{\partial\theta_0}\right)^2\right\} \tag{17}$$

One also finds

$$\left.\frac{\partial\bar{y}}{\partial\theta}\right|_0 = -\text{Tr}\left\{\left(k_0^{-1}\frac{\partial k_0}{\partial\theta_0}\right)^2\right\} \tag{18}$$

Equations (16)–(18) may now be substituted into equations (10) and (11) to yield the desired statistics of θ_{ML}.

$$\bar{\theta}_{\text{ML}} = \theta_0 \tag{19}$$

$$D_0^2(\theta_{\text{ML}}) = \text{Tr}\left\{\left(k_0^{-1}\frac{\partial k_0}{\partial\theta_0}\right)^2\right\}^{-1} \tag{20}$$

Equation (19) shows that the maximum likelihood estimate is unbiased under our assumptions.

4. Cramèr–Rao Lower Bound, General Theory

A relatively simple lower bound on mean square error is given by the Cramèr–Rao inequality [2], [3]. For an unbiased estimate $\hat{\theta}$ it asserts

$$D^2(\hat{\theta}) \geq -\left\{ \overline{\frac{\partial^2}{\partial \theta^2} \ln p(\mathbf{x}/\theta)} \right\}^{-1}_{\theta=\theta_0} \tag{21}$$

From equation (9)

$$\overline{\frac{\partial^2}{\partial \theta^2} \ln p(\mathbf{x}/\theta)} = \frac{\partial \bar{y}}{\partial \theta} \tag{22}$$

It now follows from equations (17), (18) and (20) that the right side of equation (21) is equal to $D^2(\theta_{\mathrm{ML}})$. In other words, the lower bound is realizable (with a maximum likelihood estimator). In statistical terminology, the maximum likelihood estimator is efficient under the assumptions of our analysis. This is not surprising, for we required normality of $y(\theta)$, a condition closely akin to the large sample size assumption under which the maximum likelihood estimator is asymptotically efficient according to classical theory.

The Cramèr–Rao bound is easily extended to the problem of joint estimation of several parameters. One finds for the ith parameter θ_i

$$D^2(\hat{\theta}_i) \geq [J^{-1}]_{ii} \tag{23}$$

where J is the Fisher information matrix with elements

$$J_{ij} = -\overline{\frac{\partial^2}{\partial \theta_i \, \partial \theta_j} \ln p(\mathbf{x}/\boldsymbol{\theta})} \tag{24}$$

For the case of two unknown parameters one finds

$$D^2(\hat{\theta}_i) \geq \frac{1}{1 - \mu_{12}} \left\{ \mathrm{Tr}\left(k^{-1} \frac{\partial k}{\partial \theta_i} \right)^2 \right\}^{-1} \tag{25}$$

where

$$0 \leq \mu_{12} = \frac{\left[\mathrm{Tr}\left\{ k^{-1} \dfrac{\partial k}{\partial \theta_1} \, k^{-1} \dfrac{\partial k}{\partial \theta_2} \right\} \right]^2}{\mathrm{Tr}\left\{ k^{-1} \dfrac{\partial k}{\partial \theta_1} \right\}^2 \mathrm{Tr}\left\{ k^{-1} \dfrac{\partial k}{\partial \theta_2} \right\}^2} \leq 1 \tag{26}$$

Since $\left[\text{Tr}\left(k^{-1}\dfrac{\partial k}{\partial \theta_i}\right)^2\right]^{-1}$ is the lower bound on $D^2(\hat{\theta}_i)$ when no other

parameter is unknown, the quantity $1/(1-\mu_{12})$ clearly measures the degradation in the estimate due to presence of another unknown parameter related to θ_i.

5. Bearing and Range Estimation

We now apply the techniques sketched in the previous sections to the two central problems of target location, the estimation of bearing and range. They can be treated jointly since both bearing and range affect the received data only through the sensor-to-sensor delay of the received signal wavefront.

The analysis begins with equation (13). We recall that $k = P + Q$, P being a function of θ while Q is not. For the time being we think of θ as bearing or range, not as a vector quantity. After extensive algebraic manipulations one finds

$$y(\theta) = \sum_k \left\{ h_k \mathbf{x}_k^* Q_k^{-1} V_k i\omega_k T V_k^* Q_k^{-1}\mathbf{x}_k - h_k f_k \frac{\partial G(\omega_k)}{\partial \theta} \times \right.$$
$$\left. \times |1^* V_k^* Q_k^{-1}\mathbf{x}_k|^2 - \frac{\partial G(\omega_k)}{\partial \theta} f_k \right\} \qquad (27)$$

Here $f_k = N(\omega_k)h_k$ and T is a weighting matrix whose elements are

$$t_{ij} = \frac{\partial \tau_i}{\partial \theta} - \frac{\partial \theta_j}{\partial \theta} \qquad (28)$$

The remaining symbols are defined in equations (5), (6) and (7). Equation (27) yields the estimator structure shown in Fig. 2. For ease in discussion the diagram has been subdivided into four major blocks labelled 1-4. We note

(a) Block 1 is identical with the corresponding elements of the optimal detector (Fig. 1).
(b) Blocks 3 and 4 include the gain factor $\partial G(\omega)/\partial \theta$. If the array gain does not vary with θ in the region of interest (locally isotropic noise) the block diagram reduces to the single channel combination of 1 and 2 .
(c) Block 4 does not depend on the received data. It compensates for bias, using statistical knowledge about the noise field.
(d) The output of block 3 includes a DC component (because of the square) as well as a fluctuating component, which are subtracted from the main channel output.
(e) Bearing and range estimation differ only in the weighting matrix T.

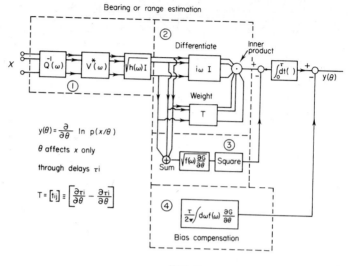

FIG. 2.

We now turn our attention to the special situation of spatially incoherent noise, $Q = I$. In that case $G(\omega) = M$ so that $\partial G/\partial \theta = 0$ and the block diagram reduces to Fig. 3a.

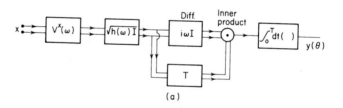

(a)

FIG. 3a.

Fig. 3b shows the block diagram of the conventional split beam tracker for comparison purposes. Aside from the weighting matrix T, the two differ in only one respect. The inner product operation multiplies the weighted output of every sensor with that of every other sensor. The split beam tracker multiplies together only the outputs of sensors in opposite halves to the array. By proper definition of T we can incorporate this feature into Fig. 3a. Thus when

$$t_{ij} = \begin{cases} 1 & i = 1 \ldots \dfrac{M}{2}, \quad j = \dfrac{M}{2} + 1 \ldots M \\ 0 & \text{otherwise} \end{cases}$$

$$\tag{29}$$

Fig. 3a becomes identical with Fig. 3b. From this point of view the split beam tracker is simply a maximum likelihood estimator with suboptimal weighting matrix.

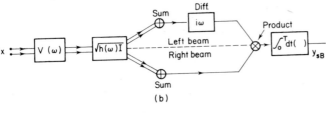

FIG. 3b.

The mean square error of the maximum likelihood estimator can be computed from equation (20). For the case of spatially incoherent noise one finds (after considerable algebra)

$$D_0^2(\theta_{\mathrm{ML}}) = \sum_k \left\{ S(\omega_k) h_k \, \omega_k^2 \, \mathrm{Tr}(TT^*) \right\}^{-1} \tag{30}$$

We note that equation (30) still holds for bearing as well as range estimation.

(a) Bearing estimation

We now specialize our results to the case of bearing estimation. In the interest of algebraic simplicity we confine our attention to linear arrays, using Z_n to designate the position of the nth sensor. If c is the velocity of sound and γ designates the signal bearing relative to the array axis (the Z axis) the T matrix for bearing estimation has the elements (for a far field source)

$$t_{ij} = \frac{\cos \gamma}{c} (Z_i - Z_j) \tag{31}$$

Thus the ideal weights vary linearly with the distance between sensors.

It is now a simple matter to calculate the efficiency of the split beam tracker. We have already seen that the maximum likelihood estimator reaches the Cramèr–Rao lower bound under our long observations time assumptions. We need therefore only divide the mean square error of the split beam tracker [4] by equation (30). One finds that the efficiency is 100% for 2 sensors and declines rapidly to an asymptotic level of 75% as the number of sensors increases. Equation (31) suggests a simple procedure for improving the efficiency: inverse shading, weight varying linearly with distance from the centre of the array. The procedure is still suboptimal but it yields an efficiency in excess of 95% for any number of sensors ($\geqslant 2$) as long as the signal-to-noise ratio exceeds 0 db (See Fig. 4).

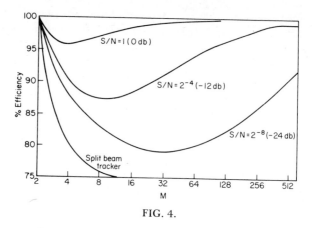

FIG. 4.

Equation (30) also gives interesting insights into the role of array geometry in determining system performance. All geometrical properties are contained in the factor $Tr(TT^*)$. If we picture the array as a set of equal masses at the various sensor locations, it is possible to identify $Tr(TT^*)$ with the moment of inertia of the configuration about an appropriate axis [6]. For targets in the same horizontal plane as the centre of gravity of the array, the moment of inertia is that of the array projected onto the horizontal plane, about a line joining the centre of gravity with with the target location. For a linear array this moment of inertia is obviously highest in the broadside direction and zero in the endfire direction. The most effective sensors, again very reasonably, are those located farthest from the line joining the centre of gravity of the array with the target.

The preceding results on bearing estimation have all assumed spatially incoherent noise fields. The split beam tracker was not optimal, but its efficiency was sufficiently high so that no really major improvements were possible. The situation can be very different when the noise field includes strong spatially coherent components. Computations for such situations tend to be very tedious, but one extreme case has been analysed in detail [6]: A noise field consisting of strong plane wave interference plus spatially incoherent noise. One obtains the following results.

(1) Split beam tracker performance is seriously degraded by a strong interference. There is both increased fluctuation (due to interference power) and a bias ($\partial G/\partial \theta \neq 0$) introduced into the estimate. The bias can often be more serious than the increased fluctuation error, even when the interference is relatively remote from the target in bearing.

(2) If the interference is far from the target in bearing (much more than one beam-width) it can be removed relatively easily by null steering [6] in each half of the split beam tracker. The performance of the null steered

split beam tracker is close to the Cramèr–Rao bound and not significantly worse than that of a tracker operating in an interference-free environment. In terms of Fig. 2, null steering is an operation closely related to the matrix filter $Q^{-1}(\omega)$ in block 1. It removes both bias and fluctuation error due to the interference and makes blocks 3 and 4 unnecessary.

(3) When interference and target are almost coincident in bearing, bias becomes the dominant problem. Null steering is no longer useful because it subtracts too much of the signal component. A simple split beam tracker with bias compensation (essentially block 4 in Fig. 2) can now provide near optimal performance [6].

(4) Only when the separation of target and interference is of the order of a beamwidth does block 3 offer any significant possibilities of improvement.

(b) *Range estimation*
Once again we confine ourselves to linear arrays. The T matrix for range estimation has the elements

$$t_{ij} = -\frac{\cos^2 \gamma}{2cr^2} (Z_i^2 - Z_j^2) \tag{32}$$

With incoherent noise one can then use equation (30) to determine the minimum (realizable) mean square error in range. In particular, for an array of length L with $M \gg 1$ equally spaced sensors one finds

$$[\mathrm{Tr}\,(TT^*)]^{-1} = \frac{45c^2 r^4}{2L^4 M^2 \cos^4 \gamma} \tag{33}$$

The corresponding expression for bearing estimation is

$$[\mathrm{Tr}\,(TT^*)]^{-1} = \frac{6c^2}{M^2 L^2 \cos^2 \gamma} \tag{34}$$

Thus the range estimate is critically dependent on the true range, while the bearing estimate is not (except for the range dependence of the signal to noise ratio). The range estimate is also more critically dependent on array aperture than the bearing estimate. All of this is qualitatively reasonable because the range measurement is in a sense little more than a triangulation procedure.

There is a strong element of artificiality in regarding range estimation—as we have done thus far—as a single parameter problem. A more realistic position would be to assume that both range and bearing are unknown *a priori*. In accordance with equation (25) the Cramèr–Rao bound on estimation error is then the single parameter bound multiplied by the degradation factor $(1 - \mu_{12})^{-1}$. For the case of incoherent noise one obtains from equation (26)

$$\mu_{12} = \frac{\{\mathrm{Tr}(T_1 T_2^*)\}^2}{\mathrm{Tr}(T_1 T_1^*)\, \mathrm{Tr}(T_2 T_2^*)} \tag{35}$$

where T_1 and T_2 are the weighting matrices for bearing and range respectively. Working with a linear array of equally spaced sensors, $r \gg L$, and choosing the origin of coordinates at one end of the array [as in equation (33)] one finds from equation (35)

$$(1 - \mu_{12})^{-1} = \frac{16M^2 - 30M + 11}{M^2 - 4} \tag{36}$$

Thus range estimation is not possible with only two sensors. For $M > 2$ the degradation factor lies between 12 and 16, indicating that lack of *a priori* bearing information reduces the range accuracy by an order of magnitude and vice versa. On the other hand, if one chooses the origin at the centre of the array one finds that $\mu_{12} = 0$ for $M > 2$. Range and bearing estimation are now decoupled.

From a purely formal point of view the origin dependence is introduced by equation (32). It is a simple matter to demonstrate that choice of the origin affects equation (33) in precisely the same manner as it does $(1 - \mu_{12})$. Hence the estimate accuracy of range with unknown bearing is, in fact, origin independent. On the other hand equation (31), the counterpart of equation (32) for bearing estimation, is independent of the origin so that the estimate of bearing with unknown range does vary with the choice of origin. It is most accurate with the origin at the centre of the array. This conclusion has a certain intuitive appeal: with the origin at the centre of the array any curvature of the wavefront affects delays to sensors on opposite sides of the origin in symmetrical fashion. With the origin located elsewhere, asymmetrical delays couple the estimated bearing to the range estimate.

6. Other Parameters

The results of this paper are readily extended to the estimation, separately or jointly, of many other parameters. The computations tend to be tedious and have been carried out only for certain special cases of particular interest. Among these is bearing estimation with a linear array in an environment consisting of a strong plane wave interference plus spatially incoherent noise. Various combinations of signal power, interference power, and interference bearing were considered as additional unknown parameters [6]. Very broadly speaking, one finds that the additional nuisance parameters have little effect on the Cramèr–Rao bound for bearing as long as signal and interference are separated by at least one beamwidth. With separations of less than a beamwidth the situation is different. Here lack of prior knowledge concerning signal power still only has a relatively minor effect on bearing accuracy. Uncertainty concerning interference power or interference bearing, on the other hand, can drastically increase the lower bound on bearing error. It appears reasonable to attribute this phenomenon to the bias effect of the interference which becomes very pronounced when the angle between target and interference diminishes to much less than a beamwidth.

Specific results have also been obtained for the maximum likelihood estimator of source extent [5]. As one might expect, the estimation error depends strongly on the angle subtended by the target when viewed from the receiving array. Expressions for error have been derived, but they are too complex algebraically to warrant discussion here.

ACKNOWLEDGEMENT

The authors wish to acknowledge support in the general area of sonar signal processing from ONR (prime contract with General Dynamics/Electrical Boat under the SUBIC program) and more recently, from the Naval Underwater Systems Center, New London Laboratory.

REFERENCES

1. Lehman, E. L. (1959). "Testing Statistical Hypotheses", Ch. 3, Wiley.
2. Rao, C. R. (1945). Information and accuracy attainable in the estimation of statistical parameters *Bull. Calcutta Math. Soc.* 37, 81.
3. Cramèr, H. (1951). "Mathematical Methods of Statistics", Ch. 32, Princeton University Press, Princeton.
4. MacDonald, V. H. and Schultheiss, P. M. (1969). Optimum passive bearing estimation in a spatially incoherent noise environment. *JASA* 46.
5. Bangs, W. J. (1971). "Array Processing with Generalized Beam-Formers", Ph.D. Dissertation, Yale University.
6. MacDonald, V. H. (1971). "Optimum Bearing Estimation with Passive Sonar Systems", Ph.D Dissertation. Yale University.

DISCUSSION

W. S. Liggett: I notice that the mean square error of the optimum detector can go to infinity near endfire. Can one not do better than that?

Answer: The problem arises because of the unbiased nature of the estimator. Near endfire one cannot work with an unbiased estimator. The best biased estimator should minimize the total mean square error, composed of both bias error and variance. If one minimizes that, one finds that the optimum bias is almost zero except within about a beamwidth of endfire. Thus, except in this region, the unbiased estimator is essentially optimal.

R. B. Gilchrist: A good procedure for achieving the effect of inverse shading and maximizing moment of inertia would appear to be to divide the sensors into 2 groups as far away from the centre of the linear array as possible.

Answer: This would indeed be a good procedure. The only limitation is the need to keep the noise independent from sensor to sensor.

Miss E. A. Killick: Can you point to illustrations where this type of thinking leads to different conclusions than a more conventional point of view?

Answer: Ultimately the conclusions have to check, of course. What we are after in this type of analysis is to gain insight into the structure and performance of optimal processors, to see whether the devices arrived at by more intuitive means could be improved significantly, and if so, how this might be done.

S. W. Autrey: Would the results for white noise carry over to isotropic noise?

Answer: If the sensor to sensor spacing is small enough so that noise coherence becomes significant, the specific computational results would no longer hold. However, for isotropic noise the resultant improvement would be of the super-directive variety and we know that theoretical performance gains of this type tend to be largely fictitious in practice. Practically, therefore, I would expect little change.

R. E. White: You said that uniqueness of the maximum likelihood estimate depends on the signal-to-noise ratio. Could you set a numerical limit on the signal to noise ratio which would make the estimate unique?

Answer: I hesitate to do so because the number would depend on too many other parameters and could therefore vary greatly from case to case.

A Recent Trend in Adaptive Spatial Processing for Sensor Arrays: Constrained Adaptation†

NORMAN L. OWSLEY

Naval Underwater Systems Center. New London, Connecticut, U.S.A.

1. Introduction

The theory of optimum processing for sensor arrays which has been developed [1, 2], presents the theoretically optimum structure with respect to maximizing spatial signal detection performance. In view of the complexity of implementing the optimum structure, a valid question is, "What is a practical sub-optimum realization of the well known optimum array processor?" Naturally, this question occurs when any truly optimum system is considered in terms of the design engineer's hardware and financial limitations. The intent of this paper, therefore, is to indicate that,

(a) a class of sub-optimal (i.e. less than fully adaptive) adaptive system can in many cases give near optimum array performance with a reasonable hardware requirement, and

(b) sub-optimal adaptive array control algorithms should be properly constrained and configured within the context of the signal and interference environment for a specific array. These two points are not, of course, mutually exclusive. The emphasis herein is placed on adaptive realization techniques for the general multi-channel processor, with applications discussed in terms of specific examples.

The concept of less than fully optimum, adaptive sensor array processing for spatial noise rejection has been proposed in various forms by several authors (see for example, References [3], [4]). The first example discussed in this paper introduces the class of linearly constrained, gradient descent least mean squared algorithms to the area of auxiliary array (or sub-array) adaptive processing. The second example discussed deals with the application of orthogonal component (i.e. eigenvector) analysis in terms of the Karhunen–Loeve expansion. Applications of this technique have been

† This work was conducted at the Naval Underwater Systems Center, New London, Connecticut, U.S.A. and was supported by the Naval Ship Systems Command (NAVSHIPS 901), Project No. SF11-121-110.

591

useful for signal processing analysis [5, 6] and in the more general areas of pattern recognition and dimensionality reduction [7, 8]. In particular a class of quadratically constrained adaptive filter control algorithms are introduced which can be applied to adaptive sensor array processing.

Section 2 of this paper gives a brief introduction to constrained multi-channel filtering for both linear and quadratic filter constraints. The adaptive realization of the constrained multi-channel filter control algorithms is also presented. Section 3 gives two examples of constrained adaptive array control wherein the emphasis is placed on maximizing performance while minimizing the number of adaptive channels required. As mentioned, the techniques discussed in this section are adaptive auxiliary arrays and orthogonal beam formation. Section 4 contains a summary of "current events" in adaptive array control and a prediction of things to come.

2. Multi-channel Adaptive Filtering with Linear and Quadratic Constraints

Consider the K-channel array processor shown in Fig. 1. The inputs $\{x_i/i = 1, 2, \ldots, K\}$ are assumed to be zero mean complex random processes for which the channel-to-channel correlation properties are known in terms

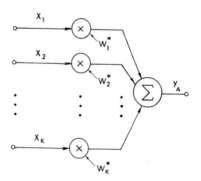

FIG. 1. K-channel array processor.

of the correlation matrix R at frequency f. The data input vector $\mathbf{X}$ and filter weight vector $\mathbf{W}$ are given respectively by

$$\mathbf{X}^T = [x_1 x_2 \ldots x_K] \tag{1}$$

and

$$\mathbf{W}^T = [w_1 w_2 \ldots w_K] \tag{2}$$

where the superscript T indicates matrix transpose. Superscripts $*$ and H will be used to indicate matrix complex conjugate and complex conjugate transpose operations respectively. Complex scalars will be denoted by lower case letters. Vector and matrix quantities are given by upper case letters with vectors in bold. The channel-to-channel correlation matrix, R, is therefore given by the expected value

$$R = E\{\mathbf{XX}^H\}. \tag{3}$$

(a) Linear constraints
The multi-channel filter weight vector $\mathbf{W}$ is said to be linearly constrained if

$$0 = f_i - \mathbf{C}_i^H \mathbf{W} \tag{4}$$

where

$$f_i = f_{Ri} + jf_{Ii} \tag{5}$$

and $j = \sqrt{-1}$. Notice that the number of linear, orthonormal constraints on $\mathbf{W}$ must be less than the dimension of $\mathbf{W}$ to have any remaining degrees of freedom available for adaptation. This type of constraint is shown in Fig. 2 for the real valued 2-dimensional weight vector case subject to a single constraint.

In Fig. 3, suppose the objective is to minimize the mean squared error power, $E\{|\epsilon|^2\}$, with respect to $\mathbf{W}$ while subjecting $\mathbf{W}$ to the linear constraint (4). The mean squared error is

$$E\{|\epsilon|^2\} = E\{|y - y_A|^2\} \tag{6}$$

$$= E\{|y|^2\} - 2\mathrm{Re}\{\mathbf{W}^H\mathbf{P}\} + \mathbf{W}^H R \mathbf{W} \tag{7}$$

where $\mathbf{P} = E\{\mathbf{X}y^*\}$.

The method of undetermined multipliers will be used to formulate the constrained minimization problem. Accordingly, a single constraint

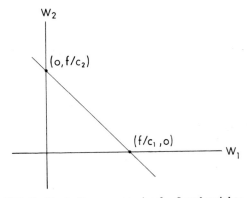

FIG. 2. Single linear constraint for 2 real weights.

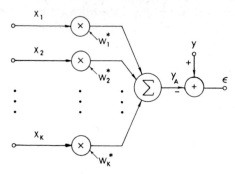

FIG. 3. Multi-channel mean squared error (MSE) filter.

criterion function $\phi_L(W)$ is formed by appending the constraint equation (4) to the mean squared error in the form

$$\phi_L(\mathbf{W}) = E\{|y|^2\} - 2\operatorname{Re}\{\mathbf{W}^H\mathbf{P}\} + \mathbf{W}^H R\mathbf{W} + \lambda_R(2f_R - [\mathbf{W}^H\mathbf{C} + \mathbf{C}^H\mathbf{W}]) +$$
$$+ \lambda_I(2f_I + j[\mathbf{W}^H\mathbf{C} - \mathbf{C}^H\mathbf{W}]) \tag{8}$$

where for the single constraint considered the constraint index i is deleted. The undetermined multipliers for the real and imaginary parts of the constraint (4) are λ_R and λ_I where a complex multiplier $\lambda = \lambda_R + j\lambda_I$ can be defined. Thus, an extremum of equation (8) is found by solving the set of K simultaneous linear equations given by the K dimensional gradient

$$\frac{\delta\phi_L(\mathbf{W})}{\delta\mathbf{W}} = -2\mathbf{P} + 2R\mathbf{W} - 2\lambda^*\mathbf{C} \tag{9}$$

$$= 0 \tag{10}$$

to yield

$$\mathbf{W} = R^{-1}[\mathbf{P} + \lambda^*\mathbf{C}]. \tag{11}$$

The constraint equation (4) can be applied to equation (11) to give

$$\lambda = (f^* - \mathbf{P}^H R^{-1}\mathbf{C})/(\mathbf{C}^H R^{-1}\mathbf{C}). \tag{12}$$

The 2-dimensional, real weight vector case is illustrated in Fig. 4. The mean squared error (MSE) surface is a quadratic in W_1 and W_2 and as such has a unique minimum. However, because of the constraint, the solution lies on the intersection of the constraint and squared error surfaces at the "cradle" point (W_1', W_2').

An adaptive algorithm for determining W by the method of constrained steepest descent can be obtained directly from equation (9) by replacing

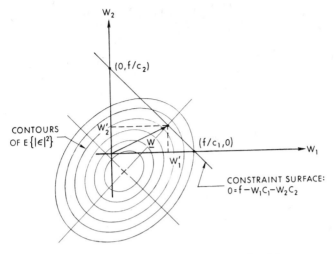

FIG. 4. Constrained MSE solution for 2 real weights.

all quantities by instantaneous sampled values at time t_n [9]. This gives an estimate of the gradient

$$\frac{\delta\phi_L(\mathbf{W})}{\delta\mathbf{W}}\bigg|_{t_n} = \frac{\delta\phi_L(\mathbf{W}(n))}{\delta\mathbf{W}(n)} \tag{13}$$

which can be used to "adjust" the filter vector according to the gradient descent algorithm ($\mu < 0$)

$$\mathbf{W}(n+1) = \mathbf{W}(n) + \mu\frac{\delta\phi(\mathbf{W}(n))}{\delta\mathbf{W}(n)} \tag{14}$$

$$= \mathbf{W}(n) + G(\mathbf{X}(n)\mathbf{X}^H(n)\mathbf{W}(n) - \mathbf{X}(n)y^*(n) - \lambda^*(n)\mathbf{C}) \tag{15}$$

where $G = 2\mu$ is the gradient step size and the stochastic estimate $R = \mathbf{X}(n)\mathbf{X}(n)^H$ is utilized. By a direct application of the constraint $0 = f - \mathbf{C}^H\mathbf{W}(n+1)$, equation (15) can be solved for $\lambda^*(n)$ and substituted back into itself to give

$$\mathbf{W}(n+1) = P[\mathbf{W}(n) + G(\mathbf{X}(n)\mathbf{X}^H(n)\mathbf{W}(n) - \mathbf{X}(n)y^*(n))] + f\mathbf{C} \tag{16}$$

where P is a matrix given by $P = [I - \mathbf{C}\mathbf{C}^H]$ and I is a $K \times K$ identity matrix. It is worth noting that in equation (16) the term

$$\mathbf{W}_u(n+1) = [\mathbf{W}(n) + G(\mathbf{X}(n)\mathbf{X}^H(n)\mathbf{W}(n) - \mathbf{X}(n)y^*(n))] \tag{17}$$

$$= [\mathbf{W}(n) - G\mathbf{X}(n)\epsilon^*(n)] \tag{18}$$

is an updated filter weight vector not subject to the linear constraint of interest. The matrix $P = [I - CC^H]$ projects $\mathbf{W}_u(n + 1)$ into a new vector which is orthogonal to $\mathbf{C}$ and the term $f\mathbf{C}$ translates this orthogonally constrained vector so as to satisfy the desired constraint. This sequence of updating operations is shown for the 2-dimensional case in Fig. 5. A proof of convergence of the mean of $\mathbf{W}(n + 1)$ as $n \to \infty$ for the real data vector case is given by Frost [10].

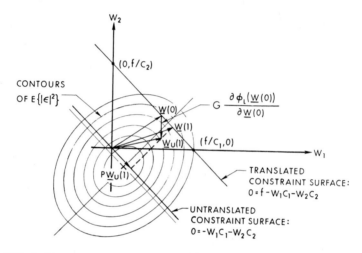

FIG. 5. Update sequence for 2 real weights subject to a linear constraint.

(b) Quadratic constraints

The filter weight vector $\mathbf{W}$ is said to be quadratically constrained if

$$0 = g_i - \mathbf{W}^H Q_i \mathbf{W} \tag{19}$$

where for our applications Q_i is assumed to be a $K \times K$ Hermitian matrix. For a multi-channel filter processor the expected output power is given by

$$E\{|y|^2\} = E\{\mathbf{W}^H \mathbf{X}\mathbf{X}^H \mathbf{W}\} \tag{20}$$

$$= \mathbf{W}^H E\{\mathbf{X}\mathbf{X}^H\}\mathbf{W} \tag{21}$$

$$= \mathbf{W}^H R \mathbf{W}. \tag{22}$$

Suppose it is desired to "extremalize" the expected output power equation (22) with respect to $\mathbf{W}$ and subject to the constraint

$0 = g - \mathbf{W}^H Q \mathbf{W}$. A criterion function for a single quadratic constraint employing the method of undermined multipliers can therefore be stated as

$$\phi_q(\mathbf{W}) = \mathbf{W}^H R \mathbf{W} + \lambda_q(g - \mathbf{W}^H Q \mathbf{W}). \tag{23}$$

The multi-dimensional gradient of equation (23) with respect to $\mathbf{W}$ is

$$\frac{\delta\phi_q(\mathbf{W})}{\delta\mathbf{W}} = 2[R\mathbf{W} - \lambda_q Q\mathbf{W}]. \tag{24}$$

The extremum value $\mathbf{W}$ is found by setting equation (24) to a K dimensional zero vector to obtain

$$R^{-1}Q\mathbf{W} = \lambda_q^{-1}\mathbf{W} \tag{25}$$

where it is assumed that R has an inverse, R^{-1}. Thus, the extremum value of $\mathbf{W}^H R \mathbf{W}$ with the quadratic constraint $g = \mathbf{W}^H Q \mathbf{W}$ is obtained when $\mathbf{W}$ is an eigenvector of $R^{-1}Q$ and λ_q^{-1} is the corresponding eigenvalue.

An adaptive control algorithm based on equation (24) which continually seeks to maximize the multi-channel processor output is given by

$$\mathbf{W}(n+1) = \mathbf{W}(n) + G(\mathbf{X}(n)\mathbf{X}(n)^H\mathbf{W}(n) - \lambda_q(n)Q\mathbf{W}(n)) \tag{26}$$

$$= \mathbf{W}(n) + G(\mathbf{X}(n)y^*(n) - \lambda_q(n)Q\mathbf{W}(n)) \tag{27}$$

where G is positive if $E\{|y|^2\}$ is to be maximized [11]. In equation (26), if the constraint $\mathbf{W}^H(n)Q\mathbf{W}(n) = g$ is enforced "periodically" by magnitude scaling of $\mathbf{W}(n)$, then the estimate of λ_q obtained from

$$\mathbf{W}^H(n)R(n)\mathbf{W}(n) = \lambda_q(n)\mathbf{W}^H(n)Q\mathbf{W}(n) \tag{27}$$

$$= \lambda_q(n)g \tag{28}$$

$$= \mathbf{W}^H(n)\mathbf{X}(n)\mathbf{X}^H(n)\mathbf{W}(n) \tag{29}$$

$$= |y(n)|^2 \tag{30}$$

is perfectly legitimate (Note: $R(n) = \mathbf{X}(n)\mathbf{X}(n)^H$). The 2-dimensional adaptation sequence indicated above is shown in Fig. 6.

For generality, multiple linear constraints can be superimposed on $W(n)$ in addition to the quadratic constraint. These linear constraints can be written in matrix form as

$$0 = \mathbf{F} - C^H\mathbf{W} \tag{31}$$

where $\mathbf{0}$ and $\mathbf{F}$ are M dimensional vectors of zeros and constraint constants $\{f_i/i = 1, 2, \ldots, M\}$ respectively. The matrix C has the vector $\mathbf{C}_i$ as its ith

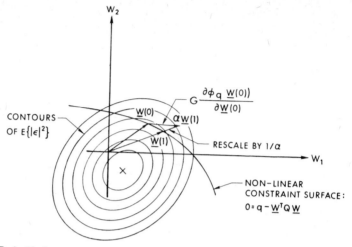

FIG. 6. Update sequence for 2 real weights subject to a quadratic constraint.

column [see equation (4)]. The set $\{C_i/i = 1, 2, \ldots, M\}$ must be a set of M orthonormal constraint vectors. The adaptive control algorithm now becomes

$$W(n + 1) = [I - CC^H] [W(n) + G(X(n)y^*(n) - \lambda_q(n)QW(n))] + CF$$

$$(32)$$

where $\lambda_q(n) = |y(n)|^2$. Clearly, the constraint $C^HW(n + 1) = F$ is upheld in equation (32) and the quadratic constraint $q = W^H(n)QW(n)$ is imposed by rescaling.

An alternate approach to quadratically constrained adaptive control is to linearize the quadratic constraint surface in the region about $W(n)$. Thus, defining the vector $Q(n) = QW(n)$ the update algorithm would incorporate the projection matrix ($Q(n)$ normalized)

$$P(n) = [I - Q(n)Q^H(n)]$$

$$(33)$$

which could be used in the same sense as for the linearly constrained case. This approach can be visualized as an attempt to approximate the convex constraint surface $\psi: 0 = g - W^H(n)QW(n)$ in the neighbourhood of $W(n)$ with a linear surface which is tangent to ψ at $W(n)$. It should be evident that the precise manner in which to implement the control algorithms for quadratically constrained multi-channel adaptive filters has various realizations and that much work remains in this area to select the best approach [12].

3. Sensor Array Adaptive Processing

This section gives two examples of how constrained multi-channel filter control can be applied to the practical problem of sensor array processing for signal detection in the presence of spatial interference. The emphasis is on the reduction of adaptive processor channels required versus a maximum of system performance.

(a) Auxiliary array processing
Figure 7 gives the general configuration for a sensor array beamformer with adaptive auxiliary array processing. The particular auxiliary array selection

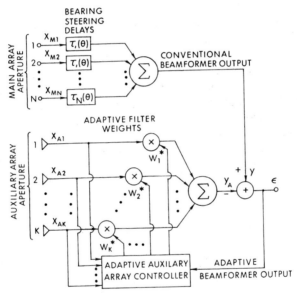

FIG. 7. Sensor array beamformer with auxiliary array adaptive processing $(K \leqslant N)$.

of interest here is a sub-array of sensors in the main array aperture. The auxiliary array filter weight vector $W^T = [w_1 w_2 \ldots w_K]$ is controlled so as to minimize the MSE, $E\{|\epsilon|^2\}$. However, the vector W is constrained so that a beam formed by the auxiliary array has zero response in the direction, θ, the main array beam is steered. Thus, if the beam steering vector

$$D^T(\theta) = [e^{-j2\pi f\tau_1(\theta)} \, e^{-j2\pi f\tau_2(\theta)} \ldots e^{-j2\pi f\tau_N(\theta)}] \qquad (34)$$

is defined, then the constraint

$$0 = D^H(\theta)S^T W \qquad (35)$$

must be upheld. In equation (35) S is a $K \times N$ sub-array selection matrix defined by

$$X_A = SX_M \tag{36}$$

where, with reference to Fig. 7,

$$X_M = [x_{M1} x_{M2} \ldots x_{MN}] \tag{37}$$

and

$$X_A = [x_{A1} x_{A2} \ldots x_{AK}]. \tag{38}$$

The auxiliary array aperture processor can clearly be equated to the linearly constrained multi-channel filter controller discussed in the previous section. An example of auxiliary (sub-) array processing will be given in terms of the processed beam output (ϵ) signal-to-noise background ratio. A line array consisting of 24 equally spaced elements is considered for a signal of interest arriving at an angle of 90° relative to the array. The first case considers a point source (i.e. far field) interference at 83° relative to the array. The signal and interference are uncorrelated and the noise background is uncorrelated from sensor to sensor. Figure 8 gives the processor output signal-to-noise ratio for conventional (time delay-and-sum) beamforming and auxiliary array beamforming with and without the

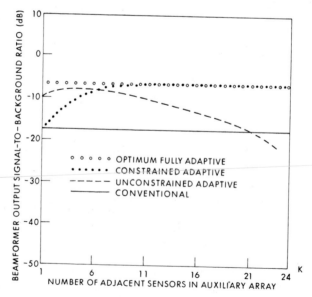

FIG. 8. Beamformer output signal-to-background versus number of sensors in auxiliary array.

linear signal preservation constraint. The auxiliary array variable is the number of adjacent elements in the sub-array. The first important result here is that linearly constrained auxiliary array processing of approximately 25% of the sensors in the array is nearly (within 2 dB) as effective in cancelling interference as if all the elements were processed. With 50% of the sensors processed the sub-array system is essentially optimum in performance. Secondly, unconstrained processing is somewhat better than constrained for a small aperture (few adaptive degrees of freedom) auxiliary array because the strict enforcement of the linear signal preservation constraint uses up one degree of freedom. This limitation could be minimized by increasing the aperture of the auxiliary array by spatially sparse element selection. Note however, that as the number of sensors increases the unconstrained processor cancels an increasing amount of signal because of more directivity to the auxiliary array beam pattern. Figure 9 gives the

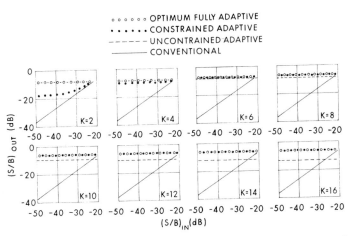

FIG. 9. Beamformer output signal-to-background ratio $(S/B)_{OUT}$ as a function of input (sensor level) signal-to-background ratio (dB), $(S/B)_{IN}$.

signal to back-ground ratio at the beamformer output versus the sensor element signal-to-background noise ratio for various numbers of adjacent auxiliary array sensors. The array gain can be obtained from these curves and is given by $A = (S/B)_{OUT} - (S/B)_{IN}$.

(b) *Orthogonal beam processing*
The inter-hydrophone correlation matrix, R, can be written in terms of its orthogonal components (eigenvectors) $\{M_i/i = 1, \ldots, N\}$ and eigenvalues $\{\lambda_i/i = 1, \ldots, N\}, (\lambda_i > \lambda_{i+1})$

$$R = \sum_{i=1}^{N} \lambda_i M_i M_i^H. \tag{39}$$

The expected power output for a conventional beam steered at angle θ is therefore given by

$$B(\theta) = E\{|y(\theta)|^2\} \tag{40}$$

$$= E\{|\mathbf{D}^H(\theta)\mathbf{X}|^2\} \tag{41}$$

$$= \mathbf{D}^H(\theta)R\mathbf{D}(\theta) \tag{42}$$

$$= \mathbf{D}^H(\theta) \sum_{i=1}^{N} \lambda_i \mathbf{M}_i \mathbf{M}_i^H \mathbf{D}(\theta) \tag{43}$$

$$= \sum_{i=1}^{N} \lambda_i |\mathbf{D}^H(\theta)\mathbf{M}_i|^2 \tag{44}$$

$$= \sum_{i=1}^{N} \lambda_i B_i(\theta) \tag{45}$$

where $B_i(\theta) = |\mathbf{D}^H(\theta)\mathbf{M}_i|^2$ is termed the ith orthogonal beam output power at frequency f [13]. Figure 10 gives $B_1(\theta)$ and $B_2(\theta)$ for a 24-element line

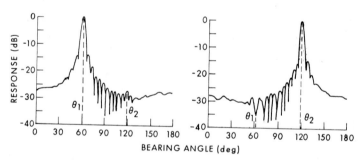

FIG. 10. Orthogonal beam angular response patterns $B_1(\theta)$ and $B_2(\theta)$.

array with statistically uncorrelated plane waves arriving at $\theta_1 = 61°$ and $\theta_2 = 119°$. The signal-to-noise ratios at the sensor level are 0 dB and -20 dB for the signals at θ_1 and θ_2 respectively. The noise is spatially uncorrelated. Notice that the predominant spatial information $(\theta_i, i = 1, 2)$ for each of the signals is "clustered" in a single orthogonal beam response. This separation of spatial signal arrival angle is relatively independent of the sensor element level signal-to-noise ratio and is the reason for considering this technique.

The algorithms for an adaptive realization of the orthogonal beam processor are of the linearly and quadratically constrained class discussed in the previous section. This relationship is obvious in view of the following theorem:

Theorem. max $\{W^H R W\} = \lambda_i$ for $W = M_i$.
where $W^H W = 1$, $W^H M_j = 0$, $j = 1, 2, \ldots, i - 1$.   (46)

In equation (46) $W^H W = 1$ and $\{W^H M_j = 0 / j = 1, \ldots, i - 1\}$ constitute the quadratic and linear constraints on W respectively.

4. Summary

Two applications for constrained multi-channel adaptive filter control to sensor array processing have been introduced. Adaptive auxiliary array processing is an accepted technique for minimization of hardware requirements while still obtaining near optimum spatial noise rejection. The effectiveness of the auxiliary array technique for a given number of auxiliary channels and sensor location selection of course depends on the geometry of the spatial noise environment. For this reason, auxiliary array selection must be considered within the context of the spatial interference characteristics. Hardware development for auxiliary array processing is straightforward and is currently advancing. Orthogonal beam processing, on the other hand, has a less firm theoretical base for sensor array processing than auxiliary array processing. The question of orthogonal (eigenvector) component coupling with regard to spatially correlated signals and/or noise has yet to be fully considered. However, adaptive algorithms for realizing the orthogonal beam technique can be easily programmed on a flexible system for adaptive sensor array processing. This experimental approach, it is felt, will provide useful information on the utility of the orthogonal beam technique.

REFERENCES

1. Bryn, F. (1962). Optimal signal processing of three-dimensional arrays operating on Gaussian signals and Noise. *J. Acoust. Soc. Am.* **34**, No. 3, 289–297.
2. Mermoz, H. (1965). "Adaptive Filtering and Optimal Utilization of An Antenna", Institute Polytechnique, Grenoble, France.
3. Anderson, V. C. (1969). Rejection of coherent arrival(s) at an array, *J. Acoust. Soc. Am.* **45**, No. 2, 406–411.
4. Cox, H. (1969). "Array Processing Against Interference", Proceedings of the Symposium on Information Processing, Purdue University, pp. 453–464.
5. Van Trees, H. L. (1968). "Detection, Estimation and Modulation Theory", John Wiley and Sons, Inc., New York.
6. Fukunaga, K. and Koontz, W. L. G. (1969). "Eigenvalues and Eigenvectors in Signal Analysis", Purdue University, Department of Electrical Engineering, Report No. TR-EE69-34.

7. Watanabe, S., Lambert, P., Kulikowski, C., Buxton, J. and Walker, R. (1967). Evaluation and selection of variables in pattern recognition. *In* "Computer and Information Sciences—II, J. T. Tou, Ed., pp. 91-122, Academic Press, New York.

8. Tou, J. T. and Heyden, R. P. (1967). Some approaches to optimum feature selection. *In* "Computer and Information Sciences—II", (J. T. Tou, Ed.), pp 57-89, Academic Press, New York and London.

9. Owsley, N. L. (1969). "A Constrained Gradient Search Method with Application to Adaptive Sonar Arrays", U.S. Naval Underwater Sound Laboratory Technical Document No. 2242-207-63.

10. Frost, O. L. (1970). "Adaptive Least Squares Optimization Subject to Linear Equality Constraints", Stanford University Ph.D. Thesis.

11. Owsley, N. L. (1971). "Source Location with an Adaptive Antenna Array", U.S. Naval Underwater Systems Center Technical Document No. NL-3015.

12. Rosen, J. B. (1964). The gradient projection method for non-linear programming; Part II: Non-linear constraints. *SIAM* 9, No. 4, pp. 514-532.

13. Kneipfer, R. R. (1970). "An Eigenvector Interpretation of an Array's Bearing Response Pattern", Naval Underwater Systems Center Report No. 1098.

DISCUSSION

H. Mermoz (paraphrased): You mentioned the application of null steering to near-field interference by using quadratically constrained adaptive minimization. How can this be?

Answer (paraphrased): "Near field" interference was a bad choice of words on my part. Instead, I should have said "partially coherent" interference for which the spatial correlation does not extend completely across the array. For this partially coherent interference, the correlation matrix for the array can not be expressed in dyadic form, which, in turn, can therefore not be reduced to a linear constraint. However, since a partially coherent interference cannot be reduced to a dyadic, the beam-former output power due to this interference must be expressed as a general quadratic form involving a correlation matrix of full rank. This quadratic form can then be used to suppress the interference by application of a quadratic constraint. For implementation of this constraint, I refer you to the paper.

M. Mermoz (paraphrased): Do you agree that an adaptive system cannot simultaneously be fully adaptive in both space and time?

Answer (paraphrased): Yes, an adaptive system can be fully spatially optimum/ adaptive for a signal from a given direction. However, to be temporally optimum the spectrum of the signal must be known *a priori* and therefore it follows that the post beamformer temporal processor cannot be fully adaptive and fully optimum simultaneously.

P. Moose (paraphrased): Could you expand on the technique of integrating multiple eigenvectors into the bearing angle display function you discussed?

Answer: First, if K eigenvectors are to be used, then K algorithms must be operational, that is, one for each desired eigenvector. This is scheme operates sequentially by feeding the estimate of each eigenvector into the subsequent eigenvector estimation algorithms where it is used to form a linear (orthogonality) constraint. In addition to these orthogonality constraints a quadratic (magnitude) constraint is used to normalize the eigenvectors. Thus, for example, the Nth eigenvector algorithm uses $(N-1)$ linear orthogonality constraints and one quadratic magnitude constraint.

Enhancement of Antenna Performance by Adaptive Processing†

G. BIENVENU and J. L. VERNET

Thomson-CSF D.A.S.M. Cagnes-sur-mer, France

1. Introduction

In this paper, experimental results obtained with two adaptive space–time processors are presented. One of them is applicable to active detection and the other, to estimation. In the first case the temporal form of the signal is known whilst in the second, it is unknown. In both cases, one assumes that the signal and the noises are uncorrelated and the signal direction is known.

The optimal antenna processors in stationary medium for the usual criteria of active or passive detection and estimation have been treated previously in the literature and a few are given in Section 2. These processors depend upon the noises received on the sensors only through the inverse of their space–time correlation matrix R. If R is unknown and moreover slowly variable, as is often the case in the sea, the same processor can be used by permanently replacing the correlation matrix by its estimation. These processors are referred to as adaptive processors. This procedure when applied directly is cumbersome since the matrix R is almost always of large dimensions; hence, its inversion entails a voluminous calculation. In the systems presented, the filters are estimated iteratively by using algorithms that avoid the inversion of R. These algorithms are analogous to those used by Widrow [1].

The performance attained by adaptive processing with regard to a processing in which the matrix R is assumed to be an identity matrix (independent noises between sensors and white in the bandwidth used) depends on the nature of the noise field. The performance is good if the noise field comprises directional noise sources. It is nevertheless limited by the convergence speed of the adaptive algorithm which should be greater than that of the noise field variation. They have been measured by simulating on computer the two types of processors. Synthesized noise

† This study has been supported by: "Direction des Recherches et Moyens d'Essais" contract No. DRME 289/71.

fields and real noise fields obtained by recording the sea noises received through hydrophones, have been used.

2. Optimum Space–Time Processing in Stationary Medium

In the following, the reception bandwidth is assumed to be limited so that the signals received can be sampled. The optimum processors are presented in a digital form. The optimization is carried out within a limited period of time T which, in the case of active detection, coincides with the duration of the signal, and which in the case of estimation, has a range of duration comparable to that of the space–time correlation functions of received noises.

(a) Notations and description of a space–time digital processor
The antenna possesses K sensors and therefore the space–time filter has K inputs. The signals issued from the sensors are sampled and delayed so that the antenna is steered in the direction of the signal. Therefore, when the signal is present, it is identical at all inputs. Figure 1 shows a scheme of a space–time digital processor. It possesses K filters with L taps.

$-x_1(t), \ldots, x_k(t), \ldots, x_K(t)$ are K input signals

$-\vec{X}_1, \ldots, \vec{X}_k, \ldots, \vec{X}_K$ are K L-dimensional vectors representing the sampled signal present in the filters.

$-\vec{X}$ is a vector with $(K \times L)$ components composed of the K vectors $\vec{X}_K$

$$\vec{X}^T = [\vec{X}_1^T, \ldots, \vec{X}_k^T, \ldots, \vec{X}_K^T]$$

$-W$ is a vector with $(K \times L)$ components formed by the K vectors $\vec{W}_k$ representing the weight of each filter.

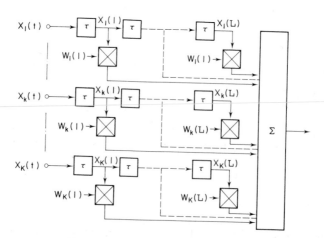

FIG. 1. Digital space–time processor.

$-M\vec{S}$ is a vector with $(K \times L)$ components of K signal vectors $\vec{S}$.

$$(M\vec{S})^T = [\vec{S}^T, \ldots, \vec{S}^T, \ldots, \vec{S}^T]$$

M is a column matrix of $K(L \times L)$ identity matrix.

When there is no signal, the vector $\vec{X}$ is made up of the noises vector: $\vec{X} = \vec{B}$.

When the signal is present, one has $\vec{X} = \vec{B} + M\vec{S}$, R is the space–time correlation matrix of noises alone; $R = E(\vec{B}\vec{B}^T)$. It is symmetric and is supposed to be inversible, therefore, it is positive definite. R_x is equal to $E(\vec{X}\vec{X}^T)$.

The vectors are represented by column matrices. $\vec{X}^T$ is the transpose of $\vec{X}$.

(b) Optimum processors [2–9]

The optimum antenna processors for different detection and estimation criteria have already been given in various papers. Three of these will be reviewed here:

Active detection. The signal $\vec{S}$ is known. Whether one uses the maximum likelihood ratio criterion in assuming that the noises form a set of jointly Gaussian zero-mean processes, or the matched filter criterion in supposing that the correlation matrix of noises is known, one obtains the same solution for the optimum digital processor $\vec{W}_{MF}$:

$$\vec{W}_{MF} = \alpha R^{-1} M\vec{S}$$

Passive detection. The noises are assumed to form a set of jointly Gaussian zero-mean processes, the signal is a Gaussian zero-mean process. The optimum digital processor is composed of a processor W_{PD}:

$$W_{PD} = R^{-1} MA$$

which delivers an L-dimensional vector $\vec{Y}$, and of a scalar product $\vec{Y}^T\vec{Y}$. A is an $(L \times L)$ matrix such that $AA^T = (R_S^{-1} + M^T R^{-1} M)^{-1}$ where $R_S = E(\vec{S}\vec{S}^T)$.

Estimation of an unknown signal. Only the direction of the signal is known. The noises are assumed to form a set of jointly Gaussian zero-mean processes. The optimum digital processor W_E is obtained by using the maximum likelihood criterion:

$$W_E = R^{-1} M(M^T R^{-1} M)^{-1}.$$

The same processor is obtained by minimizing the variance of an unbiased linear estimate.

(c) Optimum spatial processing
Several authors [4, 6, 8] have mentioned that all the optimum antenna
processors are composed of a common space-time processor independent
of the criterion, followed by a time processor that depends on the criterion.
The common processing is represented by the term $R^{-1}M$. From K input
signals with L time samples, it gives a signal with L time samples. One can
also adopt as common space processing the estimator of an unknown
signal W_E which then plays the role of an optimal antenna common to all
the processors giving the best estimation of the signal that comes from the
chosen direction. It minimizes the power due to noise sources which have
other directions. It generalizes therefore the notion of antenna. Moreover,
it possesses the special property of being realizable from the space–time
correlation matrix of noises alone, R, or else from the matrix obtained
from noises and the signal, $R + MR_{SS}M^T$. This property is especially
interesting for the adaptive realization of this processing.

3. Adaptive Processing [1]

In order to realize adaptive processing, one must estimate the correlation
matrix R of *noises alone.* As the estimation of this matrix is carried out
by using the signals issued from the sensors, and one does not know *a
priori* if the signal is present or not, one is unable to say if the estimation
obtained is really that of the matrix of noises alone, or if the matrix of
the signal has been added. However, this problem can be resolved for two
very useful processors. They are the maximum likelihood estimator,
because of the aforementioned property, and the active detection processor
in the case where the variation period of the noise field is wide compared
to the duration of the signal used. Indeed, the time constant of adaptation
can then be chosen big enough so that the signal does not upset the filters.
Two adaptive processors corresponding to each case will now be described.

4. Adaptive Spatial Processor (Estimation) [10, 11]

The estimator defined by the relation $W_E = R^{-1}M(M^TR^{-1}M)^{-1}$ estimates
L time samples of the signal all at the same time. In a real system, one will
estimate only one signal sample at each sampling time. One will therefore
realize only one column of the matrix W_E. If one estimates, for example,
the ith component of the signal vector, the processor is given by the ith
column of W_E. Thus $\vec{W}_E = R^{-1}M(M^TR^{-1}M)^{-1}\vec{\Delta}$, where $\vec{\Delta}$ is a vector with
L zero components, except the ith which is equal to 1. One obtains
therefore a sequential estimation of the signal. The adaptive algorithm is
obtained by using a property of the optimum filter. $\vec{W}_E$ minimizes the

variance of the output noise $\vec{W}_E^T R \vec{W}_E$, and gives an unbiased estimation of the ith sample of $\vec{S}$, thus

$$E[\vec{W}_E^T(\vec{B} + M\vec{S})] = \vec{W}^T M\vec{S} = s(i), \quad \text{and therefore } M^T \vec{W}_E = \vec{\Delta}.$$

The method of the projected gradient is then used

$$\vec{W}_{n+1} = \vec{W}_n - \epsilon[I - M(M^T M)^{-1} M^T] R \vec{W}_n$$

and the first vector $\vec{W}_0$ is such that $M^T \vec{W}_0 = \vec{\Delta}$. Since the corrective vector is such that $M^T[I - M(M^T M)^{-1} M^T] R \vec{W}_n = 0$, hence $M^T \vec{W}_n = \vec{\Delta}$ for all n. Therefore, all the vectors $\vec{W}_n$ obtained satisfy the constraint. The algorithm is obtained by replacing the R matrix by its instantaneous estimation $\vec{X}_n \vec{X}_n^T$. It is written as: $\vec{W}_{n+1} = \vec{W}_n - \epsilon[I - M(M^T M)^{-1} M^T] \vec{X}_n \vec{X}_n^T \cdot \vec{W}_n$.

It can be shown that if the successive input vectors $\vec{X}_n$ are independent and Gaussian, and if ϵ is small enough, $\vec{W}_n$ converges in the mean towards the optimum value $\vec{W}_E$, and that its variance can be made as small as desired by decreasing the value of ϵ. However, the weaker ϵ becomes, the slower will be the convergence.

An upper bound of ϵ which enables the variance of $\vec{W}_n$ to converge is given by 2/3 trace (R).

In fact, the simulation of the estimator shows that one must give ϵ an even weaker value in order to obtain a negligible variance. This variance of $\vec{W}_n$ creates at the output, a noise which increases with the total output power of the optimal antenna $\vec{W}_E^T R \vec{W}_E$. For a given noise field, it increases therefore with the signal, but is limited to a value so that the performance obtained is never inferior to that of the processing where R is replaced by the identity matrix.

In order to use this processor in a medium where R, and as a result trace (R), varies, ϵ should be estimated at each iteration. To accomplish this, ϵ_n is made to equal

$$\epsilon_n = \frac{k}{\vec{X}_n^T \vec{X}_n}$$

where k is a constant and $\vec{X}_n^T \vec{X}_n$ an instantaneous estimation of trace (R). The influence of the variations of ϵ_n is negligible.

5. Adaptive Processor for Active Detection

The processor to be realized is given by $\vec{W}_{MF} = \alpha R^{-1} M\vec{S}$.

In an adaptive system, it is necessary to fix the constant α so that the vector towards which $\vec{W}_n$ should converge can be defined. There are several ways to achieve this. One has chosen to prescribe the following constraint upon $\vec{W}_{MF}$: $\vec{W}_{MF}^T M\vec{S} = \beta = \text{constant}$.

$\vec{S}$ is a fixed vector, since it is a copy of the emitted signal. This relationship therefore implies that the correlation peak is proportional to the input signal. If the signal received is: a $\vec{S}$, the correlation peak equals: a β. Thus $\vec{W}_{MF}$ can be written:

$$\vec{W}_{MF} = \beta \, \frac{R^{-1}M\vec{S}}{\vec{S}^T M^T R^{-1} M\vec{S}}.$$

$\vec{W}_{MF}$ is calculated iteratively in accordance with the algorithm:

$$\vec{W}_{n+1} = \vec{W}_n + \epsilon \left\{ \left[g(\beta - \vec{W}_n^T M\vec{S}) + \frac{\vec{S}^T M^T \vec{X}_n \vec{X}_n^T \vec{W}_n}{\vec{S}^T M^T M\vec{S}} \right] M\vec{S} - \vec{X}_n \vec{X}_n^T \vec{W}_n \right\}.$$

The corrective vector has the form $(\alpha_n M\vec{S} - \vec{X}_n \vec{X}_n^T \vec{W}_n)$ where α_n is a scalar which tends towards α if the algorithm converges. The constraint is satisfied owing to the term $g(\beta - \vec{W}_n^T M\vec{S})$. With the same hypotheses as the preceding ones for the vectors $\vec{X}_n$, $\vec{W}_n$ converges in the mean towards $\vec{W}_{MF}$ with a variance which can be made as small as desired by decreasing ϵ, provided that ϵ is small enough.

In order to ensure that the convergence has a small variance, a value equivalent to that which is mentioned in paragraph 4 should be given to ϵ at each iteration: $\epsilon_n = k/\vec{X}_n^T \vec{X}_n$. The value of g should be inferior to: $\vec{S}^T M^T M\vec{S}/\vec{X}_n^T X_n$.

6. Performance of the Adaptive Processors—Simulation Experiments

The two adaptive processors have been simulated on computer. Their performance characteristics have been measured by comparing the results obtained with those given by the processors in which the R matrix is replaced by the identity matrix. $\vec{W}'_E = M(M^T M)^{-1}\vec{\Delta} = M\vec{\Delta}/k$ for the estimation

$$\vec{W}'_{MF} = \alpha M\vec{S} \quad \text{for the active detection.}$$

They are therefore equivalent to the classical antenna processing (addition of input signals with, in the active detection case, a correlation between the signal and the copy of the signal). The gain given by the adaptive processors is therefore solely due to the R matrix. When the noises are white and independent between sensors (omnidirectional noise), there is no gain. However, when there are also directional noise sources, it can attain some important values, and this is all the more so when the noise sources are stronger in relationship to the omnidirectional noise. One can demonstrate that if the optimization time, therefore the length of the filters, is infinite, the optimum processing has the property of eliminating $(K - 1)$ independent directional noise sources if there is not any omnidirectional noise (K is the number of antenna sensors). If the optimization time has the same range of duration as that of the space–time cross-correlation functions of the noises received on the sensors, one obtains a quasi-elimination of the noise sources.

(a) Description of the simulation

The simulated antenna is linear and possesses 10 sensors spaced half a wavelength (for 4.6 kHz) apart. The 3 dB beamwidth is equal to 15° for the central frequency (3 kHz). The signals received are filtered in the bandwidth from 2 to 4 kHz. They are sampled at twice the maximum frequency, that is at 10 kHz.

In the spatial processor, the filters comprise 32 memory cells, whereas the length of the cross-correlation functions is estimated at 30 sampling intervals. In active detection processor, the signal is a 10 ms rectangular pulse linearly frequency modulated from 2 kHz to 4 kHz ($BT = 20$). It consists of 100 sampling points. The number of filter memory cells has therefore been fixed at 100.

In the two cases, the direction of the signal is orthogonal to the antenna. Two types of noise fields have been used:

(i) simulated fields: they are composed of white and independent noises between the sensors, and directional noise sources with different powers and directions.

(ii) real fields: they consist of real noises recorded at sea with an antenna having characteristics identical to those of the simulated antenna.

In each type of processing, the gains of the adaptive and classical processors are equal for the signal. So it is possible to measure the gain produced by the adaptive processors by comparing the output noise powers.

(b) Simulation results

The simulation results are shown in Figs 2 to 7. In the spatial processor, the signal is a white Gaussian noise when the noises are simulated; it is a sine wave at 3 kHz when the noises are real. The initial vectors are equal to $M\vec{\Delta}/k$ for the spatial processor and the null-vector for the active detection processor. Figures 2 to 5 show the gains of the adaptive processors with respect to the classical processors for the following noise fields (the gain is measured by the ratio of the output noise powers of the two processors, the curves are traced on 10 000 iterations).

Figure 2: The noise field is composed only of a directional noise source which is a white Gaussian noise. Its direction is equal to 21°. The gain attains 21 dB in active detection. In spatial processing, the gain attained depends on the signal-to-noise ratio as has been specified by the theory. It is equal to 17 dB when there is no signal; 14 dB for a signal-to-noise ratio of −16 dB at the input, and 13.5 dB at the output; 11 dB for a

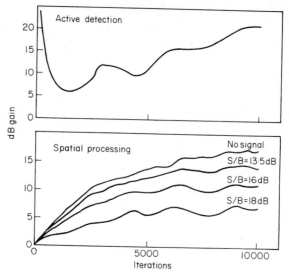

FIG. 2. One directional noise source (21°).

signal-to-noise ratio of −10 dB at the input and 16 dB at the output. 7 dB for a signal-to-noise ratio of −4 dB at the input and 18 dB at the output.

Figure 3: The noise field consists of the same noise source and omnidirectional noise with a power that is 10 times weaker. The gain attains 4 dB for the active detection. In spatial processor, it is equal to

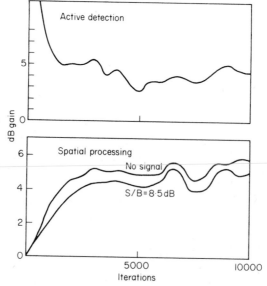

FIG. 3. Omnidirectional noise (power = 1) and one directional noise source (21°; power = 10).

5.5 dB without signal and 5 dB with a signal-to-noise ratio of −11 dB at the input and 8.5 dB at the output.

One notes a decrease of performance due to the omnidirectional noise.

Figure 4: The noise field is composed of the omnidirectional noise and the same directional noise source with a direction equals to 10°, (therefore it is in the main lobe) and a power thrice as great as that of the omnidirectional noise. The gain attains 11 dB in active detection.

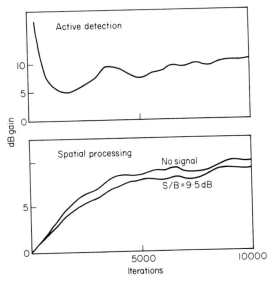

FIG. 4. Omnidirectional noise (power = 1) and one directional noise source (10°; power = 3).

In spatial processing, it is equal to 10 dB without signal, and 9.5 dB with a signal to noise ratio which equals −4 dB at the input and 9.5 dB at the output. This experiment shows that the "directivity" of an adaptive antenna depends upon the noise field.

Figure 5: The noise field consists of omnidirectional noise and of three directional noise sources which are independent white Gaussian noises. They possess powers that are 10 times higher than that of the omnidirectional noise and their directions equal 21°, 33°, 47°.

The gain attains 9 dB in active detection.

In spatial processing, it equals 7.5 dB without signal, and 7 dB for a signal-to-noise ratio of −13 dB at the input and 8 dB at the output.

The adaptation time varies in all of these cases from 3000 iterations to 10 000 iterations (that is from 0.3 to 1 second).

Figures 6 and 7 show the output signals of the processors for two types of real noise field. The noises have been recorded without any apparent directional noise sources.

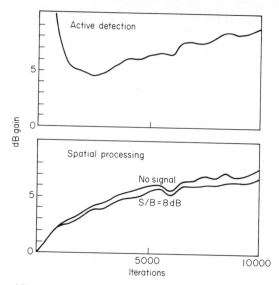

FIG. 5. Omnidirectional noise (power = 1) and three directional noise sources (21°, 33°, 47°; powers = 10).

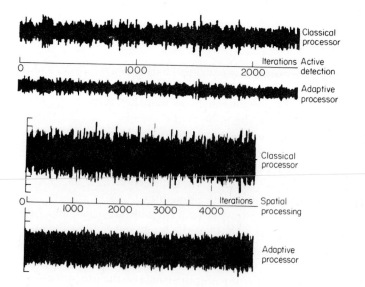

FIG. 6. Real noises.

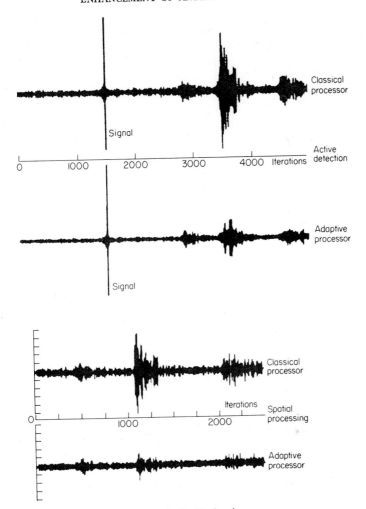

FIG. 7. Real noises.

In Fig. 6, the noises are stationary. In both processors the gain varies from 4 to 6 dB. In the case of a spatial processor, the output signal-to-noise ratio is equal to 10 dB (−4.5 dB at the input).

In Fig. 7, the noises show strong impulses. They are greatly reduced by the adaptive processors. When no impulse is present, the gain varies in the two processors from 5 to 6 dB; in the case of the spatial processor, the output signal-to-noise ratio is equal to 8 dB (−6 dB at the input). In the active detection case, the signal has been added—it does not increase the output noise variance.

7. Conclusion

A major problem in adaptive antenna processing is the estimation of the correlation matrix of noises alone. It can be solved in active detection if the signal duration is small compared to the period of variation of the noise field. In this case, the adaptation time constant can be several times the signal duration. In listening, a processor for the estimation of an unknown signal can be achieved. It may be regarded as the purely spatial processor common to the useful criteria; it generalizes the notion of antenna.

The efficiency of the two adaptive processors has been shown by the simulation results when the noise field comprises directional noise sources. It is due to the spatial elimination property of optimal antenna processing. The signal has a negligible effect in the case of active detection. For the spatial processing, its contribution remains slight for small output signal to noise ratios—8 dB to 10 dB. For very great ones, the performance drops to that of the classical processing.

REFERENCES

1. Widrow, B., Mantey, P. E., Griffiths, L. J. and Goode, B. B. (1967). Adaptive antenna systems. *Proc. IEEE* 55 (12), 2143–2159.
2. Bryn, F. (1962). Optimum signal processing of three dimensional arrays operating on Gaussian signal and noise. *JASA* 34 (3), 289–297.
3. Mermoz, H. (1964). Extension de la méthode du filtrage adapté au cas de plusieurs entrées pour l'optimisation de la détection des signaux faibles. Filtrage adapté et directivité. Thèse de doctorat.
4. Kelly, E. J. and Levin, M. J. (1964). Signal parameter estimation for seismometer arrays. Lincoln Laboratory Tech. Rep. 339.
5. Burg, J. P. (1964). Three dimensional filtering with an array of seismometers. *Geophysics*, 29 (5), 693–715.
6. Van Trees, H. L. (1966). "Optimum Processing for Passive Sonar Arrays", Ocean Electronic Symposium, Honolulu, Hawaii.
7. Capon, J., Greenfield, R. J. and Kolker, R. J. (1967). Multidimensional maximum likelihood processing of a large aperture seismic array. *Proc. IEEE* 55(a), 192–211.
8. Cox, H. (1968). "Interrelated Problems in Estimation and Detection" NATO advanced Study Institute on Signal Processing with Emphasis on Underwater Acoustics. Enschede, The Netherlands.
9. Macchi, C., Macchi, O. and Arques, P. Y. (1971). Généralisation de systèmes de détection de signaux certains. *Annales des Télécom.* 26 (9-10), 363–380.
10. Luenberger, D. G. (1968). "Optimisation by Vector Space Methods", J. Wiley & Sons Inc., New York.
11. Lacoss, R. T. (1968). Adaptive combining of wideband array data for optimal reception. *IEEE Trans. Geo. Elec.* GE-6 (2).

DISCUSSION

N. L. Owsley: For the spatial processor algorithm, the round-off errors in the computation will put the weight vector $\vec{W}_n$ outside of the constraint.

Answer: This is true. In fact, in our simulation, each N iterations, we compute the deviation of $\vec{W}_n$ from the constraint, and we correct $\vec{W}_n$ in accordance with this deviation. It is sufficient to take N equal to about 100 and the correction procedure is very simple.

Sensitivity Considerations in Adaptive Beamforming

HENRY COX

University of California, San Diego,
Marine Physical Laboratory,
Scripps Institution of Oceanography,
San Diego, California, U.S.A.

1. Introduction

The theory of optimum arrays became widely known in the underwater acoustics community through the work of Bryn [3] and Mermoz [17]. The relationship of their work to other detection and estimation problems has been discussed in [7]. Optimum array processing structures use detailed information about the signal and noise fields. Since this information is not known precisely in advance, one is led naturally to adaptive beamformers which continually adjust their parameters based upon on-line measurements of some kind. Since adaptive processors are continually adjusting, it is natural to question how sensitive performance is to small variations of the signal field, noise field and system parameters from their assumed or estimated values.

The question of sensitivities has been examined in the past [6, 12, 22, 24] in conjunction with "super-directive" arrays. An attempt will be made to point out the relationship of the results of this paper to those earlier results. The emphasis in this paper is on receiving arrays.

In Section 2 an introduction to the problem is provided using an intuitive approach. The performance measures of array gain and output power are used.

Section 3 presents the principal sensitivity results. The approach is to take partial derivatives of the gain and output signal power with respect to the size of signal, noise and steering perturbations.

Section 4 discusses the problem of signal suppression which arises in passive adaptive processors when measurements of signal-plus-noise are used when noise only measurements are desired. Interference rejection is also discussed.

A number of optimization problems are discussed in Section 5. Particular emphasis is given to formulations which somehow take sensitivity into

619

account. It is pointed out that the simple beamformer structure is no longer optimum in most of these problems. A more general array structure is suggested which possesses optimality properties.

Vector-matrix notation similar to that in Reference [7] will be used throughout this paper. Column vectors, i.e. matrices with only one column are designated by bold lower case letters such as $\mathbf{k}$ and $\mathbf{m}$. Other matrices are designated by bold upper case letters such as $\mathbf{P}$, $\mathbf{Q}$ and $\mathbf{R}$. The asterisk is used to denote complex conjugate transposition. Thus, $(\mathbf{k}^*\mathbf{k})$ is a scalar and $(\mathbf{mm}^*)$ is a dyad or square matrix of rank one.

2. Preliminary Discussion

The receiving array consists of an arbitrary arrangement of M sensors. When a signal is present the waveform at the output of the ith sensor is

$$x_i(t) = v_i(t) + n_i(t)$$

where $v_i(t)$ is the signal component and $n_i(t)$ is the noise component. The noise component includes both the effects of the external noise field and internal electronic noise. The outputs of all M sensors may be represented by the following vector equation:

$$\mathbf{x}(t) = \mathbf{v}(t) + \mathbf{n}(t) \tag{1}$$

The detection problem is usually formulated as deciding whether or not the signal component $\mathbf{v}(t)$ is present. Estimation problems arise when one seeks to obtain information about particular parameters of $\mathbf{v}(t)$. The relationships among a number of detection and estimation problems are discussed in Reference [7]. There it is shown that the same beamforming processor is optimum for a variety of different detection and estimation problems.

The signal vector $\mathbf{v}(t)$ and the noise vector $\mathbf{n}(t)$ are assumed to have mean values of zero and to be statistically independent of each other. Because the analysis of array processors is sometimes simplified by working in the frequency domain, it is convenient to work with the cross-spectral density matrices of $\mathbf{v}(t)$, $\mathbf{n}(t)$ and $\mathbf{x}(t)$ which are denoted $\widetilde{\mathbf{P}}(\omega)$, $\widetilde{\mathbf{Q}}(\omega)$ and $\mathbf{R}(\omega)$ respectively. Then

$$\mathbf{R}(\omega) = \widetilde{\mathbf{P}}(\omega) + \widetilde{\mathbf{Q}}(\omega) \tag{2}$$

when the signal is present. $\widetilde{\mathbf{P}}(\omega)$ and $\widetilde{\mathbf{Q}}(\omega)$ may be written as

$$\widetilde{\mathbf{P}}(\omega) = \sigma_s^2(\omega)\mathbf{P}(\omega) \tag{3}$$

$$\widetilde{\mathbf{Q}}(\omega) = \sigma_n^2(\omega)\mathbf{Q}(\omega) \tag{4}$$

where $\sigma_n^2(\omega)$ is the noise power spectral density averaged over the M sensors so that

$$\sigma_n^2(\omega) = \text{trace } \tilde{Q}(\omega)/M \tag{5}$$

and $\sigma_s^2(\omega)$ is similarly defined as the signal power (or energy) spectral density averaged over the M sensors. Thus, $P(\omega)$ and $Q(\omega)$ are normalized to have their traces equal to M, the number of sensors in the array. The quantity $\sigma_s^2(\omega)/\sigma_n^2(\omega)$ is the input signal-to-noise spectral ratio $(S/N)_{\text{in}}$.

A device which has the configuration shown in Fig. 1 will be called a

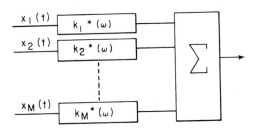

FIG. 1. General beamformer configuration.

beamformer. In general, it consists of a set of filters, one for each sensor, followed by a summation device. The transfer function of the filter on the ith sensor is $k_i^*(\omega)$. Two quantities of prime interest are the output spectral density.

$$z(\omega) = k^*(\omega)R(\omega)k(\omega) \tag{6}$$

and the array gain

$$G(\omega) = \frac{k^*(\omega)P(\omega)k(\omega)}{k^*(\omega)Q(\omega)k(\omega)} \tag{7}$$

where, from Fig. 1, the steering vector $k^*(\omega)$ is the row vector of filter transfer functions $\{k_1^*(\omega), \ldots, k_M^*(\omega)\}$. The sensitivity analysis will be concerned with determining the effects on $z(\omega)$ and $G(\omega)$ of varying $P(\omega)$, $Q(\omega)$ and $k(\omega)$ from their nominal values.

The array gain given by equation (7) is the ratio of the output signal-to-noise spectral ratio to the input signal-to-noise spectral ratio. It is easily seen to be invariant if $k^*(\omega)$ is multiplied by a scale factor. In considering sensitivity problems it is sometimes useful to normalize both the numerator and denominator of equation (7) by dividing by the magnitude squared of $k^*(\omega)$. Then equation (7) becomes

$$G(\omega) = \frac{k^*(\omega)P(\omega)k(\omega)/(k^*(\omega)k(\omega))}{k^*(\omega)Q(\omega)k(\omega)/(k^*(\omega)k(\omega))} \tag{8}$$

The numerator of equation (8) is the normalized signal response and the denominator is the normalized noise response. We shall see that the numerator and denominator of equation (8) individually or more specifically their reciprocals are of fundamental importance in sensitivity analysis.

One of the reasons for working in the frequency domain is that the signal vector $v(t)$ is usually assumed to be related to a scalar signal $s(t)$ at some source point in space by an equation of the following form:

$$v(t) = \int m(t - \tau)s(\tau)\, d\tau \qquad (9)$$

where $m_i(t)$ accounts for the propagation from the source to the ith sensor and the response of the ith sensor itself. In the ideal case of non-dispersive propagation and distortion-free sensors, $m_i(t)$ is a simple time delay $\delta(t - T_i)$. Whenever $v(t)$ is related to a scalar $s(t)$ by a known transformation $m(t)$ as in equation (9) $P(\omega)$ is simply the dyad

$$P(\omega) = m(\omega)m^*(\omega) \qquad (10)$$

where $m(\omega)$ is the Fourier transform of $m(t)$ normalized so that $m^*(\omega)m(\omega) = M$.

The noise field may be the sum of a number of components. Some components of particular significance are spatially white noise, istropic noise, and point source noise.

Spatially white noise is uncorrelated from sensor to sensor and of equal intensity at each sensor. It has the identity matrix for its normalized cross-spectral density matrix. The subscript w on $Q_w = I$ will be used to denote this type of noise component.

Another physically important noise component is isotropic noise in which the spatial density of the noise from all directions is the same. Since there is no preferred direction, the elements of the isotropic noise cross-spectral density matrix are all real. In a spherically isotropic field the cross-spectral density between two sensors separated by a distance d is $\sin(\omega d/c)/(\omega d/c)$ where c is the velocity of propagation. The symbol Q_0 will be used to denote this isotropic noise cross-spectral density matrix. Notice that Q_0 depends only on the geometry of the array and frequency. Another isotropic cross-spectral density [14] is $J_0(\omega d/c)$ which arises in two dimensional problems when the field is cylindrically isotropic.

Point source noise comes from a single point source and hence is signal-like in its spatial characteristics.

The symbol

$$Q_d = d(\omega)d^*(\omega) \qquad (11)$$

will be used to denote this type of noise component. Again the normalization $d*d = M$ is assumed. Isotropic noise is simply the limit of summing infinitesimal independent noise components of equal intensity from all directions.

Henceforth, in order to simplify the notation, the dependence of various quantities on frequency ω will usually not be shown explicitly.†

When the signal matrix has the simple form of equation (10) the expression for the array gain becomes

$$G = \frac{|k*m|^2}{k*Qk} \tag{12}$$

The numerator of equation (12) is the magnitude squared of the inner product of the steering vector k and the signal direction vector m. We may think geometrically in terms of a generalized angle γ between these vectors defined as follows:

$$\cos^2\gamma = \frac{|k*m|^2}{(k*k)(m*m)} \tag{13}$$

Substituting from equation (13) into equation (8) and using the relationship $m*m = M$ yields

$$G = M \cos^2\gamma \{k*k/(k*Qk)\} \tag{14}$$

The array gain against spatially white noise is therefore

$$G_w = M \cos^2\gamma \tag{15}$$

which is also the normalized signal response [numerator of equation (8)]. It is a function of the misalignment of the steering vector k and the signal direction vector m.

In a conventional beamformer k is chosen to be proportional to m. This choice of "matching to the signal direction" makes γ equal to zero. This maximizes the gain against spatially white noise, alias normalized signal response. The maximum value of G_w in equation (15) is M. If G_w is less than unity the array becomes more sensitive than a single sensor to spatially white noise.

The factor $\{k*k/(k*Qk)\}$ is the reciprocal of the normalized noise

---

† Many of the results expressed in general vector-matrix form may also be used directly in time-domain formulations of a variety of related problems [7].

response [the denominator of equation (8)]. It is also the gain ratio G/G_w which we shall denote by ρ. When the noise is spatially white this factor is unity. For an arbitrary noise matrix this gain ratio ρ is bounded as follows:

$$\frac{1}{\lambda_{max}} \leqslant \frac{k^*k}{k^*Qk} \leqslant \frac{1}{\lambda_{min}} \tag{16}$$

where λ_{min} and λ_{max} are respectively the smallest and largest eigenvalues of Q. Since Q is normalized to have its trace equal to M, the average size of the eigenvalues of Q is unity.

When the noise is isotropic the array gain

$$G_0 = M \cos^2 \gamma \{k^*k/(k^*Q_0 k)\} \tag{17}$$

is known as the directivity index. The gain ratio

$$G_0/G_w = k^*k/(k^*Q_0 k) \tag{18}$$

is known as the Q-factor [15, 24] or super-gain ratio [22]. We shall call ρ the generalized super-gain ratio since it involves replacing the isotropic noise matrix in equation (18) with a general noise matrix. The isotropic noise covariance matrix depends on the array geometry. For line arrays it becomes the identity matrix when elements are spaced at one-half wavelength intervals. When sensors are moved closer together than one-half wavelength some of the eigenvalues of the isotropic matrix become extremely small and ρ may become large.

"Optimum" and "adaptive" beamformers choose k to be something other than proportional to m. The underlying philosophy is to accept a reduction in normalized signal response in order to achieve an even greater reduction in normalized noise response, thereby improving the gain. Since normalized signal response and gain against spatially white noise are the same, these processors can become very sensitive to white noise if too great a reduction in normalized signal response takes place. An "optimum" steering vector k^* under a number of criteria [7] is

$$k^* = m^*Q^{-1} \tag{19}$$

Lewis and Schultheiss [14] present an interesting discussion of the gain of optimum and conventional processors and their relationship to the eigenvalues of the noise cross-spectral matrix. For the choice of k given by equation (19) the gain, white noise gain and generalized super-gain ratio become respectively

$$G = m^*Q^{-1}m \qquad (20)$$

$$G_w = \frac{(m^*Q^{-1}m)^2}{m^*Q^{-2}m} \qquad (21)$$

$$\rho = \frac{G}{G_w} = \frac{m^*Q^{-2}m}{m^*Q^{-1}m} \qquad (22)$$

The ratio (22) of course still satisfies the bound given by equation (16). McDonough [16] presents a complicated expression relating equation (21) to the ratio of the largest and smallest eigenvalues of Q. He concludes equation (21) may become very small if $\lambda_{max}/\lambda_{min}$ is large.

The theory of super-directive arrays involves achieving a high directivity index by using the optimum steering vector for isotropic noise $k^* = m^*Q_0^{-1}$. The literature on super-directive arrays is extensive [6, 12, 22, 23, 24, 25]. It is well known that high directivity index (nearly equal to M^2) can be achieved with endfire steering of arrays of closely spaced elements. In order to achieve these high gains the steering vector is made almost orthogonal to the signal direction vector and the gain against white noise becomes extremely small. For example, Gilbert and Morgan [12] report a maximum directivity index of 15.8 but a corresponding white noise gain of only 1.5 x 10^{-4} with a four element endfire array with 1/16 wavelength inter-element spacing. They also present the result that the maximum directivity index of an array averaged over all steering directions is equal to the number of elements M. To see this, we write equation (20) for isotropic noise as

$$G = \text{trace}(Q_0^{-1}mm^*) \qquad (23)$$

and notice that the average of mm^* over all directions is by definition the isotropic matrix Q_0.

Sensitivity analyses of super-directive arrays have found the super-gain ratio and the reciprocal of the white noise gain to be useful measures of sensitivity.

In this paper we shall find that $1/G_w$ and ρ play key roles in sensitivity whenever the perturbations are statistically independent from element to element.

The causes of variations of P, Q and k from their nominal values can be many and varied. They include: violations of a priori assumptions such as those concerning plane wave or point source nature of signal or interference, or isotropic nature of the noise field; imperfect measuring devices and finite measuring intervals used to estimate $\mathbf{Q}$; non-stationarity of the noise field; presence of signal in noise estimates in passive systems; position errors in sensor locations; amplitude and phase errors in analogue components; and sampling and quantization errors in digital components. In adaptive beamformers the nature of these variations is usually related to the inter-play between *a priori* assumptions and on-line measurements. Sensitivity to variations in k are especially important since the underlying philosophy of adaptive beamforming is continuing adjustment of the steering vector.

3. General Sensitivity Results

(a) Noise perturbations
Suppose that the noise field is made up of several components so that $\mathbf{Q}$ may be expressed as a sum of Hermitian matrices of the form

$$\mathbf{Q} = \alpha_1 \mathbf{Q}_1 + \alpha_2 \mathbf{Q}_2 + \cdots + \alpha_n \mathbf{Q}_n \tag{24}$$

where

$$\alpha_i \geqslant 0, \, i = 1, \ldots, n \quad \sum_{i=1}^{n} \alpha_i = 1$$

Each $\mathbf{Q}_i$ in equation (24) is non-negative definite and has its trace equal to M so that it is a legitimate normalized noise cross-spectral matrix. The parameter α_i is the relative strength of the ith component. The decomposition of $\mathbf{Q}$ into components is not unique but is sometimes useful when various physical sources of noise exist. When $\mathbf{Q}$ is given in the form of equation (24) the array gain equation (7) may be expressed as follows:

$$G = \left(\sum_{i=1}^{n} \alpha_i / G_i \right)^{-1} \tag{25}$$

where G_i is the array gain against the ith noise component, that is

$$G_i = \frac{\mathbf{k}^* \mathbf{P} \mathbf{k}}{\mathbf{k}^* \mathbf{Q}_i \mathbf{k}} \tag{26}$$

From equation (25) it is evident that a large value of α_i / G_i for any component will cause the gain G to be small. Thus a relatively weak noise component (small α_i) can limit the overall array gain if the gain against that particular component is very small. Hence choices of k which give low gain against a type of noise which is likely to be present result in systems which are sensitive to small amounts of that type of noise.

Similarly equation (6) may be rewritten as follows:

$$z = \sigma_s^2 \, \mathbf{k}^* \mathbf{P} \mathbf{k} \left[1 + \frac{\sigma_n^2}{\sigma_s^2} \sum_{i=1}^{n} \alpha_i / G_i \right] \tag{27}$$

Quantitative measures of sensitivity to changes in the noise matrix can be obtained by differentiating equations (25) and (27). Suppose that

$$Q = (1 - \alpha)Q_1 + \alpha Q_2 \tag{28}$$

The fractional sensitivity of G and z to the substitution of a little Q_2-type noise for an equal amount of Q_1-type noise are

$$\left(\frac{dG/d\alpha}{G}\right)_{\alpha=0} = 1 - (G_1/G_2) \tag{29}$$

and

$$\left(\frac{dz/d\alpha}{z}\right)_{\alpha=0} = \frac{(G_1/G_2) - 1}{1 + G_1 \sigma_s^2/\sigma_n^2} \tag{30}$$

or

$$\left(\frac{dz/d\alpha}{z}\right)_{\alpha=0} = \frac{-\left(\dfrac{dG/d\alpha}{G}\right)_{\alpha=0}}{1 + (S/N)_0} \tag{31}$$

where $(S/N)_0 = (\sigma_s^2/\sigma_n^2)G_1$ is the output signal to noise ratio. When $G_1/G_2 > 1$, the substitution of Q_2-type noise for Q_1-type noise causes a decrease in gain and an increase in output power. The gain ratio G_1/G_2 in equations (29) and (30) may be expressed as

$$\frac{G_1}{G_2} = \frac{k^*Q_2 k}{k^*Q_1 k} \tag{32}$$

In the special case of spatially white noise, equation (32) reduces to the generalized super-gain ratio

$$\frac{G_1}{G_w} = \frac{k^*k}{k^*Q_1 k} = \rho \tag{33}$$

discussed earlier. If, in addition Q_1-type noise is isotropic, equation (33) becomes the super-gain ratio or Q-factor.

(b) Signal perturbations

Similarly we may conceive of the signal field being made up of several components so that P may be expressed as

$$P = \sum_{i=1}^{n} \beta_i P_i; \quad \beta_i \geqslant 0 \; i = 1, \ldots, n; \quad \sum_{i=1}^{n} \beta_i = 1 \tag{34}$$

Each $\mathbf{P}_i$ is Hermitian and non-negative definite with trace equal to M. Then

$$G = \sum_{i=1}^{n} \beta_i G_i \tag{35}$$

and

$$z = \sigma_n^2 \, \mathbf{k}^* \mathbf{Q} \mathbf{k} \left[1 + \frac{\sigma_s^2}{\sigma_n^2} \sum_{i=1}^{n} \beta_i G_i\right] \tag{36}$$

where

$$G_i = \frac{\mathbf{k}^* \mathbf{P}_i \mathbf{k}}{\mathbf{k}^* \mathbf{Q} \mathbf{k}} \tag{37}$$

is the gain for the ith component of the signal field. The overall gain is simply the sum of the gain for each component weighted by its relative strength.

Suppose that

$$\mathbf{P} = (1 - \beta)\mathbf{P}_1 + \beta \mathbf{P}_2 \tag{38}$$

The fractional sensitivities of G and z to the substitution of a little $\mathbf{P}_2$-type signal for an equal amount of $\mathbf{P}_1$-type signal are

$$\left(\frac{dG/d\beta}{G}\right)_{\beta=0} = \frac{G_2}{G_1} - 1 \tag{39}$$

and

$$\left(\frac{dz/d\beta}{z}\right)_{\beta=0} = \frac{(G_2/G_1) - 1}{1 + \sigma_n^2/(G_1 \sigma_s^2)} \tag{40}$$

or

$$\left(\frac{dz/d\beta}{z}\right)_{\beta=0} = \frac{(S/N)_0}{1 + (S/N)_0} \left(\frac{dG/d\beta}{G}\right)_{\beta=0} \tag{41}$$

The gain ratio G_2/G_1 appearing in equations (39) and (40) may be expressed as

$$\frac{G_2}{G_1} = \frac{\mathbf{k}^* \mathbf{P}_2 \mathbf{k}}{\mathbf{k}^* \mathbf{P}_1 \mathbf{k}} \tag{42}$$

When $\mathbf{P}_1$ is the result of a point signal source with direction vector $\mathbf{m}$ and $\mathbf{P}_2$ is the identity matrix representing independent signal perturbations from sensor to sensor equation (42) becomes

$$G_2/G_1 = \mathbf{k}^* \mathbf{k}/|\mathbf{k}^* \mathbf{m}|^2 = 1/G_w \tag{43}$$

and

$$G_2 = k^*k/(k^*Qk) = \rho \tag{44}$$

which are by now familiar expressions. Low white noise gain in equation (43) means large response to a spatially incoherent signal component.

Before proceeding it is useful to consider how the situation represented by equation (38) might arise. Suppose that the signal vector suffered random perturbations so that it could be written as

$$m = \sqrt{1 - \beta}m_1 + \sqrt{\beta}\delta \tag{45}$$

where δ is a vector of random perturbations with zero mean and normalized covariance matrix $E[\delta\delta^*] = P_2$. The factor $\sqrt{1 - \beta}$ in (45) provides the normalization so that

$$E[m^*m] = \text{trace}[E(mm^*)] = M \tag{46}$$

Then

$$E[|k^*m|^2] = (1 - \beta)|k^*m_1|^2 + \beta k^* P_2 k \tag{47}$$

In this situation, the sensitivity results of equations (39) and (40) should be interpreted on an ensemble average basis.

From equation (39) it is evident that small perturbations on the average will not cause severe degradation in G or z even in $G_2 \ll G_1$. However, when $G_2 \gg G_1$ rapid increases in G and z are to be expected. Of course, any individual perturbation of the type of equation (45) may deviate significantly from this average behaviour. From equation (43) we see that when k is nearly orthogonal to m so that $1/G_w$ is large, random signal perturbations will on the average cause an increase in G and z by decreasing γ. Of course, unpredictable increases in G and z due to slight perturbations in m with no change in signal power can be a source of confusion and may be undesirable. This is especially true in processors which are supposedly constrained.

Another approach to the formulation of the signal sensitivity problem is to assume P is of the dyad form of equation (10) and to allow amplitude and phase perturbations to occur to the individual components of m, i.e.

$$m_j = m_j^0(1 + a_j) \exp(\sqrt{-1}\xi_j) \tag{48}$$

when m_j^0 is the nominal value of the jth component of m and the amplitude and phase perturbations have zero means and are assumed to be independent of each other. This approach has been taken in a recent paper by McDonough

[16] following the lead of Gilbert and Morgan [12]. When m_j is given by equation (48), the quantity $E(|k^*m|^2)$ becomes

$$E(|k^*m|^2) = \sum_{i,j} k_i^* k_j m_i^0 m_j^0 {}^* C_{ij} \tag{49}$$

where

$$C_{ij} = \begin{cases} 1 + E(a_j^2), & \text{for } i = j \\ \{1 + E(a_i a_j)\} E[\exp(\sqrt{-1}(\xi_i - \xi_j))], & \text{for } i \neq j \end{cases} \tag{50}$$

Notice that the quantity m in equation (48) is not normalized as m was in equation (46) since for m defined in (48)

$$E[m^*m] = M + \sum_j |m_j|^2 E(a_j^2) > M \tag{51}$$

As shown by McDonough, equation (49) may be simplified under the following additional assumptions which were made by Gilbert and Morgan:

(1) Amplitude perturbations are independent from sensor to sensor and of equal variance, i.e., $E[a_i a_j] = \sigma_a^2 \delta_{ij}$.
(2) Phase perturbations are Gaussian, small, independent from sensor to sensor and of equal variance, i.e., $E[\xi_i \xi_j] = \sigma_\xi^2 \delta_{ij}$, $\sigma_\xi^2 \ll 1$.
(3) m_j is a pure phasing, i.e., $|m_j|^2 = 1, j = 1, \ldots, n$.

Under the above assumptions

$$C_{ij} \begin{cases} = 1 + \sigma_a^2 & \text{for } i = j \\ \approx 1 - \sigma_\xi^2 & \text{for } i \neq j \end{cases} \tag{52}$$

and equation (49) becomes

$$E[|k^*m|^2] = (1 - \sigma_\xi^2)|k^*m^0|^2 + (\sigma_a^2 + \sigma_\xi^2)k^*k \tag{53}$$

which is almost identical to equation (47) for the corresponding case $P_2 = I$. The difference can be attributed to the normalization of equation (45).

In general equation (49) can be made to look like equation (47) by defining a matrix F with elements

$$F_{ij} = m_i^0 m_j^0 {}^* (C_{ij} - 1) \tag{52}$$

The equation (49) may be written

$$E[|k^*m|^2] = |k^*m^0|^2 + k^*Fk \tag{53}$$

Much of the sensitivity work [12, 16, 24] in the field of antenna arrays has been concerned with the quantity $E[|k^*m|^2]/|k^*m^0|^2$ when the components of k and/or m have been perturbed in amplitude and/or phase from their nominal values $\mathbf{k}^0$ and $\mathbf{m}^0$. From equation (49) it is evident that the type of result will be the same whether C_{ij} arises from perturbations of $\mathbf{m}$, $\mathbf{k}$, or some combination of both.

(c) Steering perturbations

As suggested above there are similarities between signal perturbations and perturbations in the steering vector $\mathbf{k}$. Perturbations in $\mathbf{k}$ may also be approached in two ways, analogous to equations (45) and (48) respectively. However, there is an important difference in that perturbations in $\mathbf{k}$ affect both signal and noise terms.

Following the approach of equation (45), suppose that

$$\mathbf{k} = \sqrt{1 - \epsilon}\, \mathbf{k}_1 + \sqrt{\epsilon}\, \boldsymbol{\eta} \tag{54}$$

where $\boldsymbol{\eta}$ is a vector of random perturbations with zero mean and covariance matrix $E[\boldsymbol{\eta}\boldsymbol{\eta}^*] = \mathbf{W}$ where $\mathbf{W}$ is normalized to have its trace equal to $\mathbf{k}_1^*\mathbf{k}_1$. Then expected output power is

$$\bar{z} = E[\mathbf{k}^*\mathbf{R}\mathbf{k}] = (1 - \epsilon)\mathbf{k}_1^*\mathbf{R}\mathbf{k}_1 + \epsilon\, \text{trace}(\mathbf{R}\mathbf{W}) \tag{55}$$

and the sensitivity of the expected output power is

$$\left(\frac{d\bar{z}/d\epsilon}{\bar{z}}\right)_{\epsilon=0} = \frac{\text{trace}(\mathbf{R}\mathbf{W})}{\mathbf{k}_1^*\mathbf{R}\mathbf{k}_1} - 1 \tag{56}$$

Defining a gain $\tilde{\tilde{G}}$ as follows:

$$\tilde{\tilde{G}} = E(\mathbf{k}^*\mathbf{P}\mathbf{k})/E(\mathbf{k}^*\mathbf{Q}\mathbf{k}) \tag{57}$$

we obtain

$$\tilde{\tilde{G}} = \frac{(1 - \epsilon)\mathbf{k}_1^*\mathbf{P}\mathbf{k}_1 + \epsilon\, \text{trace}(\mathbf{P}\mathbf{W})}{(1 - \epsilon)\mathbf{k}_1^*\mathbf{Q}\mathbf{k}_1 + \epsilon\, \text{trace}(\mathbf{Q}\mathbf{W})} \tag{58}$$

The sensitivity of this gain is

$$\left(\frac{d\tilde{\tilde{G}}/d\epsilon}{\tilde{\tilde{G}}}\right)_{\epsilon=0} = \frac{\text{trace}(\mathbf{P}\mathbf{W}/G_1) - \text{trace}(\mathbf{Q}\mathbf{W})}{\mathbf{k}_1^*\mathbf{Q}\mathbf{k}_1} \tag{59}$$

Simplifications again occur when the perturbations are independent from sensor to sensor and of equal variance so that $\mathbf{W} = (\mathbf{k}_1^*\mathbf{k}_1/M)\mathbf{I}$. Then equation (55) becomes

$$\bar{z} = (1 - \epsilon)\mathbf{k}_1^*\mathbf{R}\mathbf{k}_1 + \epsilon(\sigma_s^2 + \sigma_n^2)\mathbf{k}_1^*\mathbf{k}_1 \tag{60}$$

and equation (56) becomes

$$\left(\frac{d\bar{z}/d\epsilon}{\bar{z}}\right)_{\epsilon=0} = \frac{\mathbf{k}_1^*\mathbf{k}_1}{\mathbf{k}_1^*\mathbf{Q}\mathbf{k}_1}\left\{\frac{1 + (S/N)_{\text{in}}}{1 + (S/N)_0}\right\} - 1 \tag{61}$$

When $(S/N)_{in}$ and $(S/N)_0$ are both much less than unity the output power sensitivity of equation (61) is approximately equal to $\rho - 1$. When $(S/N)_{in}$ and $(S/N)_0$ are both much larger than unity, it is approximately equal to $\{(1/G_w) - 1\}$.

A similar simplification occurs in equation (58) which becomes

$$\widetilde{\widetilde{G}} = \frac{(1-\epsilon)k_1^* P k_1 + \epsilon k_1^* k_1}{(1-\epsilon)k_1^* Q k_1 + \epsilon k_1^* k_1} \tag{62}$$

and equation (59) which becomes

$$\left(\frac{d\widetilde{\widetilde{G}}/d\epsilon}{\widetilde{\widetilde{G}}}\right)_{\epsilon=0} = \frac{k_1^* k_1}{k_1^* Q k_1}\left\{\frac{1}{G_1} - 1\right\} \tag{63}$$

For the usual case of $G_1 \gg 1$ the gain sensitivity of equation (63) is approximately equal to $-\rho$.

Following the alternate approach of equation (48) and considering amplitude and phase perturbations, we may describe k as follows:

$$k_j = k_j^0(1+b_j)\exp\left(\sqrt{-1}\phi_j\right) \tag{64}$$

where k_j^0 is the nominal value of the jth component of k. The amplitude perturbation b_j and the perturbation ϕ_j have zero mean and are assumed to be independent of each other. When k is described by equation (62) the expected output power is

$$\bar{z} = E(k^* R k) = \sum_{i,j} k_i^{0*} k_j^0 D_{ij} R_{ij} \tag{65}$$

where R_{ij} is the jth component of $\mathbf{R}$ and in analogy to equation (50)

$$D_{ij} = \begin{cases} 1 + E(b_j^2), & \text{for } i = j \\ \{1 + E(b_i b_j)\}\, E[\exp\left(\sqrt{-1}\,(\phi_j - \phi_i)\right)], & \text{for } i \neq j \end{cases} \tag{66}$$

Similarly $\widetilde{\widetilde{G}}$ defined in equation (57) becomes

$$\widetilde{\widetilde{G}} = \frac{\sum\limits_{i,j} k_i^{*0} k_j^0 D_{ij} P_{ij}}{\sum\limits_{i,j} k_i^0{}^* k_j^0 D_{ij} Q_{ij}} \tag{67}$$

where P_{ij} and Q_{ij} are i, jth components of $\mathbf{P}$ and $\mathbf{Q}$ respectively. Under the following assumptions equations (63) and (65) may be simplified considerably:

(1) Amplitude perturbations are independent from sensor to sensor and of equal variance, i.e., $E[b_i b_j] = \sigma_b^2 \delta_{ij}$.
(2) Phase perturbations are Gaussian, small, independent from sensor to sensor and of equal variance, i.e., $E[\phi_i \phi_j] = \sigma_\phi^2 \delta_{ij}$, $\sigma_\phi^2 \ll 1$.
(3) Either $|k_i|^2 = |k_j|^2$ for $i, j = 1, \ldots, M$, or $Q_{ii} = P_{ii} = 1$ for $i = 1, \ldots, M$.

Then equation (65) becomes

$$\bar{z} = (1 - \sigma_\phi^2)\mathbf{k}^0 {}^*\mathbf{R}\mathbf{k}^0 + (\sigma_\phi^2 + \sigma_b^2)(\sigma_s^2 + \sigma_n^2)\mathbf{k}^0 {}^*\mathbf{k}^0 \qquad (69)$$

which closely resembles equation (60). Under the same assumptions equation (67) becomes

$$\tilde{\tilde{G}} = \frac{(1 - \sigma_\phi^2)\mathbf{k}^0 {}^*\mathbf{P}\mathbf{k}^0 + (\sigma_\phi^2 + \sigma_b^2)\mathbf{k}^0 {}^*\mathbf{k}^0}{(1 - \sigma_\phi^2)\mathbf{k}^0 {}^*\mathbf{Q}\mathbf{k}^0 + (\sigma_\phi^2 + \sigma_b^2)\mathbf{k}^0 {}^*\mathbf{k}^0} \qquad (70)$$

which closely resembles equation (62).

Finally let us note that when $\mathbf{m}$ is perturbed as in equation (48) and $\mathbf{k}$ is independently perturbed as in equation (64),

$$\frac{E[|\mathbf{k}^*\mathbf{m}|^2]}{E[\mathbf{k}^*\mathbf{Q}\mathbf{k}]} = \frac{\sum k_i^0 {}^*k_j^0 m_i^0 m_j^0 {}^*D_{ij}C_{ij}}{\sum k_i^0 {}^*k_j^0 D_{ij}Q_{ij}} \qquad (71)$$

(d) Summary

Table I presents a summary of some of the more important sensitivity results derived in this Section.

TABLE I. Summary of senstivity results.

Assumptions	Gain Sensitivity	Output Sensitivity
$G = \mathbf{k}^*\mathbf{P}\mathbf{k}/\mathbf{k}^*\mathbf{Q}\mathbf{k}$ $z = \mathbf{k}^*\mathbf{R}\mathbf{k}$	$\left(\dfrac{dG/d\epsilon}{G}\right)_{\epsilon=0}$	$\left(\dfrac{dz/d\epsilon}{z}\right)_{\epsilon=0}$
Noise Variations		
$\mathbf{Q} = (1 - \epsilon)\mathbf{Q}_1 + \epsilon\mathbf{Q}_2$	$1 - (G_1/G_2)$ (29)	$[(G_1/G_2) - 1]/(1 + (S/N)_0)$ (31)
For $\mathbf{Q}_2 = \mathbf{I}$	$1 - \rho$ (33)	$(\rho - 1)/(1 + (S/N)_0)$
Signal Variations		
$\mathbf{P} = (1 - \epsilon)\mathbf{P}_1 + \epsilon\mathbf{P}_2$	$(G_2/G_1) - 1$ (39)	$[(G_2/G_1) - 1](S/N)_0/$ $[1 + (S/N)_0]$ (41)
For $\mathbf{P}_1 = \mathbf{m}\mathbf{m}^*, \mathbf{P}_2 = \mathbf{I}$	$(1/G_w) - 1$ (43)	$[(1/G_w) - 1](S/N)_0/[1$ $+ (S/N)_0]$
Steering Variations	$\tilde{\tilde{G}}$ for G	$\bar{z}$ for z
$\mathbf{k} = \sqrt{(1 - \epsilon)}\mathbf{k}_1 + \sqrt{\epsilon}\eta$ $E(\eta\eta^*) = \mathbf{W}$, trace $\mathbf{W} =$ $\mathbf{k}^*\mathbf{k}$	$\dfrac{\mathrm{trace}(\mathbf{P}\mathbf{W}/G_1) - \mathrm{trace}(\mathbf{Q}\mathbf{W})}{\mathbf{k}_1^\dagger\mathbf{Q}\mathbf{k}_1}$ (59)	$\dfrac{\mathrm{trace}(\mathbf{R}\mathbf{W})}{\mathbf{k}_1^\dagger\mathbf{R}\mathbf{k}_1} - 1$ (56)
For $\mathbf{W} = (\mathbf{k}^*\mathbf{k}/M)\mathbf{I}$	$-\rho\{1 - (1/G_1)\} \approx -\rho$ (63)	$\rho\left\{\dfrac{1 + (S/N)_{\mathrm{in}}}{1 + (S/N)_0}\right\} - 1$ (61)

4. Interference Rejection and Signal Suppression

The problem of designing an array to be insensitive to a point source of noise has received considerable attention in the literature [1, 2, 8, 20]. In the notation of this paper, the goal is to reject a component of the noise field of the form of equation (11).

The gain of any beamformer against such an interference is

$$G_d = |k^*m|^2 / |k^*d|^2 = \cos^2\gamma / \cos^2\theta \tag{72}$$

where θ is the generalized angle between the steering vector k and the interference direction vector d. The quantity $\cos^2\theta$ is obtained by replacing m in equation (13) by d.

Perhaps, the conceptually simplest approach to interference rejection is to use a steering vector which is orthogonal to d so that $\cos^2\theta = 0$. Such a beamformer completely nulls out the unwanted interference. The quality of the null will be degraded if either k or d suffers random perturbations. Because of the symmetry of the expression

$$\cos^2\theta = |k^*d|^2 / (k^*k)(d^*d) \tag{73}$$

perturbations in k or d have similar effects on the null. Perturbations in k also affect the signal response while perturbations in d do not. The quantity $E(|k^*d|^2)$, when d suffers random perturbations, may be obtained directly from the results, equations (47) or (49) presented for $E(|k^*m|^2)$, when m underwent random perturbations. In the simplest case of independent perturbations from sensor to sensor

$$E[\cos^2\theta] = \sigma^2 / M \tag{74}$$

where σ^2 is equal to β in equation (47) or the combined variance of the amplitude and phase perturbations in equation (53). In this situation the expected gain is

$$E(G_d) = G_w / \sigma^2 \tag{75}$$

Thus if G_w is small, nulling becomes impractical.

The null-steering beamformer of Anderson [1, 2] uses the following steering vector:†

$$k^* = m^*[I - dd^*/M] \tag{76}$$

For this steering vector the white noise gain G_w is given by the following equation:

$$G_w = M[1 - |m^*d|^2 / M^2] = M[1 - \cos^2\mu] = M\sin^2\mu \tag{77}$$

---

† The matrix $[I - dd^*/M]$ is a projection operator which passes only the component of m which is orthogonal to d.

where μ is the generalized angle between the signal direction vector and the interference direction vector. Thus, the expected gain (75) against the interference will be low when the interference and signal are closely spaced such as when the interference is within the main lobe of a beam steered conventionally in the signal direction.

When an interference is added to an existing noise field the noise matrix becomes

$$\tilde{Q} = \sigma_1^2 Q_1 + \sigma_d^2 dd^* = (\sigma_1^2 + \sigma_d^2)Q \tag{78}$$

The optimum steering vector is

$$k^* = m^* Q^{-1} = [1 + (\sigma_d^2/\sigma_1^2)]m^* \left[Q_1^{-1} - \frac{Q_1^{-1} dd^* Q_1^{-1}}{d^* Q_1^{-1} d + \sigma_1^2/\sigma_d^2} \right] \tag{79}$$

Then

$$k^* d = [1 + (\sigma_d^2/\sigma_1^2)]m^* Q_1^{-1} d \left[1 - \frac{1}{1 + [d^* Q_1^{-1} d\sigma_d^2/\sigma_1^2]^{-1}} \right] \tag{80}$$

and

$$k^* m = [1 + (\sigma_d^2/\sigma_1^2)]m^* Q_1^{-1} m$$
$$\times \left\{ 1 - \left[\frac{|m^* Q_1^{-1} d|^2}{(m^* Q_1^{-1} m)(d^* Q_1^{-1} d)} \right] \left(\frac{1}{1 + [d^* Q_1^{-1} d\sigma_d^2/\sigma_1^2]^{-1}} \right) \right\} \tag{81}$$

The gain of the "optimum" beamformer is also given by equation (81).

In order to interpret equations (80) and (81) it is useful to consider the meaning of various quantities which appear in these equations:

σ_d^2/σ_1^2—interference to noise ratio at the input $(I/N)_{in}$.

$m^* Q_1^{-1} = k_1^*$—optimum steering vector in the absence of interference.

$d^* Q_1^{-1} d\sigma_d^2/\sigma_1^2 = (I/N)_{max}$—maximum possible output interference-to-noise ratio, that is, the output interference-to-noise ratio of an optimum beamformer, for Q_1-type noise, steered in the direction of the interference.

$|m^* Q_1^{-1} d|^2 = |k_1^* d|^2$—response in the interference direction of an optimum beamformer, for Q_1-type noise, steered in the signal direction.

$m^* Q_1^{-1} m$—optimum gain in the absence of interference.

The "optimum" beamformer does a tradeoff between nulling the interference and preserving gain against Q_1-type noise. This tradeoff involves the interference-to-noise ratio and the sidelobe level in the interference direction of the optimum beamformer, for Q_1-type noise, steered in the signal direction. When Q_1-type noise is spatially white so that

$Q_1 = I$, the limit of the steering vector (79) for large input interference-to-noise ratios is the Anderson null-steering processor (76) to within a constant of proportionality.

When $(I/N)_{max}$ is small the interference has little effect on the optimum steering vector which remains nearly the same as it would be in the absence of interference. When $(I/N)_{max}$ becomes large, k^*d given by equation (80) becomes small and a null develops in the direction of the interference. The factor

$$\cos^2 (\mu_{Q_1^{-1}}) = |m^*Q_1^{-1}d|^2/(m^*Q_1^{-1}m)(d^*Q_1^{-1}d) \tag{82}$$

appearing in equation (81) must be less than or equal to unity by the Schwarz inequality. It may be interpreted as the cosine squared of the generalized angle between m and d in the linear vector space in which length is defined relative to the metric Q_1^{-1}, so that $m^*Q_1^{-1}m$ is the length squared of m.

The question of signal suppression arises when the steering vector k^* is made proportional to m^*R^{-1} instead of m^*Q^{-1} and R^{-1} contains the signal vector. It is of particular concern since the nulling embodied in the R^{-1} operation usually involves measured quantities and the matching operation involved in m^* usually is based on *a priori* assumptions about the signal direction vector.

Mathematically, signal suppression may be treated as a special case of interference rejection so that the results developed above may be applied directly. In particular, we may look at the effect of the inclusion of signal in the estimator [4]

$$z = (m^*R^{-1}m)^{-1} \tag{83}$$

which is the power output of a beamformer with $k^* = m^*R^{-1}/(m^*R^{-1}m)$. This k provides a minimum variance unbiased linear estimate of the signal [4, 7]. The quantity $z^{-1} = m^*R^{-1}m$ may be obtained by replacing the initial factor $(1 + \sigma_d^2/\sigma_1^2)$ in (81) with $(1/\sigma_1^2)$ and redefining the other terms in equation (81) as follows:

d: actual (measured) signal direction vector
m: assumed (*a priori*) signal direction vector
Q_1: noise only matrix
σ_1^2: input noise level
σ_d^2: input signal level

Mismatch occurs when $m \neq d$. The effect of mismatch can be seen by examining the ratio

$$\frac{(z)_{m \neq d}}{(z)_{m=d}} = \frac{d^*Q_1^{-1}d \left[1 - \dfrac{1}{1 + (d^*Q_1^{-1}d\,\sigma_d^2/\sigma_1^2)^{-1}}\right]}{m^*Q_1^{-1}m \left[1 - \cos^2(\mu_{Q_1^{-1}}) \left(\dfrac{1}{1 + (d^*Q_1^{-1}d\,\sigma_d^2/\sigma_1^2)^{-1}}\right)\right]} \tag{84}$$

There are two distinct effects of mismatch: First is the effect of the factor $(d*Q_1^{-1}d/m*Q_1^{-1}m)$. This effect is the usual effect of mismatch discussed in Section 3 and has nothing to do with the inclusion of the signal in the matrix inversion. Second is what we shall call the anomalous signal suppression caused by the presence of $\cos^2(\mu_{Q_1^{-1}})$ in equation (84) which is a direct result of including the signal in the matrix inversion process. The anomalous signal suppression will be insignificant as long as the quantity $(d*Q_1^{-1}d\sigma_d^2/\sigma_1^2)$ is small, i.e. as long as the signal-to-noise ratio at the output of a perfectly matched optimum beamformer with $k* = d*Q_1^{-1}$ would be small. If this weak signal criterion is not satisfied, the ratio given in equation (82) will be reduced through the inclusion of the signal in the matrix inversion and the mismatch between the actual and assumed signal direction vectors.

If we define $(S/N)_{max}$ as $(d*Q_1^{-1}d\sigma_d^2/\sigma_1^2)$ then equation (84) may be simplified to the following:

$$\frac{(z)_{m \neq d}}{(z)_{m=d}} = \frac{d*Q_1^{-1}d}{m*Q_1^{-1}m} \left\{ \frac{1}{1 + (S/N)_{max} \sin^2(\mu_{Q_1^{-1}})} \right\} \tag{85}$$

where $\sin^2(\cdot) = 1 - \cos^2(\cdot)$. The factor $[1 + (S/N)_{max} \sin^2(\mu_{Q_1^{-1}})]^{-1}$ is a direct measure of the anomalous signal suppression as it affects z.

5. Optimization

There are a number of closely related optimization problems which have been or can be formulated. In this section we shall briefly sketch some of these problems with emphasis on formulations which somehow take sensitivity into consideration.

(a) Beamformers
Maximum gain and minimum variance. The problem of unconstrained array gain maximization is to choose k such that

$$G = k*Pk/k*Qk \tag{7}$$

is maximized. The solution of this problem is known [6] to be choosing k proportional to the eigenvector corresponding to the largest eigenvalue of $(Q^{-1}P)$. When $P = mm*$, the optimum k is proportional to $Q^{-1}m$ which is well known [7, 10, 17] and easily shown by direct application of the Schwarz inequality [9].

The relationship between maximizing the gain and minimizing variance under a constraint on signal response lies in that maximizing G is equivalent to minimizing the denominator of equation (7) subject of a constraint on the numerator. Since the gain in equation (7) is invariant to a scaling of k, the imposition of a constraint of the form $k*m = 1$ has no effect on G and simply determines the constant of proportionality. Thus $k = Q^{-1}m/(m^*Q^{-1}m)$ both maximizes G and satisfies the constraint $k^*m = 1$. The related problem of minimizing k^*Rk subject to the same constraint leads to $k = R^{-1}m/(m^*R^{-1}m)$.

Maximum gain subject to sensitivity constraint. As shown in Section 3 the generalized super-gain ratio ρ and the gain against white noise G_w play key roles in the sensitivity of beamformers when the perturbations are independent from sensor to sensor. The problem of maximizing the gain (7) subject to a constraint on

$$G_w = k^*Pk/k^*k \tag{86}$$

is most easily formulated as that of finding the k which minimizes $1/G$ with a constraint on $1/G_w$. That is minimizing

$$1/G + \lambda/G_w = k^*(Q + \lambda I)k/k^*Pk \tag{87}$$

where λ is a Lagrange multiplier. Finding the k which minimizes equation (87) is equivalent to finding the k which maximizes the reciprocal of (87), which in turn is an eigenvalue problem of the type discussed above. The optimum k is chosen to be proportional to the eigenvector corresponding to the largest eigenvalue of $[Q + \lambda I]^{-1}P$. This problem was first addressed by Gilbert and Morgan [12] with $P = mm^*$ so that the optimum k is equal to $[Q + \lambda I]^{-1}m$.

A related problem is to maximize G subject to a constraint on ρ. In this case we maximize

$$G + \lambda\rho = k^*[P + \lambda I]k/k^*Qk \tag{88}$$

Again we have a similar eigenvalue problem. Uzsoky and Solymar [24] examined this problem for the case of isotropic noise $Q = Q_0$. For the formulation of equation (88) to be consistent ρ must be specified within its allowable range given by equation (16). From equation (86) we see that the effect of adding the constraint on ρ is equivalent to that of adding a spatially white (spatially incoherent) component to the signal field.

Lo, Lee and Lee [15] review a number of optimization problems and present a numerical approach to the solution of the more difficult problem of maximizing G subject to a constraint on the Q-factor. They maximize

$$(|k^*m|^2/k^*Qk) + (\lambda k^*k/k^*Q_0 k)$$

All of these optimization problems are insensitive to a scaling of k. Hence the addition of a linear constraint such as $k^*m = a \neq 0$ is handled by simply scaling k.

Maximization of expected quantities. A somewhat different approach to the problem of sensitivity is to take the type and anticipated size of perturbations into account before maximizing G. Thus perturbation terms are included in the ratio to be maximized. A special case of this approach has been taken by Cheng and Tseng [6] who maximize equation (71) and present numerical results for a linear endfire array of eight dipoles. Again the magnitude of k remains free so that an additional constraint of the form $k*m = a \neq 0$ can be handled by scaling k.

Multiple linear constraints. In the preceding discussion a single linear constraint could usually be handled by using the degree of freedom of the magnitude of k which was left undetermined in the process of gain maximization. An exception arises when the constraint is a null of the form $k*d = 0$. This situation may be treated as a special case of the more general problem of multiple linear constraints. Multiple constraints may be used to reduce sensitivity to signal perturbations by keeping the gain relatively constant over a range of signal perturbations. A similar technique may be used to control sidelobes in a specific neighborhood of directions. The problem can be formulated as that of minimizing the output power z subject to the constraint $H*k = g$. Each row vector of the constraint matrix $H*$ imposes a constraint of the form $h_i^* k = g_i$. Thus, the constraint matrix $H*$ has a row for each constraint. The total number of rows must be less than the number of sensors or the problem will be overspecified. Using a Lagrange multiplier vector λ^* we may minimize

$$z = k*Rk + \lambda*[H*k - g] + [k*H - g*]\lambda \qquad (89)$$

Completing the square [7] yields

$$z = [k* + \lambda*H*R^{-1}]R[R^{-1}H\lambda + k] - \lambda*H*R^{-1}H\lambda - \lambda*g - g*\lambda \quad (90)$$

Since k appears only in the initial quadratic term of equation (90), the solution is obviously to make that term equal to zero by choosing

$$k = -R^{-1}H\lambda \qquad (91)$$

Using the constraint $H*k = g$ to eliminate λ, finally yields

$$k = R^{-1}H[H*R^{-1}H]^{-1}g \qquad (92)$$

For this choice of k the output power z becomes

$$z = g*[H*R^{-1}H]^{-1}g \qquad (93)$$

The use of an estimate for R^{-1} in equation (93) is a generalization of the estimator equation (83). One approach to maintaining signal response is to have each row vector of $H*$ be a steering vector in the neighborhood of the steering direction m and to let g be a vector with the number one as each component.

(b) General array processor
So far our discussion has centred on beamformers which have the structure of Fig. 1. A more general array processor structure is illustrated in Fig. 2. In the more general processor there are multiple outputs obtained by a

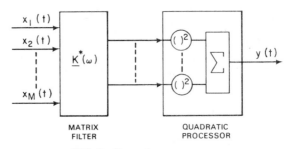

FIG. 2. General array processor.

matrix filtering of the input signals. In the processor of Fig. 2, K^* is a matrix. While the processor of Fig. 1 could be followed by a simple square-law detector and averager, the matrix filter K^* may be followed by a more general quadratic processor. It is known [7, 18] that the structure of Fig. 1 is only optimum when P is a simple dyad. Since the effect of perturbations is to destroy the dyad nature of the expectation of P we are naturally led to the consideration of the more general processor.

For example, suppose that the signal and noise were both completely incoherent from sensor to sensor, i.e., $P = Q = I$. Then all beamformers of the type of Fig. 1 will provide no gain ($G = 1$) while an incoherent combination of the sensor outputs can provide a gain of $\sqrt{M}$.

A processor which forms multiple closely spaced beams and averages output power across these beams is a special case of the general array processor.

In the general array processor the output power is defined as

$$z = \text{trace}(K^*RK) = E(y) \tag{94}$$

In order to be able to properly account for the potential of incoherent gain, we define a gain in terms of the detection index at the output of a general quadratic processor [7].

$$G = \frac{E(y \mid S + N) - E(y \mid N)}{\{E(y^2 \mid N) - E^2(y \mid N)\}^{1/2}} \left(\frac{\sigma_n^2}{\sigma_s^2}\right) \tag{95}$$

When the noise is Gaussian [7], equation (95) may be written as

$$G = \text{trace}[K^*PK] / \{\text{trace}[(K^*QK)^2]\}^{1/2} \tag{96}$$

Notice that equation (96) reduces to (7) when K is a column vector.

The problem of maximizing equation (96) has been solved [7, 9] by direct application of the following Schwarz inequality:

$$| \text{trace}(A^*B) |^2 \leqslant \text{trace}(A^*A) \, \text{trace}(B^*B) \qquad (97)$$

The optimum choice of K is

$$K = cQ^{-1}A \qquad (98)$$

where c is an arbitrary scalar constant of proportionality and $P = AA^*$. The matrix A, assumed for convenience to be of full rank, will have M rows and r columns where r is the rank of the matrix P. The maximum value of gain (96) is

$$G = \{\text{trace}[(PQ^{-1})^2]\}^{1/2} \qquad (99)$$

This maximum gain cannot be achieved by any K with less than r columns.

The problem of maximizing gain (96) determines K to within a scalar multiple. Thus, constraints of the form $\text{trace}(H^*K) = a \neq 0$ may be handled by a simple scaling as before.

The problem of minimizing $\text{trace}(K^*RK)$ subject to multiple linear constraints of the form $H^*K = L$ is readily handled by completing the square similar to the way it was done in equation (89). For details, see the development in equations (28) to (31) of [7]. The solution is

$$K = R^{-1}H[H^*R^{-1}H]^{-1}L \qquad (100)$$

for this value of K,

$$K^*RK = L^*[H^*R^{-1}H]^{-1}L \qquad (101)$$

Equations (100) and (101) are generalizations of (92) and (93). A further generalization of the estimator (83) is therefore

$$z = \text{trace} \, \{L^*[H^*R^{-1}H]^{-1}L\} \qquad (102)$$

(c) Implementation

Most of the work [5, 13, 21, 26] in adaptive beamforming has dealt with unconstrained optimization problems so that there remains much unbroken ground in the field of constrained optimization.

Some exceptions do exist. The work in antenna array optimization [6, 12, 22, 24] has been concerned with sensitivities but not concerned with appropriate algorithms for on-line adaptation. The problem of multiple linear constraints leads naturally to stochastic versions of gradient-projection type algorithms [11, 19]. This approach has also been suggested [27] for non-linear constraints. However, slow convergence may be anticipated.

6. Conclusion

General sensitivity measures have been developed for the cases of perturbations to the signal field, the noise field and the steering vector. These sensitivity measures may be used to test the practicality of particular processors. It was found that the white noise gain and the generalized supergain ratio are key parameters in determining the various sensitivities when the perturbations are independent from sensor to sensor.

The problem of anomalous signal suppression through the inclusion of the signal in the matrix inversion and subsequent mismatch has been treated as a special case of interference rejection. A simple expression for this anomalous signal suppression has been presented. The effect of signal suppression on the output power can only be significant if the signal-to-noise ratio would be large at the output of an optimum beamformer steered perfectly in the signal direction.

Various beamformer optimization problems have been considered. Constraining ρ was found to be equivalent to adding a spatially incoherent component to the signal field prior to unconstrained optimization. Constraining G_w was found to be equivalent to adding a spatially white component to the noise field.

In many important situations the simple beamformer structure is not optimum. A more general array processing configuration has been presented which provides for incoherent combination of the outputs of a number of simple beamformers. Optimization problems associated with this processor have been solved providing generalization of the results for simple beamformers.

REFERENCES

1. Anderson, V. C. (1969). DICANNE, a realizable adaptive process. *J. Acoust. Soc. Am.* **41**, (2), 398–405.
2. Anderson, V. C. and Rudnick, P. (1969). Rejection of a coherent arrival at an array. *J. Acoust. Soc. Am.* **41**, (2), 406–410.
3. Bryn, F. (1962). Optimum signal processing of three-dimensional arrays operating on Gaussian signals and noise. *J. Acoust. Soc. Am.* **34**, (3), 289–297.
4. Capon, J. (1969). High-resolution frequency-wavenumber spectrum analysis. *Proc. IEEE* **57**, 1048–1418.
5. Chang, J. H. and Tuteur, F. B. (1971). A new class of adaptive array processors. *J. Acoust. Soc. Am.* **49**, (3), 639–649.
6. Cheng, D. K. and Tseng, F.-I. (1968). Optimum spatial processing in a noisy environment for arbitrary antenna arrays subject to random errors. *IEEE Trans. Antennas Propagation* **AP-16**, 164–171.
7. Cox, H. (1968). "Interrelated Problems in Estimation and Detection I & II", Proceedings of NATO Advanced Study Institute on Signal Processing with Emphasis on Underwater Acoustics, Enschede, the Netherlands.
8. Cox, H. (1969). "Array Processing Against Interference", Proceedings of the Symposium on Information Processing, Purdue University 453–464.
9. Cox, H. (1969). Optimum arrays and the Schwartz inequality. *J. Acoust. Soc. Am.* **45**, (1), 228–232.

10. Edelblute, D. J., Fisk, J. M. and Kinnison, G. L. (1967). Criteria for optimum-signal-detection theory for arrays. *J. Acoust. Soc. Am.* **41**, (1).

11. Frost, O. L. (1970). Adaptive least squares optimization subject to linear equality constraints. Ph.D. Thesis, Stanford University.

12. Gilbert, E. N. and Morgan, S. P. (1955). Optimum design of directive antenna arrays subject to random variations. *Bell System Tech, J.* **34**, 637–663.

13. Griffiths, L. J. (1969). A simple adaptive algorithm for real-time processing in antenna arrays. *Proc. IEEE* **57**, (10), 1696–1704.

14. Lewis, J. B. and Schultheiss, P. M. (1971). Optimum and conventional detection using a linear array. *J. Acoust. Soc. Am.* **49**, (4), 1083–1091.

15. Lo, Y. T., Lee, S. W. and Lee, Q. H. (1966). Optimization of directivity and signal-to-noise ratio of an arbitrary antenna array. *Proc. IEEE* **54**, 1033–1045.

16. McDonough, R. N. (1972). Degraded performance of nonlinear array processors in the presence of data modeling errors. *J. Acoust. Soc. Am.* **51**, (4), 1186–1193.

17. Mermoz, H. (1964). "Filtrage Adapté et Utilization Optimale d'une Antenne", Proc. NATO Advanced Study Institute on Signal Processing with Emphasis on Underwater Acoustics, Grenoble. France.

18. Middleton, D. and Groginski, H. (1965). Detection of random acoustic signals by receiver with distributed elements. *J. Acoust. Soc. Am.* **38**, (5).

19. Owsley, N. L. (1973). A recent trend in adaptive spatial processing for sensor arrays: constrainted adaptation. *In* "Signal Processing". Proceedings of NATO Advanced Study Institute on Signal Processing. Academic Press, London and New York.

20. Schultheiss, P. M. (1968). Passive sonar detection in the presence of interference. *J. Acoust. Soc. Am.* **43**, (3).

21. Shor, S. W. W. (1966). Adaptive technique to discriminate against coherent noise in a narrow-band system. *J. Acoust. Soc. Am.* **39**, (1).

22. Taylor, T. T. (1955). Design of line-source antennas for narrow beamwidth and low sidelobes. *IRE Trans. Antennas Propagation* **AP-3**, 16.

23. Uzkov, A. I. (1946). An approach to the problem of optimum directive antennae design. *C. R. (Dokl) Acad. Sci. URSS* **53**, 35–38.

24. Uzsoky, M. and Solymár, L. (1956). Theory of superdirective linear arrays. *Acta Phys. Sci. Hungar.* **6**, 185–205.

25. Vanderkulk, W. (1963). Optimum processing of acoustic arrays. *J. Bri. IRE* **26**, 285–292.

26. Widrow, B., Mantey, P. E., Griffiths, L. J., and Goode, B. B. (1967). Adaptive antenna system. *Proc. IEEE* **55**, (12), 2143–2159.

27. Winkler, L. P. and Schwartz, M. (1972). Adaptive nonlinear optimization of the signal-to-noise ratio of an array subject to a constraint. *J. Acoust. Soc. Am.* **52**, (1), 39–51.

DISCUSSION

T. D. Plemons: In a practical situation we never know the matrix Q exactly. Rather, we must estimate Q based on a finite number of samples. Perhaps we could assume that our estimate consists of the real Q matrix plus a perturbation. Could we use the analysis of your paper to find out the effect of not knowing Q exactly?

Answer: There are some statistical results due to N. R. Goodman which pertain directly to the question you raise, that of using estimates in place of true cross-spectral matrices. The difficulty in using the results of my paper for this problem is that the nature of the perturbation matrix is unknown. We can easily develop a lower bound on the gain which depends only on the "size" of the perturbations. Suppose that the true Q is given by equation (28) where Q_1 is the normalized estimate and Q_2 is normalized perturbation matrix. Then using equation (25) we may write the true gain as follows:

$$G = \{[(1 - \alpha)/G_1] + [\alpha(k^*Q_2 k)/(k^*Pk)]\}^{-1} \tag{103}$$

By the Schwarz inequality equation (97)

$$k^*Q_2 k \leqslant k^*k \sqrt{\text{trace}(Q_2^2)} \tag{104}$$

Substituting from equation (104) into (103) yields the desired result

$$G \geqslant \{[(1 - \alpha)/G_1] + [\alpha\sqrt{\text{trace}(Q_2^2)}/G_w]\}^{-1} \tag{105}$$

Recall that $\sqrt{\text{trace}(Q_2^2)}$ is the usual norm of the hermitian matrix Q_2. Thus equation (105) provides a simple lower bound on the true gain in terms of the norm of the perturbation matrix and the gain against white noise G_w.

N. L. Owsley: Could you extend this analysis to treat arrays which have dead hydrophones?

Answer: Yes, but you must be very careful how you apply the results. If your beamformer is based exclusively on *a priori* information, the effect of a dead hydrophone is to reduce the magnitude of the corresponding component of k to zero. The general expressions for gain and output spectral level still apply but the results for "small" perturbations would not. However, if your beamformer is based on estimating the noise matrix through the array, an undetected dead hydrophone would badly bias the estimate. If there were no electronic noise on the output of this dead hydrophone, an "optimum" beamformer could shut down the other "noisy" sensors and listen only on the apparently quiet dead sensor. Constraining G_w to be greater than unity in the optimization procedure could prevent such a drastic result. In practice, some tests on the data could be performed to avoid these difficulties.

N. L. Owsley: Can you handle the situation in which the noise power level is tapered, that is the noise power level is different at different sensors in the array?

Answer: Yes, the results of this paper apply directly to the situation you describe. It is interesting to note that when the beamformer is a pure phasing or time delay so that $|k_i|^2$ is the same for all sensors, the output power and the gain do not depend on how the noise is distributed among the sensors. Of course it would be better to take advantage of the situation by putting greater emphasis on quieter portions of the array

P. H. Moose: Could you explain how the gain could increase if the signal became more incoherent?

Answer: I doubt that this occurs in practice. A pathological example is the "super gain" situation in which the steering vector and the signal direction vector are made nearly orthogonal resulting in extremely low gain against spatially white noise. The low gain against spatially white noise means a large response to an incoherent signal component. It is hard to imagine a physical situation in which signal perturbations would occur without the noise field also developing a non-isotropic component.

We can easily develop an upper bound on the gain in the presence of signal perturbations using the same approach which led to (105). Suppose that P is given by (38). Then (35) becomes

$$G = (1 - \beta)G_1 + \beta(k^*P_2 k)/(k^*Qk) \tag{106}$$

Then using the Schwarz inequality (97) we obtain

$$G \leqslant (1 - \beta)G_1 + \beta\rho\sqrt{\text{trace}(P_2^2)} \tag{107}$$

Equation (107) provides a simple upper bound on the gain in terms of the norm of the signal perturbation matrix $\mathbf{P}_2$ and the generalized super-gain ratio ρ. The generalized super-gain ratio in turn must satisfy (16).

N. L. Owsley: Do you see any hope of bringing the effects of quantized data into this?

Answer: I have not done that. Of course, it is a much more difficult problem. Perhaps analogous results are obtainable in special cases. It would be an interesting problem to pursue.

Adaptive Processing: Time-varying Parameters

LOREN W. NOLTE

Duke University, Durham, North Carolina, U.S.A.

1. Introduction

In many signal detection situations the signal is described by a set of parameters (e.g. amplitude, phase, waveshape) which are not precisely known to the observer. In this paper we consider signal detection problems in which parameters of either or both the signal and the noise are allowed to vary with time in an uncertain manner. The time variations are accounted for by assuming that the parameters can be modelled as a Markov stochastic process.

There are a number of approaches to this type of problem, both formal and *ad hoc*. Let us briefly consider two approaches. One popular method for attacking such problems is to form some estimate of the unknown parameters and insert these estimates in a likelihood ratio expression as if they were indeed the true parametric values. A formalized version of this approach is known as the generalized likelihood ratio philosophy and is widely applied [1, 2]. Some recent work by Jaarsma [3], however, has shown that such estimates-considered-exact may differ greatly from that set of parameters which should be inserted in the generalized likelihood ratio of the observation. This stems from the fact that the *a priori* knowledge is not properly accounted for.

Another approach to signal detection is the Bayes likelihood ratio philosophy. An early paper which clearly delineated application of this approach of decision theory to signal processing was that of Peterson, Birdsall and Fox [4]. The Bayes likelihood ratio philosophy is born of the conjecture that in most conceivable situations, the detector designer has some knowledge beyond absolute uncertainty about the unknown signal parameters. Bayesian philosophy directs that he account for this *a priori* knowledge on the parameter set θ and incorporate it in the detector design. This is accomplished by summarizing this knowledge in a probability density function $p(\theta)$. That this knowledge can be so summarized is the main tenet of classical Bayesian philosophy.

Let us next recall in a little more detail both the generalized likelihood

647

ratio (GLR) test and the Bayesian likelihood ratio. We will consider the latter first. The Bayes likelihood ratio is:

$$l(\bar{\mathbf{X}}_k) = \int l(\bar{\mathbf{X}}_k \mid \theta) p(\theta)\, d\theta \qquad (1)$$

and the hypothesis test is:

$$l(\bar{\mathbf{X}}_k) \underset{H_0}{\overset{H_1}{\gtrless}} \gamma \qquad (2)$$

where

$$l(\bar{\mathbf{X}}_k \mid \theta) = \frac{p(\bar{\mathbf{X}}_k \mid \theta, H_1)}{p(\bar{\mathbf{X}}_k \mid H_0)} \qquad (3)$$

is the conditional objective likelihood ratio, $\bar{\mathbf{X}}_k = (x_1, x_2, \ldots, x_k)$ is the vector observation, and H_1 and H_0 denote the signal plus noise and the noise hypotheses respectively. The decision threshold γ can be determined as a function of error costs and *a priori* regarding H_1 and H_0. This approach results in detectors that are optimum for minimizing the risk, minimizing error, or for that matter for most any criteria which prefers correct decisions over incorrect ones. A feature of this approach is that the processor is unstructured; that is, the physical form of the detector is not postulated beforehand but is evolved freely within the restriction of the criterion of optimality.

A test which is frequently simpler to formulate and often intuitively tempt-ing is the classical GLR test. One major feature of this test is that it explicitly incorporates estimates of the unknown signal parameters. This test compares the conditional objective likelihood ratio to a decision threshold with some estimate θ of the unknown parameter θ inserted. The GLR test is:

$$l(\bar{\mathbf{X}}_k \mid \hat{\theta}) \underset{H_0}{\overset{H_1}{\gtrless}} \gamma' \qquad (4)$$

The designer accounts for his uncertainty about θ by considering some estimate $\hat{\theta}$ as the true value. Frequently he chooses an estimator that is "well known" and which has good properties as an estimator *per se*. However, it frequently happens that his selection of an estimator is not particularly good for making good decisions as to the true hypothesis.

The basic approach we will use in this paper is the Bayes likelihood ratio one. We will show how the optimum detector for time-varying parameters can then be structured so that it resembles processors that are designed from adaptive, sequential, and generalized likelihood ratio or estimate-and-plug approaches. This approach has been used for several problems of fixed but unknown parameters [3, 5–10]. This is in contrast to starting with an adaptive, sequential or estimate-and-plug structure and hoping for optimum detection performance by proper adjustment of the variables involved.

2. Optimum Detector for Time-Varying Parameters

Let us next consider the Bayes likelihood ratio for problems with time-varying parameters. We will then evolve the structure into two forms; an adaptive one and an estimate-and-plug structure.

The two decisions possible are H_1, signal plus noise and H_0, noise alone. The total observation $\bar{X}_k$ is divided into small subobservations x_k and the value of the parameter in the ith interval associated with the jth hypothesis is denoted by θ_i^j. The parameter value is assumed to remain constant during any one subinterval but changes from interval to interval. The optimum detector forms the likelihood ratio, which is

$$l(\bar{X}_k) = \frac{p(\bar{X}_k \mid H_1)}{p(\bar{X}_k \mid H_0)} \tag{5}$$

where $p(\bar{X}_k \mid H_1)$ is the probability density function of $\bar{X}_k$ conditional to H_1 and $p(\bar{X}_k \mid H_0)$ is the probability density function of $\bar{X}_k$ conditional to H_0. Since

$$p(\bar{X}_k \mid H_j) = p(\bar{X}_{k-1} \mid H_j)p(x_k \mid \bar{X}_{k-1}, H_j); \quad j = 0, 1 \tag{6}$$

we can implement equation (5) sequentially by

$$l(\bar{X}_k) = l(\bar{X}_{k-1})l(x_k \mid \bar{X}_{k-1}) \tag{7}$$

where

$$l(x_k \mid \bar{X}_{k-1}) = \frac{p(x_k \mid \bar{X}_{k-1}, H_1)}{p(x_k \mid \bar{X}_{k-1}, H_0)} \tag{8}$$

and

$$l(\bar{X}_0) = 1 \tag{9}$$

In this form, the detector design problem becomes that of putting $l(x_k \mid \bar{X}_{k-1})$ in a practical form.

We next allow for Markov related time-varying parameters in the ith interval associated with the signal hypothesis θ_i^1 and the noise hypothesis θ_i^0. Note that θ_i^j could be a vector of parameters. Information needed for the detector design is summarized in the functions $p(\theta_i^j \mid \theta_{i-1}^j)$ and $p(\theta_1^j \mid \bar{X}_0)$ where j is 1 or 0 depending on the hypothesis. We assume that

$$p(\theta_i^j \mid \theta_{i-1}^j, \ldots, \theta_1^j) = p(\theta_i^j \mid \theta_{i-1}^j)$$

and hence $p(\theta_i^j \mid \theta_{i-1}^j)$ describes the Markov dependence of θ_i^j on the previous value θ_{i-1}^j The probability density function $p(\theta_1^j \mid \bar{X}_0)$ summarizes the designers initial or *a priori* knowledge about the parameter θ_1^j. Hatsell has shown, after a somewhat lengthy derivation, that the optimum detector design equations in sequential or adaptive form for a doubly composite hypothesis situation are [11]:

The likelihood ratio: sequential formulation

$$l(\bar{\mathbf{X}}_k) = l(\bar{\mathbf{X}}_{k-1})l(x_k \mid \bar{\mathbf{X}}_{k-1}) \tag{10}$$

$$l(x_k \mid \bar{\mathbf{X}}_{k-1}) = \frac{p(x_k \mid \bar{\mathbf{X}}_{k-1}, H_1)}{p(x_k \mid \bar{\mathbf{X}}_{k-1}, H_0)} \tag{11}$$

$$l(\bar{\mathbf{X}}_0) = 1 \tag{12}$$

where the numerator and denominator for equation (11) are obtained as follows:

Tracking detector design equations: double composite hypothesis

$$p(\theta^j_{k+1} \mid \bar{\mathbf{X}}_k) = \int p(\theta^j_k \mid \bar{\mathbf{X}}_{k-1}) \frac{p(x_k \mid \theta^j_k)}{p(x_k \mid \bar{\mathbf{X}}_{k-1}, H_j)} p(\theta^j_{k+1} \mid \theta^j_k) \, d\theta^j_k \tag{13}$$

$$p(x_k \mid \bar{\mathbf{X}}_{k-1}, H_j) = \int p(x_k \mid \theta^j_k)p(\theta^j_k \mid \bar{\mathbf{X}}_{k-1}) \, d\theta^j_k \tag{14}$$

For the single composite hypothesis situation, with $\theta^j_{k+1} = \alpha^j$ and $\theta^j_k = \beta^j$, the sequential or adaptive design equations are [11]:

Tracking detector design equations: single composite hypothesis case

$$p(\alpha \mid \bar{\mathbf{X}}_k) = p(\beta \mid \bar{\mathbf{X}}_{k-1}) \frac{l(x_k \mid \beta)}{l(x_k \mid \bar{\mathbf{X}}_{k-1})} p(\alpha \mid \beta) \, d\beta \tag{15}$$

$$l(x_k \mid \bar{\mathbf{X}}_{k-1}) = \int l(x_k \mid \beta)p(\beta \mid \bar{\mathbf{X}}_{k-1}) \, d\beta \tag{16}$$

$$l(x_k \mid \beta) = \frac{p(x_k \mid \beta, H_1)}{p(x_k \mid H_0)} \tag{17}$$

$$l(x_k \mid \bar{\mathbf{X}}_{k-1}) = \frac{p(x_k \mid \bar{\mathbf{X}}_{k-1}, H_1)}{p(x_k \mid H_0)} \tag{18}$$

A problem that has received considerable interest is that of detecting recurrent transient phenomena in background noise [5-8, 12]. To facilitate the inclusion of time uncertainty for these problems, an indicator parameter t_k is introduced where the value of t_i during the ith interval is defined by:

$$t_i = \begin{cases} 1 & \text{when transient is present in } i\text{th interval} \\ 0 & \text{when transient is absent in } i\text{th interval} \end{cases}$$

The basic sequential or adaptive design equations for the optimum detector for both time-varying parameters and time uncertainty of transient occurrences are [11]:

Tracking detector design equations: inclusion of time uncertainty

(General design equations for time uncertainty)

$$p(\alpha^j, t_{k+1} \mid \overline{X}_k) = \iint_{\beta_j \; t_k} p(\beta^j, t_k \mid \overline{X}_{k-1}) \frac{p(x_k \mid \beta^j, t_k)}{p(x_k \mid \overline{X}_{k-1}, H_j)}. \tag{19}$$

$$p(\alpha^j, t_{k+1} \mid \beta^j, t_k) \, d\beta^j \, dt_k$$

$$p(x_k \mid \overline{X}_{k-1}, H_j) = \int_{\beta_j} \int_{t_k} p(x_k \mid \beta^j, t_k) p(\beta^j, t_k \mid \overline{X}_{k-1}) \, dt_k \, d\beta^j \tag{20}$$

A case which occurs frequently enough to warrant a specialized set of design equations is that in which the indicator parameter t_i is independent of the other parameters and, in addition, is independent of t_{i-1}; i.e.

$$p(\alpha^j, t_{k+1} \mid \beta^j, t_k) = p(t_{k+1} \mid t_k) p(\alpha^j \mid \beta^j) \tag{21}$$

$$p(\alpha^j, t_{k+1} \mid \overline{X}_k) = p(t_{k+1}) p(\alpha^j \mid X_k) \tag{22}$$

and

$$p(t_{k+1} \mid t_k) = p(t_{k+1}) \tag{23}$$

Applying these equations with the assumption that

$$\int_{\xi-\epsilon}^{\xi+\epsilon} p(t_i) \, dt_i = \begin{cases} \nu & \text{for } \xi = 1 \\ (1 - \nu) & \text{for } \xi = 0 \end{cases} \qquad \epsilon > 0 \tag{24}$$

yields the following set of design equations [11]:

Design equations for signal duty-factor ν

$$p(\alpha^j \mid \overline{X}_k) = \int_{\beta^j} \frac{p(x_k \mid \beta^j)}{p(x_k \mid \overline{X}_{k-1}, H_j)} p(\beta^j \mid \overline{X}_{k-1}) p(\alpha^j \mid \beta^j) \, d\beta^j \tag{25}$$

$$p(x_k \mid \overline{X}_{k-1}, H_j) = \int_{\beta^j} p(x_k \mid \beta^j) p(\beta^j \mid \overline{X}_{k-1}) \, d\beta^j \tag{26}$$

$$p(x_k \mid \beta^j) = \nu p(x_k \mid \beta^j, t_k = 1) + (1 - \nu) p(x_k \mid \beta^j, t_k = 0) \tag{27}$$

Note that $\nu = 1$ implies periodic signals and $\nu < 1$ implies synchronously recurrent signals. The latter, for fixed but unknown parameters, has been considered in greater detail before [5–7].

The above design equations for optimum detection of signals in noise with time-varying parameters resemble those of an adaptive structure. Let us next show how the optimum detector structure can be evolved into one which resembles an estimate-and-plug or generalized likelihood ratio approach. Consider the single composite hypothesis situation. We can write the log likelihood ratio as

$$\ln l(\bar{\mathbf{X}}_k) = \ln l(\bar{\mathbf{X}}_{k-1}) + \ln \frac{l(x_k \mid \bar{\mathbf{X}}_{k-1})}{l(x_k \mid \beta)} + \ln l(x_k \mid \beta) \tag{28}$$

Using an argument analogous to that used by Jaarsma [3] we note that the left-hand side of this equation is independent of the parameter β and hence the right-hand side must also be. A convenient value to use for β is that value for which

$$\ln \frac{l(x_k \mid \bar{\mathbf{X}}_{k-1})}{l(x_k \mid \beta)} = 0 \tag{29}$$

since this will give us the desired estimate-and-plug structure while preserving the optimality of the log likelihood ratio for detection purposes. Let us denote this value of β as our estimate $\hat{\beta}$, denote equation (29) as our estimator, and plug this estimate $\hat{\beta}$ into the last term on the right-hand side of equation (28). The resultant design equations for this sequential estimate-and-plug structure are:

Tracking detector design equations: single composite hypothesis estimate-and-plug structure

$$\ln l(\bar{\mathbf{X}}_k) = \ln l(\bar{\mathbf{X}}_{k-1}) + \ln l(x_k \mid \hat{\beta}) \tag{30}$$

where the estimate $\hat{\beta}$ is obtained by solving

$$\frac{l(x_k \mid \bar{\mathbf{X}}_{k-1})}{l(x_k \mid \hat{\beta})} = 1 \tag{31}$$

Figure 1 shows a block diagram of this processor structure. We note that in general the estimator will not be a "well-known" estimator.

It is interesting to note that the optimum detector for time-varying parameters can also be formulated as a two-stage processor in a manner similar to that demonstrated by Birdsall [13] for the fixed but unknown parameter case. In this two-stage approach, the first stage is designed using some rather arbitrary a priori description $p_1(\theta_1^j \mid \bar{\mathbf{X}}_0)$ which reproduces and which is chosen primarily from the point of view of

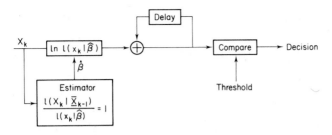

FIG. 1. Estimate-and-plug implementation of optimum Detector.

mathematical simplicity. Frequently, the first stage contains the bulk of the processing. The desired *a priori* description $p_2(\theta_1^j \mid \overline{X}_0)$ [e.g., that suggested by some physical situation] is then accounted for in a relatively simple second stage which constitutes a multiplicative, updated modifier for the output of the first stage. The details of this two-stage processor for time-varying parameters have been worked out by Hatsell [11].

3. Optimum Detection Performance: Specific Cases

Several specific optimum detectors, in adaptive and sequential forms, along with their detection performance on the ROC have been worked out (11). Cases that have been considered are time-varying signal amplitude, phase and waveshape [11] as well as time-varying noise power [14]. In addition the detection of synchronously recurrent transient signals in time-varying noise power has been considered. We only have space here to briefly mention several of these situations along with a few ROC's.

In Fig. 2 the ROC is shown for the optimum detector for a signal with time-varying amplitude. The Markov transitional statistics assumed are

$$p(\theta_k \mid \theta_{k-1}) = \frac{1}{2\pi a} \exp\left[-\frac{(\theta_k - \theta_{k-1})^2}{2a^2}\right] \qquad (32)$$

where "a" is a measure of the dispersion of θ_k with respect to θ_{k-1}. The ROC's are presented for $S_k/k\sigma_n^2 = 1$, for $k = 10$, where S_k is the average signal energy in a k-length observation interval and σ_n^2 is the variance of the noise. As the dispersion of the next value of amplitude increases, the detection performance decreases markedly.

Figure 3 shows the ROC for a synchronous-recurrent transient signal with time-varying waveshape (eight possible 3-bit waveforms). The parameter A on the graph is a measure of the rapidity with which the waveshape changes. Large A corresponds to little change in waveshape from one occurrence to the next and small A corresponds to rapid change.

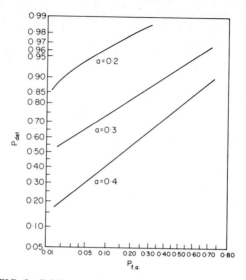

FIG. 2. ROC curves for amplitude tracking detector.

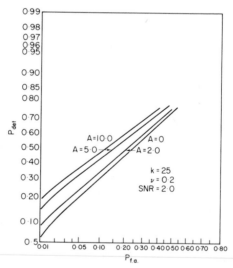

FIG. 3. ROC curves for a synchronous signal with time-varying waveshape in Gaussian noise. SNR = 2.

While the recurrence times of the transient are uncertain, $\nu = 0.2$ corresponds to an average occurrence of transients about 20% of the time.

Figure 4 shows the ROC for a synchronous-recurrent transient signal in time-varying noise power. The parameter c/α is the square root of

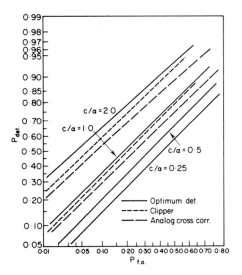

FIG. 4. ROC curves for synchronous signals in time-varying noise power.

the signal energy divided by the average noise for one sequential observation. Here $k = 10$ and $\nu = 0.2$. Several suboptimum processors are shown for comparison. The optimum detector structure for this problem can be put in the form of a limiter with a nonlinear transition region [14].

4. Summary

We have shown in this paper how optimum detectors for signals and noise with time-varying parameteters can be approached within the framework of the Bayesian likelihood ratio and have seen that we may structure these processors adaptively, sequentially, or estimate-and-plug. This approach provides both a consistent design as well as performance measure on the ROC. While a detailed treatment of a number of examples, including application to optimal array detection [15] is contained in the references, a brief summary of several cases has been presented here. (This research was sponsored by ONR, Acoustics Programs.)

REFERENCES

1. Van Trees, H. L. (1968). "Detection, Estimation, and Modulation Theory, Part I", p. 92, John Wiley and Sons, Inc., New York.
2. Helstrom, C. W. (1960). "Statistical Theory of Signal Detection", Pergamon Press.
3. Jaarsma, D. (1970). "The Theory of Signal Detectability: Bayesian Philosophy, Classical Statistics, and The Composite Hypothesis", Cooley Electronics Laboratory Technical Report No. 200, The University of Michigan, Ann Arbor, Michigan.

4. Peterson, W. W., Birdsall, T. G. and Fox, W. C. (1954). The theory of signal detectability", *IRE Trans. Inf. Theory* **PGIT-4**, 171–211.

5. Nolte, L. W. (1965). "Adaptive Realizations of Optimum Detectors for Synchronous and Sporadic Recurrent Signals in Noise", Technical Report No. 163, Cooley Electronics Laboratory, The University of Michigan, Ann Arbor, Michigan.

6. Nolte, L. W. (1968). Adaptive optimum detection: synchronous-recurrent transients. *J. Acoust. Soc. Am.* **44**, No. 1, 224–239.

7. Nolte, L. W. (1967). An adaptive realization of the optimum receiver for a sporadically recurrent waveform in noise. *IEEE Trans. Inf. Theory* **IT-3**, No. 2, 308–311.

8. Fralick, S. C. (1965). "Learning to Recognize Patterns Without a Teacher", Systems Theory Laboratory Technical Report No. 6103–10, Stanford Electronics Laboratories, Stanford, California.

9. Abramson, N. and Braverman, D. (1962). Learning to recognize patterns in a random environment. *IRE Trans. Inf. Theory* **IT-8**, 58–63.

10. Spooner, R. L. (1969). "The Theory of Signal Detectability: Extension to the Double Composite Hypothesis Situation", Technical Report No. 196, Cooley Electronics Laboratory, The University of Michigan, Ann Arbor, Michigan.

11. Hatsell, C. P. (1970). "Optimum Tracking Detectors", Technical Report No. 5, Adaptive Signal Detection Laboratory, Dept. of Electrical Engineering, Duke University, Durham, North Carolina, U.S.A.

12. Birdsall, T. G. (1964). "Likelihood Ratio and Optimum Adaptive Detection", NATO Advanced Study Institute, Grenoble, France.

13. Birdsall, T. G. (1968). "Adaptive Detection Receivers and Reproducing Densities", NATO Advanced Study Institute, Enschede, Netherlands.

14. Hatsell, C. P. and Nolte, L. W. (1971). On transient and periodic signal detection in time-varying noise power. *IEEE Trans. Aerospace Electronic Systems*, **AES-7**, No. 6.

15. Gallop, M. A. (1971). "Adaptive Optimum Array Detectors", Technical Report No. 7, Adaptive Signal Detection Laboratory, Duke University, Durham, North Carolina, U.S.A.

What is Optimality for an Adaptive Detection System?

G. VEZZOSI

Laboratoire d'Etude des Phénomènes Aléatoires,
Universite de Paris-Sud Orsay, France

1. Introduction

The noise available at the output terminals of a sonar antenna is characterized by the high temporal instability of its statistical properties. This instability results from the multiplicity of interfering noise sources, and from the random fluctuations of their respective levels. The phenomenom is neither known nor predictable. The classical techniques of signal processing, which deal with stationary or non-stationary models whose properties are known in advance, cannot afford to handle this type of situation. We need adaptive processing techniques. By adaptive processing technique is meant a method which fits automatically the surrounding noise evolution by means of a continuous control and learning from this evolution to the extent that such a learning is possible.

(a) Conditions for use of adaptive techniques
Not all kinds of noise are equally suited for adaptive processing techniques; because these methods rest upon a continuous evaluation of the noise characteristics, and such an evaluation requires a certain number of independent or weakly correlated trials from the same population, the only noises which can be processed are those which possess the local stationarity property [1]. We shall suppose this condition satisfied in the following study.

(b) Aim and scope of this paper
The literature on adaptive processing techniques is very large and diverse. Under apparently identical experimental circumstances, a great number of solutions are proposed which are based sometimes on direct estimation methods of the unknown elements and sometimes upon algorithmic methods; they frequently omit to specify the precise operation that should be done on the input data and are always presented as candidates for optimality.

Such a diversity of opinions may raise doubt upon the concept of optimality itself. It is the investigation of the detection problems of this concept with which we shall be mainly concerned in this paper. We shall show that there is not one, but two types of optimality, which lead to different structures:

(i) the mean absolute optimality, derived from likelihood ratio;
(ii) the optimality in Neyman–Pearson conditional sense, i.e. optimality with a constraint on the conditional false alarm probability.

Several examples and numerical results will be given.

2. Statement and Modelization of the Real Problem

The real problem is to detect in a continuous and adaptive manner one or more signals of unknown arrival time. The duration of each signal is a constant which is known in advance. Generally this constant is rather small. The use of sequential procedures is, therefore, superfluous and the problem becomes that of being informed of the presence of each signal at each epoch where it may end.

The shape of each signal can be completely known, or parametric with random factors, or else can be completely random, in which case the signal is a trajectory of a random process.

The noise stems from the superposition of a great number of independent microscopical contributions. This property suffices to ensure the Gaussian character (at least in the first approximation). Moreover, the macroscopical factors which determine its statistical properties vary in a random manner with time, but the time constant of this evolution is much greater than that of the noise fluctuations. This property suffices to ensure a certain local stationarity.

The signal is additive and independent of the noise. In particular, the presence of a signal does not alter the statistical properties of the noise.

(a) *Modelization.*

From these statements, it emerges that the problems arising from the adaptive and continuous detection of one or several signals are threefold. One can distinguish schematically:

(a) The questions related to the diversity of the signals to be detected: we are normally in a multiple hypothesis context.

(b) The questions related to the continuous character of the detection: the successive intervals from which the operator has to decide about the presence of a signal are not disjoined. Accordingly, there exist unavoidable correlations between the successive decisions. An optimal procedure should take these correlations into account.

(c) The questions related to the non-stationarity of the noise and to adaptivity: How conciliate are the adaptive and detection processes? What should be the receiver memory? At the basis of every decision there is an observation which is viewed as an element of a sample space governed by some probabilistic laws. But what shall we call a sample in the present situation, and what about the probabilistic law, i.e. the noise model?

The first and second questions are not peculiar to adaptive methods. They prevail even when the noise characteristics are known and immutable. Generally one disregards them, because they lead to a formidable receiver complexity and a negligible performance gain [2].

Therefore, we shall suppose that there is only one type of signal to be detected, and that this signal is alone and ends at the time t_0 considered (whatever t_0 from $-\infty$ to $+\infty$).

On the other hand, the third question is peculiar to adaptive problems. For the choice of an observation interval, two principles command attention.

Firstly, to define optimality, we require a model of the noise. But we know nothing whatever about the mechanisms of the noise instability. All that we know is that the noise is stationary on time intervals of length equal to a local stationarity range τ_s. Therefore we are forced to limit the observation interval to a length near to τ_s.

Secondly, a well-known principle in detection theory prescribes the use of the whole signal duration T to detect in the best conditions. Therefore the observation interval must be at least equal to T.

At this stage we see a possible contradiction. The two above principles are conflicting when $\tau_s < T$. In this case, we must use an interval of range T, but we are unable to specify a model of the noise in this interval, and we cannot define what is meant by an "optimal procedure".

Therefore we shall suppose that $\tau_s > T$. The noise is then stationary in each observation interval, and this simplification allows a theoretical treatment.

3. The Two Types of Optimality for Adaptive Detection Systems

Let us consider some observation interval $[t_0 - \tau_s, t_0]$, from which the presence of a signal localized in the sub-interval $[t_0 - T, t_0]$ is to be tested. Without loss of generality, we may shift this interval into $[0, \tau_s]$, and consider that the signal is present everywhere in $(0, \tau_s)$, but is equal to zero from 0 to $\tau_s - T$.

The observation $x(t)$ can be written in the form:

$$x(t) = n(t) + \eta s(t) \quad t \in (0, \tau_s)$$

where $n(t)$ is a sample from a stationary random process $N(t)$, whose correlation function $r(\tau)$ is unknown, and η and $s(t)$ are merely parameters, which represent respectively the indicator of the signal presence and the signal shape itself.

The problem is to decide whether $\eta = 0(H_0)$ or $\eta = 1(H_1)$. Thus we are looking for a decision rule, which can be viewed as a partition of the observation space $(\mathscr{X}, \mathscr{A})$ in two disjoined and complementary regions $\mathscr{X}_0$ and $\mathscr{X}_1$.

This is a classical composite hypothesis testing problem, where the parameter set can be put conveniently into the form:

$$(\Theta, \mathscr{C}) = (H, \mathscr{H}) \times (P, \mathscr{R}) \times (\Sigma, \mathscr{S})$$

Each point of the parameter set, i.e. each triplet $(\eta, r(\cdot), s(\cdot))$ defines a possible probability law for the observation. However, it is worth noticing that the three parameters $\eta, r(\cdot), s(\cdot)$ play quite different functions—while η can be considered as a "decisional parameter", because its knowledge allows it to take the decision without error, $r(\cdot)$ and $s(\cdot)$, on the other hand are nuisance parameters, because they tell nothing about the signal presence.

Optimality is defined by reference to the aggregate of the situations that we can encounter.

At each trial the true value of the parameter $\theta = (\eta, r(\cdot), s(\cdot))$ vary. Therefore $\eta, r(\cdot), s(\cdot)$ must be considered as random elements, taking their values in the sets $(H, \mathscr{H})$, $(P, \mathscr{R})$ and $(\Sigma, \mathscr{S})$ respectively, and governed by probabilistic measures $\mu_H, \mu_P,$ and μ_Σ. These random elements are independent because of a previous assumption (noise and signal are independent).

The three measures $\mu_H, \mu_P,$ and μ_Σ are almost unknown (except perhaps for the signal measure μ_Σ) but they exist, and this existence suffices to ensure the possibility of an unambiguous comparison of all strategies (by means, for example, of the mean detection probability versus the mean false alarm probability).

(a) *The absolute mean optimality*
The knowledge of μ_R and μ_Σ reduces the composite hypotheses $H_0(\eta = 0)$ and $H_1(\eta = 1)$ to simple ones. Then, it is well known that the mean absolute optimality is reached by comparing the likelihood ratio (L.R.) to a fixed threshold.

In the present situation the observation $x(t)$ is a function, and the L.R. must be obtained by some limiting operation. This can be carried out either from a temporal discretization, or from a decomposition of the observation space by use of basis functions (Karhunen–Loeve expansion, for example). We shall use the first technique.

Let $\{t_i, i = 1, 2, \ldots\}$ be a dense sequence of points in $(0, \tau_s)$, let $\vec{x} = \{x(t_i), i = 1, \ldots, n\}$, $\vec{S} = \{s(t_i), i = 1, \ldots, n\}$ be the sample values of $x(t)$ and $s(t)$ on the first n points of the sequence $\{t_i\}$, and let

$$R = [r(t_i - t_j)] i, j = 1, \ldots, n$$

be the corresponding correlation matrix.

Let $\mu_P^{(n)}$ and $\mu_\Sigma^{(n)}$ be the restriction of μ_P and μ_Σ on P_n and Σ_n.

$$L(x) = \lim_{n \to \infty} \frac{\int_{P_n} d\mu_P^{(n)} \int_{\Sigma_n} \frac{1}{\sqrt{\det R}} \exp\{-\frac{1}{2}(\vec{x} - \vec{S})^\tau R^{-1}(\vec{x} - \vec{S})\} \, d\mu_\Sigma^{(n)}}{\int_{P_n} \frac{1}{\sqrt{\det R}} \exp -\frac{1}{2}\vec{x}^\tau R^{-1}\vec{x} \, d\mu_P^{(n)}} \tag{1}$$

The L.R. $L(x)$ gives the optimal procedure, but it does not afford control of the conditional false alarm probability. Such a control is obtained by receivers of the following category.

(b) *The optimality in the Neyman–Pearson conditional sense*
For practical purposes it is often desirable to control the false alarm probability under every circumstance [such a property is always required in classical composite hypothesis testing problems [3]; but then the parameter has a fixed unknown value for each possible trial, and prior distributions are meaningless].

Mathematically, the condition is expressed by restricting the class of rivalling receivers to those which satisfy the constraint:

$$\alpha(\mathscr{X}_1|_{r(\cdot)}) \leqslant \alpha, \qquad r \in P \tag{2}$$

It is not obvious that such strategies actually exist, nor that they exist for every value of the size α.

To the extent that the class satisfying (2) is not empty, we can always associate to each of its elements a mean probability detection $\beta(\mathscr{X}_1)$. This possibility allows us to define a total preordering in the space of strategies, and the extremal element for this ordering defines the optimal strategy in the Neyman–Pearson conditional sense.

It is worth noting that there is no general rule to determine this optimal strategy (contrary to the previous optimality, which is based on the L.R. computation). The precise determination must be done in each particular case.

4. The Asymptotic Case

However, there exists a situation where the precise operations carried out by these two optimal receivers are quite simple and can be written explicitly: it is the case where the correlation function $r(\cdot)$ is known at each trial, either because the observation bears enough information to allow a perfect estimation of $r(\cdot)$ under both hypothesis, or because the observer is told the true value of $r(\cdot)$.

This situation is referred as the "asymptotic case", and is discussed in the present and following sections. We suppose throughout the sequel that the conditional detection problem (i.e. the problem where $r(\cdot)$ and $s(\cdot)$ have fixed values for all trials) is non-singular, whatever $r(\cdot) \in P$ and $s(\cdot) \in \Sigma$. Under these conditions the following statements are valid:

(1) The L.R. (1) becomes independent of the prior measure μ_P, and equals the conditional likelihood ratio (this is a well-known result since von Mises—see Neyman [4]).

(2) Because $r(.)$ is known at each trial, it is possible to conceive a receiver which maximizes the conditional probability detection $\beta(\mathscr{X}_1|_{r(\cdot)})$ under the constraint $\alpha(\mathscr{X}_1|_{r(\cdot)}) \leqslant \alpha$ for each function $r(\cdot)$; this receiver computes the conditional L.R., but compares this quantity to a threshold variable at each trial and adjusts to satisfy the constraint, which then reduces to an equality. Because this receiver maximizes the conditional detection probability, it maximizes automatically the mean detection probability, and hence is optimal in the Neyman–Pearson sense. Moreover, this receiver saturates the constraint, and therefore is a constant false-alarm probability (CFAP) receiver.

Thus we see that the two types of optimal receivers both compute the conditional L.R.

$$L(x) = \lim_{n \to \infty} \frac{\displaystyle\int_{\Sigma_n} \exp\{-\tfrac{1}{2}(\vec{x} - \vec{S})^T R^{-1}(\vec{x} - \vec{S})\}\, d\mu_\Sigma^{(n)}}{\exp\{-\tfrac{1}{2}\vec{x}^T R^{-1}\vec{x}\}}$$

$$= \lim_{n \to \infty} \int_{\Sigma_n} \exp\{\vec{x}^T R^{-1}\vec{S} - \tfrac{1}{2}\vec{S}^T R^{-1}\vec{S}\}\, d\mu_\Sigma^{(n)},$$

but differ essentially in the threshold setting rule:

(i) For the type I optimal receiver, the threshold is set to a fixed value;
(ii) For the type II optimal receiver, the threshold is set to a variable value, adjusted according to the desired false alarm probability.

The derivation of the corresponding structures is straightforward: the conditional likelihood ratio is available in many textbooks [2, 5]. One just has to compute the threshold setting rule for the type II receiver. Table I displays the corresponding results for some archetypal situations.

(a) Comments on Table I

Deterministic signal. The observation $x(t)$ is a vectorial process $\{x_i(t), i = 1, 2, \ldots, n\}$, and $r(\tau)$ is a matrix correlation function $\{r_{ij}(\tau); i; j = 1, \ldots, n\}$. The signal is supposed to be identical on the n sensors [this assumption

TABLE I

	Type I receiver (optimal in absolute sense)	Type II receiver (optimal in Neyman-Pearson sense)
Deterministic signal. array of n sensors	$\displaystyle\sum_{1\,i}^{n} \int_0^{\tau_s} x_i(t) g_i(t)\, dt \underset{H_0}{\overset{H_1}{\gtrless}} t_1 + \frac{d^2(r)}{2}$	$\displaystyle\sum_{1\,i}^{n} \int_0^{\tau_s} x_i(t) g_i(t)\, dt \underset{H_0}{\overset{H_1}{\gtrless}} t_2 \times d(r)$
Signal of unknown amplitude. array of n sensors	$\displaystyle\int_0^\infty \exp\left\{ \mu \sum_{1}^{n}\!\int_i \int_0^{\tau_s} x_i(t) g_i(t)\, dt - \frac{\mu^2}{2}\, d^2(r) \right\}$ $\cdots q(\mu)\, d\mu \underset{H_0}{\overset{H_1}{\gtrless}} t_1$	$\displaystyle\sum_{1\,i}^{n} \int_0^{\tau_s} x_i(t) g_i(t)\, dt \underset{H_0}{\overset{H_1}{\gtrless}} t_2 \times d(r)$
Signal of unknown amplitude and phase	$\displaystyle\int_0^\infty \exp\left\{ -\frac{\mu^2}{2}\, d^2(r) \right\} I_0\left\{ \mu \cdot \sqrt{\sum_{1}^{2}\!\int_i \left[\int_0^{\tau_s} x(t) g_i(t)\, dt \right]^2} \right\}$ $\cdots q(\mu)\, d\mu \underset{H_0}{\overset{H_1}{\gtrless}} t_1$	$\displaystyle\sum_{1}^{2}\!\int_i \left[\int_0^{\tau_s} x(t) g_i(t)\, dt \right]^2 \underset{H_1}{\overset{H_0}{\gtrless}} t_2 \times d^2(r)$
Random Gaussian signal	$\displaystyle k(r) \cdot \int_0^{\tau_s}\!\int_0^{\tau_s} x(s) h(s,t) x(t)\, ds\, dt \underset{H_0}{\overset{H_1}{\gtrless}} t_1$	$\displaystyle k(r) \cdot \int_0^{\tau_s}\!\int_0^{\tau_s} x(s) h(s,t)\, ds\, dt \underset{H_0}{\overset{H_1}{\gtrless}} t_2(r)$

does not restrict the generality (1)]. The functions $g_i(t)$ are the solutions of the system of integral equations:

$$\sum_{1j}^{n} \int_{0}^{T_s} r_{ij}(t - s)g_j(s) \, ds = s(t) \quad i = 1, \ldots, n$$

and $d^2(r)$ is the positive functional:

$$d^2(r) = \sum_{1i}^{n} \int g_i(t)s(t) \, dt$$

Generally one solves these equations by letting the observation interval be infinite, and using convolution.

Signal of unknown amplitude. The preceding remarks are still valid. Now $s(t)$ is a normalized version of the signal.

Note that the type I receiver implementation needs the knowledge of the prior law $q(\mu)$ of the signal amplitude. On the other hand, the type II receiver does not require this knowledge. Its threshold variation rule can be achieved by an AGC circuit at the end of the chain.

Signal of unknown amplitude and phase. Now the signal has the form

$$S(t) = \mu[S_1(t) \cos \phi + S_2(t) \sin \phi]$$

where ϕ is a random variable uniformly distributed between 0 and 2Π. The functions $g_i(t)$, $i = 1, 2$, are the solutions of the integral equations:

$$\int_{0}^{T_s} r(t - s)g_i(s) \, ds = S_i(t) \quad i = 1, 2$$

Moreover, we have supposed that the functions $r(\tau)$, $S_1(t)$, $S_2(t)$ are such that

$$\int_{0}^{T_s} g_1(t)S_1(t) \, dt = \int_{0}^{T_s} g_2(t)S_2(t) \, dt = d^2(r)$$

$$\int_{0}^{T_s} g_1(t)S_2(t) \, dt = \int_{0}^{T_s} g_2(t) S_1(t) \, dt = 0$$

These hypotheses are absolutely necessary if we want to obtain the classical envelope detector structure, as it appears on the Table (note that these conditions are fulfilled in case of white noise and orthogonal $S_1(t)$ and $S_2(t)$).

Random Gaussian signal. The integral kernel $h(s, t)$ is the solution of the integral equation:

$$\int_0^{\tau_s} \int_0^{\tau_s} r(s - u)h(u, v)[r(v - t) - \Gamma(v, t)] \, du \, dv = \Gamma(s, t)$$

where $\Gamma(v, t)$ is the covariance function of the signal process. The functional $k(r)$ is the limit:

$$k(r) = \lim_{n \to \infty} \det [R_n [R_n + \Gamma_n]^{-1}]$$

For the type II receiver, the threshold setting rule $t_2(r)$ is not available in explicit form, because of the intractability of the expression of the conditional false alarm probability in this case.

5. Some Numerical Results

The computation of the mean ROC curves is very complex in general. One has to average the conditional ROC curves with respect to the prior distribution of the correlation function $r(\tau)$; this requires a probability measure on an abstract space of correlation functions, and represents a rather formidable task. Under these circumstances, the best that we can do is to suppose that the correlation function is indexed by some finite numbers of real parameters.

We study in this section the simplest case where these parameters are reduced to a random positive multiplicative factor a. Then, the uncertainty on the noise reduces to the value of its mean power, and $r(\tau)$ can be written:

$$r(\tau) = a\rho(\tau), \rho(0) = 1.$$

This leads to an interesting problem which has been studied extensively [6, 7, 8, 9]. By means of a whitening operation and the use of the strong law of large numbers, it is possible to construct a perfect estimation of the unknown mean power a, without any more information than the knowledge of the finite observation $x(t)$, $t \in (0, \tau_s)$, and of the normalized correlation function $\rho(\tau)$.

The following results are drawn from Reference [9], and deal with the case where the signal is known with a possible unknown amplitude factor (analogous results could be settled when the signal is of unknown phase and amplitude, or when it is completely random).

The two receivers are presented in Table II.

Table II

	Type I receiver (optimal in absolute sense)	Type II receiver (optimal in Neyman–Pearson sense)
Deterministic signal	$$\dfrac{\displaystyle\int_0^{\tau_s} x(t)g(t) - \dfrac{d^2}{2}}{a} \underset{H_1}{\overset{H_0}{\gtrless}} t_1$$	$$\dfrac{\displaystyle\int_0^{\tau_s} x(t)g(t)\,dt}{\sqrt{a}} \underset{H_1}{\overset{H_0}{\gtrless}} t_2$$
Signal of unknown amplitude	$$\int_0^{\infty} \exp\left\{ \mu \dfrac{\displaystyle\int_0^{\tau_s} x(t)g(t)\,dt}{a} - \mu^2 \dfrac{d^2}{2a} \right\}$$ $$\ldots q(\mu)\,d\mu \underset{H_1}{\overset{H_0}{\gtrless}} t_1$$	$$\dfrac{\displaystyle\int_0^{\tau_s} x(t)g(t)\,dt}{\sqrt{a}} \underset{H_1}{\overset{H_0}{\gtrless}} t_2$$

It is worth noticing that the CFAP optimal receiver uses a simple AGC circuit, while the absolute optimal receiver uses an over-AGC (when the noise becomes very strong, the test statistic becomes smaller).

(a) Receiver evaluation

When the signal is of unknown amplitude, it is difficult to evaluate the ROC curve of the mean absolute optimal receiver. Therefore, it seems preferable in this case to compute an upper bound of these characteristics. Such an upper bound can be obtained by considering the observer which knows at each trial the amplitude value of the possible signal.

The ROC curves are displayed in Reference [9]. From them emerges the following conclusions:

(i) as one might expect, the type I receiver, which is optimal in the absolute mean sense, ranks above the type II receiver,

(ii) however, this superiority is rather weak, especially at low false alarm probabilities.

Therefore, there is no objection to using the type II receiver in practical situations.

6. The Case Where the Correlation Function is not Perfectly Estimatable

As we have shown previously, the asymptotic case leads to an easy and straightforward problem. Unfortunately, the situation is not so clear when there are uncertainties on the true correlation function of the noise. Rather than bring some vague elements to a general theory, we shall confine ourselves to a problem which, curiously, is completely reducible—the one where the uncertainty on the noise sums up to the value of its mean power.

The previous section dealt with the case where this parameter was perfectly estimatable: however, there are several ways to prevent such a possibility. One can for example add a white noise component to the observation, or one can discretize the observation and consider only a finite number n of samples. We shall use this second method; moreover we shall suppose that the signal is known, with a possible unknown amplitude factor.

Then the observation becomes:

$$\vec{x} = \sqrt{a}\vec{n} + \eta\mu\vec{S}$$

where

$$E\vec{n}\vec{n}^T = |\rho_{ij}| = \rho, \quad \rho_{ii} = 1 \quad i, j = 1, \ldots, n$$

and a is a random variable of law $p(a)$.

The following statements hold:

(i) among all tests $\mathscr{X}_1$ satisfying $\alpha(\mathscr{X}_1 | a) \leqslant \alpha$, the test which maximizes the conditional detection probability $\beta(\mathscr{X}_1 | a, \mu)$ is the Student one;

(ii) this test is CFAP.

Thus the type II optimal receiver is the Student test.

Table III gives the decision rules for the two types of optimal receivers.

The computation of the mean ROC curves for different values of n ($n = 500, 200, 100, 50, 20, 10$) has been done, when the signal is of known amplitude. The curves are not displayed here but the following conclusions hold:

(i) the degradation of the characteristics when n decreases is approximately of the same order for the two types of receivers;

(ii) in the range of law false alarm probabilities (10^{-5}), this degradation becomes perceptible below $n = 100$ (the relative loss in the detection probability for $\alpha = 10^{-6}$ is 10% at $n = 100$, 20% at $n = 50$, 50% at $n = 20$, 80% at $n = 10$).

Table III

	Type I receiver (optimal in absolute sense)	Type II receiver (optimal in Neyman–Pearson sense)
Signal of unknown amplitude	$$\dfrac{\displaystyle\int_0^\infty\int_0^\infty \frac{1}{a^{n/2}}\exp\left\{-\frac{1}{2a}(\vec{x}-\mu\vec{S})^T\rho^{-1}(\vec{x}-\mu\vec{S})\right\}p(a)q(\mu)\,da\,d\mu}{\displaystyle\int_0^\infty \frac{1}{a^{n/2}}\exp-\frac{1}{2a}\vec{x}^T\rho^{-1}\vec{x}\,p(a)\,da} \overset{H_0}{\underset{H_1}{\lessgtr}} t_1$$	$$\frac{\vec{x}^T\rho^{-1}\vec{S}}{\sqrt{\dfrac{1}{n}\vec{x}^T\rho^{-1}\vec{x}}} \overset{H_0}{\underset{H_1}{\lessgtr}} t_2$$

7. Conclusion

This paper has investigated the concept of optimality for adaptive detection systems. We have shown the possibility of two types of optimality: type I, which requires the whole prior information, is an absolute optimality; type II is a constrained optimality.

In some particular cases, it appears that the ROC curves are close together. It would remain to show that this property is quite general.

ACKNOWLEDGEMENT

The author wishes to express his sincere thanks to Dr. H. Mermoz for much helpful and valuable advice and comment.

BIBLIOGRAPHY

1. Mermoz, H. (1971). "Antennes de Détection Optimales et Adaptatives. Théorie et Applications". Collection Technique et Scientifique du CNET.
2. Helstrom, C. W. (1968). "Statistical Theory of Signal Detection," Pergamon Press, 2nd edition.
3. Lehmann, E. L. (1958). "Testing Statistical Hypothesis", J. Wiley & Sons, New York.
4. Neyman, J. (1962). Two breakthroughs in the theory of statistical decision making. *Rev. Int. Stat. Inst.* **30**, 1, 11-27.
5. Van Trees, H. L. (1968). "Detection, Estimation and Modulation Theory", Parts I, and III, J. Wiley & Sons, New York.
6. Spooner, R. L. (1968). On the detection of a known signal in a non Gaussian noise process. *JASA* **44**, 142.
7. Scharf, L. L. and Lytle, D. W. (1971). Signal detection in Gaussian noise of unknown level: An invariance application. *IEEE* **IT-17, (4)**.
8. Picinbono, B. and Vezzosi, G. (1970). Détection d'un signal certain dans un bruit non stationnaire et non gaussian. *Annales des Télécommunications* **25**, 11-12.
9. Vezzosi, G. and Picinbono, B. (1972). Détection d'un signal certain dans un bruit sphériquement invariant. Structure et comparaison de différents récepteurs. *Annales des Télécommunications* **27**, 3-4.

DISCUSSION

T. G. Birdsall: The case when you do not know the noise level but you do know the signal exactly was studied by Spooner in his Thesis on a likelihood ratio basis, and he did not come up to either of the structures you mentioned; the optimal test was more like a classical F-test, and its implementation did not make any use of AGC circuits. *Answer*: Spooner set up this kind of F-test by using a reproducing prior density for the noise level (if A means the noise level, Spooner assumed that $1/A$ has a χ^2 distribution). Such a result is no longer valid when A has an arbitrary probability distribution.

Anyway, the mentioned structures are not the exact optimal ones, but rather have been derived by inferring the asymptotic case and therefore are only approximately optimal. The aim was just to try to legitimatize the so-called estimate and plug method and to show up the necessity of an adaptive threshold setting rule when using this method.

Digital Logarithmic Normalization of Sonar Signals: Serial Processing

C. VAN SCHOONEVELD

Physics Laboratory NDRO,
The Hague, The Netherlands.

1. Introduction

We consider a method for processing the doppler output signals of a tone pulse echo ranging system (Fig. 1). The aim of the method is to remove the strongly time varying character of the reverberation before presenting the signals to the threshold of an alarm extraction device.

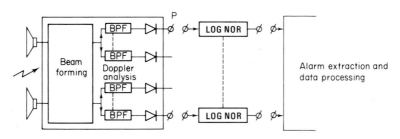

FIG. 1. Block diagram tone pulse receiver.

The sonar receiver consists essentially of beamforming, narrowband filtering and envelope detection. Each of the sonar beams is divided into a number of adjacent doppler channels. In each channel, a target echo must be detected against a joint background of noise and reverberation.

Often, the input reverberation can be regarded as a multitude of overlapping echoes, reflected by densely distributed scatterers. Consequently, the reverberation in a particular doppler channel is a random time function, characterized by a Rayleigh distributed envelope with a gradually decreasing strength (Fig. 2). Typical decreases are 50 to 70 dB in 10 sec, but due to a great and rapid variability in conditions, the precise law of decay cannot be accurately predicted. In addition, the decrease is not necessarily monotonous. Local short duration increases of reverberation are sometimes superimposed on the gradual decrease.

671

Somewhere during the sonar range, the reverberation decreases below the noise level in the channel under consideration. Normally, the noise is stationary over the remainder of the range, although its level is subject to a long-time variability.

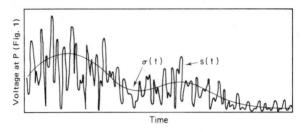

FIG. 2. Typical output of doppler channel.

The extraction of an alarm in one of the channels comes down to testing whether a voltage excursion is "significantly" greater than the background in its immediate vicinity. One way to realize this test is to transform the non-stationary time function (Fig. 2) into a stationary one (i.e. constant first and second order moments with time), which can then be presented to a threshold.

It is the task of the blocks labelled "LOG NOR" in Fig. 1 to execute this transformation. Since each doppler channel is affected in its own particular way by the reverberation, each channel must be treated in its own right. Although this requirement does not preclude the processing in one channel to be partly controlled by the signals in adjacent channels, we shall restrict the present discussion to a process without channel coupling, as depicted in Fig. 1.

2. Logarithmic Normalization

The principle of logarithmic normalization is well known [1, 2]. The signal $s(t)$ (Fig. 2) is assumed to be the product of two components: (1) a rapidly fluctuating but stationary process $f(t)$, and (2) a slowly varying time function $\sigma(t)$ which represents the original signal's time dependent r.m.s. value:

$$s(t) = f(t) \cdot \sigma(t). \tag{1}$$

A target echo, if present, is part of the rapidly fluctuating process $f(t)$. By taking the logarithm of equation (1) we replace the multiplication by an addition:

$$x(t) = r(t) + p(t). \tag{2}$$

For convenience, we use the 20 log convention to obtain the new variables in dB's:

$$x(t) = 20 \log s(t), \quad r(t) = 20 \log f(t), \quad p(t) = 20 \log \sigma(t). \tag{3}$$

The slowly varying non-stationary component $p(t)$ (henceforth referred to as "the trend") is now present as a time-varying mean value, on which the rapidly fluctuating component $r(t)$ is superimposed by addition.

Thus, the trend $p(t)$ can be removed from $x(t)$ with a linear high-pass filter (Fig. 3). Its output is a close approximation of the desired component $r(t)$. There is no need to reconvert $r(t)$ to $f(t)$ for alarm extraction purposes

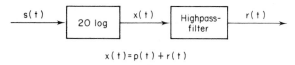

$$x(t) = p(t) + r(t)$$

FIG. 3. LOG NOR block diagram

since comparison of $f(t)$ with a threshold λ is equivalent to comparison of $r(t)$ with a threshold $20 \log \lambda$.

Figure 4 is a recording of the logarithmic signal $x(t)$, obtained experimentally at sea (100 msec tone pulse).

The trend $p(t)$ can be estimated from Fig. 4 by visual inspection. By examining a large number of recordings, a graphical ensemble of trend functions is obtained, which roughly defines their statistical properties.

As for the rapid component $r(t)$, its probability distribution can be obtained by starting from the assumed Rayleigh distribution of the original process $f(t)$ (equation (1)). The standard deviation turns out to be:

$$\overline{[(r - \bar{r})^2]}^{1/2} = 5.55 \text{ dB}. \tag{4}$$

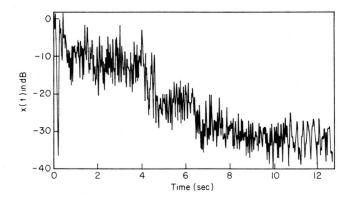

FIG. 4. Experimental recording of $x(t)$.

Starting from this kind of information and, of course, from knowledge of the doppler channel's bandwidth, we shall now consider a suitable high-pass filter.

3. High Speed Digital Processing

The proposed normalization method presupposes a possibility to process the raw sonar signals (Fig. 1) without appreciable non-linear distortion over the full dynamic range. This can be achieved with digital techniques.

Indeed, we shall assume that digital processing has been applied in the first block of Fig. 1 and that the logarithmic doppler signal $x(t)$ is available as a sequence of digital samples. Accordingly, the desired highpass processing of $x(t)$ is also realized digitally.

The LOG NOR operation must be applied on-line and in real time to all doppler channels in the system. Preferably, one hardware unit is used in time multiplex. Evidently, this requirement necessitates restricting the amount of computation to an absolute minimum.

This consideration plays a major role in this paper. Rather than striving for some "optimum" system we shall discuss a simple system which still performs reasonably well.

4. A Recursive Digital Highpass Filter

The sequence of input data to the filter is

$$x(n) = r(n) + p(n); \quad n = 0, 1, 2, \ldots, \tag{6}$$

where $x(n)$ denotes the nth sample. Observe that the input starts at $n = 0$. The $r(n)$ process is the desired rapidly fluctuating signal and $p(n)$ is the smoothly varying trend.

The design of a suitable highpass filter is reduced to the design of a corresponding lowpass filter by adopting the method of Fig. 5. Essentially, the highpass filter is obtained as "one minus a lowpass filter".

At time n, the lowpass filter produces an estimate of the trend value at a previous time $n - L$. Its output is $\hat{P}(n) = \hat{p}(n - L \mid n)$, where n is the actual time and $n - L$ is the validity instant of the estimate. The lowpass

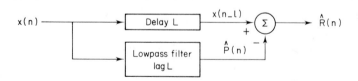

FIG. 5. Block diagram of digital highpass filter.

filter's output is subtracted from the delayed input, $x(n - L)$, to obtain an estimate of the desired component $r(\ )$, which is also valid at time $n - L$:

$$\hat{R}(n) = x(n - L) - \hat{P}(n),$$

where

$$\left.\begin{array}{l}\hat{P}(n) = \hat{p}(n - L \mid n) = \text{estimate of } p(n - L) \\[2mm] \hat{R}(n) = \hat{r}(n - L \mid n) = \text{estimate of } r(n - L)\end{array}\right\}. \tag{7}$$

The estimation lag L is a constant integer. It is introduced because a proper choice of L greatly reduces the estimation errors (see Section 5).

The lowpass filter is designed according to the method of "weighted least squares polynomial curve fitting". Normally, the method is applied to a semi-infinite data sequence [3]. A slight modification allows the application to a finite data sequence, which expands step by step as new signal samples are added. This approach results in a simple recursive algorithm, resembling a Kalman filter.

At time $i = n$ a straight line $\hat{p}(i \mid n)$(1st degree polynomial) is fitted† to the available data $x(0), \ldots, x(i), \ldots, x(n)$:

$$\hat{p}(i \mid n) = \hat{p}_0(n) + (i - n) \cdot \hat{p}_1(n); \quad i = 0, \ldots, n. \tag{8}$$

Figure 6 illustrates the procedure. The coefficients $\hat{p}_0(n), \hat{p}_1(n)$ are determined by requiring that equation (8) is the best fit in a weighted least squares sense:

$$\sum_{i=0}^{n} [x(i) - \hat{p}(i \mid n)]^2 \cdot q^{n-i} = \text{minimum}; 0 < q < 1. \tag{9}$$

The weighting function q^{n-i} (Fig. 6) puts the greatest stress on recent observations. Old observations are exponentially weighted down. The effective length of the weighting window is determined by the weighting function's time constant:

$$N = (1 - q)^{-1}. \tag{10}$$

After solving the coefficients p_0, p_1 from equations (8) and (9), the delayed trend estimate is obtained by extrapolating along the estimated straight line into the past (Fig. 6):

$$\hat{P}(n) = \hat{p}_0(n) - L \cdot \hat{p}_1(n); \quad L = \text{estimation lag}. \tag{11}$$

---

† At time $n = 0$ a proper initialization must be applied since a straight line fit requires at least 2 input data.

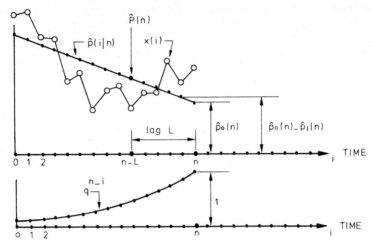

FIG. 6. Weighted least squares fitting of a straight line to the data $x(n)$.

The curve fitting procedure is repeated for each new data point $x(n)$, $x(n+1)$, ... A recursive algorithm can be obtained in the following way.

Substitution of equation (8) into equation (9), differentiation to $\hat{p}_0, \hat{p}_1$ and equating to zero yields two linear equations for $\hat{p}_0, \hat{p}_1$:

$$\begin{pmatrix} m_0(n) & m_1(n) \\ m_1(n) & m_2(n) \end{pmatrix} \begin{pmatrix} \hat{p}_0(n) \\ \hat{p}_1(n) \end{pmatrix} = \begin{pmatrix} y_0(n) \\ y_1(n) \end{pmatrix} \tag{12}$$

where

$$\left. \begin{aligned} m_j(n) &= (-1)^j \sum_{k=0}^{n} k^j q^k \\ y_j(n) &= (-1)^j \sum_{k=0}^{n} k^j q^k \cdot x(n-k) \end{aligned} \right\} . \tag{13}$$

Solution of equation 12 gives

$$\begin{pmatrix} \hat{p}_0(n) \\ \hat{p}_1(n) \end{pmatrix} = M^{-1}(n) \left(\begin{pmatrix} m_2(n) & -m_1(n) \\ -m_1(n) & m_0(n) \end{pmatrix} \begin{pmatrix} y_0(n) \\ y_1(n) \end{pmatrix} \right) . \tag{14}$$

where

$$M(n) = m_0(n)m_2(n) - m_1^2(n)$$

Equation (14) essentially solves the problem. The quantities $y_0(n), y_1(n)$ are obtained by convolution of the input $x(i)$ with an impulse response $(-1)^j k^j q^k$. These convolutions can be cast into recursive form. Next, the coefficients $\hat{p}_0(n), \hat{p}_1(n)$ are obtained as linear combinations of $y_0(n)$,

$y_1(n)$. Observe that the coefficients in the linear combinations are time dependent. Consequently, our filter is a time varying filter.

However, the system can be further simplified by applying a coupling procedure due to Levine [4]. Equation (12) is written down for time $n - 1$:

$$\begin{pmatrix} m_0(n-1) & m_1(n-1) \\ m_1(n-1) & m_2(n-1) \end{pmatrix} \begin{pmatrix} \hat{p}_0(n-1) \\ \hat{p}_1(n-1) \end{pmatrix} = \begin{pmatrix} y_0(n-1) \\ y_1(n-1) \end{pmatrix}. \tag{15}$$

Next, we express $y_j(n-1)$, $m_j(n-1)$ in $y_j(n)$, $m_j(n)$ by means of relations derived from equation (13).

$$\left. \begin{aligned} y_j(n-1) &= q^{-1} \cdot \sum_{k=0}^{j} \binom{j}{k} \cdot [y_k(n) - x(n) \cdot 0^k] \\ m_j(n-1) &= q^{-1} \cdot \sum_{k=0}^{j} \binom{j}{k} \cdot [m_k(n) - 0^k] \end{aligned} \right\} \tag{16}$$

where $0^k = 1$ if $k = 0$ and $0^k = 0$ if $k \neq 0$.

Equation (16) is substituted into equation (15) and rows and columns are manipulated until the matrix regains the same form as in equations (12):

$$\begin{pmatrix} m_0(n) & m_1(n) \\ m_1(n) & m_2(n) \end{pmatrix} \begin{pmatrix} \hat{p}_0(n-1) + \hat{p}_1(n-1) \\ \hat{p}_1(n-1) \end{pmatrix} =$$

$$= \begin{pmatrix} y_0(n) - x(n) + \hat{p}_0(n-1) + \hat{p}_1(n-1) \\ y_1(n) \end{pmatrix}. \tag{17}$$

Subtraction of equation (17) from equation (12) produces two linear equations from which $\hat{p}_0(n)$, $\hat{p}_1(n)$ are solved. The result is:

$$\left. \begin{aligned} \begin{pmatrix} \hat{p}_0(n) \\ \hat{p}_1(n) \end{pmatrix} &= \begin{pmatrix} \hat{p}'_0(n) \\ \hat{p}'_1(n) \end{pmatrix} + \begin{pmatrix} \alpha(n) \\ \beta(n) \end{pmatrix} \cdot [x(n) - \hat{p}'_0(n)] \\ \text{where} \\ \begin{pmatrix} \hat{p}'_0(n) \\ \hat{p}'_1(n) \end{pmatrix} &= \begin{pmatrix} 1 & 1 \\ 0 & 1 \end{pmatrix} \begin{pmatrix} \hat{p}_0(n-1) \\ \hat{p}_1(n-1) \end{pmatrix} \quad \text{and} \quad \begin{pmatrix} \alpha(n) \\ \beta(n) \end{pmatrix} = M^{-1}(n) \cdot \begin{pmatrix} m_2(n) \\ -m_1(n) \end{pmatrix} \end{aligned} \right\} . \tag{18}$$

The delayed estimate $\hat{P}(n)$ is given by equation (11). Combining equations (18), (11) and (7) we obtain the block diagram of Fig. 7.

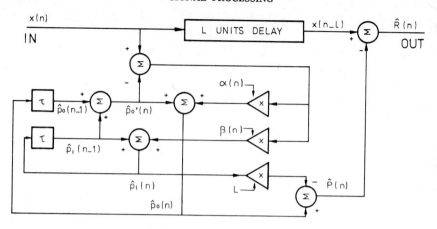

FIG. 7. Digital highpass filter.

The filter has a well known structure. The quantities $\hat{p}_0'(n)$, $\hat{p}_1'(n)$ are predictions of the expected trend, derived from the previous estimates $\hat{p}_0(n-1)$, $\hat{p}_1(n-1)$ by 1-step extrapolation along a straight line. The updated estimates $\hat{p}_0(n)$, $\hat{p}_1(n)$ are obtained by correcting the predictions with amounts proportional to the difference $[x(n) - \hat{p}_0'(n)]$ between actual observation and prediction. The proportionality constants, $\alpha(n)$ and $\beta(n)$, are time dependent. They can be precomputed by means of equation (13) and can be stored in a read-only memory.

The filter's time varying character is easily understood. For small n-values, the number of input data contributing to the output increases with each new input $x(\ )$. Accordingly, the filter operates in an "expanding memory mode". When time increases (moderate n-values) the old input data are gradually weighted down (Fig. 6). After sufficient time has elapsed (large n) the entire weighting function is effectively at work and the filter now operates in a "constant memory mode". The time variability has disappeared and the filter has become stationary. Concurrently, the constants $\alpha(n)$, $\beta(n)$ in equation (18) have converged to a limit value.

The highpass filter of Fig. 7 requires only 3 multiplications per new data point. The amount of storage is determined by the value we select for L. This in turn follows from a consideration of the estimation errors.

5. Estimation Errors

In general, the estimate $\hat{R}(n)$ will differ from its desired value $r(n-L)$. The difference is the estimation error. Since the highpass output $\hat{R}(n)$ is obtained by subtraction of $\hat{P}(n)$ from $x(n-L)$ (equation (7)), the error in $\hat{R}(n)$ equals the error in $\hat{P}(n)$, apart from the sign. Thus, we shall concentrate on the errors in $\hat{P}(n)$.

When both input components are present, i.e. when $x(n) = p(n) + r(n)$, we have

$$\hat{P}(n) \equiv \hat{p}(n - L \mid n) = \{p(n - L) + a(n, L)\} + b(n, L), \qquad (19)$$

where

$$
\begin{aligned}
p(n - L) + a(n, L) &= \text{the L.P. filter's response to an input } x(n) = p(n) \\
b(n, L) &= \text{the L.P. filter's response to an input } x(n) = r(n)
\end{aligned}
$$
$$(20)$$

$a(n, L)$ is the "tracking error" and $b(n, L)$ is referred to as the "noise error". Ideally, both should be zero.

(a) Noise error
The input consists only of the rapid component: $x(n) = r(n)$ (equation (20)). We assume that $r(n)$ has zero mean and is a white process:

$$\overline{r(n)} = 0 \quad \overline{r(n)r(k)} = \sigma^2 \cdot \delta_{nk}. \qquad (21)$$

Actually, the $r(n)$ process is definitely non-white when considered over its full bandwidth. Consequently, a significant correlation between neighbouring samples does exist. An intuitive justification for the white approximation lies in the fact that the lowpass filter is designed to suppress the rapidly varying process $r(n)$. Hence, its transfer function differs from zero only for very low frequencies. On this limited frequency interval the $r(n)$ process may safely be assumed to be white. The only requirement is that we correct the numerical value of σ^2 in equation (21) when replacing the coloured process by a white one.

Under the assumption of equation (21), the estimates $\hat{p}_0, \hat{p}_1$ also have zero mean. Their second order moments are found by substitution of equation (21) in equation (13) and by application of equation (14):

$$
\begin{pmatrix}
\overline{\hat{p}_0^2(n)} \\[2mm]
\overline{\hat{p}_0(n)\hat{p}_1(n)} \\[2mm]
\overline{\hat{p}_1^2(n)}
\end{pmatrix}
= \sigma^2 \cdot (s_0 s_2 - s_1^2)^{-2} \cdot
\begin{pmatrix}
s_2^2 S_0 - 2s_1 s_2 S_1 + s_1^2 S_2 \\[2mm]
s_1 s_2 S_0 - (s_0 s_2 + s_1^2) S_1 + s_0 s_1 S_2 \\[2mm]
s_1^2 S_0 - 2s_0 s_1 S_1 + s_0^2 S_2
\end{pmatrix}
$$
$$(22)$$

where

$$
\left.
\begin{aligned}
s_j &\equiv z_j(n, q) \\[2mm]
S_j &\equiv z_j(n, q^2)
\end{aligned}
\right\}
\quad \text{and} \quad z_j(n, t) = \sum_{k=0}^{n} k^j t^k. \qquad (23)
$$

The sums of equation (23) are easily calculated by using the recurrent relation

$$z_{j+1}(n, t) = t \cdot \frac{d}{dt} z_j(n, t).$$

Observe that s_j, S_j, and, therefore, the second order moments of equation (22), are n-dependent.

Equation (22) shows that $\hat{p}_0(n)$ and $\hat{p}_1(n)$ are correlated. This is the first reason for the introduction of the estimation lag L. By forming a linear combination of $\hat{p}_0$ and $\hat{p}_1$, as in equation (11), its variance can be reduced with respect to the variance of $\hat{p}_0$. Provided that L is positive and is restricted to integer values, the linear combination has the significance of an extrapolation into the past along the estimated straight line $\hat{p}(i \mid n)$ (equation (8)).

Substitution of equation (22) into equation (11) shows that the noise error variance depends quadratically on the lag L:

$$\overline{a^2(n, L)} \equiv \overline{\hat{P}^2(n)} = \overline{\hat{p}_0^2(n)} - \overline{2\hat{p}_0(n)\hat{p}_1(n)} \cdot L + \overline{\hat{p}_1^2(n)} \cdot L^2. \qquad (24)$$

Consequently, a careful choice of the lag L reduces the noise error. Ideally, the lag should be time-dependent since the second order moments in equation (24) are time-dependent. In our simple system, however, we select a constant L-value which gives reasonable overall performance.

Equation (24) is the output variance of the lowpass filter when the input consists of an $r(n)$ process with variance σ^2. When equation (24) is computed under the condition $\sigma^2 = 1$, it assumes the significance of a Variance Reduction Factor. Its square root is the Standard Deviation Reduction Factor:

$$\text{SDRF} = [\overline{a^2(n, L)}]^{1/2}, \quad \text{with } \sigma^2 = 1. \qquad (25)$$

The SDRF is shown in Fig. 8 (upper two plots) as a function of the lag L and the time n, respectively. The plots were computed for $q = 0,975$. The corresponding effective duration of the weighting function (equation (10)) is $N = 40$. Both plots show the influence of the lag on the SDRF. In particular shortly after starting the filter procedure (e.g. $n = 50$), a proper lag can greatly reduce the filter's noise output.

(b) Tracking error

In the present section we assume the input to consist of trend alone: $x(n) = p(n)$. Combining equations (11) and (14) we get

$$\hat{P}(n) = M^{-1} \cdot [(m_2 + Lm_1)y_0 - (m_1 + Lm_0)y_1] \qquad (26)$$

where M, m_j and y_j are time-dependent (equations (13) and (14)).

We shall derive the tracking error under the condition that the input trend $p(n)$ is a polynomial of arbitrary degree G:

$$p(i) = \sum_{g=0}^{G} p_g \cdot (i - n)^g; \quad i = 0, \ldots, n. \tag{27}$$

The actual trend value at the validity instant of estimation, $i = n - L$, is given by:

$$p(n - L) = \sum_{g=0}^{G} (-1)^g \cdot p_g \cdot L^g. \tag{28}$$

Next, we derive an expression for the estimate at time $i = n - L$. Substitution of equation (27), via equation (13), into equation (26) yields:

$$\hat{P}(n) = (s_0 s_2 - s_1^2)^{-1} \cdot \sum_{g=0}^{G} (-1)^g \cdot p_g \cdot \{(s_2 s_g - s_1 s_{g+1}) - L(s_1 s_g - s_0 s_{g+1})\}. \tag{29}$$

Subtraction of equation (28) from (29) gives the tracking error $b(n, L)$:

$$b(n, L) = \sum_{g=0}^{G} p_g \cdot \epsilon_g(n, L)$$

where

$$\epsilon_g(n, L) = (-1)^g \cdot A(n) \cdot \{B_g(n) - L \cdot C_g(n)\} - L^g \tag{30}$$

and

$$A(n) = (s_0 s_2 - s_1^2)^{-1} \quad B_g(n) = s_2 s_g - s_1 s_{g+1}$$
$$C_g(n) = s_1 s_g - s_0 s_{g+1}$$

According to equation (30) the tracking error consists of contributions from each term in the Gth degree input polynomial. The weight of each term is given by $\epsilon_g(n, L)$. By substitution of equation (23) one can easily check that

$$\epsilon_0(n, L) = \epsilon_1(n, L) \equiv 0; \quad \text{all } n, L. \tag{31}$$

Consequently, the terms of degree $g = 0$ and $g = 1$ in the input polynomial do not contribute to the tracking error. This is hardly surprising since the estimation procedure is based on fitting a straight line to the input data. The first term contributing to the tracking error is the second degree term. Its weight is given by

$$\epsilon_2(n, L) = A(n) \cdot [B_2(n) - L \cdot C_2(n)] - L^2. \tag{32}$$

Again, the weight is a quadratic function of the lag L. This is the second reason for introducing a lag. The second degree tracking error can be reduced by selecting a proper lag.

Actually, we can even select an L-value such that $\epsilon_2(n, L) = 0$. However, this again implies the use of a time-dependent lag, a refinement which complicates the system, as remarked before.

Figure 8 (lower 2 plots) presents $\epsilon_2(n, L)$ as a function of L and n. Again, the curves are valid for $q = 0.975$, or $N = 40$. The sensitivity of the second degree tracking error to the lag is evident.

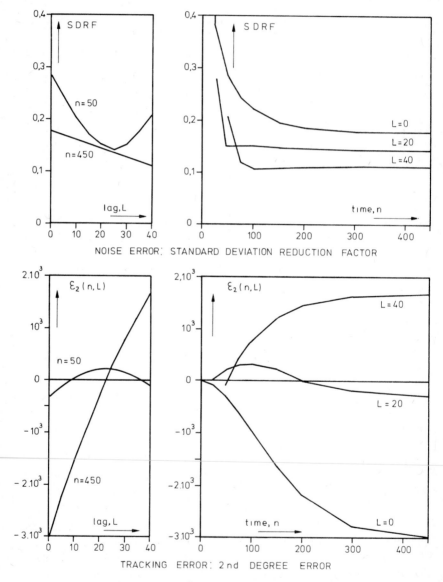

FIG. 8. Noise and tracking error for $q = 0.975 \sim N = 40$.

6. Results

In preceding sections, the structure of the filter has been established and expressions for the estimation errors have been derived. The final task is to select values for the parameters q or N (equations (9) and (10)) and L (equation (11)). Normally, they will be selected such that the estimation errors are minimized. In this context, a general remark is made.

The expressions for the noise error (equations (24) and (25)) are quite useful for this purpose. They provide direct quantitative insight. On the other hand, the expressions for the tracking error (equations (30) and (32)) should be used only as a qualitative guideline. One reason is that the input trend is approximated by a Gth degree polynomial. This is not always a good approximation. Further, the joint effect of second, third and higher degree tracking errors should be taken into account when $G > 2$. Finally, the tracking errors increase with increasing time n (Fig. 9), whereas the co-efficients of higher degree terms in actual input trends usually decrease with increasing time. The possibility of counterbalancing these effects should be exploited.

Apart from estimation errors, other considerations can play a role when selecting parameter values. For example, the effect of round off errors when using fixed point arithmetic with limited word length, might limit the useful range of q-values. Another limitation could be the storage required in a multi-doppler channel system for the delay lines of Fig. 7. Although the various channels can share the same arithmetic unit on a time-multiplex basis, each channel requires its own storage of L words.

One way to solve the problem is to resort to a trial and error method, applying the guidelines of the previous sections, and using representative realizations $p(n)$ from the experimentally obtained trend ensemble and artificially generated fluctuation processes $r(n)$. Some results obtained in this way are shown in Figs 9, 10, 11 and 12.

Each figure consists of 3 plots, each one representing the input (lower trace) and the output (upper trace) of the highpass filter. Plot A shows a filter input consisting of trend plus rapid fluctuations. Plots B and C relate to inputs consisting of trend alone and of fluctuations alone, respectively. (In plot C the input process $r(t)$ has been biased to -40 dB in order to separate both traces.)

The signals in the graphs correspond to reverberation as it would appear at the output of a doppler channel (Fig. 1), matched to a 100 msec tone pulse echo. Hence, the bandwidth of $r(t)$ is roughly 10 Hz. A sample rate of 33 Hz has been used. The horizontal scale comprises 512 signal samples, corresponding to a sonar range of ≈ 15 sec or ≈ 11 kilometre. Vertical scales are in dB. At time $n = 100$ and $n = 300$ an artificial echo with a local signal-to-reverberation ratio of 12 dB has been injected into the $r(t)$ process.

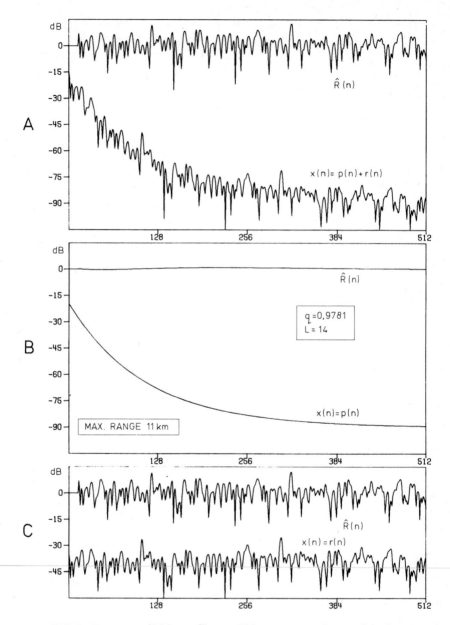

FIG. 9. Response of highpass filter to 100 msec tone pulse reverberation.

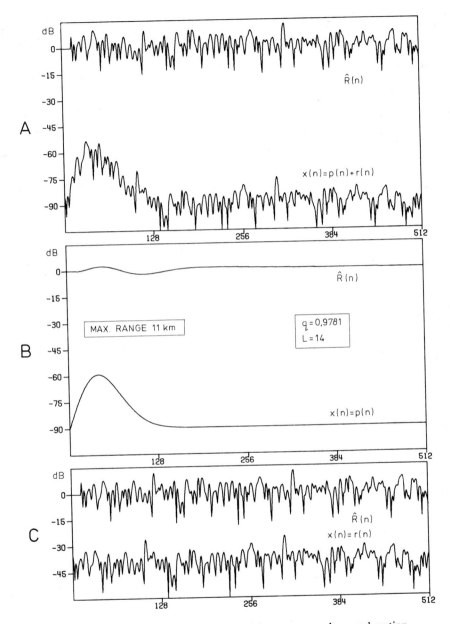

FIG. 10. Response of highpass filter to 100 msec tone pulse reverberation.

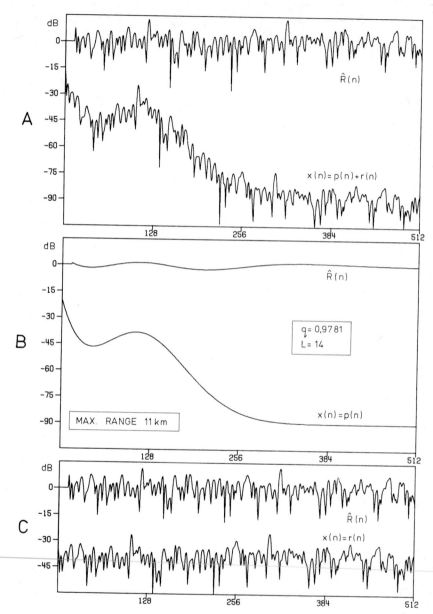

FIG. 11. Response of highpass filter to 100 msec tone pulse reverberation.

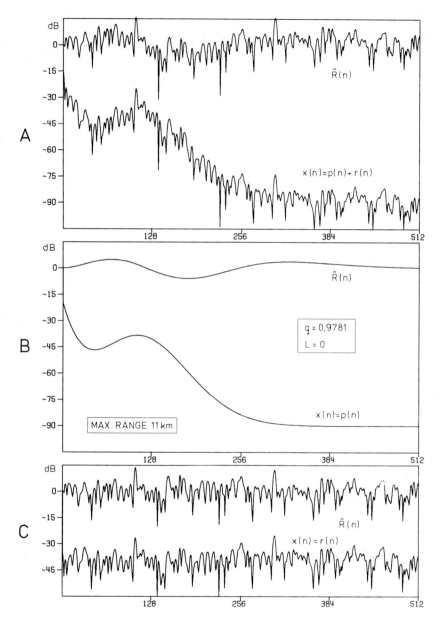

FIG. 12. Response of highpass filter to 100 msec tone pulse reverberation.

Figures 9, 10 and 11 present 3 different trend functions. In Fig. 9 the trend decreases very smoothly with range. Consequently, the tracking errors are negligible. The more violent oscillations of the trends in Figs. 10 and 11 give rise to larger tracking errors. The filter parameters used in Figs 9, 10 and 11 are $q = 0.9781$ and $L = 14$. Figure 12 shows the same input as Fig. 11, the difference being that the lag has been reduced to zero: $L = 0$. Comparison with Fig. 11 immediately shows the improved performance when applying a suitable lag.

REFERENCES

1. Hansen, V. Gregers. (1965). Logarithmic radar receiver using pulselength discrimination. *IEEE Trans AES* AES-1, No. 3.
2. Winder, A. A. (1968). "Dynamic Range Compression and Normalization in Sonar Receivers", Proceedings NATO Advanced Study Institute on Signal Processing, August 1968, Twente Institute of Technology, the Netherlands.
3. Morrison, N. (1969). "Introduction to Sequential Smoothing and Prediction", McGraw-Hill, New York.
4. Levine, N. (1961). A new technique for increasing the flexibility of recursive least squares data smoothing. *Bell System Tech. Jl.* XL, No. 3.
5. de Vries, F. P. Ph. (1973). Digital logarithmic normalization of sonar signals; batch processing. *In* "Signal Processing", Proceedings NATO Advanced Study Institute on Signal Processing, August 1972. Academic Press, London and New York.

DISCUSSION

I. G. Liddell: Given the variations in level from one channel to the other, do you think you could combine them in a profitable way?

Answer: Yes, although the benefit will depend on the channels considered. Those which are in the noise limited region right from the beginning of the range can certainly be combined. The advantage is that you get more degrees of freedom for averaging. For frequency channels in the reverb limited region the combination is more difficult and probably less profitable because of the differences in level between the channels.

J. J. Dow: Did you ever consider a process in which an additive noise term is present, besides the reverberation with multiplicative rms value?

Answer: No. I wonder whether this model is relevant to the situation described in my paper. The doppler channels are matched to the spectrum of the transmitted pulse. Therefore, they are as narrow as possible. Additive noise has approximately the same spectral shape as the reverberation in the channel and can be considered a part of it. Thus, in the narrow channel no difference is made between reverberation and noise.

H. Cox: Is taking the log essential to your system? Have you considered leaving it out?

Answer: No, we did not consider this. The first thing we want is to get a constant variance output.

A. Leisterer: How do you optimize the lag?

Answer: This depends on a lot of considerations. A few things about this matter are said at the end of the paper. Mainly, it is done by a trial and error method, using the analytically obtained results for the estimation errors as a guideline.

R. S. Thomas: Did you already measure the statistics of your system using real sea data?

Answer: Not yet; we are on the verge of doing this.

R. S. Thomas (Comment): We did something like this and found a weak stationarity, using this sort of system.

P. Raj: Did you use *a priori* information about the decreasing trend in your Kalman-like filter?

Answer: No, although this can be done of course. Actually, you could supply even more *a priori* information to the system, like for instance prior knowledge about the standard deviation of the fluctuating component (5.5 dB, see paper). It is my feeling that providing this information would not improve the performance very much. If, on the other hand, it would, one must be very certain of this information because its introduction into the system would remove some of the robustness of the least squares method.

J. W. R. Griffiths: Could normalization be done in an earlier stage of the processing chain, with the aim of reducing the dynamic ranges concerned?

Answer: Yes, this is possible. But this comes down to adaptive notch filtering or pre-whitening in the broadband input signal. This is rather difficult in itself and it requires some sort of spectrum analysis in order to control the notch filter or pre-whitener. Since the system contains already a spectrum analyser (for doppler analysis), it is better to use this first and normalize afterwards.

T. G. Birdsall: How many multiplications are required for obtaining the logarithm of a binary number?

Answer: We use an approximative method which uses no multiplications at all. It comes down to reorganizing the bits in the binary word. This operation can be done very rapidly.

C. N. Pryor, T. G. Birdsall: How many bits are used for representing the logarithmic variable?

Answer: (After some confusion). At present, we use 8 bits for the logarithmic variable, covering a range of approximately 100 dB. Hence, the quantization is in steps of roughly $100.2^{-8} = 1/4$ dB. As for the trend estimation filter, which is the important part of the digital highpass filter, a much coarser quantization is allowed for its input (say 5 bits) since the quantization errors will be averaged out. (However, it will then be difficult to check the performance of the system with smooth trend-like input signals).

Digital Logarithmic Normalization of Sonar Signals: Batch Processing

F. P. Ph. DE VRIES

Physics Laboratory NDRO,
The Hague, The Netherlands

1. Introduction

In a companion paper [1] a method for digital logarithmic normalization of sonar signals has been described. The method consists of taking the logarithm of a rectified doppler signal and of removing the slowly varying trend function $p(n)$ by means of a time varying digital highpass filter. Since it is intended for real time application to a great number of doppler channels in a multi-beam tone pulse sonar, the main emphasis has been on computational simplicity. A recursive highpass filter was obtained which processed the input signal samples as soon as they became available (serial processing) and which required only 3 multiplications per output signal sample. Figures 9, 10 and 11 of paper [1] give a visual impression of the filter's performance.

Because of its inherent simplicity this filter has several drawbacks. Some of them are:

(1) The impulse response of the filter always has a finite effective duration. Consequently the output is only based on the input signal within a relatively narrow moving window. (This is the disadvantage of any filter operating in real time.) The reason is the necessity to incorporate damping in the recursive filter loops.

(2) Shortly after starting the filter procedure only a very limited number of input signal samples is available for processing. Thus, the filter produces significant estimation errors at the beginning of the sonar range. With increasing time the number of input samples gradually increases and the estimation errors decrease accordingly to a stable limit value.

(3) Because of the required simplicity, the filter must be restricted to a low order (actually only 2 poles were used).

The purpose of the present paper is to obtain insight into the loss of performance caused by the above mentioned restrictions. Therefore, we

shall now remove the requirements of real-time operation and of computational simplicity and we shall only retain the requirement of using a linear
method. A striking feature of the input signals to the highpass filter (viz.
plots, A, Figs. 9, 10, 11, 12 of Reference [1]) is that upon visual inspection
the echoes are immediately localized. This is partially due to correlation
between trend values at widely separated time instants. This correlation
cannot be exploited by filters based on a narrow moving window. Therefore,
we shall now resort to a filter procedure which uses all samples of the whole
signal. Clearly, this procedure is not very suitable for the practical implementation since it would require storage of the signal over its entire
duration before processing. For a great number of doppler channels this
leads to a prohibitive memory capacity. However, the prime object of this
paper is to investigate how much can be improved upon the method
explained in Reference [1].

We shall adopt the method of filtering in the frequency domain, i.e.
apply a Discrete Fourier Transform (DFT) to the input sequence $x(n)$,
multiply with a suitable filter transfer function and finally apply an
Inverse Discrete Fourier Transform (IDFT) to obtain the desired highpass
output. In this way we have complete control over the transfer function.
There is no need to restrict the transfer function to a low order filter.

2. The Filter Method

The problem is to estimate the rapid fluctuations $r(n)$ from a signal $x(n)$
which is an additive combination of a slowly varying trend $p(n)$ and of
the rapid fluctuations $r(n)$, $(x(n) = p(n) + r(n); n = 0, 1, \ldots, N - 1)$. Thus,
the filter used for extraction of $r(n)$ from $x(n)$ will essentially be a highpass
one. The $x(n)$ process is the logarithmic output of one doppler channel of
a tone pulse echo ranging system. The trend $p(n)$ corresponds to the varying r.m.s. value of the original signal and $r(n)$ corresponds to its stochastic
fluctuations. The situation is depicted in Fig. 1.

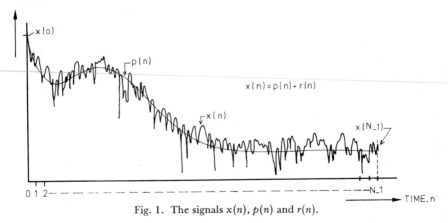

Fig. 1. The signals $x(n)$, $p(n)$ and $r(n)$.

The filter method essentially consists of a transformation of $\{x(n)\}$ to the frequency domain yielding the spectral values $\{X(r)\}$. The actual filter process consists of multiplying the spectral values $X(r)$ by the filter's transfer function $H(r)$ the result being $\hat{R}(r)$. An inverse transformation of $\hat{R}(r)$ then gives the desired estimates $\hat{r}(n)$ of the rapid fluctuations (see Fig. 2, the heavily ruled part of the diagram).

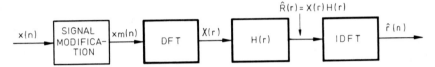

FIG. 2. Block diagram of the filter method.

The filter process corresponds to a convolution of the input signal $x(n)$ with the filter impulse response $h(n)$, where $h(n)$ is the IDFT of $H(r)$. As illustrated in Fig. 1, there are large discontinuities at the beginning and the end of the input signal ($x(0)$ and $x(N-1)$ in Fig. 1). The discontinuities are mainly caused by the behaviour of the trend component $p(n)$. These discontinuities will give rise to transient oscillations in the output signal, superimposed on the desired output component $r(n)$. The result is a distortion of the output, particularly near the edges $n = 0$ and $n = N - 1$. In order to suppress the transients, we shall modify the input signal in a suitable manner before starting the filter procedure. The modification is realized in the first block of the diagram of Fig. 2. In the sequel we shall treat in detail this signal modification and the determination of the transfer function of the filter.

3. Signal Modification

In order to remove the discontinuities at $n = 0$ and $n = N - 1$ from the input signal $x(n)$, we apply a method which is sometimes used for the numerical determination of the Fourier coefficients of a periodic function [2].

Since the input signal $x(n)$ is represented as a series of discrete spectral samples, the result of the filter procedure is a cyclic convolution of the original input signal $x(n)$ and the filter impulse response $h(n)$.

$$\hat{r}(n) = \sum_{1=0}^{N-1} x^1(n-1)\, h(1)$$

where $h(k)$ ($k = 0, 1, \ldots, N - 1$) is the IDFT of $H(r)$:

$$H(r) = \frac{1}{N} \sum_{n=0}^{N-1} h(n) \exp\left(-\frac{j2\pi}{N} nr\right) \quad r = 0, 1, \ldots, N - 1$$

and where $x^1(n)$ is a periodic repetition of the original signal $x(n)$:

$$x^1(n + kN) = x(n); \quad k = 0, \pm 1, \pm 2, \ldots$$

Consequently the two discontinuities at the beginning and the end of the original signal $x(n)$ can be treated as only one discontinuity in the periodic signal $x^1(n)$ (see Fig. 3).

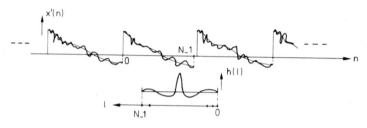

FIG. 3. Cyclic convolution.

The procedure of signal modification consists of subtracting suitable polynomials from the signal. For instance, subtraction from $x(n)$ of a straight line, with end values which equal the end values of $x(n)$, yields a modified signal $x_1(n)$ which is continuous at the edges (Fig. 4). The next step would be the subtraction of a parabola resulting in a signal which is continuous in the first derivatives too. In this manner one can proceed. However, this method requires the knowledge of the signal and its derivatives at the edges of the signal's interval. This requires an analytic expression for the signal. In practice such an expression is absent. Only the values of the signal at the edges are known allowing the subtraction of the straight line. However, there is another method to compensate for the unequal first derivatives.

From the signal $x_1(n)$ $(n = 0, 1, \ldots, N - 1)$ obtained after subtraction of the straight line, another signal $x_m(n)$ can be derived by continuing $x_1(n)$ with the negative of its mirror image (Fig. 4):

$$x_m(n) = x_1(n); \quad n = 0, 1, \ldots, N - 2$$
$$x_m(N - 1) = 0$$
$$x_m(N - 1 + n) = -x_1(N - 1 - n); \quad n = 1, 2, \ldots, N - 1$$

In this way, an odd symmetry signal of double length is obtained which has a continuous first derivative (and, naturally, a continuous zeroth derivative) at the edges when repeated with twice the original period $2N - 1$.

Summarizing, the signal modification consists of two steps: (1) subtraction of a straight line, (2) incorporation of odd symmetry by mirror imaging.

Observe that the first step of the modification (the subtraction of the straight line) removes a part of the trend function by subtraction rather than by filtering.

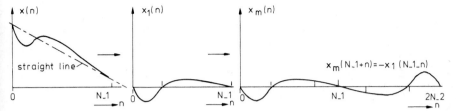

FIG. 4. Signal modification.

To determine the straight line we must use the values $x(0)$ and $x(N-1)$ since the trend values $p(0)$ and $p(N-1)$ are not at our disposal. The difference between the desired straight line based on $p(0)$ and $p(N-1)$ and the actual line based on $x(0)$ and $x(N-1)$ has to be removed by filtering.

It is interesting to note that the signal modification described above corresponds closely to a classical method for sonar signal normalization by means of gain control amplifiers, namely "Time Varied Gain". For example, subtraction of a straight line from the logarithmic output data $x(n)$ is equivalent to multiplying the pre-logarithmic signal with an exponential time function implemented by varying the gain of a controlled amplifier.

4. Determination of the Filter Transfer Function

Consider the situation depicted below.

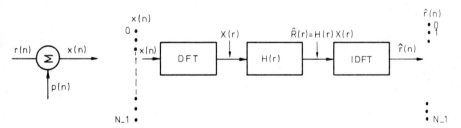

FIG. 5. Signal model and block diagram of the filter method.

The problem is to extract from the signal $x(n) = r(n) + p(n)$ $(n = 0, 1, \ldots, N-1)$ the rapid fluctuations $r(n)$, the result of the filtering is called $\hat{r}(n)$. The mean value of both trend $p(n)$ and rapid fluctuations $r(n)$ are zero.

$$\left.\begin{array}{c} \overline{r(n)} = 0 \\ \overline{p(n)} = 0 \end{array}\right\} \quad n = 0, 1, \ldots, N-1$$

$r(n)$ and $p(n)$ are statistically independent

$$\overline{p(n)r(k)} = 0, n = k = 0, 1, \ldots, N-1$$

The filter's transfer function $H(r)$ has to be determined to do the extraction in the best possible way. The criterion of goodness chosen is that the r.m.s. value of the error is as small as possible, thus

$$E\left[\sum_{n=0}^{N-1}\left\{r(n) - \hat{r}(n)\right\}^2\right] \quad \text{has to be a minimum.}$$

The averaging is done over r and p.

For the transfer function $H(r)$ we find:

$$H(r) = \frac{\overline{R(r)R^*(r)}}{\overline{P(r)P^*(r)} + \overline{R(r)R^*(r)}}$$

where $R(r)$ and $P(r)$ are the Fourier transforms of respectively $r(n)$ and $p(n)$ and the asterisk denotes complex conjugation.

We see that $H(r)$ is the ratio of the power spectrum of the desired component $r(n)$ and the power spectrum of $r(n)$ and $p(n)$.

The actual input signal to the DFT is the modified signal $x_m(n)$. This signal can be written as

$$x_m(n) = r(n) + p(n) - s_p(n) - s_r(n)$$

$r(n)$ are the desired rapid fluctuations, $p(n)$ is the trend, $s_p(n)$ is the straight line having end values $s_p(0)$ and $s_p(N-1)$ which equal the end values of the trend respectively $p(0)$ and $p(N-1)$, and $s_r(n)$ is the straight line caused by the in general non-zero values of $r(0)$ and $r(N-1)$.

The power spectrum of $r(n)$ is denoted by $\Phi_r(r)$, $(\Phi_r(r) = \overline{R(r)R^*(r)})$ and the spectra of $p(n) - s_p(n)$ and $s_r(n)$ are denoted by respectively $\Phi_p(r)$ and $\Phi_s(r)$.

The trend $p(n)$ and the rapid fluctuations $r(n)$ are independent. The line $s_r(n)$ is completely determined by the end values $r(0)$ and $r(N-1)$ of the fluctuations so $s_r(n)$ is independent of $p(n) - s_p(n)$ and, because N is large, is highly independent of $r(n)$.

Hence the transfer function is:

$$H(r) = \frac{\Phi_r(r)}{\Phi_r(r) + \Phi_p(r) + \Phi_s(r)}$$

The power spectrum $\Phi_r(r)$ is numerically determined by averaging over a large number of sample functions $r(n)$. The spectrum $\Phi_r(r)$ is analytically determined using $r(0) = r(N-1) = 0$ and $\overline{r^2(0)} = \overline{r^2(N-1)} = 5.55$ (see Reference [1], page 673).

The method described above has been applied to the following situations:

(1) From the same representative realizations $p(n)$ as used in Reference [1], the energy spectrum of each of them is numerically computed. The result is used as $\Phi_p(r)$ to give $H(r)$. Thus each specific trend is filtered with its own transfer function.

Clearly, this method requires full *a priori* knowledge of the input trend. In actual practice this knowledge is not available, therefore, the corresponding results must be regarded as a limit to what can be obtained in practice.

(2) One and the same transfer function $H(r)$ is used, taking into account $\Phi_r(r)$, $\Phi_s(r)$ and the average spectrum of specific trend spectra $\Phi_p(r)$. This method does not require *a priori* knowledge of the individual trend. It is based exclusively on the average trend behaviour. Therefore, the method connects more closely to reality. On the other hand the performance is expected to decrease with respect to method 1.

5. Results

As mentioned before, two filter methods are considered, one based on the knowledge of the energy spectrum of the trend on hand, the other on the average trend behaviour. The filter used in the former case is called "specific" the latter "fixed".

For reasons of comparison, the same trend functions and rapid fluctuations realization are used and the plots have the same format as in [1]. Figures 7, 8 and 9 consist of 3 plots; each plot represents the input (lower trace) and the output (upper trace) of the highpass filter. Plots A, B and C show the filter inputs respectively consisting of trend plus rapid fluctuations, trend only and fluctuations only. (In plot C the input sequence $r(n)$ has been displaced by -40 dB in order to separate both traces.)

The signals in the graphs correspond to reverberation as it would appear at the output of a doppler channel (Fig. 1, [1]), matched to a 100 msec tone pulse echo. Hence, the bandwidth of $r(n)$ is roughly 10 Hz. A sample frequency of 33 Hz has been used. The horizontal scale comprises 512 samples, corresponding to a sonar range of about 15 sec. or $\approx$11 kilometres. Vertical scales are in dB. At time $n = 100$ and $n = 300$ an artificial echo with a local signal to reverberation ratio of 12 dB has been injected into the $r(t)$ process.

Figure 6 consists of 3 plots. Each plot shows three curves. The curves plotted represent the Fourier coefficients $\{20 \log_{10} (\text{magnitude})\}$.

The horizontal axis is a part of the frequency axis, only the first 64 of the 512 frequency samples are shown.

One curve shows the computed coefficients of the input signal (trend plus rapid fluctuations). Another curve shows the transfer function of the highpass filter. The three plots correspond, from top to bottom, to Figs. 7, 8 and 9 respectively.

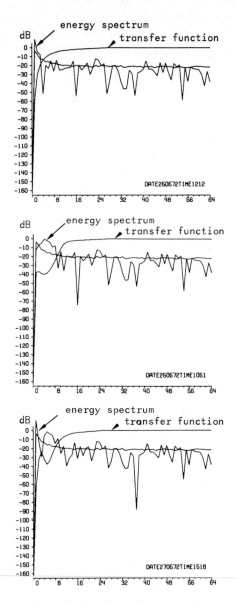

FIG. 6. Energy spectra and transfer functions, from top to bottom, corresponding to, respectively, Figs. 7, 8 and 9.

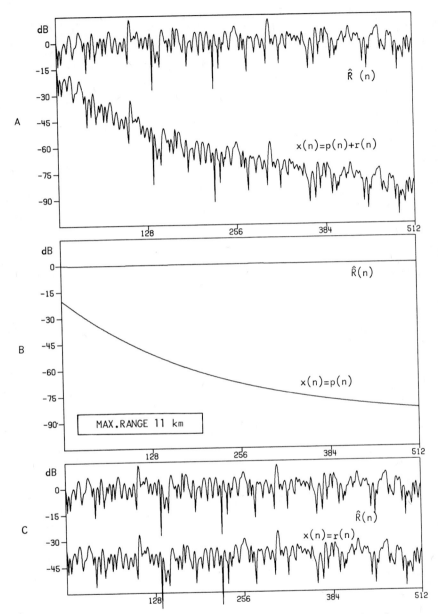

FIG. 7. Response of highpass filter to 100 msec tone pulse reverberation.

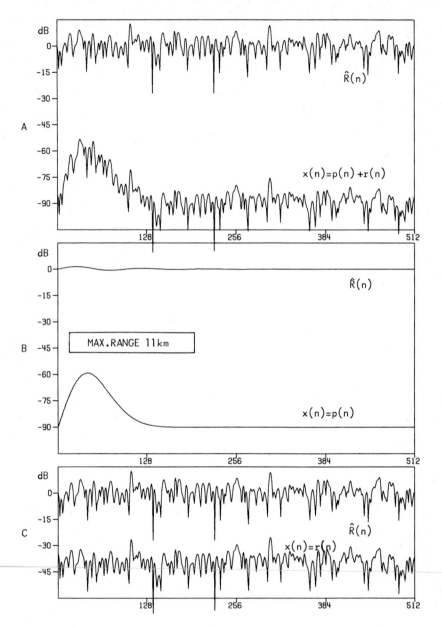

FIG. 8. Response of highpass filter to 100 msec tone pulse reverberation.

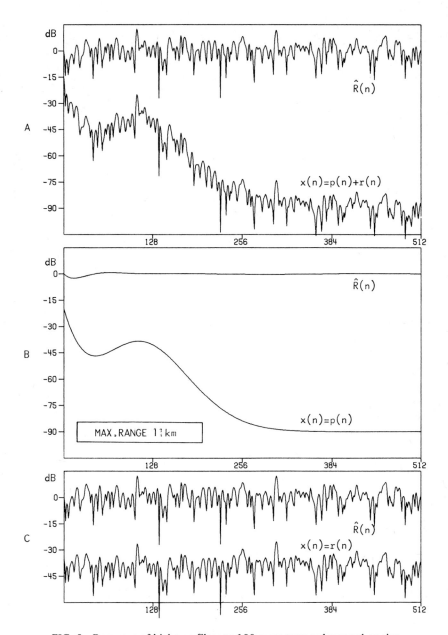

FIG. 9. Response of highpass filter to 100 msec tone pulse reverberation.

The plots shown refer to the specific filters, based on too much *a priori* knowledge.

Figures 7, 8 and 9 are to be compared with Figs. 9, 10 and 11 in [1].

Comparing the serial- and batch-processed filtering results, we see no marked difference between the two methods with respect to the rapid fluctuations (plots C). The results of filtering the smooth trend of Fig. 7 (9 Reference [1]) also show little difference. However, the residues of the batch filtered trends of Figs 8 (10 Reference [1]) and 9 (11 Reference [1]) are much smaller than the serial processed ones. Compare the echoes, in plots A, at $n = 100$ in Fig. 8 (10 Reference [1]) where the residue of the serial processed trend has a negative sign.

Note that the recursive filter used in [1] is "fixed", whereas the filtering results shown here are obtained with "specific" filters. However, the results obtained with the batch processing method using a fixed filter were only slightly different from those using specific filters. The differences are so small that the results are not even shown.

REFERENCES

1. Van Schooneveld, C. (1973). Digital logarithmic normalization of sonar signals: serial processing. *In* "Signal Processing", Procedures of NATO Advanced Study Institute on Signal Processing, August 1972, Academic Press, London and New York.
2. Hamming, R. W. (1962). "Numerical Methods for Scientists and Engineers", McGraw-Hill Book Company, Inc., New York.

DISCUSSION

P. L. Stocklin: In the examples of trend functions just shown to us, we saw trends consisting of variations superimposed on a monotonic decreasing function. These variations are confined to the beginning of the sonar range. If these variations would extend on the whole range do you expect that the batch processing method still performs better compared with the serial processing method.

Answer: We have seen that the batch processing method filters out better the unwanted trend, especially the residue is smaller where the trend varies around the monotonic decrease. Extension of the variations into the sonar range does not alter the power spectrum of the trend as far as the bandwidth is concerned. So I expect that the batch processing method still performs better than the serial processing method.

B. D. Smith: Did you consider storage from ping-to-ping of the signal. In that case you could average the signals and subtract the so obtained trend function from the signals.

Answer: In the present paper, ping-to-ping storage was not considered. The aim was to gain insight into how much the performance of serial processing is improved if we drop the simplicity constraints, but maintain the other limitations.

As for the merits of ping-to-ping storage, we are not yet sure whether the trend-to-trend correlation is sufficiently great under various conditions (lack of platform stabilization, etc).

P. L. Stocklin: Have you compared the detectability losses, due to the trend estimation errors, of both methods.

Answer: Yes, but in a simplified way. Apart from large tracking errors caused by violent trend oscillations on a limited interval of the range axis, the estimation error for the serial method has a standard deviation of $\sigma = 0.75$ dB. For the batch method this is only slightly less (the advantage of batch processing being primarily a further suppression of large anomalous tracking errors mentioned above).

To determine the detectability loss due to imperfect normalization, we have considered the system to be equivalent to a linear communication system with an *a priori* uncertainty σ in the overall attenuation between transmitter and receiver. Assuming a target with ping-to-ping Rayleigh scintillation, it is then easy to compute the ROC belong to the case of uncertain signal and reverberation levels and to compare it to the ROC for a perfect normalizer (i.e. $\sigma = 0$). Loss of detectability can then be defined as the required increase of target strength in order to compensate for the effect of imperfect normalization.

For $\sigma = 0.75$ dB (i.e. serial processing) we found losses between 0.2 and 0.5 dB, depending on the desired false alarm probability. For batch processing this would be proportionally less.

T. G. Birdsall: The odd modified signal has a pure sine transform. Could you use this knowledge in the computation of the FFT.

Answer: Yes, at the first glance one would say that the odd modified signal has two times as much points and so one has to do a 2N points FFT. However the number of degrees of freedom of the odd modified signal is the same as that of the original signal. So one expects that a modified FFT can be done with about the same computational effort. This is indeed the case.

In the *IEEE Transactions on Audio and Electroacoustics* of June 1969, vol. **AU-17**, an algorithm is described that can be applied to an even sequence. Along the same lines one can construct an algorithm for an odd sequence. Such an algorithm is used.

Normalization and Optimal Processors

W. HILL, G. H. ASH and J. R. KENEALLY

*Admiralty Underwater Weapons Establishment,
Portland, Dorset, England*

1. Introduction

In an active system receiver the original requirement for normalization
was to reduce the dynamic range at the output such that it could be presented to an operator visually on a display or aurally on an audio channel.
Thus in an active sonar system the signal would not overload under high
reverberation conditions at close range yet it would still maintain the full
dynamic range under noise limited conditions at long range. Typically the
range of control required could be 80 dB over a few seconds in time. Similarly in a passive receiver only a limited output dynamic range is required
although the input can change over 40 dB. Depending on the type of system
the change can occur over a few seconds or over several hours.

The primary aim, therefore, was to present to the operator a more or
less even representation of the background against which the target would be
detectable as an increase in signal level. Naturally variations in the output
signal occurred even when only the background was present, owing to its
random nature. These excursions would occur randomly in time and over
the displayed area if successful normalization had been achieved. Thus the
devices used to normalize the signal have often been referred to as Constant
False Alarm Rate (CFAR) devices.

A number of different methods have been devised to normalize signals.
These may be classed as: 1. Automatic Gain Control (AGC) amplifiers;
2. Time Varying Gain (TVG) amplifiers; 3. Logarithmic (Log) amplifiers;
4. Clippers; 5. Variable Threshold systems.

Inevitably normalization devices depart from the ideal and distort the
output. Not only does the mean level of the output fail to stay constant
but distortion also occurs, and these are often undesirable in an amplitude
detection system. These aspects are briefly considered in a review of the
method in the first part of the paper.

When discussing the question of normalization, however, it is relevant to
consider not only how successful the various methods are, but also how

well the concept of normalization fits the requirements of an optimum detector. In the second part of the paper Decision Theory is invoked to examine the validity of the CFAR concept. It is shown that under some conditions a CFAR detector is optimal, but that in other practical cases this is not so. The amount a CFAR system departs from optimum will depend on the detailed statistics of the background and the wanted echoes, and also on the decision criteria.

2. Normalization Techniques

(a) AGC amplifiers

Traditionally AGC amplifiers have been variable gain amplifiers whose gain is set by a control voltage obtained from the output. To generate this voltage the output of the amplifier is detected and averaged. By making the averaging time for increasing signals longer than for decreasing signals it is possible to make an echo stand out above the background with little distortion while at the same time coping with the overall rapid reduction in input level. It will not however cope with reverberation (clutter) bursts or impulsive noise which will be passed like echoes and registered as false alarms unless further steps are taken to discount them.

This is the simplest form of AGC system and does not require any particular law of control voltage versus gain. The original amplifiers used variable μ valves and transistor amplifiers have been built using the variation of gain obtained by changing the collector current or the collector emitter voltage. Often a more useful technique is to use a variable attenuator, e.g. photo resistors, field effect transistors, junction diodes as analogue attenuators often with a precise law of control voltage versus gain, or a digital attenuator using MOS transistor switches where any law of control against gain may be obtained. Finally pulse width modulation with low pass filtering may be used as a variable gain system although this technique tends to have a limited frequency response and dynamic range.

The basic principle of AGC systems is that the power of the incoming signal is estimated from an averaged length of signal and then used to correct to constant power output. Errors are incurred in the averaging process. The estimate of power is in error because the averaging time is short. In addition the sample used corrects the signal occurring later in time and this can be important when the input power is varying rapidly. AGC systems can also have an inherent instability in complex systems unless care is taken in their design.

Further problems occur in practical systems in which many data channels are processed and displayed simultaneously. As the marking rate on displays varies appreciably with only small changes in threshold, small circuit variations from channel to channel can cause difficulty in achieving a reasonably uniform rate for all of them.

(b) TVG amplifiers

TVG amplifiers are similar to AGC amplifiers except that the gain is changed according to a fixed law instead of adjusting to the input power. In this form the amplifier will require a defined law of control voltage versus gain. It is evident that TVG amplifiers on their own would only be successful when the law of power variation is a constant function of time. Often TVG is combined with AGC, the TVG taking out the gross changes of level and the AGC compensating for the departure of the output from the expected law.

(c) Log amplifiers

The effect obtained with log amplifiers is similar to normal AGC operation with rms detection. Large input changes corresponding to gain changes in the AGC system are converted into dc levels and the signal component can be filtered off by means of a high pass filter. A following antilog amplifier can reduce the amplitude distortion of the signal component and the dc level is available as a measure of the background level. Differential time constants of the filter circuit can allow for rapid pulses as in the normal system.

A logarithmic amplifier can be obtained by summing the output from several stages of gain. For small signals only the last stage will make a significant contribution to the output and there will be a linear relationship between input and output. As the input level rises the last stage will saturate and the output voltage will be dependent on the penultimate stage added to the fixed amount from the saturated last stage. In this way an approximate logarithmic response is obtained as a series of straight lines. An integrated circuit working on this principle can be obtained with a dynamic range 80 dB, a linearity of less than ±0.5 dBV and bandwidth dc to 40 MHz. Alternatively if only an output of the envelope is required the signal may be precision rectified before being applied to a dc log amplifier. Again the logarithmic relationship can be approximated to a series of straight lines or the logarithmic current to voltage relationship of a diode or saturated emitter junction of a transistor can be used. Several hybrid integrated circuits are available using transistor emitter junctions with a dynamic range of input current over 7 decades.

The log amplifier works on the principle that the probability distribution of the output voltage which results from a Gaussian input signal is independent of the power of the input signal. No practical amplifier can maintain its logarithmic characteristic down to zero input as the output would then be $-\infty$. Below a certain input level the amplifier degenerates into a linear amplifier. It has been shown that for the output distribution to be independent of the input power the logarithmic characteristic must be maintained to 20 dB below the minimum rms noise level. This increases the dynamic range requirement of the amplifier by a further 20 dB.

The disadvantages of the log amplifier are similar to the AGC amplifier. Furthermore it compresses the envelope of the signal so that after detection the echo amplitude is suppressed relative to the AGC system. As already stated this can be restored by using an antilog amplifier at the expense of further complexity.

(d) Clippers

A clipper amplifier or zero crossing detector may be considered as a normal-ization device. A constant power is available at the output of the clipper irrespective of the power at the input. The way the output power divides between the input signals has been studied and under high sine wave signal-to-noise conditions a clipper can increase the signal-to-noise ratio. Clipped systems have received a lot of study as they are particularly relevant to digital implementation. Even in a multilevel digital system where the input is encoded into a binary number, at the extremes when either the encoder is overloaded or the input is so small that only the sign stage operates then the system becomes a two state or clipped system.

Many types of clipping amplifier have been designed in the past either using diodes to perform the clipping or relying on the limiting properties of the active device itself, valve or transistor. Nowadays integrated circuit comparator amplifiers are available which make ideal clippers. Dynamic ranges of greater than 80 dB can be obtained from a single comparator while preserving the mark to space ratio at the output to better than 5% for sine wave inputs.

In a clipped system the absolute amplitude information is destroyed and the only information retained is the time of the zero crossings of the input signal. However there is a relationship between the signal to noise ratio at the output to the signal to noise ratio at the input and further processing makes use of this and also the phase of the information.

The advantages of clipped system are simplicity and relevance to digital implementation. The disadvantages are that the absolute amplitude of the information is destroyed and up to 2 dB processing gain is lost for low signal to noise ratios.

(e) Variable threshold systems

All the normalization systems so far considered have attempted to produce a constant output. Before detection criteria are applied some form of threshold is used and only the events greater than the threshold are con-sidered, leading to a fixed false alarm rate. Another approach to the problem is to allow the output to vary linearly with the input and to count the number of events above a controlled threshold. It is then possible to vary the threshold level such that the false alarm rate is constant. Such an approach could be implemented by using a digital computer having adequate storage capacity, although it could lead to problems of dynamic range in the input stages of the system. Its effectiveness would rely on the variation of the input power with time changing only slowly from transmission to transmission.

3. Optimality

Having discussed briefly the various methods of normalization, we shall now consider how closely a perfect CFAR device satisfies the conditions required for an optimum processor. In the present examination the condition of optimality will be determined by an application of Decision Theory.

We shall suppose that the total information obtained by the sonar (or radar) on any one transmission is contained in a large number of independent samples, each of which defines the condition in a certain volume of the information space (e.g. in the range, bearing, doppler dimensions). A wanted echo (target) can appear in any cell. If x_i is the magnitude of the sample obtained in cell i in any given transmission, then we can denote by $P_i(x_i/T); P_i(x_i/N)$ the probabilities of obtaining the value x_i in cell i given that it respectively contains a target or only the background noise (or reverberation). The suffix i given to P indicates that these probabilities are specific to the cell i; the probability of obtaining the same value x_i in other cells could be different.

A simple analysis of the problem using Decision Theory asserts that we should decide on the basis of the evidence x_i that a target was present in cell i if:

$$\frac{P_i(x_i/T)}{P_i(x_i/N)} \geqslant \frac{CN_iP_i(N)}{CT_iP_i(T)} \tag{1}$$

where $CT_i; CN_i$ are the costs of misclassifying targets and noise respectively, and $P_i(T); P_i(N)$ are the respective prior probabilities of a target or only noise being in cell i. Again the suffices indicate that these quantities can vary from cell to cell.

The ratio $P_i(x_i/T)/P_i(x_i/N)$ is called the Likelihood Ratio and can be denoted by $L_i(x_i)$.

One difficulty in applying Decision Theory to practical problems is in determining the appropriate values for the prior probabilities and the costs. Nevertheless it is often possible to infer from design parameters based on other criteria what are reasonable value for ratios, such as that on the right-hand side of equation (1). Generally it is assumed that the same performance is required in all cells, and thus in the present case we shall assume a constant value K for the ratio $CN_iP_i(N)/CT_iP_i(T)$. Thus equation (1) now becomes

$$L_i(x_i) = \frac{P_i(x_i/T)}{P_i(x_i/N)} \geqslant K \tag{2}$$

as the criterion for deciding that a target is present.

The aim of a normalization process is to make a transformation of the variable x_i into a variable $y_i = y_i(x_i)$, so that the probability distribution of y_i is independent of i, i.e. $P_i(x_i/N) \rightarrow P_i(y_i/N) = P(y_i/N)$ for all i. Thus we can now set a threshold, Y, at the output of the device, and the probability of the sample, y_i, in any cell exceeding it will be the same for all cells.

For this threshold Y to satisfy the requirements of Decision Theory,

$$\frac{P_i(Y/T)}{P(Y/N)} = K, \quad \text{for all } (i)$$

This implies that $P_i(Y/T)$ must be independent of i.

This will be satisfied if the distribution $P_i(x_i/T)$ is constant over the range of values of x_i for which $y_i = Y$, all i. This will occur if this range of x_i is small and if $P_i(Y/T)$ is near its maximum. It will occur over a large amplitude range if the signal strength of the target is very indeterminate, as this will cause the probability density function to be almost constant over a large interval.

A normalization device will also satisfy the criterion if the function $P_i(y_i/T)$ is independent of i. Such a situation arises when the signal-to-background ratio of the target is constant, and can occur over limited range intervals in some reverberation limited conditions.

In many situations, however, the signal-to-background ratio varies appreciably, even over relatively small range intervals. For example, this can occur when the background rises rapidly and fades again over a short interval due to the intrusion of reverberation from the sea bed; the target echo, however, meanwhile stays sensibly constant.

In a similar manner, when a bank of filters is used in a doppler receiver, the background power is much higher in the zero doppler channel than it is in the outer channels because of reverberation. The signal from the target, however, will have the same energy in each channel and thus a variation in the signal-to-background ratio will occur across the filter bank.

It will be instructive to consider the case of a filter bank in a little more detail. Let us assume that the probability distribution of the noise or reverberation background is Gaussian, and that the target superimposes a signal of constant amplitude.

The probability distributions are then:

$$P_i(x_i/N) = (\sqrt{2\pi}\sigma_i)^{-1} \, \exp -\tfrac{1}{2}(x_i/\sigma_i)^2$$

$$P_i(x_i/T) = (\sqrt{2\pi}\sigma_i)^{-1} \, \exp -\tfrac{1}{2}\left(\frac{x_i - s_i}{\sigma_i}\right)^2$$

These can be changed by normalizing to give

$$P(y_i/N) = (2\pi)^{-\frac{1}{2}} \, \exp -\tfrac{1}{2}y_i^2 \tag{3}$$

$$P_i(y_i/T) = (2\pi)^{-\frac{1}{2}} \, \exp -\tfrac{1}{2}(y_i - s_i/\sigma_i)^2 \tag{4}$$

where $y_i = x_i/\sigma_i$

Then for a CFAR device, if a threshold Y_c is set, the probability of false alarm in any channel is

$$P_{FA} = \int_{Y_c}^{\infty} (2\pi)^{-\frac{1}{2}} \, \exp -\tfrac{1}{2}y^2 \, dy$$

Let us assume that the signal-to-background ratio in the centre channel has the minimum value (s_0^2) and that the ratio increases on either side. If s_i^2 is the signal-to-noise ratio in channel i, s_i may be represented by $s_i = \alpha_i s_0$, where $\alpha_i \geqslant 1$. Equation (4) then becomes:

$$P_i(y_i/T) = (2\pi)^{-\frac{1}{2}} \, \exp -\tfrac{1}{2}(y_i - \alpha_i s_0)^2 \tag{5}$$

and the Decision Theory criterion may be expressed as

$$\ln L_i(y_i) = y_i \alpha_i s_0 - \tfrac{1}{2}(\alpha_i s_0)^2 \geqslant \ln K$$

from equations (2), (3) and (5) on substituting $y_i = y_i(k)$

Thus

$$y_i \geqslant \frac{\ln K}{\alpha_i s_0} + \tfrac{1}{2}(\alpha_i s_0) \tag{6}$$

This equation sets the threshold for channel i. If we let Y_0 be the threshold in the centre channel, we have from equation (6)

$$\ln K = s_0/2(2Y_0 - s_0)$$

and thus the threshold for channel i, Y_i, is

$$y_i = Y_0/\alpha_i + s_0/2(\alpha_i - 1/\alpha_i)$$

We may now use these threshold values to determine the probabilities of detection and false alarm in each channel. In the example shown in Fig. 1,

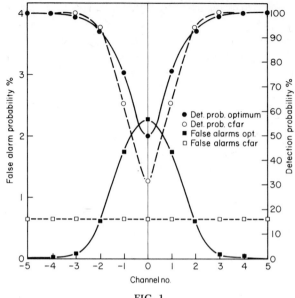

FIG. 1.

it has been assumed that the threshold Y_0 has been set at 2 (6 dB rel. to
noise power), s_0 is set at 2 (so that a probability of detection of 0.5 is
obtained), and the signal to background ratio increases by 3 dB per channel
from the central channel outwards

The CFAR case has been shown for the same average false alarm
probability as the Decision Theory case for the 11 channels.

It can be seen that in this example the Decision Theory criterion gives
a large variation of false alarm rates across the channels, and achieves a
significant improvement over the CFAR detector in the probability of
detection in the centre channel. Great significance should not be attached
to this one example, of course. The results that are obtained depend very
largely on the conditions assumed and on the statistical distributions of
the background and of the target. Nevertheless they do indicate a general
result for cases where the probability distribution function $P_i(y_i/T)$ has a
positive slope. (This will often occur for probabilities of detection greater
than 0.5.) Under these conditions the Decision Theory detector causes the
false alarm probabilities to decrease as the signal-to-background ratio (and
thus the probability of detection) increases.

In this analysis the assumption was made that the ratio $CN_iP_i(N)/Cr_iP_i(T)$
was constant for all the channels. It represents the usual assumption that a
similar performance is desirable in each information cell. It is possible,
however, to put forward arguments which would imply that the ratio
should be varied over the cells. For example it seems unreasonable to

assume the same prior probability for the presence of a target in two range bearing cells, one of which is at twice the range of the other. A more plausible assumption is that the probability at the longer range should be twice that at the shorter range. Other considerations might lead to varying the cost between cells. This aspect of optimality has been rarely considered in the past and merits further study.

One of the many simplifications made in the example was that the wanted echo would be contained wholly within one cell. It is usual, however, for it to spread over a number of cells in the range, bearing and doppler dimensions. The relative amplitudes of the signals in each of those cells contains shape and position information that could be of value in the detection process. This is distorted or destroyed by normalization when the process is carried out in each channel independently. The loss of information incurred represents a further departure of normalization devices from an optimal processor.

4. Discussion

The emphasis in the previous two sections has been to examine the possible deficiencies of Constant False Alarm Rate devices. From the discussion it might appear that the best sonar (or radar) system would be one based on the Decision Theory criterion. Indeed it is true that a detection device based precisely on this criterion would preserve all the information required for making a decision. In such a circumstance it would hardly be necessary to provide a display of raw data. A simple device which indicated range etc. when a threshold was exceeded would be sufficient.

The reason for displaying the raw sonar information to a man is to allow for a constantly changing situation. Due to variations in propagation loss, temperature structure, surface and bottom conditions, the background signals are constantly changing. In addition, the types of target that are to be detected might also change. Man, with his inherent ability to adapt to a constantly changing environment, is able to deal with such situations. It cannot be expected that mechanical devices can perform as well as a man over the whole range of conditions that are likely to be encountered.

If it is accepted that a display of raw sonar or radar data is required, then it is necessary to present to the man information in a form most easily assimilated by him. This might not be one based on an optimum processor. It is possible that a presentation which gives widely varying false alarm rates over the screen, as discussed in the example, would not be as good as one which gives a more or less uniform picture.

It is important, however, to present to the man all the information he needs to make a decision. CFAR devices destroy information, and to that extent they make the task of detection more difficult than it need be. A possible solution would be to present on the display a uniform picture by using a CFAR device, but supplying in addition other relevant information, such as the areas in which the background level is high.

5. Summary

Constant False Alarm Rate devices have a number of deficiencies. Not only do practical devices depart from the design objectives, but even perfect ones discard information and present the man with a varying decision criterion. Processors based on a Decision Theory criterion should in theory perform better than normalization devices. They are, however, difficult or impossible to implement to a degree which would allow for the constantly varying situation that exists in practice. Simplifying assumptions made in their design might present a picture on a display which was not matched to the man, and performance could be lost. It is still, therefore, an open question what criteria should be used in the design of practical systems.

DISCUSSION

C. van Schooneveld: In this paper the optimal processor has a high false alarm rate for low signal to noise ratios as the threshold is brought close to the noise level. This is in direct contradiction to the method of Mr Vezzosi in his paper "What is Optimality for an Adaptive Detection System?" which uses over AGC and has the effect of increasing the threshold for low signal to noise ratios.

M. Mermoz (in reply): Vezzosi's model is only valid when the signal is completely known including the amplitude. It is not valid in other cases, as for example the conditions in Hill's paper.

J. J. Dow (comment): Clippers can produce losses of much greater than 2 dB when these are multipath effects.

Answer: Agreed, but they do have the advantage of extreme simplicity.

J. J. Dow (comment): The effects of normalization applied at IF and after detection are not discussed in the paper. The effects obtained are dependent on the transmitted signal. For example for a tone pulse normalized at IF the output of the detector is not stationary in the frequency due to the change of background spectrum in the reverberation limited and noise limited conditions. For an FM pulse the output after detection however is substantially stationary.

G. Vezzosi: In the example given in the paper the input was normalized by dividing the root mean square of the noise. How is this done in practice?

Answer: This was just an illustration to show how a constant false alarm rate device is not always optimum. No particular consideration has been given to its practical implementation.

Pattern Recognition

I. G. LIDDELL

*Admiralty Underwater Weapons Establishment,
Portland, Dorset, England*

1. Introduction

The problem of classifying patterns occurs in a great many fields, e.g. Medical (classification of blood cells, cancer cells, etc.); Communications (automatic speech and alpha-numeric character recognition); Meteorology (recognition of weather patterns); Criminal Science (classification of finger prints) and Military Science (scene analysis, automatic radar and sonar target recognition, etc.).

Despite the considerable attention devoted to these and a great many other problems in recent years, progress has been very slow. The number of schemes which have been attended by sufficient success to merit their adoption in practical circumstances are very few indeed. In the main such schemes are those in which the design has rested heavily upon a combination of experience and intuition on the part of the designer. In certain cases success is possible because the patterns themselves are so distinctive that the problem becomes relatively trivial. Occasionally this situation arises because the form of the patterns themselves may be controlled by the designer, as for example in the case of character sets for Bankers' Cheques. The great mass of problems, however, present the designer with considerably more difficulty. Even some of the (apparently) simpler problems, such as recognition of typewritten characters, have not been attended by a great deal of success. The percentage of correctly recognized characters seldom rises above 90, which is not satisfactory for most practical applications, and frequently even these scores are only made possible by the exercise of considerable ingenuity on the part of the designer. In this sense it is fair to describe Pattern Recognition as still being more an art than a science.

Broadly speaking two quite distinct approaches to Pattern Recognition may be traced: an analytic approach in which investigators have attempted to make use of Statistical Decision Theory, Discriminant Analysis, etc., in order to formulate classification procedures; and an intuitive approach, exemplified by the Perceptron and a great many other adaptive or learning machines which are loosely modelled upon the processes of human perception and learning. The development of these two approaches has been largely separate and in terms of results there is probably little to choose between them. Nevertheless in the longer term, the future of Pattern Recognition will depend upon the strength and adequacy of the theoretical framework which can be erected to support it, and it seems unlikely that a satisfactory generalized recognition device will be achieved along the lines of presently conceived trainable machines. A generalized machine must be adaptive in a very broad sense, e.g. if the problem is to classify the letters of the alphabet, the decision that a particular example is the letter "O" has to be made without regard to the size of the letter. The problem may, however, be to classify circles of different diameter, which clearly requires a different process. As another example, the problem may be to recognize aircraft parked on an airfield. In this case the recognition device must be insensitive to rotation of the pattern. Returning to the problem of recognizing alphabetic characters, it may be impossible to effect correct recognition unless the patterns are constrained not to rotate. Conceptually, this degree of flexibility is possible, but in practice its achievement is formidable, and it would appear to be more appropriate at the present time to concentrate upon understanding ways of solving particular pattern recognition problems.

(a) *The nature of the problem*
The unknown pattern may be specified as an N-dimensional vector or measurement set:

$$X_N = (x_1, x_2, x_3, \ldots, x_N).$$

The problem of Pattern Recognition is to determine to which of Q distinct classes of patterns this particular example belongs. The solution to this problem lies in estimating the probability distribution which each class has on its members and in determining a decision rule which provides the best classification—typically by minimizing the total error rate.

A considerable amount of effort has been directed towards the solution of this problem, but comparatively little towards the systematic study of the input pattern vector itself. The vector X_N is chosen on an intuitive basis and may be affected by various practical constraints such as the need to operate in real time, or to use a particular type of sensor with limited resolving power. Taken together, these various factors give rise to a reduction in the amount of discriminant information in the pattern vector, in comparison with that contained in the source pattern. The importance of making a careful choice of pattern vector space cannot be too strongly emphasized, for the considerable loss of discriminant power which may result from an arbitrary choice cannot be regained no matter how sophisticated the recognition process is made. In a number of cases Man's own considerable pattern recognition capabilities may be put to good use in order to assess whether a particular choice of pattern vector has resulted in an unacceptably high information loss. Although useful, any such assessment tends to be qualitative and does not fill the need for a rational approach to the choice of the measurement space and for methods of comparing the relative information loss associated with different choices.

For the sake of clarity in subsequent discussion the term "Pattern Description" will be applied to that part of the problem associated with the choice of a suitable pattern vector measurement space, as shown in Fig. 1. The term "Pattern Recognition" or "classification" has been reserved for the subsequent discriminant stage, in which the measurements associated with a particular vector are used to ascribe the pattern to one of Q classes, and the overall process is referred to as "Pattern Analysis". In Fig. 1 the Pattern Description stage has been divided into two steps, the first of which is to produce a pattern vector which (ideally) contains all the information present in the source pattern. Thereafter this vector may be subjected to a process of dimensionality reduction or preprocessing, because of practical constraints placed upon processing time and storage etc., which produces a modified pattern vector for input into the classifier.

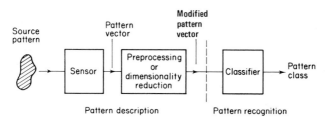

FIG. 1. Schematic pattern analyser.

2. Pattern Recognition (Classification)

In this section the operation of the final stage shown in Fig. 1 will be considered. It is assumed that a suitable modified pattern vector, Y_M, has been determined, where:

$$Y_M = (y_1, y_2, y_3, \ldots, y_M)$$
$$= F(x_1, x_2, x_3, \ldots, x_N)$$

where $(x_1, x_2, \ldots, x_N)$ is the pattern vector, X_N, before modification.

The problem we shall consider is to classify the vector Y_M as belonging to one of two possible classes. Baye's Decision Rule may be used to assign the pattern to Class 1 if:

$$\frac{P_1(Y_M)}{P_2(Y_M)} \geqslant \frac{C_2}{C_1} \cdot \frac{P_2}{P_1} \tag{1}$$

where $P_1(Y_M)$ is the probability distribution of Y_M for Class 1, $P_2(Y_M)$ is the probability distribution of Y_M for Class 2, P_1 is the prior probability of observing a member of Class 1, P_2 is the prior probability of observing a member of Class 2, C_1, is the cost of misclassifying a member of Class 1, C_2 is the cost of misclassifying a member of Class 2, and the costs of correct classification are assumed to be zero.

In many cases neither P_1 nor P_2 are known accurately. Moreover, the costs of misclassification may be virtually impossible to assess, or may vary. Such is certainly the case in sonar and radar problems, so that it is common to find the composite quantity $C_2 P_2/C_1 P_1$ determined as a subjective decision threshold.

If $P_1(Y_M)$ and $P_2(Y_M)$ may be expressed as M-dimensional multivariate Gaussian distribution [1], each takes the form:

$$P(Y_M) = \frac{1}{(2\pi)^{M/2}|K|^{\frac{1}{2}}} \exp\left[-\tfrac{1}{2}(Y_M - \overline{Y}_M)^T K^{-1}(Y_M - \overline{Y}_M)\right] \tag{2}$$

where $\overline{Y}_M$ is a vector of component means and K is the covariance matrix.
Taking logarithms gives:

$$\log P(Y_M) = -\tfrac{1}{2}(Y_M - \overline{Y}_M)^T K^{-1}(Y_M - \overline{Y}_M) + \text{const.} \tag{3}$$

and the resulting Baye's decision surface is a hyperquadric function determined by

$$\log \frac{P_1(Y_M)}{P_2(Y_M)} = \log \frac{C_2 P_2}{C_1 P_1} \tag{4}$$

i.e.,

$$\text{const} + \tfrac{1}{2}(Y_{M2} - \overline{Y}_{M2})^T K_2^{-1}(Y_{M2} - \overline{Y}_{M2}) -$$

$$-\tfrac{1}{2}(Y_{M1} - \overline{Y}_{M1})^T K_1^{-1}(Y_{M1} - \overline{Y}_{M1}) = \log \frac{C_2 P_2}{C_1 P_1}.$$

A more general result for Q classes, than that implied by equation (1) is to allocate the pattern to the kth class, when:

$$\sum_{\substack{i=1 \\ i \neq k}}^{Q} P_i P_i(Y_M) C_{ki} < \sum_{\substack{i=1 \\ i \neq j}}^{Q} P_i P_i(Y_M) C_{ji} \tag{5}$$

where $j = 1, \ldots, k - 1, k + 1, \ldots, Q$ and C_{ki} is the cost of wrongly allocating a member of the ith class to the kth class.

In many practical situations the form of the probability distributions governing the pattern vectors is unknown. Sebestyen and others [1, 2], have used Discriminant Analysis techniques, which are found to lead to results which are identical with the Bayesian Decision Theory Approach, in the general case of unequal covariance matrices for the various pattern classes. In the special case of equal covariance matrices, there exists a similar equivalence between the Decision Theory result and Fisher's Linear Discriminant [3]. Fisher used a linear function of the measurements,

$$Y = \sum_{i=1}^{M} a_i y_i \qquad (6)$$

which maximized the square of the difference between the mean values of Y for two classes divided by the variance of Y, which is assumed equal for the two classes.

A great deal of interest has centred around linear methods because of the obvious computational advantages. In general, poor results have been obtained, even when the classes do not overlap, unless the patterns are well separated in the measurement space and the dimensionality of the pattern vector is considerably greater than the number of classes. In the idealized case where each class is represented by a single point in the measurement space, perfect dichotomization of the space may be achieved by linear methods, provided that the number of classes is at most one more than the number of dimensions. In other words, any single class or group of classes may be separated linearly from the remainder. As the classes begin to be distributed about mean values, as shown in Fig. 2, rather than to exist as points, then a stage is reached, depending upon both the inter- and intra-cluster separation of the measurements, when linear separation is no longer possible, as shown in Fig. 2(c). If there exist unused measurement dimensions, or alternatively the dimensionality may be increased, a linear separation may still be possible provided that the patterns occur in "general position" in that space. Otherwise, non-linear methods become necessary.

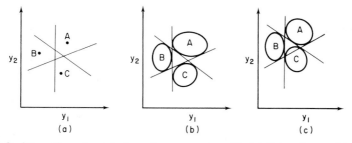

FIG. 2. Separation of $m + 1$ classes in m space ($m = 2$). (a) Point classes; (b) and (c) Distributed classes.

In situations where linear separation is not possible, even though the classes do not overlap, a variety of non-linear techniques have been established. One such method is effected by the Nearest Neighbour Classifier, in which each new pattern is tested against a set of exemplars ("typical" known patterns) from each class. The rule is to assign the new pattern to the class whose nearest exemplar lies closer than the nearest exemplars of other classes. In other words the pattern vector Y_M is classified as belonging to category A if:

$$\min_{\text{over } u} \left[\sum_{i=1}^{M} (y_i - a_{ui})^2 \right] < \min_{\text{over } v} \left[\sum_{i=1}^{M} (y_i - b_{vi})^2 \right]$$

$$< \min_{\text{over } w} \left[\sum_{i=1}^{M} (y_i - C_{wi})^2 \right] \quad \dots \text{ etc.} \tag{7}$$

for all Q classes and where y_i is the measured value of the new pattern in the ith dimension, and a_{ui}, b_{vi} are the corresponding values of each exemplar in each class. Apart from the obvious problems of storage, the method has the disadvantage of being very sensitive to "stray" samples amongst the sets of "typical" exemplars.

This difficulty is overcome by more sophisticated techniques such as the method of potential functions [4]. The principle of potential functions is explained by the simple two-dimensional representation in Fig. 3. In the figure, exemplars of two classes have been represented as positive and negative charges in the measurement space. Given this, a boundary may be computed which is the locus of points of zero potential in the space.

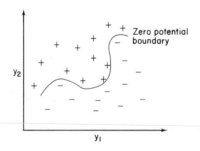

FIG. 3. Method of potential functions.

3. Pattern Description

The discussion so far has attempted to outline very briefly a small selection of Pattern Classification techniques. It has already been pointed

out, however, that the successful application of these techniques in practical circumstances will depend upon the amount of relevant information
carried in the vector which describes the original pattern. It is clearly
unsatisfactory blindly to try every possible classification process upon data
in which the necessary discriminating information has been thrown away
(or was, perhaps, never present). Yet such an approach has not infrequently
been adopted, resulting in considerable waste of resources. Despite the
difficulties involved, it remains the author's firm belief that in future a
great deal more effort will have to be directed towards the Pattern Description problem, together with the associated problems of evaluating the
information content of data and of reducing dimensionality.

A possible avenue of approach towards the solution of some of these
problems might be the application of such techniques as Principal Component Analysis [5], which is used in multivariate statistics in order to reduce
dimensionality. Figure 4 illustrates the aims of the technique in a two-
dimensional situation. The points shown, which represent a set of measurements in y_1, y_2 space, exhibit perfect positive correlation in Fig. 4(a). It is
possible therefore to define a single new variable, z_1, which completely
specifies the behaviour of the data. This variable is termed the first
principal component. In Fig. 4(b) a slightly more realistic scatter is shown,
in which the points no longer lie on a straight line but still exhibit a high
degree of correlation.

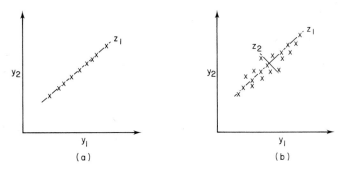

FIG. 4. Principal component analysis.

The first principal component, z_1, is now determined by finding the
coefficients of

$$z_1 = a_1 y_1 + a_2 y_2 \tag{8}$$

such that the variance of z_1 is maximized.

A second principal component, z_2, may then be defined, orthogonal to
z_1, i.e.,

$$z_2 = b_1 y_1 + b_2 y_2 \tag{9}$$

Plus Correlation $(z_1, z_2) = 0$ and Variance (z_2) is a maximum.

Treatment of problems involving more dimensions is a straightforward extension of the above method. In the rather trivial case shown, it could be that the second principal component accounts for so little of the observed variance that a satisfactory approximation is to use the first principal component alone to describe the data, thereby reducing the dimensionality from 2 to 1. When many dimensions are involved, a good approximation can often be obtained by using the first five or so principal components.

Needless to say the possibility of a significant simplification of this kind is most attractive, but there is a penalty to be paid in the sense that a great deal of storage and computation is required to carry out the analysis in the large number of dimensions involved in practical Pattern Description problems. The principal components are determined by examining the variance–covariance matrix of the set of measurements, and computing its eigen vectors and eigen values. The former define the coefficients of the linear combinations of the original measures which comprise the principal components, whilst the latter define the contribution of each principal component towards the total variance. In most practical pattern recognition problems the number of dimensions involved is large and may in some cases exceed 10^6. The problems of handling such large matrices are considerable, but in those cases which are feasible the potential gain to be made by effecting an enormous reduction in dimensionality may be well worthwhile. These difficulties may, of course, be eased by reducing the number of dimensions of the original pattern vectors by decreasing resolution. The danger of doing this is the familiar one of under resolving leading to a loss of vital discriminant information.

In order to effect a complete pattern analysis using this technique, some form of discriminant analysis must follow. One technique is to perform the Principal Component Analysis upon a mixture of data from all the pattern classes, followed by a Multivariate Discriminant Analysis upon those components which account for the majority of the scatter.

A further problem in Pattern Description may arise because the actual number of distinct classes of patterns is unknown. The patterns due to received echoes and noise from a variety of sources observed on a sonar display represent a good example of this situation. Although it is convenient to divide the data into the two classes of "Target Echoes" and "Background Events", the latter undoubtedly consists of a number of undefined sub-classes, and in consequence there is a very real danger of choosing a highly biased subset of background data upon which to base the Pattern Analysis. The number of background events which may be observed in quite short periods of time and covering only a very limited spectrum of operating conditions is very great and the time required for detailed examination of all of them would be prohibitive, and possibly wasteful.

To alleviate the problem, the technique of Cluster Analysis [6] may be applied, which seeks to partition the measurement space in some optimum sense. Consider for example the simple single and two dimensional problems illustrated in Fig. 5.

In the single dimensional case, the data may be imagined to be partitioned into two groups. By partitioning in all possible ways and calculating each

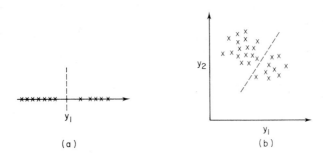

FIG. 5. Clustering patterns in one and two dimensions.

time the sum of the variance of the resulting pairs of groups (weighted by the number of members in each group), one particular partition will be found which minimizes this sum. This partition, shown as a dashed line in the figure, is that which gives optimum separation in the sense of maximizing the distance between the centres of gravity of the two groups. Figure 5(b) shows a similar situation in two dimensions. As the number of dimensions increases, however, the computational problem mounts very quickly. Once again, the dimensionality reduction possible by means of Principal Component Analysis may be of considerable aid.

In situations where it is felt that the unknown data is likely to be well behaved in the sense of forming closely knit clusters which are well separated from one another, the so-called "Chain" method of clustering may be adopted [7]. In the application of this technique the distance between the first and second samples is computed and the second becomes part of the first cluster if this distance is less than a predetermined minimum value. Failing this the second sample forms the basis of a second cluster. The process continues with successive samples, each one being added to an existing cluster or forming the first member of a new cluster if the minimum distance criterion is not satisfied.

(a) *Additional information*

In many practical applications, the information contained in a single pattern may be insufficient to obtain a low enough error rate even though the pattern vector may contain all the discriminant information in the source pattern. Such is the case in character recognition where it is clear that almost error free classification is achieved by man, largely because he can take into account the relationship between adjacent patterns, i.e. context. In sonar and radar, relationships between patterns in the time domain, i.e. tracking, may be used in order to improve the confidence with which events may be classified as target or non-target.

Other problems which are relevant to consider in the Pattern Description stage, relate to questions of invariance to translation, scale and rotation, which were touched upon in the introductory remarks. Invariance to translation is required in almost all applications, i.e., whether the patterns in question are alpha-numeric characters, blood cells, radar echoes, finger prints, etc., the performance of the classifier is required to remain unchanged regardless of their position in the field of view. In practice this is frequently achieved by "manual" centring (although automatic techniques have been evolved) e.g. centre of gravity, or edge registration.

As far as invariance to scale and rotation are concerned, the requirements will vary according to the task. As with translation, both manual and automatic techniques may be employed. Spiral scanning techniques and manual rotation of exemplars are examples of techniques which help to produce rotational invariant results. Invariance to scale change may be aided by using relationships between the ratios of scale related measures in the Pattern Description.

4. Testing

In view of the somewhat ad hoc procedures adopted in the search for satisfactory methods of Pattern Analysis, a considerable amount of trial and error is frequently necessary in most practical applications. Because of this adequate test procedures are of particular importance, and represent a part of Pattern Analysis which is often responsible for tying up considerable resources, both in terms of manpower and time.

Perhaps the simplest method is to measure the mean percentage of patterns misclassified, and in many circumstances (e.g. automatic recognition of numerals) this technique may prove adequate. Despite its simplicity, however, there always exists the danger that a single, overall figure may hide the fact that the performance against a particular class or classes is poor. More particularly, in situations where the number of patterns in each class is expected to be dissimilar and when the costs of misclassifying the various classes are not identical, the application of this technique can be highly misleading. In these circumstances a technique which may be of use is the Confusion Matrix, in which the number of events correctly classified and misclassified in each class, is shown in a matrix form. The number x, in element ij, of the matrix is incremented each time a pattern belonging to the ith class is classified as belonging to the jth class. Thus, the numbers x_{ij}, where $i = j$ which form the diagonal, represent the correctly recognized patterns in each class, and the remainder, where $i \neq j$, represent the misclassifications.

Another very useful technique may be borrowed from the field of communications, in the form of the Receiver-Operator Characteristic (ROC). In this method a graph is constructed of the probability of correctly classifying class A against that of misclassifying members of all other classes as belonging to class A. This device is particularly useful in problems which are effectively two class problems, i.e. those in which only two classes exist or in which one class is required to be separated from a number of others which need not be differentiated. The recognition of sonar echoes, mentioned earlier, and the recognition of cancer cells are good examples of problems which can usefully exploit this technique. In each case the numbers of patterns in the class of particular importance, wanted targets and cancerous cells respectively, are many orders of magnitude less than those which comprise the combined classes of non targets and healthy cells. In both cases the costs of misclassification are not identical. It is clearly much less disastrous to classify a healthy cell as cancerous, than vice versa. Likewise, the price to be paid for failing to classify a target correctly is likely to be higher than that for classifying a background event as a submarine.

A further attraction of the ROC stems from the fact that comparison of a variety of different methods of classification is excellently summarized on it. The performance of a random classifier is represented as a diagonal line passing through the origin and point 1, 1. Better than random performance is indicated by a curve lying above this line. In situations when more than two classes must be distinguished it is possible to draw a series of ROCs showing the probability of correctly classifying each particular class against the probability of members of the remaining classes being misclassified. Given a significant number of classes, however, it may rapidly become more difficult to assess their overall meaning.

In order that the estimates of error rates may be made to a satisfactorily high degree of confidence it is clearly necessary that the test patterns should be representative. Where possible, random sampling of the classes should be used, and even then it is the author's opinion that it is dangerous to follow the line adopted by some investigators, of using the same "representative" sample of data both for designing the process and subsequently testing it. It has already been mentioned in the context of sonar data that a finite, representative sample of background data may be difficult to obtain because of the probable sub-structure of that data in the form of a set of undefined classes.

Other problems arise because the Pattern Description stage is sensitive, or over-sensitive to phenomena like translation, magnification and rotation. Different problems put different constraints upon these properties and the performance of a given scheme will depend upon how well these constraints have been met.

5. Conclusion

An attempt has been made to present some of the underlying principles of Pattern Analysis. There are many gaps and whilst more emphasis has been placed on those techniques which have an established theoretical basis, the author is the first to realize that both the state of understanding of the overall problem and of the existing theory is very inadequate.

Particular stress has been laid upon the requirement for a more logical approach to the problems of Pattern Description, including dimensionality reduction. Even though solutions which are of general application may seem unlikely to be obtained for a very long time, any attempt to quantify and rationalize the approach to particular problems or sets of similar problems can only be beneficial at the present time.

On the question of generality, it is the view of the author that the construction of a generalized type of learning machine capable of emulating the pattern analysing capabilities of man, which so many investigators have sought after, remains at present a relatively remote prospect. Different pattern recognition tasks have different problems associated with them: One such task may require recognition without regard to rotation; another may solely be concerned with looking for rotation; some problems require a size invariant classification, etc. A device which seeks to classify hand-written letters, cannot be expected to compete with a man unless, like the man, it uses context, both in the sense of likely groups of adjacent letter, and in the sense of (often) having some prior knowledge of the kinds of words that may be encountered, e.g. in determining postal addresses the human operator knows by virtue of the task that certain types of words (towns, counties, etc.) are going to be encountered. Whilst a general purpose learning machine cannot be discounted in principle, it is clear that such a device would be likely to be many orders of magnitude more complex than the most sophisticated schemes attempted so far. In consequence it would seem far more profitable to progress more slowly and logically, by trying to understand basic principles, and continuing to build systems aimed at solving particular problems.

REFERENCES

1. Nilsson, N. J. (1965). "Learning Machines", McGraw-Hill, New York.
2. Sebestyen, G. S. (1962). "Decision Making Processes in Pattern Recognition", Macmillan.
3. Fisher, R. A. (1952). "Contributions to Mathematical Statistics", J. Wiley and Sons, New York.

4. Arkadev and Braverman (1967). "Teaching Computers to Recognise Patterns", Academic Press, New York and London.
5. Morrison, D. F. (1967). "Multivariate Statistical Methods", McGraw-Hill, New York.
6. Sokal, R. R. and Sneath, P. H. A. (1963). "Principles of Numerical Taxonomy", Freeman.
7. Bonner, R. E. (1962). On Some Clustering Techniques, *IBM J. Res. Dev.* 6, 353-359.

DISCUSSION

M. G. Vezzosi pointed out the considerable degree of similarity between the Statistical Decision Theory Approach to Pattern Recognition and that of Multiple Hypothesis Testing, and went on to enquire whether the problem might be approached from a Neyman-Pearson rather than Baysian standpoint.

In reply, the author stated that the latter approach had been used as an example only. A number of approaches to the problem are possible and it was not the intention of the paper to postulate an optimum solution to the problem of discriminating between a number of classes. Rather, the intention had been to emphasise the fact that the major problems lay in describing the input pattern vector in such a way that an unacceptably high information loss was not incurred.

Professor T. G. Birdsall commented on the fact that many problems of automatic pattern recognition still appear to stem from a lack of experience of the nature of the patterns involved, mentioning sonar targets in particular.

The author agreed, but stated that the evolution of techniques for both evaluating the information content of patterns and defining descriptors which are independent of man's intuition would be highly desirable. Nevertheless, it would seem that present approaches to any particular problem must of necessity rely upon man's experience of the patterns involved.

Professor P. Stocklin made reference to the concept of signal processing as a sequential process, beginning with energy detection, followed by estimation and classification, and wondered whether a similar view might be appropriate for Pattern Recognition.

It was the view of the author that such an approach would be likely to be essential in most cases. Although the use of as much information as possible at the earliest stage in the processing chain is desirable (a view which was confirmed by Mr E. R. Thomas), the nature of the process itself might preclude the use of all information. For example, in the sonar case, shape in the most general sense should include the time domain, i.e. multi ping information. In practice, single ping clues would be used first, and a likelihood ratio determined for a given event, and this would be modified by information from subsequent pings etc.

Post Detection Information Processing

ROBERT M. LAUVER

Bell Laboratories,
Whippany, New Jersey, U.S.A.

1. Introduction

A significant challenge for future oceanographic research will be to devise methods for analysing large bodies of data. As in outer space, improved inner space sensor technology now allows for the acquisition of massive amounts of data spanning temporal and spatial dimensions. In fact, modern technology now provides a plethora of data which at times staggers the researcher's will to attack it.

However, modern technology, with the advent of fast minicomputers utilizing mass storage memories and good interactive display devices, is providing a solution to the problem of analysing large quantities of data. The solution involves the employment of automated techniques to glean from the masses of raw data the essentials for review and analysis by the researcher. In fact, because of the quantities of data involved, much data reduction is now accomplished in *real time.*

Instrumental in achieving the feasibility of modern day data reduction has been the progress made over the past decade in information-processing techniques that can be implemented on improved computers. As evident in papers presented at this and past conferences, great progress has been made in signal detection and processing techniques. However, signal detection is only a necessary first step. The next question is, "What do I do with the signals now that I have detected them?"

This paper looks at the problem of post-detection information processing from a systems vantage point by formulating a model of an information processing system, providing an overview of available processing techniques, and projecting the impact on the system when included within the framework of the system is the end user, the researcher.

2. Statement of the Problem

The basic problem considered in this paper, as shown in Fig. 1, is to operate on a set of raw data input, using all available *a priori* information,

A PRIORI INFORMATION

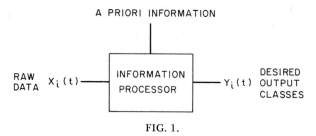

FIG. 1.

so as to produce a finite set of outputs, the desired output classes along with appropriate descriptive information concerning each member of each output class, and to keep this set of descriptors current as a function of time.

Because of the real-time constraint, input data is limited to the past and present. However, the output may be in the form of estimates of past or present events as well as predictions of future events.

This process may be further subdivided as shown in Fig. 2.

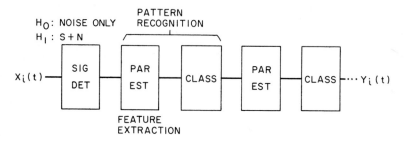

FIG. 2.

The initial process, shown as signal detection, is a process whereby the presence of possibly interesting events in the data set is indicated. Here, emphasis is placed on maximizing the probability of detection for a given risk function. The remaining processes form a chain of signal characterization and classification functions, in which each stage reduces the dimensionality of the process and the information content. In principle, at least, it would be desirable to carry all information to the last step of the process before the final decision takes place, but, in practice, hardware constraints (computer capacity and memory) dictate that data reduction takes place as early as possible.

The disciplines which deal with the analysis of this process are found in the areas of pattern recognition and multivariate data analysis. The next section of this paper will review, in tutorial fashion, some of the available processing techniques. In particular, emphasis will be placed on the requisite *a priori* information assumed by the technique.

3. Review of Pattern Recognition Techniques: [13, 18, 22]

The process of pattern recognition, as shown in Fig. 2, classically consists of two fundamental processes: (1) Feature Extraction; (2) Classification. Of the two, greater attention has been paid to the problem of classification than to feature extraction.

(a) Feature extraction
This process—also known as preprocessing, filtering or prefiltering, feature or measurement extraction, or dimensionality reduction—is usually the first stage of the pattern recognition process. The primary purpose of feature extraction is to select from the input data that "essential" information necessary for classification with an acceptable probability of error, while reducing the dimensionality of the data (and the information content) in order to reduce the processing load. Many approaches to feature selection have been intuitive, relying strongly on the designer's knowledge of the problem being solved to select that appropriate subset of the data deemed most appropriate for classification. Theoretical guidance [12] to feature extractor design has centred on techniques to rank order a preselected set of features or combinations thereof according to some criteria of optimality. Some examples are:

Feature space transformation. This technique [6, 26] employs a linear transformation of the data. Also known as Karhunen-Loeve analysis and principal factors, the technique, given a set of features $x_i, i = 1 \ldots n$ of a pattern vector $X(x_1 \ldots x_n)$ requires the solution of the eigenvalue equation

$$(\Sigma - \lambda I)W = 0 \tag{1}$$

where Σ is the convariance matrix of the features. The basis vectors of the transformed space consist of the eigenvectors of equation (1). Dimensionality is reduced to dimension M by taking only the eigenvectors associated with the M largest eigenvalues.

Information theoretic approach. Information theoretic approaches have been suggested for evaluating feature effectiveness. Both the divergence and the average information have been considered in rank-ordering features according to effectiveness.

The divergence [17] between two pattern classes ω_i and ω_j is defined as:

$$J_{ij} = E[L/\omega_i] - E[L/\omega_j] \tag{2}$$

where

$$L = \log \left[p(X/\omega_i)/p(X/\omega_j) \right]. \tag{3}$$

For Gaussian density functions with equal *a priori* probability of occurrence of the classes and equal covariance matrices, the probability of misclassification of an optimal classifier can be shown to be

$$e_{ij} = \int_{\frac{\sqrt{J_{ij}}}{2}}^{\infty} \frac{1}{\sqrt{2\pi}} \exp \left[-\frac{y^2}{2} \right] dy. \tag{4}$$

The probability of misclassification is a monotonically decreasing function of J_{ij} and therefore the magnitude of J_{ij} may be taken as a measure of the effectiveness of a set of features i.e. feature set X_k is considered more effective than feature set X_l if $J_{ij}(X_k) > J_{ij}(X_l)$. For more than two pattern classes the criterion of maximizing the minimum divergence or the expected divergence between any pair of classes has been suggested.

An alternate technique [16] for selecting effective features is to employ the average (mutual) information between a feature and the pattern classes. If a feature x_l can take on n_l possible values, denoted by $x_l(k)$ $k = 1, \ldots, n_l$, then the average information between x_l and the set of classes $\Omega(\omega_1 \ldots \omega_N)$ is given by

$$I_l = \sum_{i=1}^{N} \sum_{k=1}^{n_l} p[\omega_i, x_l(k)] \log \frac{p[\omega_i, x_l(k)]}{p[\omega_i] p[x_l(k)]} \tag{5}$$

Thus each feature may be rank ordered according to its mutual information content with the set of classes.

While the above examples illustrate techniques that give guidance for feature extractor design, it is generally recognized that the selection from the input data of the "essential" information necessary for classification is still one of the principal problems of pattern recognition.

(b) Classification

The problem of classification is to obtain data transformations that operate on patterns to produce decisions as to their class membership. If the pattern vectors $X(x_1 \ldots x_n)$ are viewed as points in an n dimensional space, the problem of classification may be viewed geometrically as a problem of mapping the points of the pattern vector feature space into the class category numbers or alternately dividing the feature space into separate regions where associated with each region (pattern class ω_i, $i = 1 \ldots N$) is a discriminate function $g_i(X)$. The discriminate function is such that if the input pattern vector X is in class ω_i, the value of $g_i(X)$ must be the largest. That is, for all $X \in \omega_i$

$$g_i(X) > g_j(X) \quad j = 1 \ldots N \quad i \neq j \tag{6}$$

Thus, in the feature space, the boundary partition (called the decision boundary) between regions associated with class ω_i and class ω_j, respectively is expressed by:

$$g_i(X) - g_j(X) = 0 \tag{7}$$

Techniques for classification depend greatly upon the amount of *a priori* information assumed available to the designer. When sufficient *a priori* information is available, classifiers may be designed that perform perfect classification or are optimized to produce a minimum probability of misclassification. However, rarely in practice is sufficient *a priori* information available. Four classes of available *a priori* information will be considered. These are:

Class I. Known statistical distributions. In this case, the statistical problem is sufficiently well understood that the conditional density functions $p(X/\omega_i)$ for all classes are known. In addition, the *a priori* probabilities of occurrence of a class $p(\omega_i)$ are known. Deterministic situations are considered a subset of this case.

Class II. Known statistical distributions except for parameters. In this case parameters of known distributions such as the mean or variance of a Gaussian distribution are unknown.

Class III. Unknown statistical distribution. In this case, both the functional form of the distribution and the parameters are unknown.

Class IV. Unknown number of classes. In each of the above cases, the number of pattern classes was assumed to be known. In this case, the distributions, parameters and number of pattern classes are unknown.

As less is known about the process, more reliance must be placed upon inferring parameters or structure from the data itself. In this case, inference is made from a group of patterns called the training set. Two types of training sets may be available. Either the training set contains a set of labels identifying the "true" class of the members, yielding techniques called "learning with a teacher" or "supervised learning" or the true classes are unknown yielding techniques called "learning without a teacher" or "unsupervised learning". When learning techniques are employed, the resultant classifier is then tried on a separate testing set of patterns to measure its generalizing ability. For supervised learning situations, a potential pitfall appears in that it is possible to overdesign the classifier by over-tuning to the training set at the expense of performance on the test set.

Examples of classification techniques follow.

Class I. Known statistical distributions. The simplest approach for pattern recognition is probably the deterministic approach of "template-matching", where a set of templates or prototypes, one for each pattern class, is stored. The unknown input pattern is compared with the template of each class and the classification is based on a preselected matching or similarity criterion. The templates may be exact representations of the classes or some derived feature such as the mean of the class.

Statistical classification techniques are built upon the theory of statistical decision theory [28].

The simplest case is the two-class problem where the conditional densities $p(X/\omega_1)$ and $P(X/\omega_2)$ and the *a priori* probabilities of the classes $p(\omega_1)$ and $p(\omega_2)$ are known. The maximum *a posteriori* probability criterion yields the decision rule: choose $X \in \omega_1$ if:

$$\frac{p(\omega_1/X)}{p(\omega_2/X)} > 1 \tag{8}$$

and choose ω_2 otherwise. By the use of Bayes rule

$$p(A/B) = \frac{p(B/A)p(A)}{p(B)} \tag{9}$$

an alternate form of equation (9) is choose ω_1 if:

$$\frac{p(X/\omega_1)}{p(X/\omega_2)} \geqslant \frac{p(\omega_2)}{1 - p(\omega_2)} \tag{10}$$

otherwise choose ω_2.

The probabilities $p(X/\omega_i)$ are called likelihood functions and the ratio $p(X/\omega_1)/p(X/\omega_2)$ is called the likelihood ratio. When cost functions can be associated with the various decisions, the decision rule is to choose ω_1 if:

$$\frac{p(X/\omega_1)}{p(X/\omega_2)} \geqslant \frac{p(\omega_2)}{[1 - p(\omega_2)]} \frac{(c_{12} - c_{22})}{(c_{21} - c_{11})} \tag{11}$$

where c_{ij} is the cost associated with choosing ω_i when actually ω_j is true. This class of classifiers is called Bayes classifiers.

For the multiclass case, Bayes rule becomes

$$p(\omega_i/X) = \frac{p(X/\omega_i)p(\omega_i)}{\sum\limits_{j=1}^{N} p(X/\omega_j)p(\omega_j)} \tag{12}$$

and the decision rule becomes choose that ω_i which maximizes $p(X/\omega_i)p(\omega_i)$.

Class II. Known statistical distributions except for parameters. Techniques in this class are often referred to as parametric learning techniques. When the true class of the members of the training set is known, iterative Bayes estimation techniques [1, 14] are used to estimate from the data the value of an arbitrary parameter vector θ. The Bayes estimator is

$$p(\theta/X_1 \ldots X_n) = \frac{p(X_n/\theta)p(\theta/X_1 \ldots X_{n-1})}{p(X_n/\theta)p(\theta/X_1 \ldots X_{n-1}) \, d\theta}. \tag{13}$$

The likelihood function may then be calculated from

$$p(X_{n+1}/X_1 \ldots X_n, \omega_i) = \int p(X_{n+1}/X_1 \ldots X_n, \omega_i, \theta)p(\theta/X_1 \ldots X_n, \omega_i) \, d\theta. \tag{14}$$

This technique is particularly useful when the functional form of $p(\theta/X_1 \ldots X_n)$ is independent of n, such as with the Gaussian distribution. Densities for which this holds true are called reproducing densities.

When the classification of the members of the training set are unknown, the computation of $p(\theta/X_1 \ldots X_n)$ is difficult because the sequence of occurrence of the classes ω_i is unknown [25]. The density $p(\theta/X_1 \ldots X_n, \omega_n^\alpha)$ could be computed where ω_n^α is an assumed sequence of occurrences of the classes ω_i; i.e.

$$\omega_n^\alpha: \langle \omega_{\langle}^1), \omega_{\langle}^2), \ldots, \omega_{\langle}^n) \rangle$$

and $\omega_{\langle}^1$), is a member of $\Omega(\omega_1 \ldots \omega_N)$. The complete density can then be computed from

$$p(\theta/X_1 \ldots X_n) = \sum_{\alpha=1}^{N^n} p(\theta/X_1 \ldots X_n \omega_n^\alpha)p(\omega_n^\alpha/X_1 \ldots X_n). \tag{15}$$

This technique is impractical since N^n alternatives must be considered after n observations, therefore approximations must be made.

One technique [23], is to implement the optimum detector and let the decisions of the detector constitute the estimate of the sequence in which the classes occurred. Such a detector is called a decision-directed detector. Percentages of error for high signal-to-noise ratios approach those of taught receivers.

Another technique [7] is to use the optimum detector for known parameters, but utilize estimates of any unknown parameters to replace these parameters in the optimum rule. In this case, the estimate of the unknown parameter must be made from the mixture distribution,

$$p(X) = \sum_{i=1}^{N} p(\omega_i)p(X/\omega_i) \tag{16}$$

where the component category distributions $p(\omega_i)$ and $p(X/\omega_i)$ are known except for the unknown parameter θ.

Class III. Unknown statistical distribution. Techniques in this class are often referred to as nonparametric learning techniques.

When the distribution functions used to determine the decision boundaries are unknown, one technique is to assume a functional form for the decision boundary and then estimate the parameters from a known training sample. A particularly simple form is a linear boundary which gives rise to the class of techniques called linear classifiers [19]. In this case, an unknown pattern X is assigned to class ω_1 if

$$g(X) = X \cdot W = \sum_{i=1}^{n} x_i W_i > \theta \tag{17}$$

and to class ω_2 otherwise, where W is a set of weight vectors and $g(X)$ is a linear discriminate function.

A special case of the linear classifier is the minimum distance classifier where the N pattern categories are characterized by a set of points $P_1 \ldots P_N$. The distance of a received pattern vector from a point P_i, such as the mean of the class which may be estimated from the training set, is then

$$|X - P_i|^2 = (X - P_i) \cdot (X - P_i) = X \cdot X - 2X \cdot P_i + P_i \cdot P_i \tag{18}$$

The minimum distance classifer then selects the largest of the expressions

$$X \cdot P_i - \tfrac{1}{2} P_i \cdot P_i \quad \text{for} \quad i = 1 \ldots N \tag{19}$$

The discriminate functions for this case are given by

$$g_i(X) = X \cdot P_i - \tfrac{1}{2} P_i \cdot P_i \quad \text{for} \quad i = 1 \ldots N \tag{20}$$

This dot product type of operation is sometimes called template matching, correlation detection, and matched filtering.

In general, training of linear classifers is accomplished by adjusting the weight vectors of the discriminate function in an error-correcting fashion when an erroneous decision has been made on the training sample. Three rules for correction have been suggested:

(1) Fixed increment rule: adjust by a fixed positive number.
(2) Absolute correction rule: adjust by the smallest integer that places the incorrectly classified sample in the correct class region.
(3) Fractional correction rule: adjust by a fractional amount of rule 2.

For linearly separable sets, it can be shown that these techniques converge to the correct solution.

Another technique [8] in this class is the "nearest neighbour decision rule", where patterns in the test set are assigned to the class of the pattern closest (in terms of an arbitrary distance function) to it as determined from the training set. For the large sample case, it can be shown that the probability of error is less than twice the Bayes probability of error.

For the unsupervised learning situation, several techniques have been suggested. An example [20] is the construction of histograms from the unclassified training samples. The histogram can then be used to estimate parameters for a decision rule, i.e., possible picking modes of the density as estimates of the means.

Class IV. Unknown number of classes. This class represents the minimal amount of available *a priori* information and thus the most difficult case to cope with. Unfortunately, this class is also fairly often encountered in underwater acoustics.

In this case, training must be of the unsupervised type since, if the classes of the training set were known and were truly representative of the output classes, the number of classes would have to be known.

Techniques [2] which deal with this class of problem are known as cluster analysis, numerical taxonomy, or mode seeking.

With little or no *a priori* information available, classification is based upon inferences obtained from the data. Here the problem is to sort a set of pattern vectors into subsets, where the subsets are as "similar" as possible. These subsets, then, are called clusters, where it is assumed that pattern vectors with like features derive from the same class and therefore will "cluster" together in feature space. However, in this case, there is not a defined standard objective performance criterion such as probability of misclassification for guidance. Rather, class separation is based upon a distance measure such as similarity for intraclass members of a set or dissimilarity for interclass members of sets. Common criteria in clustering are the sum of squared distances within a cluster and the sum of distances of the cluster centres from the overall average, where each cluster centre is weighted by the number of objects in the cluster (see also [21]).

A number of clustering techniques are available. Some examples are:

(1) *Total enumeration.* Conceptually, one could evaluate all possible partitions of a set of objects into clusters [11]. Total enumeration is generally computationally feasible only for a very small number of objects.

(2) *Hierarchical techniques.* Widely used in numerical taxonomy are agglomerative and diversive hierarchical clustering schemes. Agglomerative procedures [5, 15] start with N objects regarded as N single-element clusters. The first step is the clustering of two objects which optimize some criterion. These objects are then regarded as a single

object and their relation to every other object is adjusted, thus reducing
the order of the data set by one. This completes the first step of the
clustering process. The process continues until, after $N - 1$ steps, all the
objects belong to a single cluster.

Divisive procedures [9] start with all N objects grouped together into a
single cluster. These objects are then partitioned into two clusters by
optimizing some criterion function. Each cluster so formed is similarly
split into two more clusters and the process continues until each cluster
contains exactly one element.

(3) *Iterative techniques.* Nonhierarchical schemes have also been used
in clustering a fixed set of patterns [3]. Principally, iterative schemes
begin with an arbitrary set of exhaustive and mutually exclusive clusters
and successively improve the set of clusters by transferring patterns
from one cluster to another until no further improvement is possible.

(4) *Interactive clustering.* This technique [4] is used with an interactive
computer system connected to a graphic display. Here the researcher
has a variety of data displays, a data manipulation capability, and
interaction with the clustering routines.

4. Man Machine Interaction

The foregoing has reviewed some of the techniques for implementing the
processes of Fig. 2. However, in analysing large quantities of data, processing is frequently not left to machine techniques alone but involves a
continuous interaction of the researcher with the machine process during
the course of the processing. Thus, Fig. 2 may be expanded to a more
complex system that includes the human operator, as in Fig. 3. In this
case, provision is made for interaction with the individual subprocesses,
allowing the operator to change the machine decision according to con-
textual review of previous machine decisions, finer grained examination of
the input raw data, or perhaps his whim.

Human operators are also information processors which possess both

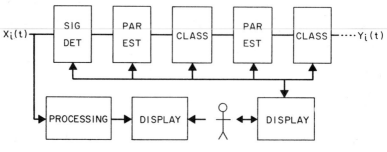

FIG. 3.

good and bad characteristics when compared to machine processors. Man's ability as a pattern recognizer in certain situations has been well established. However, his capacity to process ever-increasing amounts of data and to deal with ever-increasing dimensionality of the data is limited, as is his ability to cope continuously with long temporal stretches of data without fatigue. Thus in the future, man may constitute to a higher and higher degree the limiting factor in man-machine performance.

An example [24] of the operator's impact on total system performance is given below to illustrate the interaction effect of operator performance and machine performance. A simple model is shown in Fig. 4. Here C_i and M_i represent machine classifications on a set of data, C_i being the set of

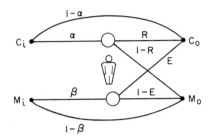

FIG. 4.

those correctly classified and M_i being those incorrectly classified. These may also be viewed as the results of a clustering technique, C_i being correct clusters and M_i incorrect clusters. It is assumed that a percentage of these decisions are reviewed by the operator and that he changes the machine's decisions, perhaps based on examination of the raw data. Thus, he reviews α of the correct clusters and β of the incorrect clusters. Next, the operator is viewed as a decision device characterized by two parameters R and E. R is the percentage of correct clusters that he confirms to be correct and $(1 - R)$ is the percentage of correct clusters that he incorrectly classifies. Similarly, E is the percentage of incorrect clusters that he corrects and $(1 - E)$ is the percentage of incorrect clusters that he incorrectly classifies. The output classes C_0 and M_0 are the final results of the system. The question is, "What performance improvement did the operator bring to the system?" From the figure it can be seen that the criterion for the operator to increase the performance of the system is

$$\frac{E}{1 - R} > \frac{\alpha C_i}{\beta M_i} = G \tag{21}$$

where G is the ratio of good to bad machine clusters that he reviews. This condition is plotted in Fig. 5 as a function of E and R. For the operator to improve the system performance, his operating point (E, R) must be

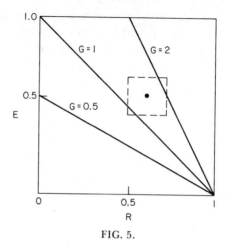

FIG. 5.

above the machine operating curve G. Some experimental results have shown human operating points falling within the dotted square shown on the diagram. A representative point ($E = 0.5$, $R = 0.6$) has been chosen as an example. In this case, for the operator not to change the system performance, G must be equal to 1.25; i.e., the ratio of good to bad machine clusters reviewed by the operator is 1.25. For a higher ratio, the operator degrades system performance, for a lower ratio, he improves system performance. This can better be seen in Fig. 6, where the percentage improvement in system performance, defined to be

$$\% \text{ Improvement} = \frac{C_0 - C_i}{C_i} = (R - 1) + \frac{E}{G} \qquad (22)$$

is shown as a function of G.

This simple example has shown that the impact of a human operator

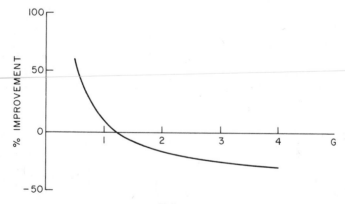

FIG. 6.

upon an information processor's performance cannot be ignored. If the operator is viewed as an imperfect information processor then, depending upon the relative effectiveness of the man versus the machine, it is possible that he will degrade the total system performance if care is not taken in the design of the man/machine interface.

5. Truth

The final consideration taken up in this paper concerns the criteria by which system performance is judged. The ultimate criterion is experimental verification, which requires testing with data where "truth" is known and can be compared with system output. For nonsupervised learning situations, it may be possible to identify samples for testing purposes which are not sufficiently representative to be used as training samples.

In underwater acoustics, where performance must be judged on the basis of large masses of data, truth will become an ever-increasingly difficult entity to define.

This problem may be viewed from an information theoretic point of view (see Fig. 7). The desired output T, which is a functional transformation of the source data X; i.e., $T = D(X)$, is defined to be truth. System evaluation then consists of comparing the output of the system under test, Z, with truth to determine system performance. However, in practice truth is not available to us and must be estimated from the input data Y. Thus, a truth processor is employed, which in practice is a computer assisted man laboriously post-analysing a segment of the data to determine an estimate of truth $\hat{T}$.

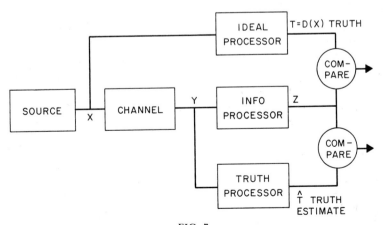

FIG. 7.

Studies [27] have shown that, independent of the information-preserving properties of the data transformation $D(\ )$, the reduction of the entropy of the output U; i.e. the reduction of the uncertainty concerning truth, achievable through the use of a feed-forward estimating information processor (truth processor) is always bounded from above by $I(X; Y)$ where

$$I(X; Y) = \int_{-\infty}^{\infty} dX \int_{-\infty}^{\infty} dY\, P(X; Y) \ln \left[\frac{P(X; Y)}{P_x(X)P_y(Y)} \right] \qquad (23)$$

is the acoustic channel transmittance. Thus we are faced with the prospect that even with the best truth processor available, it may be impossible to completely determine truth. In practice, to determine what is truth is further compounded by the fact that the analyst or operator is not an optimal information processor or "truth processor", as shown in the previous section. Therefore, as real-time information processors improve in capability, evaluation will more and more involve the problem of resolving differences between the processor under test and the current truth estimate to determine new "truth". Thus "truth" for system evaluation purposes will become a bootstrap operation whose definition depends, in part, on the output of the system under test.

6. Conclusion

The following observations have been made in this paper.

(1) Underwater acoustics research offers an interesting challenge for the use of large computer-interactive information-processing systems.
(2) These systems will combine signal processing, information processing and man-machine interactive techniques.
(3) The choice of applicable information processing technique is governed by the available *a priori* information.
(4) Study and optimization of these techniques should take into account the interface with a human operator.
(5) The operator may become the limiting factor in man-machine systems and have the potential for degrading performance if his limitations are not taken into account.
(6) Truth for system evaluation purposes will increasingly be influenced by the system under test.

REFERENCES

1. Abramson, N. and Braverman, D. (1962). Learning to recognize patterns in a random environment. *IRE Trans. Information Theory* **IT-8**, 58–63.
2. Ball, G. H. (1965). Data analysis in the social sciences: What about the details? *Proc. Fall Joint Computer Conf.* 533–559.
3. Ball, G. H. and Hall, D. J. (1966). ISODATA, an iterative method of multivariate

analysis and pattern classification. Presented at the International Communications Conf., Philadelphia, U.S.A.

4. Ball, G. H. and Hall, D. J. (1970). Some implications of interactive graphic computer systems for data analysis and statistics. *Technometrics* **12**, No. 1, 17–31.

5. Bonner, R. E. (1961). On some clustering techniques. *IBM J. Res. Devl.* **8**, 22–32.

6. Chien, Y. T. and Fu, K. S. (1962). On the generalized Karhunen-Loeve expansion. *IEEE Trans. Inform. Theory* **13**, No. 2, 518–520.

7. Cooper, D. B. and Cooper, P. W. (1964). Adaptive pattern recognition and signal detection without supervision. *IEEE International Conv. Rec.* pt. I.

8. Cover, T. M. and Hart, P. E. (1967). Nearest neighbor pattern classification. *IEEE Trans. Information Theory* **IT-13**, 21–27.

9. Edwards, A. W. F. and Cavalli-Sforza, L. L. (1965). A method for cluster analysis. *Biometrics* **21**, 372–375.

10. Fralick, S. C. (1967). Learning to recognize patterns without a teacher. *IEEE Trans. Information Theory* **IT-13**, 57–65.

11. Fortier, J. J. and Solomon, H. (1966). Cluster procedures. *In* "Multivariate Analysis" (Paruchur. R. Krishnaiah, ed.), pp. 493–506, Academic Press, New York and London.

12. Fu, K. S., Min, P. J. and Li, T. J. (1970). Feature selection in pattern recognition. *IEEE Trans. Systems Science Cybernetics* **SSC-6**, No. 1, 33–39.

13. Ho, Y. C. and Agrawala, A. K. (1968). On pattern classification algorithms; introduction and survey. *Proc. IEEE*, **56**, No. 12, 2101–2114.

14. Keehn, D. G. (1965). A note on learning for Gaussian properties. *IEEE Trans. Information Theory* **IT-11**, 125–132.

15. Lance, G. N. and Williams, W. T. (1967). A general theory of classificatory sorting strategies, I. Hierarchical Systems. *The Computer Journal* **9**, No. 4, 373–380. *Also* II, Clustering Systems. *The Computer Journal* **10**, No. 3, 271–277.

16. Lewis, P. M. (1962). The characteristic selection problem in recognition systems. *IRE Trans. Information Theory* **IT-8**, No. 2, 171–178.

17. Marill, T. and Green, D. M. (1963). On the effectiveness of receptors in recognition systems. *IEEE Trans. Information Theory* **IT-9**, No. 1, 11–17.

18. Nagy, G. (1968). State of the art in pattern recognition. *Proc. IEEE* **56**, No. 5, 836–862.

19. Nilsson, N. J. (1965). "Learning Machines", McGraw-Hill, New York.

20. Patrick, E. A. and Hancock, J. C. (1966). Nonsupervised sequential classification and recognition of patterns. *IEEE Trans. Information Theory* **IT-12**, 362–372.

21. Rogers, D. J. and Tanimoto, T. T. (1960). A computer program for classifying plants. *Science* **132**, 1115–1118.

22. Rosen, C. A. (1967). Pattern classification by adaptive machines. *Science* **156**, No. 3771, 38–44.

23. Scudder, H. J. (1965). Adaptive communications receivers. *IEEE Trans. Information Theory* **IT-11**, 167–174.

24. Seagren, M. K. Internal communication.

25. Spragins, J. (1966). Learning without a teacher. *IEEE Trans. Information Theory* **IT-12**, 223–229.

26. Watanabe, S. (1965). Karhunen–Loeve expansion and factor analysis, theoretical remarks and applications, Information Theory, Statistical Decision Functions, Random Processes, Trans. 4th Prague Conf. 635–660.

27. Weidemann, H. L. and Stear, E. B. (1970). Entropy analysis of estimating systems. *IEEE Trans. Information Theory* **IT-16**, No. 3, 264–270.

28. Whalen, A. D. (1971). "Detection of Signals in Noise", Academic Press, New York and London.

DISCUSSION

J. J. Dow: In the figure illustrating the percent improvement that the operator brings to the system, is there any way of knowing how well the machine is currently performing? If so, it may provide guidance as to which areas are candidates for further effort.

Answer: There is no general answer to the question. Each specific problem must be viewed individually. There are cases where the machine is better than the man and conversely, cases where the man is better than the machine.

J. J. Dow: So the answer is that we are all over the curve.

Answer: Correct.

W. S. Liggett: Your discussion of system performance evaluation makes a case for tests under controlled conditions so that you can determine truth from the source.

Answer: That is correct and controlled condition experiments will play an important role in system evaluation. However, there is always the question of how general and representative the controlled experiment is and the answer is that controlled experiments usually represent only a subset of the conditions against which the system has been designed to operate.

I. G. Liddell: (Comment) I agree with the point that there is a lot to be learned about how man performs. However, an even greater problem arises when you try to put the man and the machine systems together. The good things that the man can do and the good things that the machine can do become distorted when you try to put them together. One must consider possible changes in man's performance when confronted with decisions made by the computer and the fact that he may react differently to the same data when the computer is part of the process than he did before.

Three-dimensional Displays

D. J. CREASEY and J. A. EDWARDS

University of Birmingham, Birmingham, England

1. Introduction

Commonly, in echo-ranging systems, only two-dimensional (2-D) data is determined; these being usually the range and bearing of each detected target. In some situations however, a knowledge is required of the three-dimensional (3-D) coordinates of each target—as for example in air-traffic control. In displaying such data, the display system may or may not seek to provide a 3-D illusion. Where the target field is relatively simple and changing slowly, satisfactory results can be obtained by displaying the information on two 2-D displays. In other applications, information regarding the third dimension may, with advantage, be recorded in code on a 2-D to give a single display. However in complex or quickly changing situations, the task of the operator may be simplified considerably by displaying the data on an illusory 3-D display.

Illusory displays of the contact-analogue type have proved useful for aircraft and submarine control [1, 2], but little attention has been given to the use of such displays for portraying real target fields. Holographic techniques are of course available for image-reconstruction in three-dimensions although most of such applications reported refer to optical systems in which information is not presented at the time of collection. When applied to echo-ranging systems, holographic techniques are insufficiently developed for real time applications, and they will not be considered.

The type of display considered in this paper is that which seeks to display the range of a target, or part of a target, by utilizing one or more of the visual depth cues [3, 4] which are the indicators of depth employed when viewing everyday scenes. In other words these cues are the items of evidence used by an observer's eye–muscle–brain system to perceive his environment [5].

For work involving the relatively quick estimation of distance in a complicated situation, rather than the precise measurement of distance, the most suitable signal processor seems to be the human brain. Situations of

745

this sort arise, for example in vehicle-control, and displays for these situations will be described. Displays involving the normal processes of perception are also of interest in certain applications requiring target recognition. Preliminary work in this area will also be discussed.

2. 3-D Perception

The perception of depth appears to be very much a process of the brain rather than one of simple vision. It seems that one's brain accepts a construction of a scene which is the most likely considering the available cues in the light of previous experience. Previous experience is of considerable importance and is the basis of many optical illusions.

The appropriate cues are described in detail in the relevant literature and will be mentioned only briefly here. It is convenient to divide these cues into two classes: monocular and binocular. The monocular cues are relevant to the perception of objects at a distance and to the understanding of pictures. It is of significance that, given adequate experience, we are capable of accepting a two-dimensional encoded representation of three dimensions and of perceiving depth while yet being aware that no depth exists.

The monocular cues are useful in displays having a familiar structure especially if the displayed targets can be focused at a distance. The binocular cues are relevant to the perception of close objects. These stereoscopic cues can be particularly strong and are useful where the display contains no familiar structure. In order to make use of the valuable stereoscopic effects, the range of distant targets must be scaled appropriately.

The monocular cues include object recognition, perspective, the casting of shadows, parallax, the interposition of a near object in front of a far one, brightness and clarity. The most important of these is probably recognition but perspective provides a very strong cue for observers with appropriate experience and the cues of parallax and interposition are valuable especially when motion is involved. Accommodation, which is the adjustment of the focus of the eye is also a monocular cue although probably not a particularly informative one. The binocular cues include convergence of the eyes and the disparity of the images recorded on each retina. Stereoscopic displays utilizing these cues can provide a very convincing illusion of depth.

In situations where the observer is moving relative to his environment, as is the case in vehicle-control applications, depth information is provided for all objects with a relative velocity component across his line of sight by a velocity gradient which exists between the observer and his horizon. Distant objects appear almost stationary whereas near objects appear to move quickly. This effect is, of course, a valuable aid to the perception of the observer's own motion. The perception of the component of motion along the observer's line of sight will depend on the rate of change of angular size of the object in view in monocular displays but will involve convergence and disparity in stereoscopic presentations.

A cue, sometimes available to display designers which is not naturally available, is that of the order in which the scene is constructed. In echo-ranging systems there may be an observable time delay between the presentation of targets suitably spaced in range. Further, the temporal introduction of a ground plane may serve to emphasize other depth cues which may be present.

3. Applications

Specific situations in which the authors and their colleagues have found the application of these principles to be of benefit are described briefly below.

(a) P.P.I.-based perspective display

One application has involved the display of P.P.I. information obtained from a short range mobile X-band radar, the maximum range being 100 metres. This radar was built for evaluation as an aid to drivers of, for example, airfield emergency vehicles in conditions of poor visibility. A similar application might be the navigation of marine craft in confined waters.

In complex cluttered environments, interpretation of the two dimensional P.P.I. obtained from the radar was found to be difficult. In simple, well-structured, environments the general arrangement of targets could be seen easily enough but the rapid assessment of the progress of the radar vehicle with respect to the environment was not easy. Some improvement was obtained by the use of a distorted display giving a perspective effect such as is described below. Even with such a display, the information presented to the driver was inadequate for confident progress except in familiar, well-structured surroundings.

For the purposes of this type of display, which relies on geometrical perspective, the targets are assumed to be situated on a flat horizontal "ground plane", this being the plane of the P.P.I. The observer is assumed to be at height H above this plane viewing the targets through a vertical transparent screen in the "picture plane" which is positioned at a distance D before him.

The apparent position of any target can then be specified with respect to axes in either the ground plane or the picture plane. If the axes are chosen as shown in Fig. 1, where the picture reference axes are formed by the horizon and a central vertical line, the ground coordinates (x, y)

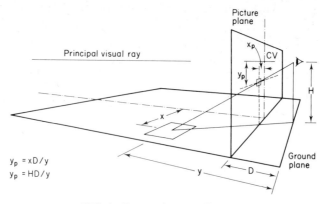

FIG. 1. Perspective coordinates.

can be transformed into picture coordinates (x_p, y_p) using the following relationships:

$$x_p = xD/y \quad y_p = -HD/y$$

If the P.P.I. deflection signals are resolved into analogues of the ground coordinates, the perspective deflection waveforms can be obtained by the implementation of these equations using analogue computer techniques. Further, if the P.P.I. scan signals consist of a range scanning waveform $f(t)$ modulated by the sine and the cosine of the aerial rotation angle θ, then over a small sector, the perspective deflection coordinates are approximately as given below:

$$x_p \doteq k_1 D \theta \quad y_p = -k_2 HD/f(t)$$

These equations describe a distorted "B-scan" display in which the range deflection is non-linear.

The targets are portrayed by intensity modulation of the C.R.T. In addition it is helpful in some circumstances, especially when estimating speed, if the target echo signal is made to give a supplementary vertical deflection as in an "A-scan". If this is done then the significant echoes should be hard limited and the supplementary deflection attenuated as a function of range in order to preserve the perspective.

Experience with the display suggests that an improvement results provided that enough targets are present and that structure of the target field is recognizable. Tests using a simulator in which these conditions were met gave good results. For example, subjects were shown a "road" having an apparent width of 3 m as seen from a height of about 1 m at a picture plane distance of 150 mm. The display aperture was approximately 100 mm. It was found that stationary targets on the road could be estimated with reasonable accuracy without practice. Typically, at 20 m range, distance ahead could be estimated to within 3 m and lateral position to within 1 m.

Subjects were also shown a similar display in which targets moved along the road towards the observer, this phenomenon being readily accepted as indicating the observer's speed. The performance of untrained observers at estimating speed was poor, being typically 20 km/h in error at 50 km/h. The addition of the video signal to the vertical deflection signal gave a significant improvement.

(b) Range-bearing displays in sonar

The target-field data is contained in the three space variables, range R, bearing angle, θ, and evaluation angle ϕ, together with the intensity of the received signal. Hudson [6] has shown that by keeping the elevation angle constant and very small, it is possible to produce a 2-D display which gives a 3-D illusion. Hudson's display plotted range against bearing. Firstly the display was range gated, and the target field scanned in bearing at this particular range. The display produced a horizontal raster upon which a vertical deflection was superimposed; this deflection being proportional to signal intensity. At the end of each bearing scan, the range gate and the vertical deflection of the raster were both appropriately incremented. The sonar beam used was directional in both the bearing and elevation axes. The display so formed when a 200 mm metal disc target was used is shown in Fig. 2. The pronounced scattering from the front edge and back surface can be seen together with a slight scattering in the centre.

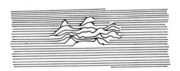

FIG. 2. Tracing of a range/bearing sonar display of a disc showing the front edge and far side with slight scattering from the centre.

(c) Bearing-elevation perspective displays

A perspective display of the features on an undulating surface can be obtained by drawing a series of sections through the surface [6]. If the undulating surface is the sea floor then the display is effectively a series of narrow beam echo-sounder tracings of the sea floor taken on near parallel courses and placed side by side. Figure 3 shows a relief display of a cylinder and two stones on a sand floor, the total area of the display being about 0.1 m². To obtain this display, the transmit/receiver transducer was placed above the sand floor and the display raster formed was the θ, ϕ raster of the polar coordinates. Superimposed upon this raster was a deflection in the ϕ direction which was inversely proportional to the range of the target surface. To obtain this a negative going ramp generator was started when the transmitter pulsed, and this ramp waveform was sampled when the first major

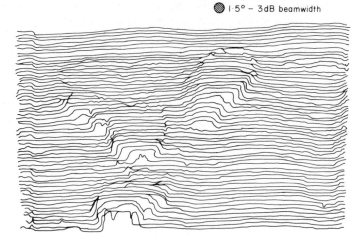

FIG. 3. Tracing of the relief display of cylinder and two stones on a sand-floor. The cylinder is the middle of the three objects.

echo was received. The transducer was made to scan very slowly in the θ direction, several pulses being transmitted per beamwidth. Consecutive heights of the sampled ramp waveform were smoothed and the resulting smoothed samples used to deflect the raster in the ϕ direction.

By suitable choice of the deflection sensitivity, the display so formed can be made to appear as a perspective picture of the surface viewed from an angle. Furthermore, if suitable proportions of the θ and ϕ raster deflections are added to each other the perspective effect on the basic raster can be altered and a parallax effect induced.

The technique is unsuitable should mid-water targets exist which do not extend downwards to the viewed surface. For example a spherical target suspended above the sea-floor would give a display which looked like a post connected to the sea-floor. Also the target strength parameter has been lost in the display of Fig. 3. This display was made on an x–y plotter which did not possess any intensity-modulation facility. This could be rectified by using a cathode-ray tube display in which case not only could the perspective and parallax cues be incorporated, but also shading. Furthermore, there is no reason why a stereo pair of displays should not be produced to give the added cues of stereoscopy and convergence.

(d) Stereoscopic displays

The stereoscopic effect has been investigated in another type of situation. Here of interest were the positions of mid-water targets relative to each other and to the bottom and surface. The sonar used was of the phase sampling type, which dispenses with amplitude information. This uses the phase of an incoming plane wave measured on a sectioned array to determine the bearing of a target [7]. In some biological studies it was noticed that targets often disappeared from the display having moved out of the sonar sector in the non-scanned direction. A second complete sonar was produced so that target movements could be observed in all three dimensions. The two receiving arrays were placed on mutually perpendicular axes and a single transmitting array insonified the volume of interest. Because of the binary nature of each signal processor, each processor was able to recognize only a single target in a single range annulus on any one scan [8]. The outputs of the two signal processors were originally displayed on two separate B-scan displays. While this was satisfactory for the observation of stationary fields, operators had the greatest difficulty in following moving targets on both B-scans simultaneously and it was decided to attempt a stereoscopic display presentation for real time viewing.

It was assumed that a target echo processed and accepted by the elevation-scan sonar would also be present in the bearing-scan signal processor. The control waveforms of the two signal processors were synchronized and the threshold of one processor was relaxed so that the threshold of the other processor controlled the collection of data. Thus the processors were made to process data from the same target simultaneously.

In a normally sighted person, each eye receives a slightly different view of a scene. The brain combines the two images and judges depth information from the difference in the two images [9]. The two images differ most for objects near the eye, and least for distant objects. For very distant objects, the lines of sight are nearly parallel and there is little stereoscopic effect. The stereoscopic threshold is measured in terms of the parallatic angles subtended by two targets to the eyes, see Fig. 4. For normal viewing this

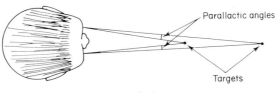

FIG. 4.

threshold depends upon the background lighting level: in poor lighting conditions it is of the order of 6 sec., while it approaches 1 sec for good lighting conditions [10].

The range coordinate in a stereoscopic display can be made to differentially deflect the two θ, ϕ displays which form the stereo-pair. This deflection is in the θ direction. For a target on zero bearing, at a range R, the differential angular deflection between the operator's eyes spaced a distance x apart should be 2 arctan $(x/2R)$. This approximates to x/R when $x \ll R$.

There are a number of ways of presenting the stereo-pair of images, one image being viewed by each eye. The optical images may be differentially polarized and the operator can view the displays through spectacles, each lens being appropriately polarized [11]. Alternatively the stereo-pair may be colour-coded, one red and one green. The observer then views the images with the appropriate colour filter in front of each eye. This technique has been tried, but the persistence of colour cathode-ray tubes is insufficient for sonar and radar use [12]. The method found to be most successful is to provide two completely different θ, ϕ displays and to use mirrors, prisms and lenses to present one display to each eye.

Using this last technique, it is possible to position the imaged displays, one directly in front of each eye. Targets at infinity will not be differentially deflected and the eyes will be in their relaxed positions for such targets. The deflection waveform should be dependent not only upon the range but also the bearing of any target. However quite satisfactory results have been obtained using a lateral deflection which decreases linearly with increasing range, zero deflection occurring at the longest range of interest. This gives a small range distortion, but even at short ranges where the distortion is a maximum it is not noticeable.

A rapid assessment of the value of such a display was made by testing 22 subjects with six different test patterns, each pattern with 3 targets at different ranges, bearings and elevation angles. Although the number of subjects was too small for statistical significance, only two subjects experienced zero stereoscopic effect, and all but 5 subjects were able to judge the relative positions of three targets correctly in over 50% of the tests.

To date there has been no subjective system evaluation using real sonar data, but this is about to be done. The positioning of single targets relative to the background formed by the bottom and surface of lakes has been determined correctly by those who have tried the display system using recorded sonar data.

4. Conclusions

The design of 3-D displays is, of course, a man-machine interface problem and, as such, it involves considerations of human performance as well as of equipment design. Human performance is variable and appropriate data, when available, will be of a statistical nature; it is likely that no illusory 3-D display will be compatible with all possible operators. Even a well designed display will be unrealistic unless the detail provided by the echo-ranging system is comparable with the detail obtainable by direct vision.

In any case radar and sonar images usually differ from those obtained optically. Nevertheless the picture decoding capabilities of most trained operators permit the use of unrealistic displays in which the information coding is familiar. Learning is an important factor in operator performance.

Displays which rely on geometrical perspective may be useful in applications such as vehicle control in well-defined channels. Apart from possibly easing display interpretation, these displays have the advantage over plan displays of a better selection of displayed information as a function of range. The display at shorter ranges is expanded to give better lateral and range resolution, and the position of an observer with respect to the target field is made obvious. Appreciation of movement, both target movement and observer movement with respect to the displayed environment is more natural.

Stereoscopic displays are a form of perspective display to which another strong depth cue has been added, and they can provide an effective method of presenting echo-ranging data. Such displays are ineffective for some 5% of the population [12], and difficulty may be experienced by untrained operators in fusing the separate images. Further the apparent range interval over which images can be fused may vary considerably from subject to subject. This last difficulty may be resolved, depending upon the application, by allowing each subject to adjust the stereo-deflection sensitivity.

ACKNOWLEDGMENTS

The authors would like to thank their many colleagues in the Department of Electronic and Electrical Engineering, University of Birmingham for many helpful discussions.

REFERENCES

1. Balding, G. H. and Susskind, C. (1960). Generation of artificial electronic displays with application to integrated flight instrumentation. *I.R.E. Trans. ANE*, p. 92 (September).
2. McLane, R. C. and Wolf, J. D. (1967). Symbolic and pictorial displays for submarine control. *IEEE Trans.* **HFE-8**, No. 2, 148.
3. Vernon, M. D. (1962). "The Psychology of Perception", Pelican Books.
4. Gregory, R. L. (1966). "Eye and Brain", Weidenfeld and Nicholson.
5. Gregory, R. L. (1970). "The Intelligent Eye", Weidenfeld and Nicholson.
6. Hudson, J. E. (1970). New types of high resolution sonar displays. IERE Conference on Electronic Engineering in Ocean Technology. *IERE Conf. Proc.* No. 19, 121.
7. Nairn, D. (1968). Clipped-digital technique for the sequential processing of sonar signals. *JASA* **44**, 5.
8. Creasey, D. J. and Braithwaite, H. B. (1969). Experimental results of a sonar system with a digital signal processing unit. *Applied Acoustics* **2**, p. 39.
9. Morgan, C. T., Cook III, J. S., Chapanis, A. and Lund, M. W. (1963). "Human Engineering Guide to Equipment Design", p. 60, McGraw-Hill, New York.

10. Berry, R. N., Riggs, L. A. and Duncan, C. P. (1950). The relation of vernier and depth discriminations to field brightness. *J. Exp. Psychology* **40**, 346 [61].

11. Holland, G. E. and Emmons, B. W. "Comparative Tests on 3-D Displays". U.S. Air Force, Air Material Command, Cambridge Field Station, Report 1-50.

12. See Reference 9, p. 87.

13. Haber, R. N. (Ed.) (1968). "Contemporary Theory and Research in Visual Perception", Holt Rinehart and Winston, New York.

14. Brown, R. (1969). Three dimensional television. *Wireless World* **May 1969**, 206.

15. Rawson, E. G. (1969). Vibrating varifocal mirrors for 3-D imaging. *IEEE Spectrum*, September 1969, 37–43.

16. Ciccotto, D. (1969). "Cathode ray tube having image forming elements in displaced parallel planes", Patent U.S.A. 3478242, 13 May 1966, (published 11 November 1969).

17. Jacobson, A. D. (1970). Requirements for holographic displays. *Proc. Soc. Inf. Display (U.S.A.)* **11**, No. 2, 82–89.

18. Perkins, W. J. (1971). Computer displays in 3-D. *Electron. Aust.* **33**, No. 1, 22–23.

19. McDermott, J. (1971). Holographic displays in real time move from dream to development. *Electron. Des. (U.S.A.)* **19**, No. 15, 28.

20. Herman, S. (1971). Principles of binocular 3-D displays with application to television. *J. Soc. Motion Pic. and T. V. Eng. (U.S.A.)* **80**, No. 7, Pt. 1, 539–544.

21. Lewis, J. D., Verber, C. M. and McGhee, R. B. (1971). A true three-dimensional display. *IEEE Trans.* **ED-18**, No. 9, 724–732.

22. Ortony, A. (1971). A System for stores viewing. Conference on Displays, Loughborough September 1971, IEE Conference Publication No. 80, pp. 225–232.

23. Elbourn, R. W. (1971). "H.A.P.P.I. (Height and Plan Position Indicator)" Conf. on Displays, Loughborough, September 1971, IEE Conference Publication No. 80, pp. 239–248.

24. Keats, A. B. (1971). "A Three Dimensional Display of Multi-parameter Data", Conf. on Displays, Loughborough, September 1971, IEE Conference Publication No. 80, pp. 177–181.

25. Fuller, K. L. (1971). AVOID, a short-range high-definition radar. *Philips Tech. Rev.* **32**, No. 1, 13–19.

DISCUSSION

P. L. Stocklin: In a complicated situation have you tried to add the perspective cue in your stereoscopic system? Further have you tried this system in a target detection situation?

Answer: An approximate geometric perspective cue exists on a single B-scan display. Since the stereo-pair is formed directly from two B-scans, the stereoscopic display naturally possesses the perspective cue. We are about to try the system in target detection situations.

C. Van Schooneveld: Can colour be used as an added cue?

Answer: Yes colour can be added as an extra cue either as a "natural" cue or in a colour coding scheme. However most sonar situations require a long persistence display and you are limited in the persistence of the phosphors used for colour cathode-ray tubes. We have tried using red and green coding for stereo-pairs but found this unsatisfactory for practical situations because of the lack of persistence.

J. W. R. Griffiths: In the particular digital signal processor you use to form your stereo-display, there is a fundamental limitation that only one target can exist in a single range cell. Could a stereo-system be formed using two within-pulse sector scanning sonars, and not produce too many ambiguities?

Answer: The essential requirement to produce the stereo-display is that one target only should exist in a single range cell, or that *each* bearing-elevation cell should possess unambiguous data. If the scanning system can be designed with sufficient bandwidth, then correct data can be obtained by scanning in both bearing and elevation during a time no longer than the pulse width. From a technological point of view a 2-D beamformer would be preferable to a within-pulse sector scanning sonar.

The advantage of the single digital system where only one detected target can exist per range annulus is that the data rate is considerably reduced. The disadvantage is that when multiple targets exist in the same range annulus, there may be an ambiguity in the displayed position of a target due to mutual interference during the hard-limiting process. If interference is so excessive to cause this ambiguity, then it will usually cause the threshold to fail and the ambiguous data would not be displayed.

Some Recent Results With a Long Range Side Scan Sonar

M. L. SOMERS

National Institute of Oceanography, Wormley, Surrey, England

1. Introduction

Side scan sonar is a technique for investigating the seabed, in conjunction with other geophysical methods, by recording and displaying bottom reverberation of shallow angles of incidence. The sonar transducer is designed to have a horizontal beamwidth of 1.5 to 2.5 degrees and a somewhat larger vertical beamwidth, about 10 to 15 degrees. It is mounted on a platform of sufficient stability with respect to roll and yaw and the platform is towed at right angles to the main beam. Pulses of sound are transmitted at regular intervals and between transmissions reverberation is recorded. The main beam is depressed from the horizontal so as to strike the bottom at a grazing angle. The pulses are transmitted in a bandwidth governed by the down range resolution required. The physical movement of the transducer platform combined with this down range resolution produces a scanning action which is mapped as a two-dimensional optical density pattern on a suitable recording paper.

The technique was pioneered by the National Institute of Oceanography in 1958 by Tucker and Stubbs [1] as a result of work originally reported by Chesterman *et al.* [2]. These early devices had a range of about 1 km and were suitable for continental shelf areas where the depth of water is 200 m or less. The Tucker and Stubbs sonar operated at a carrier frequency of 37 kHz and substantially the same equipment is still in use today. Meanwhile many developments have taken place in other laboratories and equipment has been commercially available for a number of years mainly from the U.S.A. but also from Western European countries. All the commercially available equipments are of the shallow water type with operating frequencies between 30 kHz and 100 kHz. Records are normally displayed on either wet or dry electrosensitive paper, both of which, under favourable conditions, can display a signal dynamic range of about 20 dB. Development work in this field is tending towards large area coverage by means of a mosaic collection of displays. For this purpose various distortions and aberrations have to be removed from the records. One example of this approach was described by Hopkins [3] who uses a cathode ray tube controlled by navigational and yaw information. Another development was described by Cholet *et al.* [4] who use a special lens for anamorphic scale changing.

2.　Work at the National Institute of Oceanography

Interest at N.I.O. centres primarily on the deep oceans and this is true also of geophysical and geological investigations. This being so it was natural that the possibility of extending side scan sonar techniques to the deep oceans should be investigated. Preliminary studies were undertaken in 1964 and development was put in hand in 1966. The first sea trials of the new sonar known by the acronym G.L.O.R.I.A. were successfully carried out in May/June 1969 and these were followed by the first season of operations in July and August of that year. Since then extensive cruises have taken place in the summers of 1970 and 1971. The areas covered are illustrated by the track charts of Fig. 1. The system was described by Rusby [5]. It is not necessary here to go into the detailed acoustic and engineering considerations involved in extending side scan sonar practice to the deep ocean, but it is worth pointing out that very little advantage is to be gained by gradual extension of sonar range. This is because side scan geometry is such that the depth to range is fairly restricted and the ocean is quite sharply divided into deep and shallow regions. At the time of the description in Reference [5] the basic parameters of the G.L.O.R.I.A. sonar were as follows: operating frequency 6.4 kHz, horizontal beam angle 2.7°, vertical beam angle 15°, acoustic pulse length 30 mS, acoustic power 50 kw. The transducer is mounted in a towed vehicle for four main reasons, namely to obtain a quieter operating environment, to increase the acoustic cavitation threshold, to obtain some de-coupling in yaw from the parent ship and to obtain more favourable propagation conditions. The maximum slant range works out at 22 Km, with a pulse repetition period of 30 seconds. The towed vehicle has active steering to assist in damping short term yaw and the received beam is stabilized to the direction of transmission. Using the system with these parameters and a wet electro-sensitive paper for display, records such as Fig. 2 were obtained. This shows the mouth of a submarine canyon near Capri. The recorder was a twin helix machine with the shorter helix arranged to give a true scale picture. The longer helix achieves greater resolution at the cost of a 3:1 distortion. Figure 2b shows another replay of nearly the same data but this time a contour map of the area has been included to make the interpretation easier. Note that the range scale is 12 Km, this is because acoustic conditions in the Mediterranean generally disallow unreflected propagation to the bottom at 20 Km. Other features worth noting are

FIG. 1. Map of G.L.O.R.I.A. coverage.

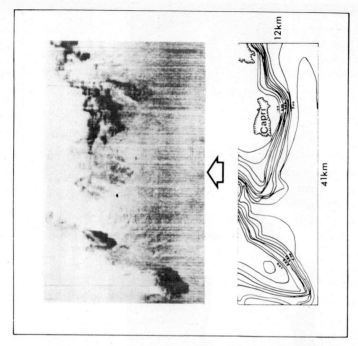

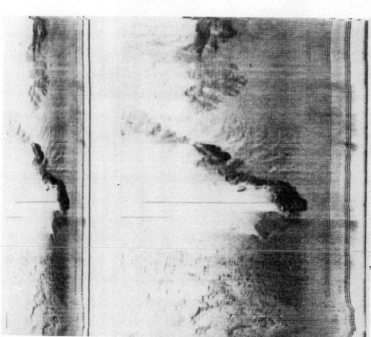

FIG. 2.   Showing a submerged canyon near Capri. Left: Twin helix replay: right: single replay with contour map and scale.

the very high target strengths associated with the sidewalls and their pronounced acoustic shadows and the double bottom echo and its triplet structure. These displays were obtained as it happens by replaying tape recorded information at a speed up ratio of 4:1, wet paper recorders are much less satisfactory working in real time in this application.

3. Recent Developments

The development of the G.I.O.R.I.A. system was a very complex exercise in engineering and seamanship, and it is fair to say that these difficulties absorbed a disproportionate part of the effort. For this reason the signal processing and display tended to be neglected. However in the winter of 1970/71 it was possible to give these aspects more attention and two developments of great importance were carried out. Firstly, a linear processor for pulses of large bandwidth-time product was obtained and secondly, a photographic display system for off-line reproduction of records was procured. The linear processor is a correlator kindly lent by the Admiralty Research laboratory, and since its installation the pulse in regular use has been a 4 second linear FM sweep of 100 Hz bandwidth. As will be appreciated this has resulted in a dramatic improvement in signal-to-noise ratio, the major source of noise being the ship. Figure 3 is an example of a record taken with the linear correlator displayed on wet paper. The scales are the same as Fig. 2 and in fact this is again the Mediterranean but farther East being a portion of the Mediterranean ridge near Crete. The dynamic range of signal available to the display from this record was in excess of 35 dB, but as has been mentioned the display can only accept 20 dB, and this accounts for the rather disappointing showing of this record with its undoubtedly better signal to noise ratio.

The photographic display is a standard photographic picture receiver used by newspaper offices to receive pictures by wire or radio. It has however been modified to enable it to replay G.L.O.R.I.A. records, the modifications being to the gearing and drive circuitry. It is still necessary to accept a distorted picture from this machine and in order to construct a mosaic chart some form of anamorphic scale changing is required. The recorder uses a crater tube as a modulated light source to expose photographic paper which is loaded in sheets on a revolving drum. The crater tube carriage is mounted on a lead screw and moves parallel to the drum axis. The paper loading, unloading and processing are quite automatic. Anamorphic scale changing can be carried out by photographing a moving object (record) on to moving film through a slit. This process was described by Moore [6] and, independently, much later by Honick [7]. Cholet [4] mentions another method of anamorphosis using a cylindrical lens, but the author is not yet in a position to compare the respective merits

of the two methods. The photographic recorder is capable of dealing with
the full dynamic range of the tape-recorded signal, and results in a great
improvement in the appearance of the display. This is illustrated in
Fig. 4 which shows the same ground as Fig. 3 but reproduced by the new
technique. It was mentioned earlier that the records are frequently required
in true scale for the purposes of constructing a mosaic, and that two

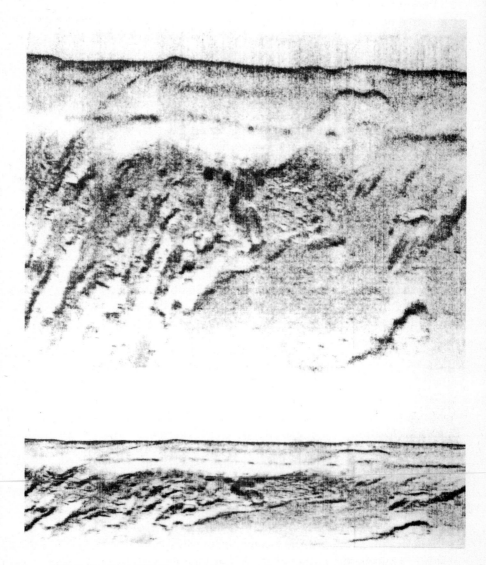

FIG. 3. Portion of Mediterranean ridge near Crete. Linear correlator replayed on
wet paper.

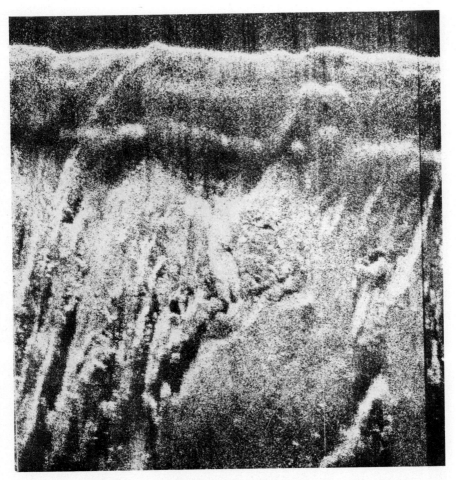

FIG. 4. Same ground as Fig. 3 but replayed on photographic recorder.

methods of anamorphosis are available. Figure 5 shows yet again the same ground but this time reduced to true scale. The upper part of the picture is the output of the linear correlator, while in parallel with this the system has a hard-clipped digital DELTIC correlator the output of which is shown in the lower part of the picture. The two systems present an interesting comparison, for although the digital correlator appears as a much flatter picture remarkably little detail is actually missing. It appears that the digital record has a much lower dynamic range except where there is a strong discontinuity. There are also some interesting echoes in mid-water in the digital record exhibiting the capture effect of the hard-clipping.

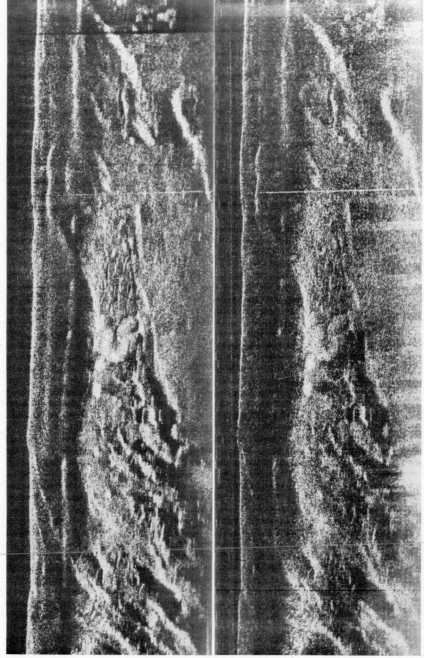

FIG. 5. The same ground as Fig. 3 again but anamorphosed to true scale and showing the linear correlator output (top) and the digital correlator output (bottom).

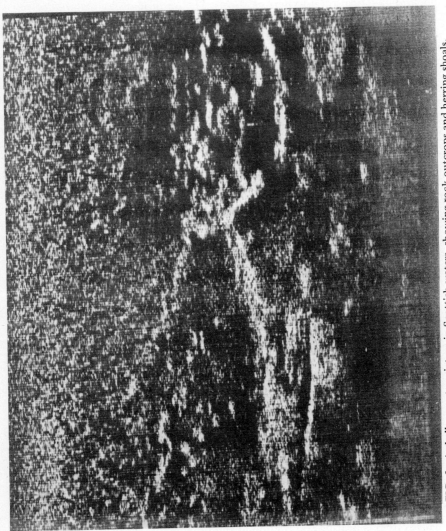

FIG. 6. A shallow water picture in Scottish waters, showing rock outcrops and herring shoals.

4. Work in Shallow Water

In September 1971 the sonar was used in Scottish waters in an experimental investigation into herring shoals. The purpose of the experiment was to go to an active herring fishing ground and to see if the sonar had anything to offer as a tool in fisheries research. Good geological control of a selected piece of ground was obtained and then successive runs were made over this ground with the sonar working at the 12 Km range in about 100 m of water. As the runs progressed and results were plotted the geological features quickly became obvious and could be picked out quite rapidly in later runs. The ground had of course been chosen partly for its absence of rocky features and considerable luck was experienced with the weather so that propagation conditions were good throughout the three days of the sonar runs. A typical run across the ground is shown in Fig. 6, the ship's tracking being across the top of the picture. The bank of rock, which is called Hawes bank, can be seen clearly breaking out of the flat shingle bottom at about half range. Note the characteristic ledges of rock just behind the boundary on the left-hand side of the picture at about 8 Km range. The linear feature of about 6 Km length in the centre at about 5 or 6 Km is a shoal of herring whose progress was plotted in successive runs. Eventually it proved possible to con a commercial purse-seiner retained for the purpose on to one end of such a shoal so that the shoal could be sampled. The results cannot be discussed in detail here, but great interest has been created in the fisheries laboratories. Figure 6 has been anamorphosed to a true scale.

5. Conclusion

A description has been given of the great improvement in quality of record which results from the much greater signal-to-noise ratio at the linear correlator output, and how this improvement can be realized in the display by using the photographic recorder. It is hoped, in the next season's trials, to take this display system to sea together with a system of anamorphic scale changing. The other line of development is an experimental system for imposing a non-linear sweep on the photographic recorder to compensate for the slant range distortion which arises from working in deep water. This system is nearing completion at the time of this Institute.

REFERENCES

1. Tucker, M. J. and Stubbs, A. R. (1961). A narrow beam echo ranger for fishery and geological investigations. *Brit. J. App. Phys* **12**, 103.
2. Chesterman, W. D., Clynick, P. R. and Strike, A. H. B. (1958). An acoustic aid to sea bed survey. *Acustica* **8**, 285.

3. Hopkins, J. C. (1970). Cathode ray tube display and correction of side scan sonar signals. Proceedings of I.E.R.E. Conference on Electronic Engineering in Ocean Technology. Swansea 1970 p. 151.
4. Cholet, J., Fontanel, A. and Grau, G. (1968). Survey of the sea bottom by means of a Side Scan Sonar. XIe Congrès International De Photogrammetrie Lausanne 1968.
5. Rusby, J. S. M. (1970). A long range side scan sonar for use in the deep sea. *Int. Hydrographic Rev.* **47**, No. 2.
6. Hilary B. Moore (1949). A camera for producing one dimensional reduction in narrow records. *Sears Found. J. Mar. Res.* 1949.
7. Honick, K. R. (1971). Anamorphosis by continuous flow photography. *J. Photographic Science* **19**, 1971.

DISCUSSION

C. N. Pryor: Do you know of any attempts to use ping to ping coherent processing as in side looking radars?

Answer: Yes, we have looked into the possibility, and the Raytheon Company have produced an exhaustive study. The main difficulty lies in the velocity of propagation. To achieve down-range resolution energy has to be transmitted as you implied in pings, and to avoid ambiguities one has to wait for the return from one ping before transmitting the next. On the other hand the pings are used to sample the phase function so there has to be a minimum density of pings along the track, which sets a maximum value for the transducer velocity. At the expense of dead bands at fixed ranges and much complexity the sampling problem can be eased by frequency multiplexing, but at the maximum multiplex ratio likely to be practicable the ship's speed would still be unacceptably slow. As well as the actual duration of the survey the problem of stabilizing the array becomes acute at the required speed of one or two knots. It is an attractive idea and we are still considering an experiment, but we feel that in the field of long range side scan sonar the technique of aperture synthesis is unlikely to be of much practical benefit.

C. Van Schooneveld: In the pictures of the bottom you have shown do you use only time variation or do you use angular information as well?

Answer: We only use time differences directly but in fact the lower limb of the main vertical lobe is used as an aid to normalizing, that is to say the operator seeks to hit the bottom at the extreme range with the peak of the main lobe. I might add that all the pictures I have shown have been normalized by time-varied gain to compensate for spreading and attenuation, and the vertical directivity adds a little to this process but that is the only way in which angular information is used.

Author Index

Numbers in brackets are reference numbers and are included to assist in locating references in the text where authors' names are not mentioned. Numbers with an asterisk indicate the pages on which references are listed in full.

A

Abramson, N., 648(9), 656*, 735(1), 742*
Ackroyd, M. H., 2(6), 4(9), 6(10), 8*
Adler, R., 470(38), 475*
Agrawala, A. K., 731(13), 743*
Alsup, J. M., 458(1), 470(1, 41), 472(41), 473*, 475*
Al-Temimi, C. A., 321(32), 322(32), 325*
Altman, L., 465(8a, 8b), 473*
Anderson, T. W., 337(1), 343*
Anderson, V. C., 591(3), 603*, 634(1, 2), 642*
Angelakos, D. J., 255(1, 2, 3), 256(3), 258*
Aoki, Y., 61*
Applications of Finite Amplitude Acoustics to Underwater Sound, 311(7), 321(7), 324*
Archbold, E., 548(5), 559*
Arkadev, 720(4), 727*
Arneson, S. H., 468(24), 472(24), 474*
Arques, P. Y., 607(9), 616*
Åstrom, K. J., 137(6), 140*
Au, K. T., 529(3), 542*
Avenhaus, E., 159(15), 162*

B

Bailey, W. H., 466(9), 472(9), 473*
Balding, G. H., 745(1), 753*
Ball, G. H., 737(2), 738(3, 4), 742*, 743*
Bangs, W. J., 589(5), 589*
Barbagelata, A., 206(4), 222*, 528(19), 543*
Barker, R. H., 463(5), 473*
Barnard G. R., 322(35), 325*
Barnett, H., 554(7), 560*
Bartram, J. F., 325*
Bass, C. A., 36(1), 39*

Bello, P. A., 225(3), 226(3), 240*
Berglund, C. N., 465(7), 473*
Berktay, H. O., 311(1, 3, 4, 6), 314(1, 22), 316(23), 319(28), 321(32), 322(32, 36, 37), 324*, 325*
Berry, R. N., 751(10), 754*
Bertheas, J., 459(2), 473*
Beyer, R. T., 311(10, 11), 317(10, 11), 324*
Bieren, K. von, 469(36), 474*
Birdsall, T. G., 348(1), 355*, 647(4), 650(12), 652(13), 656*
Blackstock, D. T., 318(25, 27), 319(25), 320(29), 325*
Blue, J. E., 311(8, 19), 314(19), 324*
Bogert, B. P., 21*
Bogner, R. E., 146(3), 149(5), 161*
Bohlin, T., 137(6), 140*
Boni, P., 222*
Boniganni, W. L., 468(31), 472(31), 474*
Bonner, R. E., 723(7), 727*, 737(5), 743*
Bown, G., 156(8), 159(8), 161*
Bracewell, R., 505(6), 521*
Braithwaite, H. B., 751(8), 753*
Bramhall, J. N., 39*
Braverman, D., 648(9), 656*, 720(4), 727*, 735(1), 742*
Breuer, D. R., 462(4), 472(4), 473*
Brillinger, D. R., 328(2), 343*
Brown, B. M., 432(2), 433(2), 434(2), 436(2), 437(2), 446*
Brown, R., 754*
Brown, W. M., 566(5), 574*
Browning, D. G., 311(13), 315(13), 324*
Bryn, F., 271(1), 278*, 591(1), 603*, 607(2), 616*, 619(3), 642*
Buckner, D. N., 360(3), 361*
Buie, J. L., 462(4), 472(4), 473*
Burg, J. P., 132(2), 136(2), 140*, 607(5), 616

Burnsweig, J., 468(24), 472(24), 474*
Buss, D. D., 466(9), 472(9), 473*
Buxton, J., 592(7), 604*
Byram, G. W., 470(44), 475*

C

Cain, G. D., 152(6), 161*
Capon, J., 607(7), 616*, 636(4), 642*
Castanet, A., 206(4), 222*, 239(5), 240*, 528(19), 543*
Cavalli-Sforza, L. L., 738(9), 743*
Chang, J. H., 641(5), 642*
Cheng, D. K., 619(6), 625(6), 637(6), 639(6), 641(6), 642*
Chernov, L. A., 271(2), 279*
Chesterman, W. D., 757(2), 766*
Chien, Y. T., 731(6), 743*
Chilov, G. E., 64(1), 69(1), 75*
Cholet, J., 758(4), 761(4), 767*
Chrestenson, H. E., 25(3), 39*
Ciccotto, D., 754*
Claiborne, L. T., 468(27), 474*
Clark, C. A., 466(15), 474*
Clay, C. S., 327(16), 344*
Clynick, P. R., 757(2), 766*
Collins, D. R., 466(9), 472(9), 473*
Collins, R. E., 518(15), 521*
Comeau, J., 255(2). 258*
Conference Record, 484(8), 487*
Conroy, J. J., 472(49), 475*
Cooley, J. W., 21*
Cooper, D. B., 735(7), 743*
Cooper, P. W., 735(7), 743*
Coquin, G. A., 467(19), 474*
Court, J. N., 471(45), 475*
Cover, T. M., 736(8), 743*
Cox, H., 331, 343*, 360(2), 361*, 591(4), 603*, 607(8), 608(8), 616*, 619(7), 620(7), 623(7), 624(7), 634(8), 636(7), 637(7, 9), 640(7), 641(7, 9), 642*
Cramer, H., 582(3), 589*
Creasey, D. J., 751(8), 753*
Crochiere, R., 158(14), 162*
Crystal, T., 149(4), 161*

D

Damon, R. W., 469(37), 475*
Davenport, W. B. Jr., 364(3), 369(3), 372*
De Bruine, R. F., 133(4), 140*
Dempster, A. P., 327(4), 328(4), 343*
Desoer, C. A., 508(12), 521*
Deutsch, R., 265(2), 270*
Diehl, R., 572(7), 574*

Diess, B., 217(5), 222*
Di Franco, J. V., 363(1), 366(1), 372*
Dow, J. J., 432(2), 433(2), 434(2), 436(2), 437(2), 446*
Dunn, H. K., 1(1), 8*
Dunn, J. R., 311(6), 324*

E

Edelblute, D. J., 637(10), 643*
Edwards, A. W. F., 738(9), 743*
Ehrman, L., 149(4), 161*
Elbourn, R. W., 754*
Ellinthorpe, A. W., 103(4), 114*
Ellis, J. H., 146(2), 161*
Emmons, B. W., 752(11), 754*
Ennos, A., 548(5), 559*
Erickson, C. W., 61*
Exhibition Reports, 479(5), 487*

F

Faure, P., 119(2), 127*, 418(2), 426*
Feature Articles, 478(4), 487*
Fettweiss, A., 158(13), 162*
Figoli, A., 206(1), 222*
Fine, N. J., 23(4), 26(5), 31(4, 5), 39*
Fishbein, M., 466(13), 473*
Fisher, L., 330(5), 343*
Fisher, R. A., 719(3), 726*
Fisk, J. M., 637(10), 643*
Fjallbrandt, T., 152(7), 161*
Flammer, C., 132(3), 140*
Fontanel, A., 758(4), 761(4), 767*
Fortier, J. J., 737(11), 743*
Fortuin, L., 239(5, 6, 11), 240*, 241*
Foster, T., 472(49), 475*
Fox, W. C., 348(1), 355*, 647(4), 656*
Fralick, S. C., 648(8), 650(8), 656*, 743*
Franck, A. R., 468(26), 474*
Freytag, R. W., 467(16), 472(16), 474*
Frost, O. L., 596(10), 604*, 641(11), 643*
Fu, K. S., 731(6, 12), 743*
Fukunaga, K., 330(6), 343*, 592)6), 603*
Fuller, K. L., 754*

G

Gabor, D., 1(2), 8*, 545(1), 559*
Gallop, M. A., 655(15), 656*
Galpin, R. K. P., 478(1), 482(6), 486*, 487*
Garner, H. L., 159(16), 162*
Gastmans, R., 239(5), 240*

Gazey, B. K., 311(6), 324*
Gelfand, I. 25(6), 39*, 64(1), 69(1), 75*
Gerard, H. M., 468(33), 474*
Gibbs, J. E., 23(7, 8, 9, 10, 11), 29(7, 8, 11, 12), 34(7), 39*, 61*
Gilbert, E. N., 619(12), 625(12), 630(12), 638(12), 641(12), 643*
Golay, M. J. E., 61*
Gold, B., 515(14), 521*
Goldfeld, S. M., 338(7), 343*
Goode, B. B., 605(1), 608(1), 616*
Goodman, N. R., 328(8), 329(8), 343*
Gottlieb, M., 472(49), 475*
Goyal, R. C., 469(35), 474*
Grace, O. D., 120(5), 128*
Grant, R., 554(7), 560*
Gratian, J. W., 467(16), 472(16), 474*
Grau, G., 758(4), 761(4), 767*
Green, D. M., 731(17), 743*
Greenfield, R. J., 607(7), 616*
Gregory, R. L., 745(4, 5), 753*
Greguss, P., 561(1), 574*
Griffiths, L. J., 605(1), 608(1), 616*, 641(13), 643*
Groginski, H., 624(18), 640(18), 643*

H

Haber, R. N., 754*
Haines, K., 545(3), 559*
Hájek, J., 403(1), 411*
Hall, D. J., 738(3, 4), 742*, 743*
Hamming, R. W., 702*
Hancock, J. C., 737(20), 743*
Hands, E., 506(9), 521*, 528(5), 542*
Hannan, E. J., 330(9), 344*
Hansen, V. Gnegers, 672(1), 688
Harabedian, A., 360(3), 361*
Harmuth, H. F., 27(13), 38(13), 39*, 48(1), 50(1), 61*
Harrington, J. V., 396(8), 398*
Harris, J. L., 468(27), 474*
Hart, P. E., 736(8), 743*
Harvey, F. K., 61*
Hastrup, O. F., 239(7), 240*
Hatsell, C. P., 650(11), 651(11), 653(11, 14), 655(14), 656*
Healy, M. J. R., 21*
Heimdal, P., 271(1), 278*
Helms, H. D., 515(13), 521*, 530(12), 538(12), 542*, 543*
Helstrom, C. W., 375(2), 398(2), 398*, 415(1), 426*, 647(2), 655*, 658(2), 662(2), 669*
Herman, S., 754*
Hersey, H. S., 530(21), 543*
Hewitt, E., 25(14), 39*

Heyden, R. P., 592(8), 604*
Hickernell, F. S., 467(18), 474*
Hildebrand, B., 545(3), 559*
Himmelblau, David M., 422(8), 426*
Hixson, E. L., 529(3), 542*
Ho, Y. C., 731(13), 743*
Hobaek, H., 311(2), 324*
Hochman, H. T., 465(6), 473*
Hogan, D. L., 465(6), 473*
Holland, G. E., 752(11), 754*
Holland, M. G., 469(34), 474*
Honick, K. R., 761(7), 767*
Hopkins, J. C., 758(3), 767*
Hovem, J. M., 229(4), 239(8), 240*
Hu, J. V., 156(9), 159(9), 161*
Huber, P. J., 405(2), 409(2), 411*
Hubley, B., 530(22), 543*
Hudson, J. E., 749(6), 753*
Huffman, D. A., 161(19), 162*
Hunsinger, B. J., 468(26), 474*
Hutchins, H. Stuart IV, 255(1), 258*

I

Idselis, M., 530(6), 542*
Igarashi, Y., 109(5), 114*
Isbell, D. E., 530(7), 542*

J

Jaarsma, D., 647(3), 648(3), 652(3), 655*
Jacobson, A. D., 754*
James, A. T., 337(10), 344*
Jayachandran, K., 331(14), 344*
Jenkins, G. M., 331(11), 344*
Johnsen, J., 98(1, 2), 114*
Johnson, C., 548(6), 560*

K

Kac, M., 378(4), 398*
Kalachev, A. I., 311(5), 321(33), 324*, 325*
Keats, A. B., 754*
Keehn, D. G., 735(14), 743*
Kelly, E. J., 418(5), 426*, 430(1), 432(1), 445*, 607(4), 608(4), 616*
Kharkevich, A. A., 2(5), 5(5), 8*
Kincaid, Thomas, G., 21*
Kinnison, G. L., 637(10), 643*
Kino, G. S., 468(29), 474*
Kluvanek, I., 36(15), 39*
Kneipfer, R. R., 602(13), 604*
Kock, W. E., 61*
Koenig, W., 1(1), 8*
Kolker, R. J., 607(7), 616*

Konrad, W. L., 311(13), 315(13), 321(34), 324*, 325*
Kooij, T., 364(4), 371(4), 372*, 421(7), 426*
Koontz, W. L. G., 592(6), 603*
Kraus, J. D., 505(3), 521*
Kraut, E. A., 468(30), 474*
Kreuzer, J. L., 61*
Kuhn, L., 470(43), 475*
Kulikowski, G., 592(7), 604*

L

LaBarre, J. B. K., 38(16), 39*
Lacoss, R. T., 608(11), 616*
Lacy, L. Y., 1(1), 8*
Lambert, P., 592(7), 604*
Lance, G. N., 737(15), 743*
Lardat, C., 468(25), 474*
Laval, R., 206(4), 222*, 224(1), 239(5, 6, 9), 240*, 241*, 528(19), 543*
Lawson, J. L., 364(2), 372*, 383(7), 398*
Leahy, D. J., 314(22), 324*
LeCam, L., 403(3), 411*
Lee, Q. H., 624(15), 638(15), 643*
Lee, S. W., 530(9), 542*, 624(15), 638(15), 643*
Lehman, E. L., 577(1), 589*, 661(3), 669*
Leith, E., 545(2), 559*
Leith, E. N., 565(4), 567(4), 574*
Leondes, C. T., 265(1), 269*
Lerner, E. C., 418(5), 426*
Levin, M. J., 3(8), 8*, 607(4), 608(4), 616*
Levine, N., 677(4), 688*
Levy, P., 25(17), 39*
Lewis, J. B., 622(14), 624(14), 643*
Lewis, J. D., 754*
Lewis, J. T., 530(21), 543*
Lewis, P. M., 732(16), 743*
Li, T. J., 731(12), 743*
Liebermann, L. J., 278(3), 279*
Liggett, W. S. Jr., 328(12, 13), 344*
Lillie, R., 554(7), 560*
Lim, J. T., 482(7), 487*
Lim, T. C., 468(30), 474*
Lines, P. D., 36(18), 39*
Lo, Y. T., 530(9), 542*, 624(15), 638(15), 643*
Lockhart, G. B., 145(1), 161*
Luenberger, D. G., 608(10), 616*
Lund, G. R., 287(2), 296*
Lunde, E. B., 268(3), 270*
Luukala, M., 468(29), 474*
Lytle, D. W., 665(7), 669*

M

Macchi, C., 607(9), 616*
Macchi, O., 607(9), 616*
McDermott, J., 754*
MacDonald, V. H., 585(4), 586(6), 587(6), 588(6), 589*
McDonough, R. N., 625(16), 630(16), 643*
McGonegal, C. A., 515(14), 521*
McGrath, J. J., 360(3), 361*
McLane, R. C., 745(2), 753*
McMahon, D. H., 469(37), 475*
McMahon, G. W., 530(22), 543*
Maerfeld, C., 468(25), 474*
Maloney, W. T., 469(37), 475*
Mantey, P. E., 605(1), 608(1), 616*
Manz, J. W., 50(2), 61*
Marcum, J. I., 79(1), 94*, 375(1), 398*
Marill, T., 731(17), 743*
Mark, W. D., 2(4), 8*
Martin, R. D., 410(4), 411*
Matsinger, B. J., 469(34), 474*
Mellen, R. H., 311(13, 18), 315(13, 18), 320(18), 321(18, 34), 324*, 325*
Merklinger, H. M., 311(15), 317(15), 318(15), 320(15, 31), 324*, 325*
Mermoz, H., 493(1, 2), 502*, 591(2), 603*, 607(3), 616*, 619(17), 637(17), 643*, 657(1), 669*
Mertz, L., 459(3), 473*
Metherell, A. F., 561(2, 3), 567(6), 574*
Middleton, D., 119(3), 120(6), 127*, 128*, 383(6), 398*, 418(3), 419(3), 426*, 624(18), 640(18), 643*
Millard, M. J., 29(12), 39*
Miller, R. L., 21*
Min, P. J., 731(12), 743*
Moffett, M. B., 311(10, 11, 17, 18), 315(17, 18), 317(10, 11), 320(17, 18), 321(17, 18, 34), 324*, 325*
Mohammed, A., 530(22), 543*
Moore, Hilary, B., 761(6), 767*
Moose, P. H., 419(6), 426*
Morgan, C. T., 751(9), 752(12), 753(12), 753*
Morgan, S. P., 619(12), 625(12), 630(12), 638(12), 641(12), 643*
Morgenthaler, G. W., 31(19), 40*
Morris, J. C., 506(9, 10), 521*, 528(5), 529(14), 542*
Morrison, D. F., 727*
Morrison, N., 675(3), 688*
Muir, I. G., 311(9), 318(9), 324*
Muir, T. G., 311(8, 16, 20), 313(16, 20), 320(29, 37), 324*, 325*

N

Nagy, G., 731(18), 743*
Nairn, D., 751(7), 753*
Naugol'nykh, K. A., 318(26), 325*
Neyman, J., 662(4), 669*
Nilsson, N. J., 718(1), 719(1), 726*, 736(19), 743*
Noll, A. Michael, 21*
Nolte, L. W., 648(5, 6, 7), 650(5, 6, 7), 652(5, 6, 7), 653(14), 655(14), 656*
Nuttal, A. H., 103(4), 114*

O

Ol'shevskii, V. V., 117(1), 118(1), 119(1), 127*, 418(4), 419(4), 426*
Olson, D. R., 330(6), 343*
Ortony, A., 754*
Overshott, K. J., 466(12), 473*
Owsley, N. L., 595(9), 597(11), 604*, 641(19), 643*

P

Paley, R. E. A. C., 27(20), 31(20), 40*
Parkes, D. J., 239(5), 240*
Patrick, E. A., 737(20), 743*
Pazzini, M., 206(1, 2, 3, 4), 222*, 239(5), 240*, 528(19), 543*
Pearl, J., 29(21), 38(21), 40*
Pedinoff, M. E., 468(33), 474*
Pekau, D., 572(7), 574*
Perini, J., 530(6), 542*
Perkins, W. J., 754*
Perzley, W., 466(13), 473*
Peterson, W. W., 161(18), 162*, 348(1), 355*, 647(4), 656*
Pichler, F., 23(22), 29(25), 31(22), 34(25), 36(23), 38(23, 24), 40*, 61*
Picinbono, B., 665(8, 9), 666(9), 669*
Pillai, K. C. S., 331(14), 344*
Pinnow, D. A., 470(40), 475*
Pitt, S. P., 120(5), 128*
Plemons, T. D., 119(4), 127*
Pollak, H. O., 132(1), 140*
Polyak, B. T., 32(26), 40*
Porcello, L. J., 566(5), 574*
Portnoy, S., 330(15), 344*
Powell, C., 470(42), 475*
Powell, R., 545(4), 559*
Preston, K. Jr., 61*, 471(46), 475*
Price, R., 139(8), 140(8), 140*

Prichard, R. L., 518(17), 521*
Proceedings of Symposium on Nonlinear Acoustics, 311(14), 318(14), 321(14), 324*
Pryor, C. N., 79(2), 86(3), 91(3), 94*, 360(1), 361*

Q

Quandt, R. E., 338(7), 343*
Quate, C. F., 468(28), 470(39), 472(48), 474*, 475*

R

Rabiner, L. R., 156(9), 159(9), 161*, 515(14), 521*, 538(20, 23), 543*
Rabinowitz, P., 530(11), 542*
Raikov, D., 25(6), 39*
Ramsay, J. F., 505(5), 521*
Rao, C. R., 582(2), 589*
Rawson, E. G., 754*
Reed, I. S., 430(1), 432(1), 445*
Reeder, H. A., 432(2), 433(2), 434(2), 436(2), 437(2), 446*
Requicha, A. A. G., 531(2), 533(2), 534(2), 536(2), 541(4), 542*
Riblet, H. J., 518(16), 521*
Rice, S. O., 378(3), 396(9), 398*
Rihaczek, A. W., 1(3), 2(3), 8*, 364(5), 371(5), 372*
Rogers, D. J., 737(21), 743*
Root, W. L., 364(3), 369(3), 372*, 430(1), 432(1), 445*
Rosen, C. A., 731(22), 743*
Rosen, J. B., 598(12), 604*
Ross, K. A., 25(14), 39*
Rubin, W. L., 363(1), 366(1), 372*
Rudnick, P., 634(2), 642*
Rumsey, V. H., 530(8), 542*
Rusby, J. S. M., 758(5), 767*

S

Sablatash, M., 159(12), 162*
Sangster, F. L. J., 478(3), 487*
Saraga, W., 478(2), 487*
Schafer, R. W., 538(23), 543*
Scharf, L. L., 665(7), 669*
Schissler, L. R., 468(23), 472(23), 474*
Schroeder, M. R., 22*
Schultheiss, P. M., 585(4), 589*, 622(14), 624(14), 634(20), 643*
Schulz, M. B., 469(34), 474*
Schwartz, L., 64(2), 75*
Schwartz, M., 641(27), 643*
Schwartz, S. C., 410(4), 411*

Schweitzer, B. P., 467(22), 474*
Scott-Scott, M., 146(3), 161*
Scudder, H. J., 735(23), 743*
Scudder, Henry J. III, 21*
Seagren, M. K., 739(24), 743*
Sebestyen, G. S., 719(2), 726*
Selfridge, R. G., 31(27), 40*
Shajenko, P., 548(6), 560*
Shanks, J. L., 157(10), 161*
Shannon, C. E., 136(5), 140*
Shelkunoff, S. A., 505(2), 521*
Shilov, G., 25(6), 39*
Shinners, S. M., 268(4), 270*
Shooter, J. A., 320(29), 322(35, 36),
    325*
Shor, S. W. W., 641(21), 643*
Showalter, A. E., 36(34), 40*
Shreider, Yu A., 32(26), 40*
Šidak, Z., 403(1), 411*
Signal Processor Speeds Code Change,
    472(47), 475*
Skolnik, M. I., 360(4), 361*, 383(5),
    398*
Slepian, A., 132(1), 140*
Smith, B. V., 311(12), 319(28), 324*,
    325*
Smith, I. R., 466(12), 473*
Smith, R. P., 529(1), 542*
Smith, R. R., 506(11), 521*
Smyth, C. N., 61*
Sneath, P. H. A., 722(6), 727*
Sokal, R. R., 722(6), 727*
Solomon, H., 737(11). 743*
Solymár, L., 619(24), 624(24), 625(24),
    630(24), 638(24), 641(24), 643*
Sondhi, M. M., 22*, 61*
Sostrand, K. A. 99(3), 114*, 224(2),
    225(2), 240*
Speiser, J. M., 468(32), 474*
Spek, G. A. van der, 364(6), 372*
Spencer, R. C., 505(4), 521*
Spooner, R. L., 648(10), 656*, 665(6),
    669*
Spragins, J., 735(25), 743*
Squire, W. D., 458(1), 470(1, 41),
    472(41), 473*, 475*
Staples, E. J., 468(27), 474*
Stear, E. B., 742(27), 743*
Steer, K., 478(3), 487*
Stern, R., 109(5), 114*
Stetson, K., 545(4), 559*
Stiefel, E. L., 530(10), 542*
Stockham, T. G. Jr., 542*
Stocklin, P. L., 348(2, 3), 356*
Strain, R. J., 465(7), 473*
Strike, A. H. B., 757(2), 766*

Stubbs, A. R., 757(1), 766*
Stutt, C. A., 3(7), 4(7), 8*
Sullivan, F. J. M. S., 411(5), 411*
Susskind, C., 745(1), 753*

T

Tanimoto, T. T., 737(21), 743*
Taylor, T. T., 619(22), 624(22), 625(22),
    641(22), 643*
Thiele, R., 286(1), 296*
Thompson, R. B., 468(28, 30), 474*
Tiemann, J. J., 466(10), 472(10), 473*
Tiersten, H. F., 467(19), 474*
Tolstoy, I., 327(16), 344*
Tomlinson, M., 160(17), 162*
Tou, J. T., 592(8), 604*
Tournois, P., 459(2), 468(25), 473*,
    474*
Tremain, D. E., 255(3), 256(3), 258*
Trotter, H. F., 338(7), 343*
Truchard, J. J., 322(35), 325*
Tseng, C. C., 467(21), 474*
Tseng, F. I., 619(6), 625(6), 637(6),
    639(6), 641(6), 642*
Tucker, D. G., 506(7, 8), 521*, 528(13),
    542*
Tucker, M. J., 757(1), 766*
Tufts, D. W., 530(21), 543*
Tukey, J. W., 21*
Tuteur, F. B., 641(5), 642*

U

Uhlenbeck, G. E., 364(2), 372*, 383(7),
    398*
Upatnieks, J., 545(2), 559*
Urick, R. J., 287(2), 290(3), 296*,
    327(17), 330(17), 344*
Uzkov, A. I., 625(23), 643*
Uzsoky, M., 619(24), 624(24), 625(24),
    630(24), 638(24), 641(24), 643*

V

Vaart, H. van de, 468(23), 472(23), 474*
Van de Linde, V. D., 411(5), 411*
Van den Bos, A., 140*
Vanderkulk, W., 625(25), 643*
Van Ness, J. N., 330(5), 343*
Van Schooneveld, C., 691(1), 692(1),
    696(1), 697(1), 702(1), 702*
Van Trees, H. L., 592(5), 603*, 607(6),
    608(6), 616*, 647(1), 655*, 662(5),
    669*
Verloren, P., 140*

Vernon, M. D., 745(3), 753*
Vezzosi, G., 665(8, 9), 666(9), 669*
Vilenkin, N. J., 25(28), 31(29), 40*
Vilenkin, N. Y., 64(1), 69(1), 75*
Voltmer, F. W., 467(17), 474*
Von Willisen, F. K., 471(45), 475*
Vries, F. P. Ph. de, 688*

W

Wahba, G., 328(18), 329(18), 344*
Waldner, M., 468(33), 474*
Walker, R., 592(7), 604*
Walsh, J. L., 27(30), 31(30), 40*
Watanabe, S., 592(7), 604*, 731(26), 743*
Watari, Ch., 31(31), 40*
Watts, D. G., 331(11), 344*
Weaver, W., 136(5), 140*
Weidemann, H. L., 742(27), 743*
Weinstein, C. J., 158(11), 161*
Weiser, F. W., 31(32), 40*
Weiss, P., 31(33), 32(33), 40*
Wertheim, W., 466(11), 473*
Westervelt, P. J., 311(10, 11, 21), 317(10, 11), 324*, 325*
Whalen, A. D., 734(28), 743*

White, R. M., 467(17, 20), 469(35), 470(20), 474*
Whitehouse, H. J., 458(1), 466(14), 468(32), 470(1, 41), 472(14, 41), 473*, 474*, 475*
Widrow, B., 605(1), 608(1), 616*, 641(26), 643*
Wijmans, W., 239(10), 241*
Wilkerson, C. D. W., 470(39), 475*
Wille, P., 286(1), 296*
Willette, J. G., 311(20), 313(20), 322(35), 324*, 325*
Williams, R. E., 469(36), 474*
Williams, W. T., 737(15), 743*
Winder, A. A., 672(2), 688*
Winkler, L. P., 641(27), 643*
Winslow, D. K., 470(39), 475*
Wolf, J. D., 745(2), 753*
Wolff, Irving, 505(1), 521*
Wolff, S. S., 411(5), 411*

Z

Zadeh, L. A., 508(12), 521*
Zeek, R. W., 36(34), 40*
Zory, P., 470(42), 475*
Zverev, V. A., 311(5), 321(33), 324*, 325*

Signal Processing

Sponsors

North Atlantic Treaty Organisation

Department of Electronic and Electrical Engineering
University of Loughborough, Loughborough, England

Submarine Signal Division
Raytheon Company, Portsmouth, Rhode Island, U.S.A.

INSTITUTE DIRECTOR

Professor J. W. R. Griffiths